LONDON MATHEMATICAL SOCIETY LECTURE NOTE SERIES

The titles below are available from booksellers, or from Cambridge University Press at
www.cambridge.org/mathematics

287 Topics on Riemann surfaces and Fuchsian groups, E. BUJALANCE, A.F. COSTA & E. MARTÍNEZ (eds)
288 Surveys in combinatorics, 2001, J.W.P. HIRSCHFELD (ed)
289 Aspects of Sobolev-type inequalities, L. SALOFF-COSTE
290 Quantum groups and Lie theory, A. PRESSLEY (ed)
291 Tits buildings and the model theory of groups, K. TENT (ed)
292 A quantum groups primer, S. MAJID
293 Second order partial differential equations in Hilbert spaces, G. DA PRATO & J. ZABCZYK
294 Introduction to operator space theory, G. PISIER
295 Geometry and integrability, L. MASON & Y. NUTKU (eds)
296 Lectures on invariant theory, I. DOLGACHEV
297 The homotopy category of simply connected 4-manifolds, H.-J. BAUES
298 Higher operads, higher categories, T. LEINSTER (ed)
299 Kleinian groups and hyperbolic 3-manifolds, Y. KOMORI, V. MARKOVIC & C. SERIES (eds)
300 Introduction to Möbius differential geometry, U. HERTRICH-JEROMIN
301 Stable modules and the D(2)-problem, F.E.A. JOHNSON
302 Discrete and continuous nonlinear Schrödinger systems, M.J. ABLOWITZ, B. PRINARI & A.D. TRUBATCH
303 Number theory and algebraic geometry, M. REID & A. SKOROBOGATOV (eds)
304 Groups St Andrews 2001 in Oxford I, C.M. CAMPBELL, E.F. ROBERTSON & G.C. SMITH (eds)
305 Groups St Andrews 2001 in Oxford II, C.M. CAMPBELL, E.F. ROBERTSON & G.C. SMITH (eds)
306 Geometric mechanics and symmetry, J. MONTALDI & T. RATIU (eds)
307 Surveys in combinatorics 2003, C.D. WENSLEY (ed.)
308 Topology, geometry and quantum field theory, U.L. TILLMANN (ed)
309 Corings and comodules, T. BRZEZINSKI & R. WISBAUER
310 Topics in dynamics and ergodic theory, S. BEZUGLYI & S. KOLYADA (eds)
311 Groups: topological, combinatorial and arithmetic aspects, T.W. MÜLLER (ed)
312 Foundations of computational mathematics, Minneapolis 2002, F. CUCKER *et al* (eds)
313 Transcendental aspects of algebraic cycles, S. MÜLLER-STACH & C. PETERS (eds)
314 Spectral generalizations of line graphs, D. CVETKOVIĆ, P. ROWLINSON & S. SIMIĆ
315 Structured ring spectra, A. BAKER & B. RICHTER (eds)
316 Linear logic in computer science, T. EHRHARD, P. RUET, J.-Y. GIRARD & P. SCOTT (eds)
317 Advances in elliptic curve cryptography, I.F. BLAKE, G. SEROUSSI & N.P. SMART (eds)
318 Perturbation of the boundary in boundary-value problems of partial differential equations, D. HENRY
319 Double affine Hecke algebras, I. CHEREDNIK
320 L-functions and Galois representations, D. BURNS, K. BUZZARD & J. NEKOVÁŘ (eds)
321 Surveys in modern mathematics, V. PRASOLOV & Y. ILYASHENKO (eds)
322 Recent perspectives in random matrix theory and number theory, F. MEZZADRI & N.C. SNAITH (eds)
323 Poisson geometry, deformation quantisation and group representations, S. GUTT *et al* (eds)
324 Singularities and computer algebra, C. LOSSEN & G. PFISTER (eds)
325 Lectures on the Ricci flow, P. TOPPING
326 Modular representations of finite groups of Lie type, J.E. HUMPHREYS
327 Surveys in combinatorics 2005, B.S. WEBB (ed)
328 Fundamentals of hyperbolic manifolds, R. CANARY, D. EPSTEIN & A. MARDEN (eds)
329 Spaces of Kleinian groups, Y. MINSKY, M. SAKUMA & C. SERIES (eds)
330 Noncommutative localization in algebra and topology, A. RANICKI (ed)
331 Foundations of computational mathematics, Santander 2005, L.M PARDO, A. PINKUS, E. SÜLI & M.J. TODD (eds)
332 Handbook of tilting theory, L. ANGELERI HÜGEL, D. HAPPEL & H. KRAUSE (eds)
333 Synthetic differential geometry (2nd Edition), A. KOCK
334 The Navier–Stokes equations, N. RILEY & P. DRAZIN
335 Lectures on the combinatorics of free probability, A. NICA & R. SPEICHER
336 Integral closure of ideals, rings, and modules, I. SWANSON & C. HUNEKE
337 Methods in Banach space theory, J.M.F. CASTILLO & W.B. JOHNSON (eds)
338 Surveys in geometry and number theory, N. YOUNG (ed)
339 Groups St Andrews 2005 I, C.M. CAMPBELL, M.R. QUICK, E.F. ROBERTSON & G.C. SMITH (eds)
340 Groups St Andrews 2005 II, C.M. CAMPBELL, M.R. QUICK, E.F. ROBERTSON & G.C. SMITH (eds)
341 Ranks of elliptic curves and random matrix theory, J.B. CONREY, D.W. FARMER, F. MEZZADRI & N.C. SNAITH (eds)
342 Elliptic cohomology, H.R. MILLER & D.C. RAVENEL (eds)
343 Algebraic cycles and motives I, J. NAGEL & C. PETERS (eds)
344 Algebraic cycles and motives II, J. NAGEL & C. PETERS (eds)
345 Algebraic and analytic geometry, A. NEEMAN

346 Surveys in combinatorics 2007, A. HILTON & J. TALBOT (eds)
347 Surveys in contemporary mathematics, N. YOUNG & Y. CHOI (eds)
348 Transcendental dynamics and complex analysis, P.J. RIPPON & G.M. STALLARD (eds)
349 Model theory with applications to algebra and analysis I, Z. CHATZIDAKIS, D. MACPHERSON, A. PILLAY & A. WILKIE (eds)
350 Model theory with applications to algebra and analysis II, Z. CHATZIDAKIS, D. MACPHERSON, A. PILLAY & A. WILKIE (eds)
351 Finite von Neumann algebras and masas, A.M. SINCLAIR & R.R. SMITH
352 Number theory and polynomials, J. MCKEE & C. SMYTH (eds)
353 Trends in stochastic analysis, J. BLATH, P. MÖRTERS & M. SCHEUTZOW (eds)
354 Groups and analysis, K. TENT (ed)
355 Non-equilibrium statistical mechanics and turbulence, J. CARDY, G. FALKOVICH & K. GAWEDZKI
356 Elliptic curves and big Galois representations, D. DELBOURGO
357 Algebraic theory of differential equations, M.A.H. MACCALLUM & A.V. MIKHAILOV (eds)
358 Geometric and cohomological methods in group theory, M.R. BRIDSON, P.H. KROPHOLLER & I.J. LEARY (eds)
359 Moduli spaces and vector bundles, L. BRAMBILA-PAZ, S.B. BRADLOW, O. GARCÍA-PRADA & S. RAMANAN (eds)
360 Zariski geometries, B. ZILBER
361 Words: Notes on verbal width in groups, D. SEGAL
362 Differential tensor algebras and their module categories, R. BAUTISTA, L. SALMERÓN & R. ZUAZUA
363 Foundations of computational mathematics, Hong Kong 2008, F. CUCKER, A. PINKUS & M.J. TODD (eds)
364 Partial differential equations and fluid mechanics, J.C. ROBINSON & J.L. RODRIGO (eds)
365 Surveys in combinatorics 2009, S. HUCZYNSKA, J.D. MITCHELL & C.M. RONEY-DOUGAL (eds)
366 Highly oscillatory problems, B. ENGQUIST, A. FOKAS, E. HAIRER & A. ISERLES (eds)
367 Random matrices: High dimensional phenomena, G. BLOWER
368 Geometry of Riemann surfaces, F.P. GARDINER, G. GONZÁLEZ-DIEZ & C. KOUROUNIOTIS (eds)
369 Epidemics and rumours in complex networks, M. DRAIEF & L. MASSOULIÉ
370 Theory of p-adic distributions, S. ALBEVERIO, A.YU. KHRENNIKOV & V.M. SHELKOVICH
371 Conformal fractals, F. PRZYTYCKI & M. URBAŃSKI
372 Moonshine: The first quarter century and beyond, J. LEPOWSKY, J. MCKAY & M.P. TUITE (eds)
373 Smoothness, regularity and complete intersection, J. MAJADAS & A. G. RODICIO
374 Geometric analysis of hyperbolic differential equations: An introduction, S. ALINHAC
375 Triangulated categories, T. HOLM, P. JØRGENSEN & R. ROUQUIER (eds)
376 Permutation patterns, S. LINTON, N. RUŠKUC & V. VATTER (eds)
377 An introduction to Galois cohomology and its applications, G. BERHUY
378 Probability and mathematical genetics, N. H. BINGHAM & C. M. GOLDIE (eds)
379 Finite and algorithmic model theory, J. ESPARZA, C. MICHAUX & C. STEINHORN (eds)
380 Real and complex singularities, M. MANOEL, M.C. ROMERO FUSTER & C.T.C WALL (eds)
381 Symmetries and integrability of difference equations, D. LEVI, P. OLVER, Z. THOMOVA & P. WINTERNITZ (eds)
382 Forcing with random variables and proof complexity, J. KRAJÍČEK
383 Motivic integration and its interactions with model theory and non-Archimedean geometry I, R. CLUCKERS, J. NICAISE & J. SEBAG (eds)
384 Motivic integration and its interactions with model theory and non-Archimedean geometry II, R. CLUCKERS, J. NICAISE & J. SEBAG (eds)
385 Entropy of hidden Markov processes and connections to dynamical systems, B. MARCUS, K. PETERSEN & T. WEISSMAN (eds)
386 Independence-friendly logic, A.L. MANN, G. SANDU & M. SEVENSTER
387 Groups St Andrews 2009 in Bath I, C.M. CAMPBELL *et al* (eds)
388 Groups St Andrews 2009 in Bath II, C.M. CAMPBELL *et al* (eds)
389 Random fields on the sphere, D. MARINUCCI & G. PECCATI
390 Localization in periodic potentials, D.E. PELINOVSKY
391 Fusion systems in algebra and topology, M. ASCHBACHER, R. KESSAR & B. OLIVER
392 Surveys in combinatorics 2011, R. CHAPMAN (ed)
393 Non-abelian fundamental groups and Iwasawa theory, J. COATES *et al* (eds)
394 Variational problems in differential geometry, R. BIELAWSKI, K. HOUSTON & M. SPEIGHT (eds)
395 How groups grow, A. MANN
396 Arithmetic dfferential operators over the p-adic Integers, C.C. RALPH & S.R. SIMANCA
397 Hyperbolic geometry and applications in quantum Chaos and cosmology, J. BOLTE & F. STEINER (eds)
398 Mathematical models in contact mechanics, M. SOFONEA & A. MATEI
399 Circuit double cover of graphs, C.-Q. ZHANG
400 Dense sphere packings: a blueprint for formal proofs, T. HALES
401 A double Hall algebra approach to affine quantum Schur–Weyl theory, B. DENG, J. DU & Q. FU
402 Mathematical aspects of fluid mechanics, J. ROBINSON, J. L. RODRIGO & W. SADOWSKI (eds)
403 Foundations of computational mathematics: Budapest 2011, F. CUCKER, T. KRICK, A. SZANTO & A. PINKUS (eds)
404 Operator methods for boundary value problems, S. HASSI, H. S. V. DE SNOO & F. H. SZAFRANIEC (eds)

London Mathematical Society Lecture Note Series: 405

Torsors, Étale Homotopy and Applications to Rational Points

Edited by

ALEXEI N. SKOROBOGATOV
Imperial College London

Shaftesbury Road, Cambridge CB2 8EA, United Kingdom

One Liberty Plaza, 20th Floor, New York, NY 10006, USA

477 Williamstown Road, Port Melbourne, VIC 3207, Australia

314–321, 3rd Floor, Plot 3, Splendor Forum, Jasola District Centre, New Delhi – 110025, India

103 Penang Road, #05–06/07, Visioncrest Commercial, Singapore 238467

Cambridge University Press is part of Cambridge University Press & Assessment, a department of the University of Cambridge.

We share the University's mission to contribute to society through the pursuit of education, learning and research at the highest international levels of excellence.

www.cambridge.org
Information on this title: www.cambridge.org/9781107616127

First published 2013

A catalogue record for this publication is available from the British Library

ISBN 978-1-107-61612-7 Paperback

Contents

List of contributors *page* vii
Preface ix

PART ONE LECTURE NOTES 1

1 **Three lectures on Cox rings** 3
J. Hausen

2 **A very brief introduction to étale homotopy** 61
T. M. Schlank and A. N. Skorobogatov

3 **Torsors and representation theory of reductive groups** 75
V. Serganova

PART TWO CONTRIBUTED PAPERS 121

4 **Torsors over Luna strata** 123
I. V. Arzhantsev

5 **Abélianisation des espaces homogènes et applications arithmétiques** 138
C. Demarche

6 **Gaussian rational points on a singular cubic surface** 210
U. Derenthal and F. Janda

7 **Actions algébriques de groupes arithmétiques** 231
P. Gille and L. Moret-Bailly

8 **Descent theory for open varieties** 250
D. Harari and A. N. Skorobogatov

9 **Homotopy obstructions to rational points** 280
Y. Harpaz and T. M. Schlank

10 **Factorially graded rings of complexity one** 414
J. Hausen and E. Herppich

11 **Nef and semiample divisors on rational surfaces** 429
A. Laface and D. Testa

12 **Example of a transcendental 3-torsion Brauer-Manin obstruction on a diagonal quartic surface** 447
T. Preu

Contributors

Ivan V. Arzhantsev *Department of Higher Algebra, Faculty of Mechanics and Mathematics, Moscow State University, Leninskie Gory 1, GSP-1, Moscow, 119991, Russia.*
arjantse@mccme.ru

Cyril Demarche *Université Pierre et Marie Curie (Paris 6), Institut de Mathématiques de Jussieu, 4 place Jussieu, 75252 Paris Cedex 05, France.*
demarche@math.jussieu.fr

Ulrich Derenthal *Mathematisches Institut, Ludwig-Maximilians-Universität München, Theresienstr. 39, 80333 München, Germany.*
ulrich.derenthal@mathematik.uni-muenchen.de

Philippe Gille *C.N.R.S. et École normale supérieure, DMA, 45, rue d'Ulm, F-75230 Paris Cedex 05, France.*
Philippe.Gille@ens.fr

David Harari *Mathématiques, Bâtiment 425, Université Paris-Sud, Orsay, 91405 France.*
David.Harari@math.u-psud.fr

Yonatan Harpaz *Einstein Institute of Mathematics, Givat Ram, The Hebrew University of Jerusalem, Jerusalem, 91904, Israel.*
harpazy@gmail.com

Jürgen Hausen *Mathematisches Institut, Universität Tübingen, Auf der Morgenstelle 10, 72076 Tübingen, Germany.*
juergen.hausen@uni-tuebingen.de

Elaine Herppich *Mathematisches Institut, Universität Tübingen, Auf der Morgenstelle 10, 72076 Tübingen, Germany.*
elaine.herppich@uni-tuebingen.de

Felix Janda *Departement Mathematik, ETH Zürich, Rämistr. 101, 8092 Zürich, Switzerland.*
felix.janda@math.ethz.ch

Antonio Laface *Departamento de Matemática, Universidad de Concepción, Casilla 160-C, Concepción, Chile.*
antonio.laface@gmail.com

Laurent Moret-Bailly *IRMAR, Université de Rennes 1, Campus de Beaulieu, F-35042 Rennes Cedex, France.*
Laurent.Moret-Bailly@univ-rennes1.fr

Thomas Preu *Institut für Mathematik, Universität Zürich, Winterthurerstrasse 190, CH-8057 Zürich, Switzerland.*
preu@math.uzh.ch

Tomer M Schlank *Einstein Institute of Mathematics, Givat Ram, The Hebrew University of Jerusalem, Jerusalem, 91904, Israel.*
Tomer.Schlank@mail.huji.ac.il

Vera Serganova *Department of Mathematics, University of California, Berkeley, CA, 94720-3840 USA.*
serganov@math.berkeley.edu

Alexei N. Skorobogatov *Department of Mathematics, South Kensington Campus, Imperial College London, SW7 2BZ United Kingdom.*

Institute for the Information Transmission Problems, Russian Academy of Sciences, 19 Bolshoi Karetnyi, Moscow, 127994 Russia.
a.skorobogatov@imperial.ac.uk

Damiano Testa *Mathematics Institute, University of Warwick, Coventry, CV4 7AL, United Kingdom.*
adomani@gmail.com

Preface

The workshop "Torsors: theory and applications" took place at the International Centre for Mathematical Sciences in Edinburgh from 10 – 14 January 2011. It was organised by Victor Batyrev and myself, and was funded by EPSRC and LMS. The aim was to bring together mathematicians who work with torsors from different perspectives such as geometric invariant theory, algebraic geometry, representation theory, non-abelian cohomology, and those who apply torsors in arithmetic geometry.

This collection contains the lecture notes of two mini-courses presented at the workshop by Jürgen Hausen and Vera Serganova, as well as the papers contributed by participants. It also contains lecture notes of a mini-course on étale homotopy by Tomer Schlank from the study group organised in Imperial College in the Autumn of 2010 (edited by Tomer Schlank and myself).

Alexei Skorobogatov
Imperial College London

PART ONE

LECTURE NOTES

1
Three lectures on Cox rings

Jürgen Hausen
Mathematisches Institut, Universität Tübingen

Introduction

These are notes of an introductory course held at the conference "Torsors: Theory and Applications" in Edinburgh, January 2011. Cox rings play an important role in arithmetic and algebraic geometry, and in fact appeared independently in these fields [12, 13, 25]. We present basic ideas and concepts, and in particular we treat the interaction with Geometric Invariant Theory. The notes are kept as a survey; for details and proofs we refer to [4].

The first lecture begins with a rigorous definition of the Cox sheaf $\mathcal{R}$ of a variety X as a sheaf of algebras graded by the divisor class group $\mathrm{Cl}(X)$. The Cox ring is then the algebra $\mathcal{R}(X)$ of global sections. After recalling the basic correspondence between graded algebras and quasitorus actions, we discuss the relative spectrum $\widehat{X} := \mathrm{Spec}_X \mathcal{R}$ of the Cox sheaf. We call $\widehat{X} \to X$ the characteristic space; it coincides with the universal torsor [12, 38] precisely for locally factorial varieties X. The first lecture ends with a characterization of $\widehat{X}$ in terms of Geometric Invariant Theory.

The aim of the second lecture is to present a machinery for encoding varieties via their Cox ring. A first input is an explicit description of the variation of not necessarily quasiprojective good quotients for quasitorus actions on affine varieties. This allows one to encode torically embeddable varieties with finitely generated Cox ring in terms of what we call "bunched rings". Specialized to the case of toric varieties [15, 18, 33], the bunched ring data correspond to fans via Gale duality. Moreover, the language of bunched rings extends basic features of techniques developed for weighted complete intersections [17, 26] to the multigraded case. We show how to read off basic geometric properties from the defining data and briefly touch upon the relations to Mori Theory.

The third lecture is devoted to varieties with torus action. Generalizing the case of a toric variety, we describe the Cox ring in terms of the action. In the case of a complexity one action, one obtains a very explicit presentation in terms of trinomial relations. Combining this with the language of bunched rings leads to a concrete description of rational complete varieties with a complexity one action, turning them into an easy accessible class for concrete computations. We demonstrate this in the case of $\mathbb{K}^*$-surfaces. Some classification results on Fano threefolds and del Pezzo surfaces based on this approach are included in the text.

I would like to thank the referee for his careful reading of the manuscript and many helpful comments.

1.1 First lecture

1.1.1 Cox sheaves and Cox rings

We work in the category of (reduced) varieties over an algebraically closed field $\mathbb{K}$ of characteristic zero. The Cox ring $\mathcal{R}(X)$ will contain a lot of accessible information about the underlying variety X, the rough idea is to define it as

$$\mathcal{R}(X) := \bigoplus_{[D] \in \mathrm{Cl}(X)} \Gamma(X, \mathcal{O}_X(D)),$$

where the grading is by the divisor class group $\mathrm{Cl}(X)$ of X. A priori, it is not clear what the ring structure should be, in particular if there is torsion in the divisor class group $\mathrm{Cl}(X)$. Our first task is to clarify this; we closely follow [4, Chap. I].

We begin with recalling the basic notions of divisors on the normal algebraic variety X. A *prime divisor* on X is an irreducible hypersurface $D \subseteq X$. The group of *Weil divisors* on X is the free abelian group $\mathrm{WDiv}(X)$ generated by the prime divisors. We write $D \geq 0$ for a Weil divisor D, if it is a nonnegative linear combination of prime divisors. To every rational function $f \in \mathbb{K}(X)^*$ one associates its *principal divisor*

$$\operatorname{div}(f) := \sum_D \operatorname{ord}_D(f)D \in \operatorname{WDiv}(X),$$

where D runs through the prime divisors of X and $\operatorname{ord}_D(f)$ is the vanishing order of f at D. The *divisor class group* $\operatorname{Cl}(X)$ of X is the factor group of $\operatorname{WDiv}(X)$ by the subgroup $\operatorname{PDiv}(X)$ of principal divisors. To every Weil divisor D on X, one associates a *sheaf* $\mathcal{O}_X(D)$ of $\mathcal{O}_X$-modules: for any open $U \subseteq X$ one sets

$$\Gamma(U, \mathcal{O}_X(D)) := \{f \in \mathbb{K}(X)^*; \ (\operatorname{div}(f) + D)_{|U} \geq 0\} \ \cup \ \{0\},$$

where the restriction map $\operatorname{WDiv}(X) \to \operatorname{WDiv}(U)$ is defined for a prime divisor D as $D_{|U} := D \cap U$ if it intersects U and $D_{|U} := 0$ otherwise. The sheaf $\mathcal{O}_X(D)$ is reflexive, i.e. canonically isomorphic to its double dual, and of rank one. Note that for any two functions $f_1 \in \Gamma(U, \mathcal{O}_X(D_1))$ and $f_2 \in \Gamma(U, \mathcal{O}_X(D_2))$ the product $f_1 f_2$ belongs to $\Gamma(U, \mathcal{O}_X(D_1 + D_2))$.

Definition 1.1 The *sheaf of divisorial algebras* associated to a subgroup $K \subseteq \operatorname{WDiv}(X)$ is the sheaf of K-graded $\mathcal{O}_X$-algebras

$$\mathcal{S} := \bigoplus_{D \in K} \mathcal{S}_D, \qquad \mathcal{S}_D := \mathcal{O}_X(D),$$

where the multiplication in $\mathcal{S}$ is defined by multiplying homogeneous sections in the field of functions $\mathbb{K}(X)$.

The sheaf of divisorial algebras associated to a finitely generated group of Weil divisors turns out to be a sheaf of normal algebras. A crucial observation is the following:

Proposition 1.2 *Let $\mathcal{S}$ be the sheaf of divisorial algebras associated to a finitely generated group $K \subseteq \operatorname{WDiv}(X)$. If the canonical map $K \to \operatorname{Cl}(X)$ sending $D \in K$ to its class $[D] \in \operatorname{Cl}(X)$ is surjective, then $\Gamma(X, \mathcal{S})$ is a unique factorization domain.*

Example 1.3 On the projective line $X = \mathbb{P}_1$, consider $D := \{\infty\}$, the group $K := \mathbb{Z}D$, and the associated K-graded sheaf of algebras $\mathcal{S}$. Then we have isomorphisms

$$\varphi_n : \mathbb{K}[T_0, T_1]_n \to \Gamma(\mathbb{P}_1, \mathcal{S}_{nD}), \qquad f \mapsto f(1, z),$$

where $\mathbb{K}[T_0, T_1]_n \subseteq \mathbb{K}[T_0, T_1]$ denotes the vector space of all polynomials homogeneous of degree n. Putting them together we obtain a graded isomorphism

$$\mathbb{K}[T_0, T_1] \cong \Gamma(\mathbb{P}_1, \mathcal{S}).$$

Example 1.4 (The affine quadric surface) Consider the two-dimensional affine variety

$$X := V(\mathbb{K}^3;\ T_1T_2 - T_3^2) \subseteq \mathbb{K}^3.$$

We have the functions $f_i := T_{i|X}$ on X and with the prime divisors $D_1 := V(X; f_1)$ and $D_2 := V(X; f_2)$ on X, we have

$$\operatorname{div}(f_1) = 2D_1, \qquad \operatorname{div}(f_2) = 2D_2, \qquad \operatorname{div}(f_3) = D_1 + D_2.$$

For $K := \mathbb{Z}D_1$, let $\mathcal{S}$ denote the associated sheaf of divisorial algebras. Consider the sections

$$g_1 := 1 \in \Gamma(X, \mathcal{S}_{D_1}), \qquad g_2 := f_3 f_1^{-1} \in \Gamma(X, \mathcal{S}_{D_1}),$$

$$g_3 := f_1^{-1} \in \Gamma(X, \mathcal{S}_{2D_1}), \qquad g_4 := f_1 \in \Gamma(X, \mathcal{S}_{-2D_1}).$$

Then g_1, g_2 generate $\Gamma(X, \mathcal{S}_{D_1})$ as a $\Gamma(X, \mathcal{S}_0)$-module, and g_3, g_4 are inverse to each other. Moreover, we have

$$f_1 = g_1^2 g_4, \qquad f_2 = g_2^2 g_4, \qquad f_3 = g_1 g_2 g_4.$$

Thus, g_1, g_2, g_3 and g_4 generate the $\mathbb{K}$-algebra $\Gamma(X, \mathcal{S})$. Setting $\deg(Z_i) := \deg(g_i)$, we obtain a K-graded isomorphism

$$\mathbb{K}[Z_1, Z_2, Z_3^{\pm 1}] \to \Gamma(X, \mathcal{S}), \qquad Z_1 \mapsto g_1,\ Z_2 \mapsto g_2,\ Z_3 \mapsto g_3.$$

We are ready to define the Cox ring. From now on X is a normal variety with $\Gamma(X, \mathcal{O}^*) = \mathbb{K}^*$ and finitely generated divisor class group $\mathrm{Cl}(X)$. The idea is to start with the sheaf $\mathcal{S}$ of divisorial algebras associated to a group K of Weil divisors projecting onto $\mathrm{Cl}(X)$ and then to identify systematically isomorphic homogeneous components $\mathcal{S}_D$ and $\mathcal{S}'_D$ via dividing by a suitable sheaf of ideals.

Construction 1.5 Fix a subgroup $K \subseteq \mathrm{WDiv}(X)$ such that the map $c\colon K \to \mathrm{Cl}(X)$ sending $D \in K$ to its class $[D] \in \mathrm{Cl}(X)$ is surjective. Let $K^0 \subseteq K$ be the kernel of c, and let $\chi\colon K^0 \to \mathbb{K}(X)^*$ be a character, i.e. a group homomorphism, with

$$\operatorname{div}(\chi(E)) = E, \qquad \text{for all } E \in K^0.$$

Let $\mathcal{S}$ be the sheaf of divisorial algebras associated to K and denote by $\mathcal{I}$ the sheaf of ideals of $\mathcal{S}$ locally generated by the sections $1 - \chi(E)$, where 1 is homogeneous of degree zero, E runs through K^0 and $\chi(E)$

is homogeneous of degree $-E$. The *Cox sheaf* associated to K and χ is the quotient sheaf $\mathcal{R} := \mathcal{S}/\mathcal{I}$ together with the $\mathrm{Cl}(X)$-grading

$$\mathcal{R} = \bigoplus_{[D]\in \mathrm{Cl}(X)} \mathcal{R}_{[D]}, \qquad \mathcal{R}_{[D]} := \pi\left(\bigoplus_{D'\in c^{-1}([D])} \mathcal{S}_{D'}\right).$$

where $\pi\colon \mathcal{S} \to \mathcal{R}$ denotes the projection. The Cox sheaf $\mathcal{R}$ is a quasicoherent sheaf of $\mathrm{Cl}(X)$-graded $\mathcal{O}_X$-algebras. The *Cox ring* is the ring of global sections

$$\mathcal{R}(X) := \bigoplus_{[D]\in \mathrm{Cl}(X)} \mathcal{R}_{[D]}(X), \qquad \mathcal{R}_{[D]}(X) := \Gamma(X, \mathcal{R}_{[D]}).$$

In general, the Cox sheaf is not a sheaf of divisorial algebras. However, the following shows that it is not too far away from being so.

Remark 1.6 Suppose that we have the same situation as in Construction 1.5. Then, for every $D \in K$, we have an isomorphism of sheaves

$$\pi_{|\mathcal{S}_D} : \mathcal{S}_D \to \mathcal{R}_{[D]}.$$

Moreover, for every open subset $U \subseteq X$, we have a canonical isomorphism on the level of sections

$$\Gamma(U, \mathcal{S})/\Gamma(U, \mathcal{I}) \cong \Gamma(U, \mathcal{S}/\mathcal{I}).$$

In particular, the Cox ring $\mathcal{R}(X)$ is isomorphic to the quotient

$$\Gamma(X, \mathcal{S})/\Gamma(X, \mathcal{I})$$

of global sections.

Proposition 1.7 *If K, χ and K', χ' are data as in Construction 1.5, then the associated Cox sheaves are isomorphic as graded sheaves.*

Example 1.8 (The affine quadric surface, continued) Look again at the two-dimensional affine variety discussed in Example 1.4:

$$X := V(\mathbb{K}^3; T_1T_2 - T_3^2) \subseteq \mathbb{K}^3.$$

The divisor class group $\mathrm{Cl}(X)$ is of order two; it is generated by $[D_1]$. The kernel of the projection $K \to \mathrm{Cl}(X)$ is $K^0 = 2\mathbb{Z}D_1$ and a character as in Construction 1.5 is

$$\chi\colon K^0 \to \mathbb{K}(X)^*, \qquad 2nD_1 \mapsto f_1^n.$$

The ideal $\mathcal{I}$ is globally generated by the section $1 - f_1$, where $f_1 \in \Gamma(X, \mathcal{S}_{-2D_1})$. Consequently, the Cox ring of X is given as

$$\mathcal{R}(X) \;\cong\; \Gamma(X,\mathcal{S})/\Gamma(X,\mathcal{I}) \;\cong\; \mathbb{K}[Z_1, Z_2, Z_3^{\pm 1}]/\langle 1 - Z_3^{-1}\rangle \;\cong\; \mathbb{K}[Z_1, Z_2],$$

where the $\mathrm{Cl}(X)$-grading on the polynomial ring $\mathbb{K}[Z_1, Z_2]$ is given by $\deg(Z_1) = \deg(Z_2) = [D_1]$.

The construction of Cox sheaves (and thus also Cox rings) of a variety can be made canonical by fixing a suitable point. We say that $x \in X$ is *factorial* if the local ring $\mathcal{O}_{X,x}$ is factorial; that means that every Weil divisor is principal near x.

Construction 1.9 Let $x \in X$ be a factorial point and consider the sheaf of divisorial algebras $\mathcal{S}^x$ associated to the group of Weil divisors avoiding x:

$$K^x \;:=\; \{D \in \mathrm{WDiv}(X);\; x \notin \mathrm{Supp}(D)\} \;\subseteq \mathrm{WDiv}(X).$$

Let $K^{x,0} \subseteq K^x$ be the subgroup consisting of principal divisors. Then, for every $E \in K^{x,0}$, there is a unique $f_E \in \Gamma(X, \mathcal{S}_{-E})$, which is defined near x and satisfies

$$\mathrm{div}(f_E) \;=\; E, \qquad\qquad f_E(x) \;=\; 1.$$

The map $\chi^x : K^x \to \mathbb{K}(X)^*$ sending E to f_E is a character as in Construction 1.5. We call the resulting Cox sheaf $\mathcal{R}^x$ the *canonical Cox sheaf of the pointed space* (X, x).

As noted in Remark 1.6, the homogeneous components of the Cox sheaf are isomorphic to reflexive rank one sheaves $\mathcal{O}_X(D)$. The following associates to every homogeneous section of the Cox ring a divisor, not depending on the choices made in Construction 1.5.

Construction 1.10 In the setting of Construction 1.5, let $D \in K$ and consider an element $0 \neq f \in \mathcal{R}_{[D]}(X)$. By Remark 1.6, there is a (unique) $\widetilde{f} \in \Gamma(X, \mathcal{S}_D)$ with $\pi(\widetilde{f}) = f$. The $[D]$*-divisor* and the $[D]$*-localization* of f are

$$\mathrm{div}_{[D]}(f) \;:=\; \mathrm{div}(\widetilde{f}) + D, \qquad\qquad X_{[D],f} \;:=\; X \setminus \mathrm{Supp}(\mathrm{div}_{[D]}(f)).$$

We now discuss basic algebraic properties of the Cox ring. Recall that a *Krull ring* is an integral ring R with a family $(\nu_i)_{i \in I}$ of discrete valuations such that for every non-zero f in the quotient field $Q(R)$, one has $\nu_i(f) \neq 0$ only for finitely many $i \in I$ and $f \in R$ if and only if $\nu_i(f) \geq 0$ holds for all $i \in I$.

Theorem 1.11 *The Cox ring $\mathcal{R}(X)$ is an (integral) normal Krull ring. Moreover, one has the following statements on localization and units.*

(i) *For every non-zero homogeneous $f \in \mathcal{R}_{[D]}(X)$, there is a canonical isomorphism*

$$\Gamma(X, \mathcal{R})_f \cong \Gamma(X_{[D],f}, \mathcal{R}).$$

(ii) *Every homogeneous unit of $\mathcal{R}(X)$ is constant. If X is complete then even $\mathcal{R}(X)^* = \mathbb{K}^*$ holds.*

Finally, we take a closer look at divisibility properties of Cox rings. In general, they are not factorial, i.e. unique factorization domains, but the following weaker property will be guaranteed.

Definition 1.12 Consider an abelian group K and a K-graded integral $\mathbb{K}$-algebra $R = \oplus_K R_w$.

(i) A non-zero non-unit $f \in R$ is *K-prime* if it is homogeneous and $f|gh$ with homogeneous $g, h \in R$ implies $f|g$ or $f|h$.
(ii) We say that R is *factorially graded* if every homogeneous non-zero non-unit $f \in R$ is a product of K-primes.

Remark 1.13 The concepts "factorially graded" and "factorial" coincide if the grading group is torsion free. As soon as we have torsion in the grading group, they may differ. For example, consider

$$R := \mathbb{K}[T_1, T_2, T_3]/\langle T_1^2 + T_2^2 + T_3^2 \rangle.$$

This is not a factorial ring, but it becomes factorially graded by the group $K = \mathbb{Z} \oplus \mathbb{Z}/2\mathbb{Z} \oplus \mathbb{Z}/2\mathbb{Z}$ when we set

$$\deg(T_1) := (1, \bar{0}, \bar{0}), \quad \deg(T_2) := (1, \bar{1}, \bar{0}), \quad \deg(T_3) := (1, \bar{0}, \bar{1}).$$

Theorem 1.14 *Suppose that the Cox ring $\mathcal{R}(X)$ satisfies $\mathcal{R}(X)^* = \mathbb{K}^*$; for example, assume X to be complete. Then $\mathcal{R}(X)$ is factorially $\mathrm{Cl}(X)$-graded. If moreover $\mathrm{Cl}(X)$ is torsion free, then $\mathcal{R}(X)$ is factorial.*

The following statement shows that divisibility in the Cox ring $\mathcal{R}(X)$ can be formulated geometrically in terms of $[D]$-divisors on the underlying variety X.

Proposition 1.15 *Suppose that the Cox ring $\mathcal{R}(X)$ satisfies $\mathcal{R}(X)^* = \mathbb{K}^*$.*

(i) *An element $0 \neq f \in \Gamma(X, \mathcal{R}_{[D]})$ divides $0 \neq g \in \Gamma(X, \mathcal{R}_{[E]})$ if and only if $\mathrm{div}_{[D]}(f) \leq \mathrm{div}_{[E]}(g)$ holds.*

(ii) *An element* $0 \neq f \in \Gamma(X, \mathcal{R}_{[D]})$ *is* $\mathrm{Cl}(X)$*-prime if and only if the divisor* $\mathrm{div}_{[D]}(f) \in \mathrm{WDiv}(X)$ *is prime.*

1.1.2 Quasitorus actions

Here we recall the correspondence between affine algebras A graded by a finitely generated abelian group K and affine varieties X coming with an action of a quasitorus H. Moreover, we discuss good quotients and their basic properties. Standard references are [8, 28, 31, 35, 39].

An *affine algebraic group* is an affine variety G together with a group structure such that the group operations are morphisms. A *morphism of affine algebraic groups* is a morphism of the underlying varieties which is moreover a group homomorphism. A *character* is a morphism $G \to \mathbb{K}^*$ to the multiplicative group of the ground field. Endowed with pointwise multiplication, the characters of G form an abelian group $\mathbb{X}(G)$.

Definition 1.16 A *quasitorus*, also called a *diagonalizable group*, is an affine algebraic group H whose algebra of regular functions $\Gamma(H, \mathcal{O})$ is generated as a $\mathbb{K}$-vector space by the characters $\chi \in \mathbb{X}(H)$. A *torus* is a connected quasitorus.

Example 1.17 The *standard n-torus* is $\mathbb{T}^n := (\mathbb{K}^*)^n$. Its characters are precisely the Laurent monomials $T^\nu = T_1^{\nu_1} \cdots T_n^{\nu_n}$, where $\nu \in \mathbb{Z}^n$, and its algebra of regular functions is the Laurent polynomial algebra

$$\Gamma(\mathbb{T}^n, \mathcal{O}) \;=\; \mathbb{K}[T_1^{\pm 1}, \ldots, T_n^{\pm 1}] \;=\; \bigoplus_{\nu \in \mathbb{Z}^n} \mathbb{K} \cdot T^\nu \;=\; \mathbb{K}[\mathbb{Z}^n].$$

To any finitely generated abelian group K one associates in a functorial way a quasitorus, namely $H := \mathrm{Spec}\, \mathbb{K}[K]$, the spectrum of the group algebra $\mathbb{K}[K]$.

Remark 1.18 The quasitorus $H := \mathrm{Spec}\, \mathbb{K}[K]$ can be realized as a closed subgroup of a standard r-torus as follows. By the elementary divisors theorem, we find generators $w_1, \ldots, w_r$ of K such that the epimorphism $\mathbb{Z}^r \to K$, $e_i \mapsto w_i$ has kernel

$$\mathbb{Z}a_1e_1 \oplus \ldots \oplus \mathbb{Z}a_se_s \;\subseteq\; \mathbb{Z}^r, \qquad a_1, \ldots, a_s \in \mathbb{Z}_{\geq 1}.$$

The corresponding morphism $H \to \mathbb{T}^r$ is a closed embedding realizing H as the kernel of $\mathbb{T}^r \to \mathbb{T}^s$, $(t_1, \ldots, t_r) \mapsto (t_1^{a_1}, \ldots, t_s^{a_s})$. In particular, we see that H is a direct product of a torus and a finite abelian group:

$$H \;\cong\; C(a_1) \times \ldots \times C(a_s) \times \mathbb{T}^{r-s}, \qquad C(a_i) \;:=\; \{\zeta \in \mathbb{K}^*;\ \zeta^{a_i} = 1\}.$$

Theorem 1.19 *We have contravariant exact functors being essentially inverse to each other:*

$$\begin{array}{rcl} \{\textit{finitely generated abelian groups}\} & \longleftrightarrow & \{\textit{quasitori}\} \\ K & \mapsto & \operatorname{Spec} \mathbb{K}[K], \\ \mathbb{X}(H) & \leftarrow & H. \end{array}$$

Under these equivalences, the free finitely generated abelian groups correspond to the tori.

Quasitori are linearly reductive in characteristic zero: every rational representation even splits into one-dimensional ones. Applying this to the representation on the algebra of global functions of an affine variety with quasitorus action, we obtain a grading by the character group in the following way.

Remark 1.20 Let a quasitorus H act on a not necessarily affine variety X. Then the algebra $\Gamma(X, \mathcal{O})$ becomes $\mathbb{X}(H)$-graded via

$$\Gamma(X, \mathcal{O}) = \bigoplus_{\chi \in \mathbb{X}(H)} \Gamma(X, \mathcal{O})_\chi,$$
$$\Gamma(X, \mathcal{O})_\chi := \{f \in \Gamma(X, \mathcal{O});\ f(h \cdot x) = \chi(h) f(x)\}.$$

We now associate in functorial manner to every affine algebra $A = \oplus_K A_w$ graded by a finitely generated abelian group K the affine variety $X = \operatorname{Spec} A$ with an action of the quasitorus $H = \operatorname{Spec} \mathbb{K}[K]$. Again, this can be made concrete.

Construction 1.21 Let K be a finitely generated abelian group and A a K-graded affine algebra. Set $X = \operatorname{Spec} A$. If $f_i \in A_{w_i}$, $i = 1, \ldots, r$, generate A, then we have a closed embedding

$$X \to \mathbb{K}^r, \qquad x \mapsto (f_1(x), \ldots, f_r(x)),$$

and $X \subseteq \mathbb{K}^r$ is invariant under the diagonal action of $H = \operatorname{Spec} \mathbb{K}[K]$ given by the characters $\chi^{w_1}, \ldots, \chi^{w_r}$. Note that for any $f \in A$ homogeneity is characterized by

$$f \in A_w \ \text{ if and only if } \ f(h \cdot x) = \chi^w(h) f(x) \text{ for all } h \in H,\ x \in X.$$

Theorem 1.22 *We have contravariant functors being essentially in-*

verse to each other:

$$\{\textit{graded affine algebras}\} \longleftrightarrow \{\textit{affine varieties with quasitorus action}\}$$
$$A \mapsto \operatorname{Spec} A,$$
$$\Gamma(X, \mathcal{O}) \leftarrow\!\shortmid X.$$

Definition 1.23 Let G be a linearly reductive affine algebraic group G acting on a variety X. A morphism $p\colon X \to Y$ is called a *good quotient* for this action if it has the following properties:

(i) $p\colon X \to Y$ is affine and G-invariant,
(ii) the pullback $p^*\colon \mathcal{O}_Y \to (p_*\mathcal{O}_X)^G$ is an isomorphism.

A morphism $p\colon X \to Y$ is called a *geometric quotient* if it is a good quotient and its fibers are precisely the G-orbits.

Remark 1.24 Let a linearly reductive group G act on an affine variety X. Then Hilbert's finiteness theorem ensures that the algebra of invariants $\Gamma(X, \mathcal{O})^G$ is finitely generated. This gives a good quotient $p\colon X \to Y$ with $Y := \operatorname{Spec} \Gamma(X, \mathcal{O})^G$.

Remark 1.25 Let $A = \oplus_K A_w$ be an affine algebra graded by a finitely generated abelian group K. Then $H = \operatorname{Spec} \mathbb{K}[K]$ acts on $X = \operatorname{Spec} A$ and we have

$$\Gamma(X, \mathcal{O})^H = A_0.$$

In order to compute A_0, choose homogeneous generators $f_1, \dots, f_r$ of A and consider the homomorphism

$$Q\colon \mathbb{Z}^r \to K, \qquad e_i \mapsto \deg(f_i).$$

Then, for any set B of generators of the monoid $\mathbb{Z}^r_{\geq 0} \cap \ker(Q)$, the algebra A_0 of invariants is generated by the products $f_1^{\nu_1} \cdots f_r^{\nu_r}$, where $\nu \in B$.

The basic properties of good quotients are that they map closed invariant subsets to closed sets and that they separate disjoint closed invariant sets. An immediate consequence is that the target space carries the quotient topology. Another application is the following statement on the fibers.

Proposition 1.26 *Let a linearly reductive algebraic group G act on a variety X, and let $p\colon X \to Y$ be a good quotient. Then p is surjective and for any $y \in Y$ one has:*

(i) *There is exactly one closed G-orbit $G \cdot x$ in the fiber $p^{-1}(y)$.*

(ii) *Every orbit* $G{\cdot}x' \subseteq p^{-1}(y)$ *has* $G{\cdot}x$ *in its closure.*

The first statement means that a good quotient $p\colon X \to Y$ parametrizes the closed orbits of the G-variety X. Using the description of the fibers one easily verifies that a good quotient is categorical, i.e. universal with respect to invariant morphisms. In particular, the quotient space is unique up to isomorphism which justifies the common notation $X /\!\!/ G$.

Example 1.27 Consider the $\mathbb{K}^*$-action $t\cdot(z_1, z_2) = (t^a z_1, t^b z_2)$ on $\mathbb{K}^2$. The following three cases are typical.

(i) We have $a = b = 1$. Every $\mathbb{K}^*$-invariant function is constant and the constant map $p\colon \mathbb{K}^2 \to \{\mathrm{pt}\}$ is a good quotient.

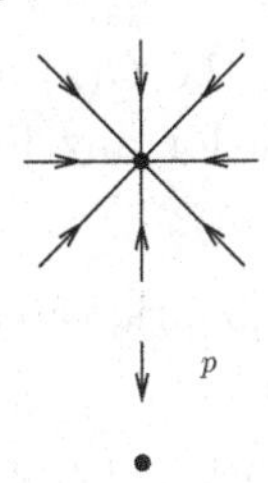

(ii) We have $a = 0$ and $b = 1$. The algebra of $\mathbb{K}^*$-invariant functions is generated by z_1 and the map $p\colon \mathbb{K}^2 \to \mathbb{K}$, $(z_1, z_2) \mapsto z_1$ is a good quotient.

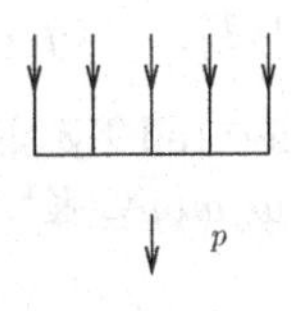

(iii) We have $a = 1$ and $b = -1$. The algebra of $\mathbb{K}^*$-invariant functions is generated by $z_1 z_2$ and $p\colon \mathbb{K}^2 \to \mathbb{K}$, $(z_1, z_2) \mapsto z_1 z_2$ is a good quotient.

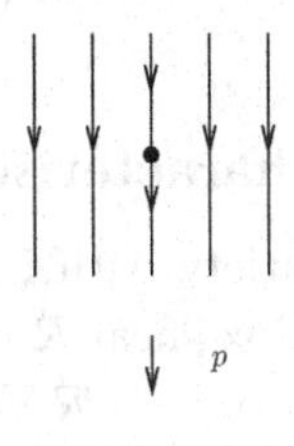

Note that the general p-fiber is a single $\mathbb{K}^*$-orbit, whereas $p^{-1}(0)$ consists of three orbits and is reducible.

Example 1.28 (The affine quadric surface as a quotient) Consider the action of the multiplicative group $C(2) = \{1, -1\}$ on $\mathbb{K}^2$ given by

$$\zeta \cdot z := (\zeta z_1, \zeta z_2).$$

This action has 0 as a fixed point and any $z \neq 0$ has trivial isotropy group. The algebra $A \subseteq \mathbb{K}^2$ of invariants is generated by

$$f_{11} := T_1^2, \qquad f_{22} := T_2^2, \qquad f_{12} := T_1 T_2.$$

The ideal of relations among them is generated by the polynomial $f_{11} f_{22} - f_{12}^2$. Consequently, with $X := V(w_1 w_2 - w_3^2) \subseteq \mathbb{K}^3$, the quotient map is

$$\pi \colon \mathbb{K}^3 \to X, \qquad z \mapsto (f_{11}(z), f_{22}(z), f_{12}(z)).$$

The fibers of π are precisely the $C(2)$-orbits. In particular, π is a geometric quotient (as it holds for any finite group action on an affine variety).

Example 1.29 (The affine quadric threefold as a quotient) Consider the action of $\mathbb{K}^*$ on $\mathbb{K}^4$ given by

$$t \cdot z := (tz_1, t^{-1} z_2, tz_3, t^{-1} z_4).$$

Note that $\mathbb{K}^*$ has 0 as a fixed point and any $z \neq 0$ has trivial isotropy group. The algebra $A \subseteq \mathbb{K}[T_1, \ldots, T_4]$ of invariants is generated by

$$f_{12} := T_1 T_2, \qquad f_{34} := T_3 T_4, \qquad f_{14} := T_1 T_4, \qquad f_{23} := T_2 T_3.$$

The ideal of relations is generated by the polynomial $f_{12} f_{34} - f_{14} f_{23}$. Thus, with $X := V(w_1 w_2 - w_3 w_4) \subseteq \mathbb{K}^4$, the quotient map is

$$\pi \colon \mathbb{K}^4 \to X, \qquad z \mapsto (f_{12}(z), f_{34}(z), f_{14}(z), f_{23}(z)).$$

The fiber over any point $0 \neq x$ is a free $\mathbb{K}^*$-orbit and hence isomorphic to $\mathbb{K}^*$, whereas the fiber over the point $0 \in X$ is given by

$$\pi^{-1}(0) = V(T_1, T_3) \cup V(T_2, T_4) \subseteq \mathbb{K}^4.$$

1.1.3 Characteristic spaces

As before, X is a normal variety with $\Gamma(X, \mathcal{O}^*) = \mathbb{K}^*$. We discuss the geometric counterpart of a Cox sheaf $\mathcal{R}$ on X, its relative spectrum. In order to obtain a reasonable object, $\mathcal{R}$ should be locally of finite type. By Theorem 1.11, this holds if the Cox ring $\mathcal{R}(X)$ is finitely generated. Moreover, one can show that $\mathcal{R}$ is locally of finite type for every $\mathbb{Q}$-factorial X.

Construction 1.30 Let $\mathcal{R}$ be a Cox sheaf on X and suppose that $\mathcal{R}$ is locally of finite type. Then the relative spectrum

$$\widehat{X} := \mathrm{Spec}_X(\mathcal{R})$$

is a quasiaffine variety. The $\mathrm{Cl}(X)$-grading of the sheaf $\mathcal{R}$ defines an action of the diagonalizable group

$$H_X := \mathrm{Spec}\ \mathbb{K}[\mathrm{Cl}(X)]$$

on $\widehat{X}$. The canonical morphism $q_X : \widehat{X} \to X$ is a good quotient for this action, and we have an isomorphism of graded sheaves

$$\mathcal{R} \cong (q_X)_*(\mathcal{O}_{\widehat{X}}).$$

We call $q_X : \widehat{X} \to X$ the *characteristic space* associated to $\mathcal{R}$, and H_X the *characteristic quasitorus* of X.

In the case of a locally factorial variety X, the characteristic space coincides with the universal torsor over X. As soon as X has non-factorial singularities the two concepts differ from each other, as we will indicate below.

Proposition 1.31 *Consider the characteristic space $q_X : \widehat{X} \to X$.*

(i) *The inverse image $q_X^{-1}(X_{\mathrm{reg}})$ of the set of smooth points is smooth, H_X acts freely there and $q_X^{-1}(X_{\mathrm{reg}}) \to X_{\mathrm{reg}}$ is an étale H_X-principal bundle.*
(ii) *For any closed set $A \subseteq X$ of codimension at least two, the inverse image $q_X^{-1}(A) \subseteq \widehat{X}$ is also of codimension at least two.*
(iii) *Let $\widehat{x} \in \widehat{X}$ be a point such that the orbit $H_X \cdot \widehat{x} \subseteq \widehat{X}$ is closed, and consider an element $f \in \Gamma(X, \mathcal{R}_{[D]})$. Then we have*

$$f(\widehat{x}) = 0 \text{ if and only if } q_X(\widehat{x}) \in \mathrm{Supp}(\mathrm{div}_{[D]}(f)).$$

We now relate properties of the H_X-action to geometric properties on X. For $x \in X$, let $\mathrm{PDiv}(X, x) \subseteq \mathrm{WDiv}(X)$ denote the subgroup of all Weil divisors, which are principal on some neighbourhood of x. We define the *local class group* of X at x to be the factor group

$$\mathrm{Cl}(X, x) := \mathrm{WDiv}(X) / \mathrm{PDiv}(X, x).$$

Obviously the group $\mathrm{PDiv}(X)$ of principal divisors is contained in $\mathrm{PDiv}(X, x)$. Thus, there is a canonical epimorphism $\pi_x : \mathrm{Cl}(X) \to \mathrm{Cl}(X, x)$. We denote by $H_{X,\widehat{x}} \subseteq H_X$ the isotropy group of $\widehat{x} \in \widehat{X}$.

Proposition 1.32 *Consider the characteristic space $q_X \colon \widehat{X} \to X$. Given $x \in X$, fix a point $\widehat{x} \in q_X^{-1}(x)$ with closed H_X-orbit. Then we have a canonical isomorphism*

$$\mathrm{Cl}(X, x) \cong \mathbb{X}(H_{X,\widehat{x}}).$$

Recall that a point $x \in X$ is factorial if and only if near x every Weil divisor is principal. We say that $x \in X$ is *$\mathbb{Q}$-factorial* if near x for every Weil divisor some multiple is principal.

Corollary 1.33 *Consider the characteristic space $q_X : \widehat{X} \to X$.*

(i) *A point $x \in X$ is factorial if and only if the fiber $q_X^{-1}(x)$ is a single H_X-orbit with trivial isotropy.*
(ii) *A point $x \in X$ is $\mathbb{Q}$-factorial if and only if the fiber $q_X^{-1}(x)$ is a single H_X-orbit.*

The first part of the following says in particular that characteristic space and universal torsor coincide if and only if X has at most factorial singularities.

Corollary 1.34 *Consider the characteristic space $q_X : \widehat{X} \to X$.*

(i) *The action of H_X on $\widehat{X}$ is free if and only if X is locally factorial.*
(ii) *The good quotient $q_X : \widehat{X} \to X$ is geometric if and only if X is $\mathbb{Q}$-factorial.*

Recall that the *Picard group* of X is the factor group of the group $\mathrm{CDiv}(X)$ of locally principal Weil divisors by the subgroup of principal divisors:

$$\mathrm{Pic}(X) \;=\; \mathrm{CDiv}(X)/\,\mathrm{PDiv}(X) \;=\; \bigcap_{x \in X} \ker(\pi_x).$$

Corollary 1.35 *Consider the characteristic space $q_X : \widehat{X} \to X$. Let $\widehat{H}_X \subseteq H_X$ be the subgroup generated by all isotropy groups $H_{X,\widehat{x}}$, where $\widehat{x} \in \widehat{X}$. Then we have*

$$\ker\big(\mathbb{X}(H_X) \to \mathbb{X}(\widehat{H}_X)\big) = \bigcap_{\widehat{x} \in \widehat{X}} \ker\big(\mathbb{X}(H_X) \to \mathbb{X}(H_{X,\widehat{x}})\big)$$

and the projection $H_X \to H_X/\widehat{H}_X$ corresponds to the inclusion $\mathrm{Pic}(X) \subseteq \mathrm{Cl}(X)$ of character groups.

Corollary 1.36 *If the variety $\widehat{X}$ contains an H_X-fixed point, then the Picard group $\mathrm{Pic}(X)$ is trivial.*

As we noted, the characteristic space of a variety X is a quasiaffine variety $\widehat{X}$ with an action of the characteristic quasitorus H_X having X as a good quotient. Our next aim is to characterize this situation in terms of Geometric Invariant Theory.

Definition 1.37 Let G be an affine algebraic group and W a G-variety. We say that the G-action on W is *strongly stable* if there is an open invariant subset $W' \subseteq W$ with the following properties:

(i) the complement $W \setminus W'$ is of codimension at least two in W,
(ii) the group G acts freely, i.e. with trivial isotropy groups, on W',
(iii) for every $x \in W'$ the orbit $G \cdot x$ is closed in W.

Definition 1.38 Let a group G act on a normal variety Y. A divisor $\sum a_D D$ is called *G-invariant* if its multiplicities satisfy $a_{g \cdot D} = a_D$ for every $g \in G$. We say that Y is *G-factorial* if every G-invariant divisor is principal.

Note that a quasiaffine variety with a quasitorus H acting on it is H-factorial if and only if its ring of functions is factorially graded by the character group $\mathbb{X}(H)$.

Remark 1.39 Let X be a normal variety with characteristic space $q_X : \widehat{X} \to X$. Then $\widehat{X}$ is H_X-factorial and $q_X^{-1}(X_{\mathrm{reg}}) \subseteq \widehat{X}$ satisfies the required properties of $W' \subseteq W$ of Definition 1.37.

Theorem 1.40 *Let a quasitorus H act on a normal quasiaffine variety $\mathcal{X}$ with a good quotient $q \colon \mathcal{X} \to X$. Assume that $\Gamma(\mathcal{X}, \mathcal{O}^*) = \mathbb{K}^*$ holds, $\mathcal{X}$ is H-factorial and the H-action is strongly stable. Then there is a commutative diagram*

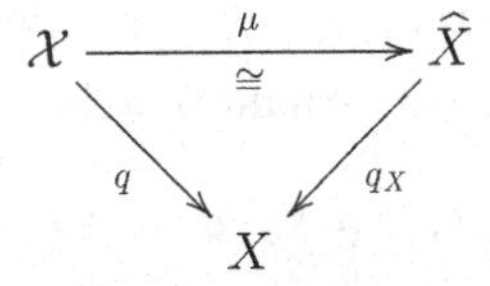

where the quotient space X is a normal variety with $\Gamma(X, \mathcal{O}_X^*) = \mathbb{K}^*$, we have $\mathrm{Cl}(X) = \mathbb{X}(H)$ and $q_X : \widehat{X} \to X$ is a characteristic space for X and the isomorphism $\mu \colon \mathcal{X} \to \widehat{X}$ is equivariant with respect to the actions of $H = H_X$.

Example 1.41 (The affine quadric surface, continued) In Example 1.28, we realized the surface $X = V(T_1T_2 - T_3^2)$ as a strongly stable quotient of $\mathbb{K}^2$ by $\mathbb{Z}/2\mathbb{Z}$. Thus, $\mathrm{Cl}(X)$ is of order two and $\mathbb{K}^2 \to X$ is a characteristic space. Moreover, from Corollary 1.36 we infer $\mathrm{Pic}(X) = 0$.

Example 1.42 (The affine quadric threefold, continued) In Example 1.29, we realized the affine threefold $X = V(T_1T_2 - T_3T_4)$ as a strongly stable quotient of $\mathbb{K}^4$ by $\mathbb{K}^*$. Thus, we have $\mathrm{Cl}(X) \cong \mathbb{Z}$ and $\mathbb{K}^4 \to X$ is a characteristic space. Moreover, from Corollary 1.36 we infer $\mathrm{Pic}(X) = 0$.

These two examples are special cases of the much bigger class of *toric varieties*, i.e., normal varieties X endowed with a torus action $T \times X \to X$ such that for some $x_0 \in X$ the orbit map $T \to X$, $t \mapsto t \cdot x_0$ is an open embedding. Toric varieties admit a complete description in terms of lattice fans; standard references are [14, 15, 18, 33]. In this picture, the Cox ring and characteristic space look as follows; see [6, 7, 13, 32].

Construction 1.43 Assume that the toric variety X arises from a fan Σ in a lattice N. The condition $\Gamma(X, \mathcal{O}^*) = \mathbb{K}^*$ means that the primitive vectors $v_1, \ldots, v_r \in N$ on the rays of Σ generate $N_{\mathbb{Q}}$ as a vector space.

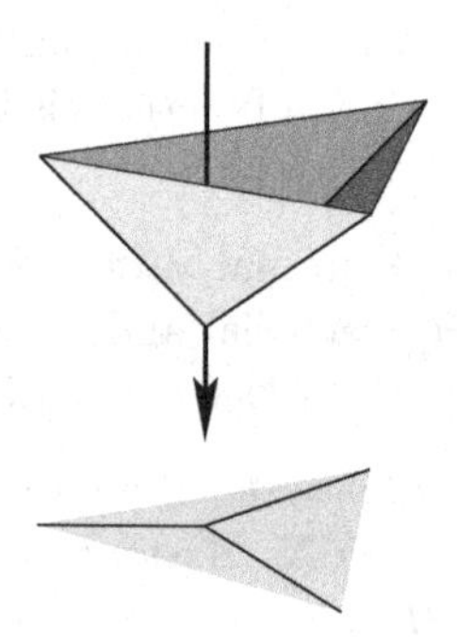

Set $F := \mathbb{Z}^r$ and consider the linear map $P\colon F \to N$ sending the i-th canonical base vector $f_i \in F$ to $v_i \in N$. There is a fan $\widehat{\Sigma}$ in F consisting of certain faces of the positive orthant $\delta \subseteq F_{\mathbb{Q}}$, namely

$$\widehat{\Sigma} \ := \ \{\widehat{\sigma} \preceq \delta;\ P(\widehat{\sigma}) \subseteq \sigma \text{ for some } \sigma \in \Sigma\}.$$

The fan $\widehat{\Sigma}$ defines an open toric subvariety $\widehat{X}$ of $\overline{X} = \mathrm{Spec}(\mathbb{K}[\delta^\vee \cap E])$, where $E := \mathrm{Hom}(F, \mathbb{Z})$. Note that all rays $\mathrm{cone}(f_1), \ldots, \mathrm{cone}(f_r)$ of the positive orthant $\delta \subseteq F_{\mathbb{Q}}$ belong to $\widehat{\Sigma}$ and thus we have

$$\Gamma(\widehat{X}, \mathcal{O}) \ = \ \Gamma(\overline{X}, \mathcal{O}) \ = \ \mathbb{K}[\delta^\vee \cap E].$$

As $P\colon F \to N$ is a map of the fans $\widehat{\Sigma}$ and Σ, i.e. it sends cones of $\widehat{\Sigma}$ into cones of Σ, it defines a morphism $p\colon \widehat{X} \to X$ of toric varieties. Note that we have $\overline{X} = \mathbb{K}^r$ and in terms of the coordinates $T_i = \chi^{e_i}$, the open subset $\widehat{X} \subseteq \overline{X}$ is given as

$$\widehat{X} = \overline{X} \setminus V(T^\sigma;\, \sigma \in \Sigma), \qquad T^\sigma = T_1^{\varepsilon_1} \cdots T_r^{\varepsilon_r}, \quad \varepsilon_i = \begin{cases} 1, & v_i \notin \sigma, \\ 0, & v_i \in \sigma. \end{cases}$$

Now, consider the dual map $P^* \colon M \to E$, where $M := \mathrm{Hom}(N, \mathbb{Z})$, set $K := E/P^*(M)$ and denote by $Q \colon E \to K$ the projection. We obtain a K-grading of the polynomial ring $\mathbb{K}[T_1, \dots, T_r]$ by setting

$$\deg(T_i) := Q(e_i) \in K.$$

This gives an action of $H := \mathrm{Spec}\, \mathbb{K}[K]$ on $\overline{X}$. The set $\widehat{X} \subseteq \overline{X}$ is invariant and $p \colon \widehat{X} \to X$ is a good quotient. Moreover,

$$W := \overline{X} \setminus \bigcup_{i \neq j} V(T_i, T_j) \subseteq \widehat{X}$$

satisfies the conditions of Definition 1.37 for this action. Thus, $p \colon \widehat{X} \to X$ is a characteristic space for X. In particular, we have

$$\mathrm{Cl}(X) \cong K, \qquad \mathcal{R}(X) \cong \mathbb{K}[T_1, \dots, T_r].$$

1.2 Second lecture

1.2.1 Variation of good quotients

Given a variety X with an action of a linearly reductive group G, the task of Geometric Invariant Theory is to describe the *good G-sets*, i.e. the invariant open subsets $U \subseteq X$ admitting a good quotient $U \to U /\!\!/ G$. In general, there will be several such good G-sets; this effect is also called "variation of good quotients". For details, see [4, Sec. III.1] and the original references [5, 9, 31].

Example 1.44 For the $\mathbb{K}^*$-action $t \cdot (z_1, z_2) = (t^a z_1, t^b z_2)$ on $\mathbb{K}^2$; as in Example 1.27, we consider the three typical cases.

(i) Let $a = b = 1$. Besides the good quotient $\mathbb{K}^2 \to \{\mathrm{pt}\}$, there is a nice geometric quotient $\mathbb{K}^2 \setminus \{0\} \to \mathbb{P}_1$, $(z_1, z_2) \mapsto [z_1, z_2]$.

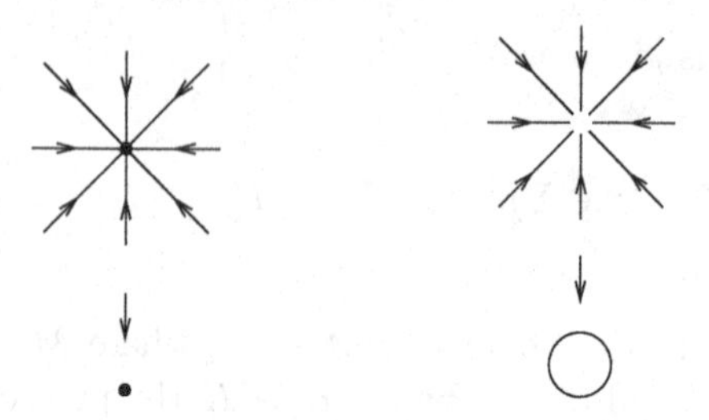

(ii) Let $a = 0$ and $b = 1$. Besides the good quotient $\mathbb{K}^2 \to \mathbb{K}$, $(z_1, z_2) \mapsto z_1$, there is a geometric quotient $\mathbb{K}^2 \setminus V(T_2) \to \mathbb{K}$, $(z_1, z_2) \mapsto z_1$.

(iii) Let $a = 1$ and $b = -1$. Besides the good quotient $\mathbb{K}^2 \to \mathbb{K}$, $(z_1, z_2) \mapsto z_1 z_2$, there are two geometric ones: $\mathbb{K}^2 \setminus V(T_i) \to \mathbb{K}$, $(z_1, z_2) \mapsto z_1 z_2$.

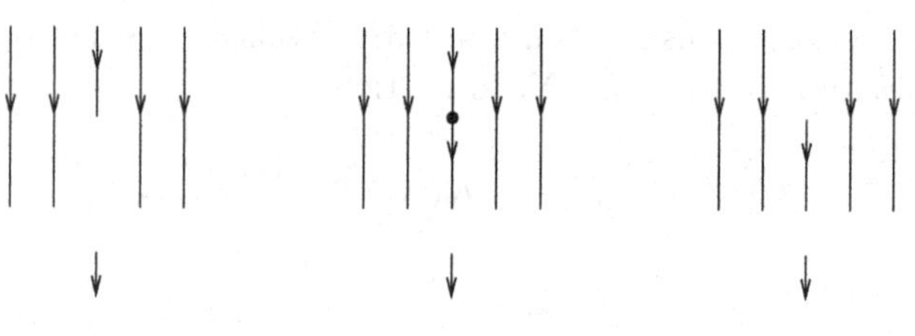

All the good G-sets occurring in this example are maximal with respect to inclusions of the following type: a subset $U' \subseteq U$ of a good G-set $U \subseteq X$ is called *G-saturated* in U if it satisfies $U' = \pi^{-1}(\pi(U'))$, where $\pi\colon U \to U /\!\!/ G$ is the good quotient. Any good G-set is G-saturated in a maximal one; for a description it is reasonable to focus on the latter ones.

Our aim is to present concrete combinatorial descriptions for quasiprojective and for torically embeddable quotients of quasitorus actions on certain affine varieties. Let us fix the setting. By K we denote a finitely generated abelian group and we consider an affine K-graded $\mathbb{K}$-algebra

$$A = \bigoplus_{w \in K} A_w.$$

Then the quasitorus $H := \operatorname{Spec} \mathbb{K}[K]$ acts on the affine variety $X := \operatorname{Spec} A$. Let $K_{\mathbb{Q}} := K \otimes_{\mathbb{Z}} \mathbb{Q}$ denote the rational vector space associated to K. Given $w \in K$, we write again w for the element $w \otimes 1 \in K_{\mathbb{Q}}$.

Definition 1.45 The *weight cone* of the K-graded algebra A is the convex polyhedral cone

$$\omega_X \ := \ \omega(A) \ = \ \mathrm{cone}(w \in K;\ A_w \neq \{0\}) \ \subseteq \ K_{\mathbb{Q}}.$$

To every point $x \in X$, we associate its *orbit cone*; this is the convex polyhedral cone

$$\omega_x \ := \ \mathrm{cone}(w \in K;\ f(x) \neq 0 \text{ for some } f \in A_w) \ \subseteq \ \omega_X.$$

In order to see that these cones are indeed polyhedral, let $f_1, \ldots, f_r$ be homogeneous generators for A and set $w_i := \deg(f_i)$. Then the weight cone ω_X is generated by $w_1, \ldots, w_r$ and the orbit cone ω_x is generated by those w_i with $f_i(x) \neq 0$. In particular, we see that ω_X is the general orbit cone and that there are only finitely many orbit cones.

Example 1.46 We determine the orbit cones of $\mathbb{K}^*$-action $t \cdot (z_1, z_2) = (t^a z_1 t^b z_2)$ on $\mathbb{K}^2$; again, we consider the three typical cases:

(i) We have $a = b = 1$. The weight cone is $\omega_{\mathbb{K}^2} = \mathbb{Q}_{\geq 0}$ and the possible orbit cones are

$$\omega_{(0,0)} = \{0\}, \qquad \omega_{(1,1)} = \mathbb{Q}_{\geq 0}.$$

(ii) We have $a = 0$ and $b = 1$. The weight cone is $\omega_{\mathbb{K}^2} = \mathbb{Q}_{\geq 0}$ and the possible orbit cones are

$$\omega_{(0,0)} = \{0\}, \qquad \omega_{(1,0)} = \mathbb{Q}_{\geq 0}.$$

(iii) We have $a = 1$ and $b = -1$. The weight cone is $\omega_{\mathbb{K}^2} = \mathbb{Q}$ and the possible orbit cones are

$$\omega_{(0,0)} = \{0\}, \quad \omega_{(1,0)} = \mathbb{Q}_{\geq 0}, \quad \omega_{(0,1)} = \mathbb{Q}_{\leq 0}, \quad \omega_{(1,1)} = \mathbb{Q}.$$

In this example we considered a subtorus action on a toric variety, and we used the fact that it suffices to determine one orbit cone for each toric orbit. This idea can be generalized using equivariant embeddings and gives the following concrete recipes for computing orbit cones.

Remark 1.47 Fix a system of homogeneous generators $\mathfrak{F} = (f_1, \ldots, f_r)$ for our K-graded algebra A. Then H acts diagonally on $\mathbb{K}^r$ via the characters $\chi^{w_1}, \ldots, \chi^{w_r}$, where $w_i = \deg(f_i)$, and we have an H-equivariant closed embedding

$$X \ \to \ \mathbb{K}^r, \qquad x \ \mapsto \ (f_1(x), \ldots, f_r(x)).$$

With $E = \mathbb{Z}^r$ and $\gamma = \mathrm{cone}(e_1, \ldots, e_r)$, we may identify $\mathbb{K}[T_1, \ldots, T_r]$

with $\mathbb{K}[E \cap \gamma]$ and thus regard $\mathbb{K}^r$ as the affine toric variety associated to the dual cone $\delta := \gamma^\vee$. For any $\gamma_0 \preceq \gamma$ and $\delta_0 := \gamma_0^\perp \cap \delta$, the following statements are equivalent:

(i) the product over all f_i with $e_i \in \gamma_0$ lies not in $\sqrt{\langle f_j;\ e_j \notin \gamma_0 \rangle} \subseteq A$,
(ii) there is a point $z \in X$ with $z_i \neq 0 \Leftrightarrow e_i \in \gamma_0$ for all $1 \leq i \leq r$,
(iii) the toric orbit $\mathbb{T}^r \cdot z_{\delta_0} \subseteq \mathbb{K}^r$ corresponding to $\delta_0 \preceq \delta$ meets X,
(iv) the intersection $\delta_0^\circ \cap \mathrm{Trop}(X)$ with the tropical variety is non-empty.

In order to determine the orbit cones, denote by $Q\colon E \to K$ the homomorphism sending e_i to w_i. Moreover, we call $\mathfrak{F}$*-faces* the $\gamma_0 \preceq \gamma$ satisfying (i). Then the orbit cones of X are precisely the images $Q(\gamma_0)$, where $\gamma_0 \preceq \gamma$ is an $\mathfrak{F}$-face.

Example 1.48 Set $K := \mathbb{Z}^2$ and consider the K-grading of $\mathbb{K}[T_1, \ldots, T_5]$ defined by $\deg(T_i) := w_i$, where w_i is the i-th column of the matrix

$$Q := \begin{bmatrix} 1 & -1 & 0 & -1 & 1 \\ 1 & 1 & 1 & 0 & 2 \end{bmatrix}.$$

The corresponding action of $H = \mathbb{T}^2$ on $\mathbb{K}^5$ leaves $X := V(T_1T_2 + T_3^2 + T_4T_5)$ invariant. The possible orbit cones ω_x for $x \in X$ are

$$\{0\}, \quad \mathrm{cone}(w_1), \quad \mathrm{cone}(w_2), \quad \mathrm{cone}(w_4), \quad \mathrm{cone}(w_5),$$

$$\mathrm{cone}(w_1, w_4), \quad \mathrm{cone}(w_2, w_4), \quad \mathrm{cone}(w_1, w_5), \quad \mathrm{cone}(w_2, w_5).$$

Definition 1.49 The *GIT-cone* of an element $w \in \omega_X$ is the (nonempty) intersection of all orbit cones containing it:

$$\lambda(w) := \bigcap_{\substack{x \in X, \\ w \in \omega_x}} \omega_x.$$

We write $\Lambda(X, H)$ for the set of all GIT-cones. The *set of semistable points* associated to a GIT-cone $\lambda \subseteq \omega_X$ is

$$X^{ss}(\lambda) \;=\; \{x \in X;\ \lambda \subseteq \omega_x\} \;\subseteq\; X.$$

Remark 1.50 Given a GIT-cone $\lambda \in \Lambda(X, H)$ and any weight $w \in \lambda^\circ$ in its relative interior, one easily checks

$$X^{ss}(\lambda) = \{x \in X;\ f(x) \neq 0 \text{ for some } f \in A_{nw},\ n > 0\}.$$

That means that $X^{ss}(\lambda)$ is the set of semistable points associated to the linearization of the trivial bundle given by the character χ^w in the sense of Mumford.

Example 1.51 We compute GIT-cones and associated sets of semistable points for the $\mathbb{K}^*$-action $t \cdot (z, w) = (t^a z, t^b w)$ on $\mathbb{K}^2$ in the three typical cases.

$$a = 1, b = 1: \quad \lambda(0) = \{0\} \quad X^{ss}(\lambda(0)) = \mathbb{K}^2,$$
$$\lambda(1) = \mathbb{Q}_{\geq 0} \quad X^{ss}(\lambda(1)) = \mathbb{K}^2 \setminus \{0\},$$

$$a = 0, b = 1: \quad \lambda(0) = \{0\} \quad X^{ss}(\lambda(0)) = \mathbb{K}^2,$$
$$\lambda(1) = \mathbb{Q}_{\geq 0} \quad X^{ss}(\lambda(1)) = \mathbb{K}^2 \setminus V(T_2),$$

$$a = -1, b = 1: \quad \lambda(0) = \{0\} \quad X^{ss}(\lambda(0)) = \mathbb{K}^2,$$
$$\lambda(-1) = \mathbb{Q}_{\leq 0} \quad X^{ss}(\lambda(-1)) = \mathbb{K}^2 \setminus V(T_1),$$
$$\lambda(1) = \mathbb{Q}_{\geq 0} \quad X^{ss}(\lambda(1)) = \mathbb{K}^2 \setminus V(T_2).$$

In the following statement, by a *quasifan* we mean a finite collection Λ of not necessarily pointed polyhedral cones in a rational vector space such that any two $\lambda_1, \lambda_2 \in \Lambda$ intersect in a common face and for $\lambda \in \Lambda$ also every face of λ belongs to Λ.

Theorem 1.52 *The collection $\Lambda(X, H) = \{\lambda(w);\ w \in \omega_X\}$ of all GIT-cones is a quasifan in $K_{\mathbb{Q}}$ having the weight cone ω_X as its support.*

(i) *For every $\lambda \in \Lambda(X, H)$, there is a good quotient $X^{ss}(\lambda) \to Y(\lambda)$ for the action of H on $X^{ss}(\lambda)$.*
(ii) *For any two GIT-cones $\lambda_1, \lambda_2 \in \Lambda(X, H)$, we have $\lambda_2 \preceq \lambda_1$ if and only if $X^{ss}(\lambda_1) \subseteq X^{ss}(\lambda_2)$ holds.*
(iii) *If $X^{ss}(\lambda_1) \subseteq X^{ss}(\lambda_2)$ holds, then there is an induced projective morphism $Y(\lambda_1) \to Y(\lambda_2)$; in particular, every $Y(\lambda)$ is projective over $Y(0)$.*

The collection $\Lambda(X, H)$ is called the *GIT-(quasi-)fan* of the H-variety X. For a diagonal H-action on $\mathbb{K}^r$, the orbit cones are cone$(\deg(T_i);\ i \in I)$, where I runs through the subsets of $\{1, \ldots, r\}$ and, thus the GIT-fan equals the Gelfand-Kapranov-Zelevinsky decomposition associated to $\deg(T_1), \ldots, \deg(T_r) \in K_{\mathbb{Q}}$.

Example 1.53 Consider the action of the standard three torus $\mathbb{T}^3$ on

$\mathbb{K}^6$ defined by $\deg(T_i) = w_i$, where $w_1, \ldots, w_6$ are defined as

$$\begin{array}{llllll} w_1 & = & (1,0,0), \quad w_2 & = & (0,1,0), \quad w_3 & = & (0,0,1), \\ w_4 & = & (1,1,0), \quad w_5 & = & (1,0,1), \quad w_6 & = & (0,1,1). \end{array}$$

Then the GIT-fan subdivides the positive orthant in $\mathbb{Q}^3$; intersecting with a suitable plane perpendicular to the line through $(1,1,1)$ gives the following picture.

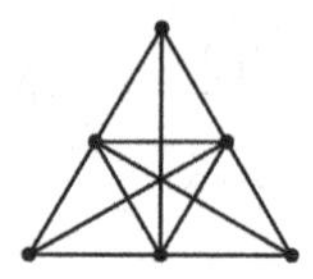

If every H-invariant divisor on X is principal, then the GIT-fan controls the whole variation of good quotients with a quasiprojective quotient space. For the precise statement let us call a good H-set $U \subseteq X$ *qp-maximal* if $U /\!\!/ H$ is quasiprojective and U is maximal with respect to H-saturated inclusion among all good H-sets $W \subseteq X$ with $W /\!\!/ H$ quasiprojective.

Theorem 1.54 *Assume that X is normal and for every H-invariant divisor on X some positive multiple is principal. Then, with the GIT-fan $\Lambda(X,H)$ of the H-action on X, we have mutually inverse order reversing bijections*

$$\begin{array}{rcl} \Lambda(X,H) & \longleftrightarrow & \{\textit{qp-maximal subsets of } X\} \\ \lambda & \mapsto & X^{ss}(\lambda) = \{x \in X;\ \lambda \subseteq \omega_x\} \\ \bigcap_{x \in U} \omega_x =: \lambda(U) & \leftarrow\!\shortmid & U. \end{array}$$

Now we look for more general quotient spaces. We say that a variety X has the *A_2-property* , if any two points $x, x' \in X$ admit a common affine open neighborhood in X. By [44], the normal A_2-varieties are precisely those that admit a closed embedding into a toric variety.

Definition 1.55 Let Ω_X denote the collection of all orbit cones ω_x, where $x \in X$. A *bunch of orbit cones* is a nonempty collection $\Phi \subseteq \Omega_X$ such that

(i) given $\omega_1, \omega_2 \in \Phi$, one has $\omega_1^\circ \cap \omega_2^\circ \neq \emptyset$,
(ii) given $\omega \in \Phi$, every orbit cone $\omega_0 \in \Omega_X$ with $\omega^\circ \subseteq \omega_0^\circ$ belongs to Φ.

A *maximal bunch of orbit cones* is a bunch of orbit cones $\Phi \subseteq \Omega_X$ which cannot be enlarged by adding further orbit cones.

Definition 1.56 Let $\Phi, \Phi' \subseteq \Omega_X$ be bunches of orbit cones. We say that Φ *refines* Φ' (written $\Phi \le \Phi'$), if for any $\omega' \in \Phi'$ there is an $\omega \in \Phi$ with $\omega \subseteq \omega'$.

Example 1.57 Consider once more the action of the standard three torus $\mathbb{T}^3$ on $\mathbb{K}^6$ discussed in Example 1.53. Here are two maximal bunches, indicated by drawing their minimal members:

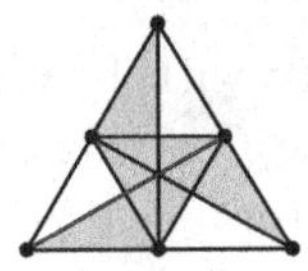
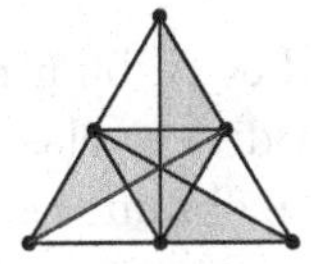

Definition 1.58 To any collection of orbit cones Φ of X, we associate the following subset of X:

$$U(\Phi) := \{x \in X;\ \omega_0 \preceq \omega_x \text{ for some } \omega_0 \in \Phi\}.$$

Conversely, to any H-invariant subset $U \subseteq X$, we associate the following collection of orbit cones

$$\Phi(U) := \{\omega_x;\ x \in U \text{ with } H \cdot x \text{ closed in } U\}.$$

By an $(H,2)$-*maximal subset* of X we mean a good H-set $U \subseteq X$ with $U /\!\!/ H$ an A_2-variety such that U is maximal with respect to H-saturated inclusion among all good H-sets $W \subseteq X$ with $W /\!\!/ H$ an A_2-variety. We are ready for the result.

Theorem 1.59 *Assume that X is normal and for every H-invariant divisor on X some positive multiple is principal. Then we have mutually inverse order reversing bijections*

$$\left\{\begin{matrix} \textit{maximal bunches of} \\ \textit{orbit cones in } \Omega_X \end{matrix}\right\} \longleftrightarrow \{(H,2)\textit{-maximal subsets of } X\}$$

$$\Phi \mapsto U(\Phi)$$

$$\Phi(U) \mapsfrom U.$$

Remark 1.60 Every GIT-chamber $\lambda \in \Lambda(X,H)$ defines a bunch of orbit cones $\Phi(\lambda) = \{\omega_x;\ \lambda^\circ \subseteq \omega^\circ\}$. These bunches turn out to be maximal and they correspond to the qp-maximal subsets of X; in particular, the latter ones are $(H,2)$-maximal. The bunches of Example 1.57 give rise to non-projective complete quotients.

1.2.2 Cox rings and combinatorics

Here we present the combinatorial approach to varieties with finitely generated Cox ring developed in [10, 19], see also [4, Chap. III]. The approach generalizes the combinatorial description of toric varieties and has many common features with methods of [17, 26] for investigating subvarieties of weighted complete spaces. The whole thing is based on the following simple observation.

Remark 1.61 Let X be a normal variety with $\Gamma(X, \mathcal{O}^*) = \mathbb{K}^*$ and finitely generated divisor class group. If the Cox ring $\mathcal{R}(X)$ is finitely generated, then we obtain the following picture

$$\begin{array}{ccccccc} \mathrm{Spec}_X \mathcal{R} & = & \widehat{X} & \subseteq & \overline{X} & = & \mathrm{Spec}\, \mathcal{R}(X) \\ & {\scriptstyle /\!\!/ H_X} & \downarrow & & & & \\ & & X & & & & \end{array}$$

where $\widehat{X} \subseteq \overline{X}$ is an open H_X-invariant subset of the H_X-factorial affine variety $\overline{X}$ and the characteristic space $\widehat{X} \to X$ is a good quotient for the H_X-action. We call the affine H_X-variety $\overline{X}$ the *total coordinate space* of X.

Thus, we see that all varieties sharing the same divisor class group K and finitely generated Cox ring R occur as good quotients of suitable open subsets of Spec R by the action of Spec $\mathbb{K}[K]$. The latter ones we just described in combinatorial terms via Geometric Invariant Theory. We now turn this picture into a combinatorial language allowing explicit computations.

Definition 1.62 Let K be a finitely generated abelian group and R a factorially K-graded affine algebra with $R^* = \mathbb{K}^*$. Moreover, let $\mathfrak{F} = (f_1, \ldots, f_r)$ be a system of pairwise nonassociated K-prime generators for R.

(i) The *projected cone* associated to $\mathfrak{F}$ is $(E \xrightarrow{Q} K, \gamma)$, where $E := \mathbb{Z}^r$, the homomorphism $Q\colon E \to K$ sends the i-th canonical basis vector $e_i \in E$ to $w_i := \deg(f_i) \in K$ and $\gamma \subseteq E_{\mathbb{Q}}$ is the convex cone generated by $e_1, \ldots, e_r$.

(ii) We say that the K-grading of R is *almost free* if for every facet $\gamma_0 \preceq \gamma$ the image $Q(\gamma_0 \cap E)$ generates the abelian group K.

(iii) We say that $\gamma_0 \preceq \gamma$ is an *$\mathfrak{F}$-face*, if the product over all f_i with $e_i \in \gamma_0$ does not lie in $\sqrt{\langle f_j;\ e_j \not\in \gamma_0 \rangle} \subseteq A$.

(iv) Let $\Omega_{\mathfrak{F}} = \{Q(\gamma_0);\ \gamma_0 \preceq \gamma\ \mathfrak{F}\text{-face}\}$ denote the collection of projected $\mathfrak{F}$-faces. An *$\mathfrak{F}$-bunch* is a nonempty subset $\Phi \subseteq \Omega_{\mathfrak{F}}$ such that

1 for any two $\tau_1, \tau_2 \in \Phi$, we have $\tau_1^\circ \cap \tau_2^\circ \neq \emptyset$,
2 if $\tau_1^\circ \subseteq \tau^\circ$ holds for $\tau_1 \in \Phi$ and $\tau \in \Omega_{\mathfrak{F}}$, then $\tau \in \Phi$ holds.

(v) We say that an $\mathfrak{F}$-bunch Φ is *true* if for every facet $\gamma_0 \preceq \gamma$ the image $Q(\gamma_0)$ belongs to Φ.

Definition 1.63 A *bunched ring* is a triple $(R, \mathfrak{F}, \Phi)$, where R is an almost freely factorially K-graded affine $\mathbb{K}$-algebra such that $R^* = \mathbb{K}^*$ holds, $\mathfrak{F}$ is a system of pairwise non-associated K-prime generators for R and Φ is a true $\mathfrak{F}$-bunch.

Construction 1.64 Let $(R, \mathfrak{F}, \Phi)$ be a bunched ring. Then Φ is a bunch of orbit cones for the action of $H := \operatorname{Spec} \mathbb{K}[K]$ on $\overline{X} := \operatorname{Spec} R$. Thus, we have the associated open set and its quotient

$$\widehat{X} := \widehat{X}(R, \mathfrak{F}, \Phi) = \overline{X}(\Phi) \subseteq \overline{X},$$

$$X := X(R, \mathfrak{F}, \Phi) := \widehat{X}(R, \mathfrak{F}, \Phi) /\!\!/ H.$$

We denote the quotient map by $p\colon \widehat{X} \to X$. Conditions 1.62 (ii) and (v) ensure that the H-action on $\widehat{X}$ is strongly stable. Moreover, every member f_i of $\mathfrak{F}$ defines a prime divisor $D_X^i := p(V(\widehat{X}, f_i))$ on X.

Theorem 1.65 *Let $\widehat{X} := \widehat{X}(R, \mathfrak{F}, \Phi)$ and $X := X(R, \mathfrak{F}, \Phi)$ arise from a bunched ring $(R, \mathfrak{F}, \Phi)$. Then X is a normal A_2-variety with*

$$\dim(X) = \dim(R) - \dim(K_{\mathbb{Q}}), \qquad \Gamma(X, \mathcal{O}^*) = \mathbb{K}^*,$$

there is an isomorphism $\mathrm{Cl}(X) \to K$ sending $[D_X^i]$ to $\deg(f_i)$, the map $p\colon \widehat{X} \to X$ is a characteristic space and the Cox ring $\mathcal{R}(X)$ is isomorphic to R.

Theorem 1.66 *Every complete normal A_2-variety with finitely generated Cox ring arises from a bunched ring; in particular, every projective normal variety with finitely generated Cox ring does so.*

Let us illustrate Construction 1.64 with two examples. The first one shows how toric varieties fit into the picture of bunched rings.

Example 1.67 (Bunched polynomial rings) Consider a bunched ring $(R, \mathfrak{F}, \Phi)$ with $R = \mathbb{K}[T_1, \dots, T_r]$ and $\mathfrak{F} := (T_1, \dots, T_r)$. Then $X(R, \mathfrak{F}, \Phi)$

is a toric variety. Its defining fan Σ is obtained from Φ via linear Gale duality:

$$
\begin{array}{ccccccccc}
0 & \longrightarrow & L_{\mathbb{Q}} & \xrightarrow{Q^*} & F_{\mathbb{Q}} & \xrightarrow{P} & N_{\mathbb{Q}} & \longrightarrow & 0 \\
 & & & & \Sigma^{\uparrow} & \xrightarrow{\delta_0 \mapsto P(\delta_0)} & \Sigma & & \\
 & & & & \uparrow {\scriptstyle \gamma_0 \mapsto \gamma_0^{\perp} \cap \delta} & & & & \\
 & & \Phi & \xleftarrow[Q(\gamma_0) \leftarrow\!\shortmid \gamma_0]{} & \Phi^{\uparrow} & & & & \\
0 & \longleftarrow & K_{\mathbb{Q}} & \xleftarrow[Q]{} & E_{\mathbb{Q}} & \xleftarrow[P^*]{} & M_{\mathbb{Q}} & \longleftarrow & 0
\end{array}
$$

Here $\Phi^{\uparrow}$ consists of those faces of the orthant $\gamma \subseteq E_{\mathbb{Q}}$ that map onto a member of Φ and $\Sigma^{\uparrow}$ of the corresponding faces of the dual orthant $\delta \subseteq F_{\mathbb{Q}}$. Note that $P \colon F \to N$ is the same map as in Construction 1.43 and the D_X^i are exactly the toric prime divisors.

Example 1.68 (A singular del Pezzo surface) Consider $K := \mathbb{Z}^2$ and the K-grading of $\mathbb{K}[T_1, \ldots, T_5]$ given by $\deg(T_i) := w_i$, where w_i is the i-th column of

$$Q := \begin{bmatrix} 1 & -1 & 0 & -1 & 1 \\ 1 & 1 & 1 & 0 & 2 \end{bmatrix}$$

Then this K-grading descends to a K-grading of the following residue algebra which is known to be factorial:

$$R := \mathbb{K}[T_1, \ldots, T_5] \,/\, \langle T_1T_2 + T_3^2 + T_4T_5 \rangle.$$

The classes $f_i \in R$ of $T_i \in \mathbb{K}[T_1, \ldots, T_5]$, where $1 \le i \le 5$, form a system $\mathfrak{F}$ of pairwise nonassociated K-prime generators of R. We have

$$E \;=\; \mathbb{Z}^5, \qquad \gamma \;=\; \operatorname{cone}(e_1, \ldots, e_5)$$

and the K-grading is almost free. Computing the $\mathfrak{F}$-faces, we see that there is one maximal true $\mathfrak{F}$-bunch Φ; it has $\tau := \operatorname{cone}(w_2, w_5)$ as its unique minimal cone.

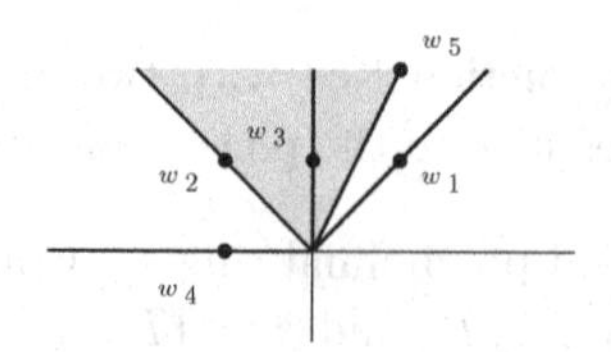

Note that τ is a GIT-cone, $\Phi = \Phi(\tau)$ holds with $\Phi(\tau)$ as in Remark 1.60 and $\widehat{X}(R, \mathfrak{F}, \Phi)$ equals $\overline{X}^{ss}(\tau)$ in $\overline{X} = V(\mathbb{K}^5; T_1T_2 + T_3^2 + T_4T_5)$. For $X = X(R, \mathfrak{F}, \Phi)$ we have

$$\dim(X) = 2, \qquad \mathrm{Cl}(X) = \mathbb{Z}^2, \qquad \mathcal{R}(X) = R.$$

Definition 1.69 Let $(R, \mathfrak{F}, \Phi)$ be a bunched ring and $(E \xrightarrow{Q} K, \gamma)$ its projected cone. The *collection of relevant faces* and the *covering collection* are

$$\mathrm{rlv}(\Phi) := \{\gamma_0 \preceq \gamma;\ \gamma_0 \text{ an } \mathfrak{F}\text{-face with } Q(\gamma_0) \in \Phi\},$$
$$\mathrm{cov}(\Phi) := \{\gamma_0 \in \mathrm{rlv}(\Phi);\ \gamma_0 \text{ minimal}\}.$$

Construction 1.70 (Canonical toric embedding) Any bunched ring $(R, \mathfrak{F}, \Phi)$ defines a bunched polynomial ring $(R', \mathfrak{F}', \Phi')$ by "forgetting the relations": If the system of generators of R is $\mathfrak{F} = (f_1, \ldots, f_r)$, set

$$R' := \mathbb{K}[T_1, \ldots, T_r], \qquad \deg(T_i) := \deg(f_i) \in K, \qquad \mathfrak{F}' := (T_1, \ldots, T_r)$$

and let Φ' be the $\mathfrak{F}'$-bunch generated by Φ, i.e. it consists of all projected faces $Q(\gamma_0)$ with $\tau^\circ \subseteq Q(\gamma_0)^\circ$ for some $\tau \in \Phi$. Then we obtain a commutative diagram, where the induced map of quotients $\imath\colon X \to Z$ is a closed embedding of the varieties X and Z associated to the bunched rings $(R, \mathfrak{F}, \Phi)$ and $(R', \mathfrak{F}', \Phi')$ respectively:

$$\begin{array}{ccccccc} \overline{X} & \supseteq & \widehat{X} & \longrightarrow & \widehat{Z} & \subseteq & \overline{Z} \\ & & \big\downarrow /\!\!/ H & & \big\downarrow /\!\!/ H & & \\ & & X & \xrightarrow[\imath]{} & Z & & \end{array}$$

By construction, Z is toric and we have an isomorphism $\imath^*\colon \mathrm{Cl}(Z) \to \mathrm{Cl}(X)$. Then, in the setting of Example 1.67, the toric orbits of Z intersecting X are precisely the orbits $B(\sigma) \subseteq Z$ corresponding to cones $\sigma = P(\gamma_0^*)$ with $\gamma_0 \in \mathrm{rlv}(\Phi)$. In particular, we obtain a decomposition into locally closed strata

$$X = \bigcup_{\gamma_0 \in \mathrm{rlv}(\Phi)} X(\gamma_0), \qquad X(\gamma_0) := X \cap B(\sigma).$$

Remark 1.71 In general, the canonical toric ambient variety Z is not complete, even if X is. If X is projective, then $\Phi = \Phi(\lambda)$ holds with a GIT-cone $\lambda \in \Lambda(\overline{X}, H)$. The toric GIT-fan $\Lambda(\overline{Z}, H)$ refines $\Lambda(\overline{X}, H)$ and every $\eta \in \Lambda(\overline{Z}, H)$ with $\eta^\circ \subseteq \lambda^\circ$ defines a projective completion of Z. For example, in the setting of Example 1.68, the two GIT-fans are:

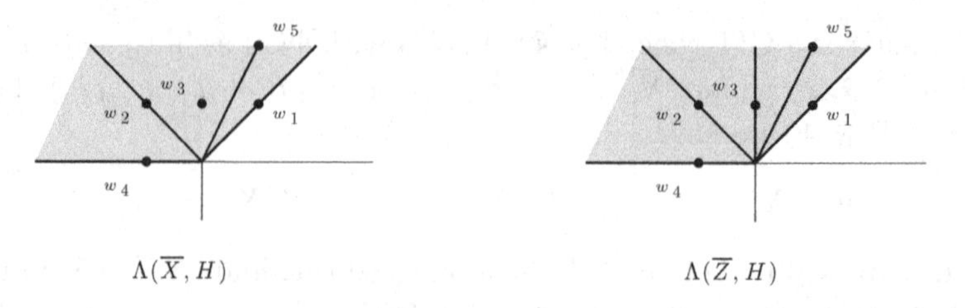

The GIT-cones cone(w_2, w_3) and cone(w_3, w_5) in $\Lambda(\overline{Z}, H)$ provide completions of Z by $\mathbb{Q}$-factorial projective toric varieties Z_1 and Z_2 and cone(w_3) gives a completion by a projective toric variety Z_3 with a non-$\mathbb{Q}$-factorial singularity.

We now indicate how to read off basic geometric properties from defining data. In the sequel, X is the variety arising from a bunched ring $(R, \mathfrak{F}, \Phi)$.

Theorem 1.72 *Consider a relevant face $\gamma_0 \in \mathrm{rlv}(\Phi)$ and a point $x \in X(\gamma_0)$. Then we have a commutative diagram*

$$\begin{array}{ccc} \mathrm{Cl}(X) & \longrightarrow & \mathrm{Cl}(X, x) \\ \cong \updownarrow & & \updownarrow \cong \\ K & \longrightarrow & K/Q(\mathrm{lin}(\gamma_0) \cap E)) \end{array}$$

In particular, the local divisor class groups are constant along the pieces $X(\gamma_0)$, where $\gamma_0 \in \mathrm{rlv}(\Phi)$. Moreover, the Picard group of X is given by

$$\mathrm{Pic}(X) \cong \bigcap_{\gamma_0 \in \mathrm{cov}(\Phi)} Q(\mathrm{lin}(\gamma_0) \cap E).$$

Theorem 1.73 *Consider a relevant face $\gamma_0 \in \mathrm{rlv}(\Phi)$ and point $x \in X(\gamma_0)$ in the corresponding stratum.*

(i) *The point x is factorial if and only if Q maps $\mathrm{lin}(\gamma_0) \cap E$ onto K.*
(ii) *The point x is $\mathbb{Q}$-factorial if and only if $Q(\gamma_0)$ is of full dimension.*

In particular, X is $\mathbb{Q}$-factorial, if and only if Φ consists of full-dimensional cones. If $\widehat{X}$ is smooth, then every factorial point of X is smooth.

Theorem 1.74 *In the divisor class group $K = \mathrm{Cl}(X)$, we have the following descriptions of the cones of effective, movable, semiample and* *ample divisors:*

$$\mathrm{Eff}(X) = Q(\gamma), \qquad \mathrm{Mov}(X) = \bigcap_{\gamma_0 \text{ facet of } \gamma} Q(\gamma_0),$$

$$\mathrm{SAmple}(X) = \bigcap_{\tau \in \Phi} \tau, \qquad \mathrm{Ample}(X) = \bigcap_{\tau \in \Phi} \tau^{\circ}.$$

Example 1.75 (The singular del Pezzo surface, continued) The variety X of Example 1.68 is a $\mathbb{Q}$-factorial surface. It has a single singularity, namely the point in the piece $x_0 \in X(\gamma_0)$ for $\gamma_0 = \mathrm{cone}(e_2, e_5)$. The local class group $\mathrm{Cl}(X, x_0)$ is cyclic of order three and the Picard group of X is of index 3 in $\mathrm{Cl}(X)$. Moreover, the ample cone of X is generated by w_2 and w_5. In particular, X is projective.

Remark 1.76 Applying Theorems 1.72, 1.73 and 1.74 to the canonical toric ambient variety Z shows that X inherits local class groups and singularities from Z and the Picard group as well as the various cones of divisors of X and Z coincide. However, factorial singularities of X are smooth points of Z, see Example 1.110 for an example. Moreover, $\mathrm{Pic}(X) = \mathrm{Pic}(Z)$ and $\mathrm{Ample}(X) = \mathrm{Ample}(Z)$ can get lost when replacing Z with a completion.

The following two statements concern the case that R is a complete intersection in the sense that with $d := r - \dim(X) - \dim(K)$, there are K-homogeneous generators $g_1, \dots, g_d$ for the ideal of relations between $f_1, \dots, f_r$. Set $w_i := \deg(f_i)$ and $u_j := \deg(g_j)$.

Theorem 1.77 *Suppose that R is a complete intersection as above. Then the canonical class of X is given in $K = \mathrm{Cl}(X)$ by*

$$\mathcal{K}_X = \sum u_j - \sum w_i.$$

Remark 1.78 (Computing intersection numbers) Suppose that R is a complete intersection and that $\Phi = \Phi(\lambda)$ holds with a full-dimensional $\lambda \in \Lambda(\overline{X}, H)$. Fix a full-dimensional $\eta \in \Lambda(\overline{Z}, H)$ with $\eta^{\circ} \subseteq \lambda^{\circ}$. For $w_{i_1}, \dots, w_{i_{n+d}}$ let $w_{j_1}, \dots, w_{j_{r-n-d}}$ denote the complementary weights and set

$$\tau(w_{i_1}, \dots, w_{i_{n+d}}) := \mathrm{cone}(w_{j_1}, \dots, w_{j_{r-n-d}}),$$
$$\mu(w_{i_1}, \dots, w_{i_{n+d}}) := [K : \langle w_{j_1}, \dots, w_{j_{r-n-d}} \rangle].$$

Then the intersection product $K_{\mathbb{Q}}^{n+d} \to \mathbb{Q}$ of the ($\mathbb{Q}$-factorial) toric variety Z_1 associated to $\Phi(\eta)$ is determined by the values

$$w_{i_1} \cdots w_{i_{n+d}} = \begin{cases} \mu(w_{i_1}, \ldots, w_{i_{n+d}})^{-1}, & \eta \subseteq \tau(w_{i_1}, \ldots, w_{i_{n+d}}), \\ 0, & \eta \not\subseteq \tau(w_{i_1}, \ldots, w_{i_{n+d}}). \end{cases}$$

As a complete intersection, $X \subseteq Z_1$ inherits intersection theory. For a tuple $D_X^{i_1}, \ldots, D_X^{i_n}$ on X, its intersection number can be computed by

$$D_X^{i_1} \cdots D_X^{i_n} = w_{i_1} \cdots w_{i_n} \cdot u_1 \cdots u_d.$$

Example 1.79 (The singular del Pezzo surface, continued) Consider once more the surface X of Example 1.68. The degree of the defining relation is $\deg(T_1T_2 + T_3^2 + T_4T_5) = 2w_3$ and thus the canonical class of X is given as

$$\mathcal{K}_X \;=\; 2w_3 - (w_1 + w_2 + w_3 + w_4 + w_5) \;=\; -3w_3.$$

In particular, we see that the anticanonical class is ample and thus X is a (singular) del Pezzo surface. The self intersection number of the canonical class is

$$\mathcal{K}_X^2 \;=\; (3w_3)^2 \;=\; \frac{9(w_1 + w_2)(w_4 + w_5)}{4}.$$

The $w_i \cdot w_j$ equal the toric intersection numbers $2w_i \cdot w_j \cdot w_3$. In order to compute them, let w_{ij}^1, w_{ij}^2 denote the weights in $\{w_1, \ldots, w_5\} \setminus \{w_i, w_j, w_3\}$. Then we have

$$w_i \cdot w_j \cdot w_3 = \begin{cases} \mu(w_i, w_j, w_3)^{-1}, & \tau \subseteq \operatorname{cone}(w_{ij}^1, w_{ij}^2), \\ 0, & \tau \not\subseteq \operatorname{cone}(w_{ij}^1, w_{ij}^2), \end{cases}$$

where the multiplicity $\mu(w_i, w_j, w_3)$ is the absolute value of $\det(w_{ij}^1, w_{ij}^2)$. Thus, we can proceed in the computation:

$$\mathcal{K}_X^2 = (3w_3)^2 = \frac{9 \cdot 2}{4}\left(|\det(w_2, w_5)|^{-1} + |\det(w_1, w_4)|^{-1}\right) = \frac{9}{2}\cdot\frac{4}{3} = 6.$$

1.2.3 Mori dream spaces

We take a closer look to the $\mathbb{Q}$-factorial projective varieties with a finitely generated Cox ring. Hu and Keel [25] called them *Mori dream spaces* and characterized them in terms of cones of divisors. In the context of normal complete varieties their statement reads as follows; by a *small birational map* we mean a rational map defining an isomorphism of open subsets with complement of codimension two.

Theorem 1.80 *Let X be a normal complete variety with finitely generated divisor class group. Then the following statements are equivalent.*

(i) *The Cox ring $\mathcal{R}(X)$ is finitely generated.*

(ii) *There are small birational maps $\pi_i \colon X \to X_i$, where $i = 1, \ldots, r$, such that each semiample cone* $\mathrm{SAmple}(X_i) \subseteq \mathrm{Cl}_{\mathbb{Q}}(X)$ *is polyhedral and one has*

$$\mathrm{Mov}(X) = \pi_1^*(\mathrm{SAmple}(X_1)) \cup \ldots \cup \pi_r^*(\mathrm{SAmple}(X_r)).$$

Moreover, if one of these two statements holds, then there is a small birational map $X \to X'$ with a $\mathbb{Q}$-factorial projective variety X'.

There are many examples of Mori dream spaces. Besides the toric and more generally spherical varieties, all unirational varieties with a reductive group action of complexity one are Mori dream spaces. Other important examples are the log terminal Fano varieties [11]. Moreover, K3- and Enriques surfaces are Mori dream spaces if and only if their effective cone is polyhedral [3, 43]; the same holds in higher dimensions for Calabi-Yau varieties [29]. Specializing to surfaces, the above theorem gives the following.

Corollary 1.81 *Let X be a normal complete surface with finitely generated divisor class group* $\mathrm{Cl}(X)$. *Then the following statements are equivalent.*

(i) *The Cox ring $\mathcal{R}(X)$ is finitely generated.*

(ii) *One has* $\mathrm{Mov}(X) = \mathrm{SAmple}(X)$ *and this cone is polyhedral.*

Moreover, if one of these two statements holds, then the surface X is $\mathbb{Q}$-factorial and projective.

The Mori dream spaces sharing a given Cox ring fit into a nice picture in terms of the GIT-fan; by the *moving cone* of the K-graded algebra R we mean here the intersection $\mathrm{Mov}(R)$ over all $\mathrm{cone}(w_1, \ldots, \widehat{w_i}, \ldots, w_r)$, where the w_i are the degrees of any system of pairwise nonassociated homogeneous K-prime generators for R.

Remark 1.82 Let $R = \oplus_K R_w$ be an almost freely factorially graded affine algebra with $R_0 = \mathbb{K}$ and consider the GIT-fan $\Lambda(\overline{X}, H)$ of the action of $H = \mathrm{Spec}\ \mathbb{K}[K]$ on $\overline{X} = \mathrm{Spec}\ R$.

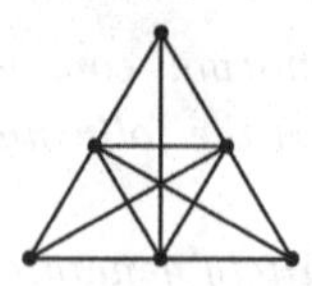

Then every GIT-cone $\lambda \in \Lambda(\overline{X}, H)$ defines a projective variety $X(\lambda) = \overline{X}^{ss}(\lambda) /\!\!/ H$. If $\lambda^\circ \subseteq \mathrm{Mov}(R)^\circ$ holds, then $X(\lambda)$ is the variety associated to the bunched ring $(R, \mathfrak{F}, \Phi(\lambda))$ with $\Phi(\lambda)$ defined as in Remark 1.60. In particular, in this case we have

$$\mathrm{Cl}(X(\lambda)) = K, \qquad \mathcal{R}(X(\lambda)) = R,$$

$$\mathrm{Mov}(X(\lambda)) = \mathrm{Mov}(R), \qquad \mathrm{SAmple}(X(\lambda)) = \lambda.$$

All projective varieties with Cox ring R are isomorphic to some $X(\lambda)$ with $\lambda^\circ \subseteq \mathrm{Mov}(R)^\circ$ and the Mori dream spaces among them are precisely those arising from a full dimensional λ.

Let X be the variety arising from a bunched ring $(R, \mathfrak{F}, \Phi)$. Every Weil divisor D on X defines a positively graded sheaf

$$\mathcal{S}^+ := \bigoplus_{n \in \mathbb{Z}_{\geq 0}} \mathcal{S}_n^+, \qquad \mathcal{S}_n^+ := \mathcal{O}_X(nD).$$

The algebra of global sections of this sheaf inherits finite generation from the Cox ring. In particular, we obtain a rational map

$$\varphi(D) \colon X \dashrightarrow X(D), \qquad X(D) := \mathrm{Proj}(\Gamma(X, \mathcal{S}^+)).$$

Note that $X(D)$ is explicitly given as the closure of the image of the rational map $X \dashrightarrow \mathbb{P}_m$ determined by the linear system of a sufficiently big multiple nD.

Remark 1.83 Consider the GIT-fan $\Lambda(\overline{X}, H)$ of the action of $H = \mathrm{Spec}\, \mathbb{K}[K]$ on $\overline{X} = \mathrm{Spec}\, R$. Let $\lambda \in \Lambda(\overline{X}, H)$ be the cone with $[D] \in \lambda^\circ$ and $W \subseteq \overline{X}$ the open subset obtained by removing the zero sets of the generators $f_1, \ldots, f_r \in R$. Then we obtain a commutative diagram:

$$\begin{array}{ccccc}
\widehat{X} & \supseteq & W & \subseteq & \overline{X}^{ss}(\lambda) \\
\big\downarrow {\scriptstyle /\!\!/ H} & & \big\downarrow & & \big\downarrow {\scriptstyle /\!\!/ H} \\
X & \supseteq & W/H & \longrightarrow & X(\lambda) \\
& \searrow_{\varphi(D)} & & & \big\| \\
& & & \dashrightarrow & X(D).
\end{array}$$

Proposition 1.84 *Let $D \in \mathrm{WDiv}(X)$ be any Weil divisor, and denote by $[D] \in \mathrm{Cl}(X)$ its class. Then the associated rational map $\varphi(D)\colon X \to X(D)$ is*

(i) *birational if and only if $[D] \in \mathrm{Eff}(X)^\circ$ holds,*
(ii) *small birational if and only if $[D] \in \mathrm{Mov}(X)^\circ$ holds,*
(iii) *a morphism if and only if $[D] \in \mathrm{SAmple}(X)$ holds,*
(iv) *an isomorphism if and only if $[D] \in \mathrm{Ample}(X)$ holds.*

Remark 1.85 The observations made so far imply in particular that two Mori dream surfaces are isomorphic if and only if their Cox rings are isomorphic as graded rings. Moreover they prove the implication "(i)⇒(ii)" of Theorem 1.80. For the other direction, one reduces finite generation of the Cox ring $\mathcal{R}(X)$ to finite generation of the semiample subalgebras $\oplus_{K \cap \lambda} \Gamma(X, \mathcal{R}_{[D]})$, where $\lambda^\circ \subseteq \mathrm{Mov}(X)^\circ$, which is given by classical results.

Recall that two Weil divisors $D, D' \in \mathrm{WDiv}(X)$ are said to be *Mori equivalent*, if there is a commutative diagram

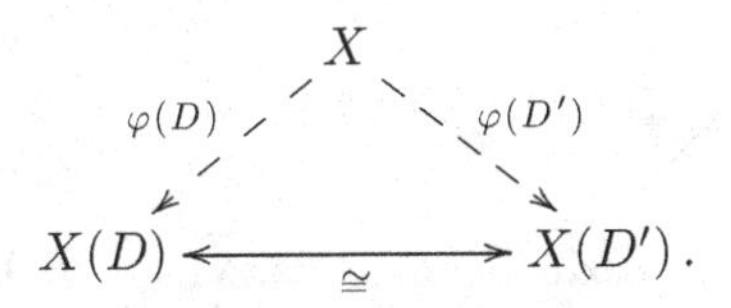

$$\begin{array}{ccc} & X & \\ \varphi(D) \swarrow & & \searrow \varphi(D') \\ X(D) & \underset{\cong}{\longleftrightarrow} & X(D'). \end{array}$$

Proposition 1.86 *For any two Weil divisors D, D' on X, the following statements are equivalent.*

(i) *The divisors D and D' are Mori equivalent.*
(ii) *One has $[D], [D'] \in \lambda^\circ$ for some GIT-chamber $\lambda \in \Lambda(\overline{X}, H)$.*

Let us have a look at the effect of blowing up and more general modifications on the Cox ring. In general, this is delicate, even finite generation may be lost. We discuss a class of toric ambient modifications having good properties in this regard; we restrict to hypersurfaces, for the general case see [19].

The starting point is a variety X_0 arising from a bunched ring $(R_0, \mathfrak{F}_0, \Phi_0)$ where the K_0-graded algebra R_0 is of the form $\mathbb{K}[T_1, \ldots, T_r]/\langle f_0 \rangle$ and $\mathfrak{F}_0$ consists of the (pairwise nonassociated K_0-prime) classes $T_i + \langle f_0 \rangle$; as usual, we denote their K_0-degrees by w_i. We obtained the canonical toric embedding via

$$
\begin{array}{ccccccc}
\overline{X}_0 & \supseteq & \widehat{X}_0 & \longrightarrow & \widehat{Z}_0 & \subseteq & \overline{Z}_0 . \\
 & & \downarrow {\scriptstyle /\!\!/ H} & & \downarrow {\scriptstyle /\!\!/ H} & & \\
 & & X_0 & \xrightarrow[\imath]{} & Z_0 & &
\end{array}
$$

The morphism $\widehat{Z}_0 \to Z_0$ is given by a map $\widehat{\Sigma}_0 \to \Sigma_0$ of fans living in lattices $F_0 = \mathbb{Z}^r$ and N_0. Let $v_1, \ldots, v_r$ be the primitive lattice vectors in the rays of Σ_0 and suppose that for $2 \leq d \leq r$, the cone σ_0 generated by $v_1, \ldots, v_d$ belongs to Σ_0. Consider the stellar subdivision $\Sigma_1 \to \Sigma_0$ at a vector

$$v_\infty = a_1 v_1 + \cdots + a_d v_d .$$

Let m_∞ be the index of this subdivision, i.e. the gcd of the entries of v_∞, and denote the associated toric modification by $\pi\colon Z_1 \to Z_0$. Then we obtain the strict transform $X_1 \subseteq Z_1$ mapping onto $X_0 \subseteq Z_0$. Moreover, we have commutative diagrams

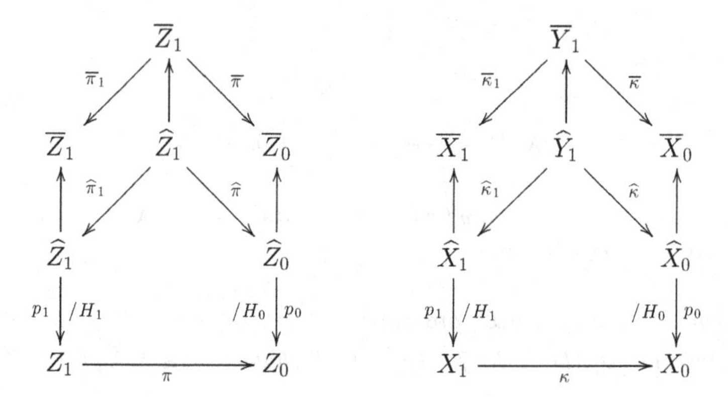

where $H_1 = \operatorname{Spec}(\mathbb{K}[K_1])$ for $K_1 = E_1/M_1$ in analogy to $K_0 = E_0/M_0$ etc. Note that the Cox ring $\mathbb{K}[T_1, \ldots, T_r, T_\infty]$ of $\overline{Z}_1$ comes with a K_1-grading. Moreover, with respect to the coordinates corresponding to the rays of the fans Σ_i, the map $\overline{\pi}\colon \overline{Z}_1 \to \overline{Z}_0$ is given by

$$\overline{\pi}(z_1, \ldots, z_r, z_\infty) \; = \; (z_\infty^{a_1} z_1, \ldots, z_\infty^{a_d} z_d, z_{d+1}, \ldots, z_r).$$

We want to formulate an explicit condition on the setting which guarantees that $\Gamma(\overline{X}_1, \mathcal{O})$ is the Cox ring of the proper transform X_1. For this, consider the grading

$$\mathbb{K}[T_1, \dots, T_r] = \bigoplus_{k \geq 0} \mathbb{K}[T_1, \dots, T_r]_k, \quad \text{where } \deg(T_i) := \begin{cases} a_i & i \leq d, \\ 0 & i \geq d+1. \end{cases}$$

Then we may write $f_0 = g_{k_0} + \cdots + g_{k_m}$ where $k_0 < \cdots < k_m$ and each g_{k_i} is a nontrivial polynomial having degree k_i with respect to this grading.

Definition 1.87 We say that the polynomial $f_0 \in \mathbb{K}[T_1, \dots, T_r]$ is *admissible* if

(i) the toric orbit $0 \times \mathbb{T}^{r-d}$ intersects $\overline{X}_0 = V(f_0)$,
(ii) g_{k_0} is a K_1-prime polynomial in at least two variables.

Note that for the case of a free abelian group K_1, the second condition just means that g_{k_0} is an irreducible polynomial.

Proposition 1.88 *If, in the above setting, the polynomial f_0 is admissible, then the Cox ring of the strict transform $X_1 \subseteq Z_1$ is*

$$\mathcal{R}(X_1) = \mathbb{K}[T_1, \dots, T_r, T_\infty]/\langle f_1(T_1, \dots, T_r, \sqrt[m_\infty]{T_\infty})\rangle,$$

where in

$$f_1 := \frac{f_0(T_\infty^{a_1} T_1, \dots, T_\infty^{a_d} T_d, T_{d+1}, \dots, T_r)}{T_\infty^{k_0}} \in \mathbb{K}[T_1, \dots, T_r, T_\infty]$$

only powers $T_\infty^{lm_\infty}$ with $l \geq 0$ of T_∞ occur, and the notation $\sqrt[m_\infty]{T_\infty}$ means replacing $T_\infty^{lm_\infty}$ with T_∞^l in f_1.

Example 1.89 Consider the polynomial $f_0 := T_1 + T_2^2 + T_3 T_4$ and the factorial algebra $R_0 := \mathbb{K}[T_1, \dots, T_4]/f_0$. Then R_0 is graded by $K_0 := \mathbb{Z}$ via the weight matrix

$$Q_0 = [2, 1, 1, 1].$$

With $\mathfrak{F}_0 := (\overline{T}_1, \dots, \overline{T}_4)$ and $\Phi_0 = \{\mathbb{Q}_{\geq 0}\}$, we obtain a bunched ring $(R_0, \mathfrak{F}_0, \Phi_0)$. The associated variety X_0 is a projective surface. In fact, there is an isomorphism $X_0 \to \mathbb{P}_2$ induced by

$$\mathbb{K}^4 \to \mathbb{K}^3, \qquad (z_1, z_2, z_3, z_4) \mapsto (z_2, z_3, z_4).$$

The canonical toric ambient variety Z_0 of X_0 is an open subset of the weighted projective space $\mathbb{P}(2,1,1,1)$. To obtain a defining fan Σ_0 of Z_0, consider the Gale dual matrix

$$P_0 := \begin{pmatrix} -1 & 2 & 0 & 0 \\ -1 & 0 & 1 & 1 \\ 0 & 1 & -1 & 0 \end{pmatrix}.$$

Let $v_1, \ldots, v_4$ denote its columns $v_1, \ldots, v_4$. According to Example 1.67, the maximal cones of Σ_0 are $P(\gamma_0^{\perp} \cap \delta)$, where γ_0 runs through the covering collection $\mathrm{cov}(\Phi_0)$. In terms of the columns $v_1, \ldots, v_4$ of P they are given as

$$\sigma_{1,2,3} := \mathrm{cone}(v_1, v_2, v_3), \qquad \sigma_{1,2,4} := \mathrm{cone}(v_1, v_2, v_4),$$
$$\sigma_{3,4} := \mathrm{cone}(e_3, e_4).$$

Now we subdivide $\sigma_{1,2,3}$ by inserting the ray through $v_\infty = 3v_1 + v_2 + 2v_3$. Note that f_0 is admissible; the g_{k_0}-term is $T_2^2 + T_3T_4$. According to Proposition 1.88, the defining equation of the Cox ring of the proper transform X_1 is

$$f_1 = \frac{T_\infty^3 T_1 + T_\infty^2 T_2^2 + T_\infty^2 T_3 T_4}{T_\infty^2} = T_\infty T_1 + T_2^2 + T_3 T_4.$$

In order to obtain the grading matrix, we have to look at the matrix with the new primitive generators as columns. In abuse of notation, we put $v_\infty = (-1,-1,-1)$ at the first place:

$$P_1 := \begin{pmatrix} -1 & -1 & 2 & 0 & 0 \\ -1 & -1 & 0 & -1 & 0 \\ -1 & 0 & 1 & -1 & 0 \end{pmatrix}.$$

The degree matrix is the Gale dual:

$$Q_1 = \begin{pmatrix} 1 & -1 & 0 & -1 & 1 \\ 1 & 1 & 1 & 0 & 2 \end{pmatrix}.$$

Thus, up to renaming of the variables, we obtain the Cox ring of the singular del Pezzo surface considered in Examples 1.68 and 1.79. In particular, we see that this del Pezzo surface is a modification of the projective plane.

1.3 Third lecture

1.3.1 Varieties with torus action

We describe the Cox ring of a variety with torus action following [22]. First recall the example of a complete toric variety X. Its Cox ring is given in terms of the prime divisors $D_1, \dots, D_r$ in the boundary $X \setminus T{\cdot}x_0$ of the open orbit:

$$\mathcal{R}(X) \;=\; \mathbb{K}[T^{D_1}, \dots, T^{D_r}], \qquad \deg(T^{D_i}) \;=\; [D] \;\in\; \mathrm{Cl}(X).$$

Now let X be any normal complete variety with finitely generated divisor class group and consider an effective algebraic torus action $T \times X \to X$, where $\dim(T)$ may be less than $\dim(X)$. For a point $x \in X$, denote by $T_x \subseteq T$ its isotropy group. The points with finite isotropy group form a non-empty T-invariant open subset

$$X_0 \;:=\; \{x \in X;\; T_x \text{ is finite}\} \;\subseteq\; X.$$

This set will replace the open orbit of a toric variety. Let E_k, where $1 \le k \le m$, denote the prime divisors in $X \setminus X_0$; note that each E_k is T-invariant with infinite generic isotropy, i.e. the subgoup of T acting trivially on E_k is infinite. According to a result of Sumihiro [40], there is a geometric quotient $q\colon X_0 \to X_0/T$ with an irreducible normal but possibly non-separated orbit space X_0/T.

Example 1.90 Consider $\mathbb{P}_1 \times \mathbb{P}_1$, the $\mathbb{K}^*$-action given with respect to inhomogeneous coordinates by $t{\cdot}(z,w) = (z, tw)$. Let X be the $\mathbb{K}^*$-equivariant blow up of $\mathbb{P}_1 \times \mathbb{P}_1$ at the fixed points $(0,0)$, $(1,0)$ and $(\infty,0)$.

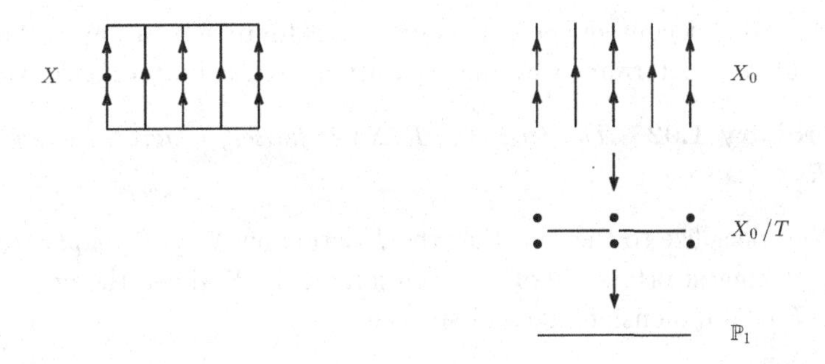

The open set X_0 is obtained by removing the two fixed point curves and the three isolated fixed points. The quotient X_0/T is the non-separated projective line with the points $0, 1, \infty$ doubled; note that there is a canonical map $X_0/T \to \mathbb{P}_1$.

As it turns out, one always finds kind of *separation* for the orbit space in our setting: there are a rational map $\pi\colon X_0/T \dashrightarrow Y$, an open subset $W \subseteq X_0/T$ and prime divisors $C_0, \ldots, C_r$ on Y such that following holds:

- the complement of W in X_0/T is of codimension at least two and the restriction $\pi\colon W \to X_0/T$ is a local isomorphism,
- each inverse image $\pi^{-1}(C_i)$ is a disjoint union of prime divisors C_{ij}, where $1 \le j \le n_i$,
- the map π is an isomorphism over $Y \setminus (C_0 \cup \ldots \cup C_r)$ and all prime divisors of X_0 with nontrivial generic isotropy occur among the $D_{ij} := q^{-1}(C_{ij})$.

Let $l_{ij} \in \mathbb{Z}_{\geq 1}$ be the order of the generic isotropy group of the T-action on the prime divisor D_{ij} and define monomials $T_i^{l_i} := T_{i1}^{l_{i1}} \cdots T_{in_i}^{l_{in_i}}$ in the variables T_{ij}. Moreover, let 1_{E_k} and $1_{D_{ij}}$ denote the canonical sections of E_k and D_{ij}.

Theorem 1.91 *There is a graded injection $\mathcal{R}(Y) \to \mathcal{R}(X)$ of Cox rings and $S_k \mapsto 1_{E_k}$, $T_{ij} \mapsto 1_{D_{ij}}$ defines an isomorphism of $\mathrm{Cl}(X)$-graded rings*

$$\mathcal{R}(X) \cong$$
$$\mathcal{R}(Y)[T_{ij}, S_1, \ldots, S_m;\ 0 \le i \le r,\ 1 \le j \le n_i] / \langle T_i^{l_i} - 1_{C_i};\ 0 \le i \le r \rangle,$$

where the $\mathrm{Cl}(X)$-grading on the right hand side is defined by associating to S_k the class of E_k and to T_{ij} the class of D_{ij}.

As a direct consequence, we obtain that finite generation of the Cox ring of X is determined by the separation Y of the orbit space X_0/T.

Corollary 1.92 *The Cox ring $\mathcal{R}(X)$ is finitely generated if and only if $\mathcal{R}(Y)$ is so.*

We specialize to the case that the T-action on X is of *complexity* one, i.e. its biggest orbits are of codimension one in X. Then the orbit space X_0/T is of dimension one and smooth.

Remark 1.93 For a normal complete variety X with a torus action $T \times X \to X$ of complexity one, the following statements are equivalent.

(i) $\mathrm{Cl}(X)$ is finitely generated.
(ii) X is rational.

Moreover, if one of these statements holds, then the separation of the orbit space is a morphism $\pi\colon X_0/T \to \mathbb{P}_1$.

The former prime divisors $C_i \subseteq Y$ are now points $a_0, \dots, a_r \in \mathbb{P}_1$ and $\pi^{-1}(a_i)$ consists of points $x_{i1}, \dots, x_{in_i} \in X_0/T$. As before, we may assume that all prime divisors of X_0 with non trivial generic isotropy occur among the prime divisors $D_{ij} = q^{-1}(x_{ij})$. Consider again the monomials $T_i^{l_i} = T_{i1}^{l_{i1}} \cdots T_{in_i}^{l_{in_i}}$, where $l_{ij} \in \mathbb{Z}_{\geq 1}$ is the order of the generic isotropy group of D_{ij}. Write $a_i = [b_i, c_i]$ with $b_i, c_i \in \mathbb{K}$. For every $0 \leq i \leq r-2$ set $j := i+1$, $k := i+2$ and define a trinomial

$$g_i := (b_j c_k - c_j b_k) T_i^{l_i} \;+\; (b_k c_i - c_k b_i) T_j^{l_j} \;+\; (b_i c_j - c_i b_j) T_k^{l_k}.$$

Theorem 1.94 *Let X be a normal complete variety with finitely generated divisor class group and an effective algebraic torus action $T \times X \to X$ of complexity one. Then, in terms of the data defined above, the Cox ring of X is given as*

$$\mathcal{R}(X) \cong \mathbb{K}[S_1, \dots, S_m, T_{ij};\ 0 \leq i \leq r,\ 1 \leq j \leq n_i] \;/\; \langle g_i;\ 0 \leq i \leq r-2 \rangle.$$

where 1_{E_k} corresponds to S_k, $1_{D_{ij}}$ corresponds to T_{ij}, and the $\mathrm{Cl}(X)$-grading on the right hand side is defined by associating to S_k the class of E_k and to T_{ij} the class of D_{ij}.

Let us discuss some applications of Theorem 1.94. As an initial application, we compute the Cox ring of a surface obtained by repeated blowing up points of the projective plane that lie on a given line; the case $n_0 = \dots = n_r = 1$ was done by other methods in [36].

Example 1.95 (Blowing up points on a line) Consider a line $Y \subseteq \mathbb{P}_2$ and points $p_0, \dots, p_r \in Y$. Let X be the surface obtained by blowing up n_i times the point p_i, where $0 \leq i \leq r$; in every step, we identify Y with its proper transform and p_i with the point in the intersection of Y with the exceptional curve. Set

$$g_i := (b_j c_k - c_j b_k) T_i \;+\; (b_k c_i - c_k b_i) T_j \;+\; (b_i c_j - c_i b_j) T_k,$$

with $p_i = [b_i, c_i]$ in $Y = \mathbb{P}_1$, the monomials $T_i = T_{i0} \cdots T_{in_i}$ and the indices $k = i+2$, $j = i+1$. Then the Cox ring of the surface X is given as

$$\mathcal{R}(X) = \mathbb{K}[T_{ij}, S;\ 0 \leq i \leq r,\ 0 \leq j \leq n_i] \;/\; \langle g_i;\ 0 \leq i \leq r-2 \rangle.$$

We verify this using a $\mathbb{K}^*$-action; note that blowing up $\mathbb{K}^*$-fixed points always can be made equivariant. With respect to suitable homogeneous coordinates z_0, z_1, z_2, we have $Y = V(z_0)$. Let $\mathbb{K}^*$ act via

$$t\cdot[z] := [z_0, tz_1, tz_2].$$

Then $E_1 := Y$ is a fixed point curve, the j-th (equivariant) blowing up of p_i produces an invariant exceptional divisor D_{ij} with a free $\mathbb{K}^*$-orbit inside and Theorem 1.94 gives the claim.

As a further application we indicate a general recipe for computing the Cox ring of a rational hypersurface given by a trinomial equation in the projective space. We perform this for the E_6 cubic surface in $\mathbb{P}_3$; the Cox ring of the resolution of this surface has been computed in [23].

Example 1.96 (The E_6 cubic surface) There is a cubic surface X in the projective space having singular locus $X^{\mathrm{sing}} = \{x_0\}$ and x_0 of type E_6. The surface is unique up to projectivity and can be realized as follows:

$$X \;=\; V(z_1 z_2^2 + z_2 z_0^2 + z_3^3) \;\subseteq\; \mathbb{P}_3.$$

Note that the defining equation is a trinomial but not of the shape of those occurring in Theorem 1.94. However, any trinomial hypersurface in a projective space comes with a complexity one torus action. Here, we have the $\mathbb{K}^*$-action

$$t\cdot[z_0, \ldots, z_4] = [z_0, t^{-3}z_1, t^3 z_2, tz_3].$$

This allows us to use Theorem 1.94 for computing the Cox ring. The task is to find the divisors E_k, D_{ij} and the orders l_{ij} of the isotropy groups. Note that $\mathbb{K}^*$ acts freely on the big torus of $\mathbb{P}_3$. The intersections of X with the toric prime divisors $V(z_i) \subseteq \mathbb{P}_3$ are given as

$$X \cap V(z_0) \;=\; V(z_0, z_1 z_2^2 + z_3^3), \qquad X \cap V(z_1) \;=\; V(z_1, z_2 z_0^2 + z_3^3),$$

$$X \cap V(z_3) \;=\; V(z_3, z_2(z_1 z_2 + z_0^2)) \;=\; (X \cap V(z_2)) \cup (X \cap V(z_1 z_2 + z_0^2)).$$

The first two sets are irreducible and both of them intersect the big torus orbit of the respective toric prime divisors $V(z_0)$ and $V(z_1)$. In order to achieve this also for $V(z_2)$, $V(z_3)$, we use a suitable weighted blow up of $\mathbb{P}_3$ at $V(z_2) \cap V(z_3)$. In terms of fans, this means to perform a certain stellar subdivision. Consider the matrices

$$P \;=\; \begin{pmatrix} -1 & 1 & 0 & 0 \\ -1 & 0 & 1 & 0 \\ -1 & 0 & 0 & 1 \end{pmatrix}, \qquad P' \;=\; \begin{pmatrix} -1 & 1 & 0 & 0 & 0 \\ -1 & 0 & 1 & 0 & 3 \\ -1 & 0 & 0 & 1 & 1 \end{pmatrix}.$$

The columns $v_0, \ldots, v_3$ of P are the primitive generators of the fan Σ of $Z := \mathbb{P}_3$ and we obtain a fan Σ' subdividing Σ at the last column v_4 of P'; note that v_4 is located on the tropical variety $\operatorname{trop}(X)$. Consider the associated toric morphism and the proper transform

$$\pi\colon Z \to Z, \qquad X' := \overline{\pi^{-1}(X \cap \mathbb{T}^3)} \subseteq Z_1.$$

A simple computation shows that the intersection of X' with the toric prime divisors of Z' is irreducible and intersects their big orbits. Moreover, $\pi\colon X' \to X$ is an isomorphism, because along X' nothing gets contracted. To proceed, note that we have no divisors of type E_k and that there is a commutative diagram

$$\begin{array}{ccc} X_0' & \longrightarrow & Z_0' \\ \big\downarrow {\scriptstyle /\mathbb{K}^*} & & \big\downarrow {\scriptstyle /\mathbb{K}^*} \\ X_0'/\mathbb{K}^* & \longrightarrow & Z_0'/\mathbb{K}^*. \end{array}$$

We determine the quotient $Z_0' \to Z_0'/\mathbb{K}^*$. The group $\mathbb{K}^*$ acts on Z' via homomorphism $\lambda_v\colon \mathbb{K}^* \to \mathbb{T}^3$ corresponding to $v = (-3,3,1) \in \mathbb{Z}^3$. The quotient by this action is the toric morphism given by any map $S\colon \mathbb{Z}^3 \to \mathbb{Z}^2$ having $\mathbb{Z}\cdot v$ as its kernel. We take S as follows and compute the images of the columns $v_0, \ldots, v_4$ of P':

$$S := \begin{pmatrix} 1 & 0 & 3 \\ 0 & 1 & -3 \end{pmatrix}, \qquad S\cdot P' = \begin{pmatrix} -4 & 1 & 0 & 3 & 3 \\ 2 & 0 & 1 & -3 & 0 \end{pmatrix}.$$

This shows that the toric divisors D_Z^1 and D_Z^4 corresponding to v_1 and v_4 are mapped to a doubled divisor in the non-separated quotient; see also [1]. The generic isotropy group of the $\mathbb{K}^*$-action along the toric divisor D_Z^i is given as the gcd l_i of the entries of the i-th column of $S\cdot P'$. We obtain

$$l_0 = 2, \qquad l_1 = 1, \qquad l_2 = 1, \qquad l_3 = 3, \qquad l_4 = 3.$$

By construction, the divisors $D_X^i := D_Z^i$ of the embedded variety $X_0' \subseteq Z_0'$ inherit the orders l_i of the isotropy groups and the behaviour with respect to the quotient map $X_0' \to X_0'/\mathbb{K}^*$. Renaming these divisors by

$$D_{01} := D_X^1, \quad D_{02} := D_X^4, \quad D_{11} := D_X^3, \quad D_{21} := D_X^0,$$

we arrive in the setting of Theorem 1.94; since D_X^2 has trivial isotropy group and gets not multiplied by the quotient map, it does not occur here. The Cox ring of $X \cong X'$ is then given as

$$\mathcal{R}(X) \cong \mathbb{K}[T_{01}, T_{02}, T_{11}, T_{21}]/\langle T_{01}T_{02}^3 + T_{11}^3 + T_{21}^2\rangle.$$

1.3.2 Writing down all Cox rings

As we observed, the Cox ring of a rational complete normal variety with a complexity one torus action admits a nice presentation by trinomial relations. Now we ask which of these trinomial rings occur as a Cox ring. First, we formulate the answer in algebraic terms and then turn to the geometric point of view; for the details, we refer to [20].

Construction 1.97 Fix $r \in \mathbb{Z}_{\geq 1}$, a sequence $n_0, \dots, n_r \in \mathbb{Z}_{\geq 1}$, set $n := n_0 + \dots + n_r$, and fix integers $m \in \mathbb{Z}_{\geq 0}$ and $0 < s < n + m - r$. The input data are

- a *sequence* $A = (a_0, \dots, a_r)$ of vectors $a_i = (b_i, c_i)$ in $\mathbb{K}^2$ such that any pair (a_i, a_j) with $j \neq i$ is linearly independent,
- an *integral block matrix* P of size $(r + s) \times (n + m)$ the columns of which are pairwise different primitive vectors generating $\mathbb{Q}^{r+s}$ as a cone:

$$P = \begin{pmatrix} P_0 & 0 \\ d & d' \end{pmatrix},$$

where d is an $(s \times n)$-matrix, d' an $(s \times m)$-matrix and P_0 an $(r \times n)$-matrix build from tuples $l_i := (l_{i1}, \dots, l_{in_i}) \in \mathbb{Z}^{n_i}_{\geq 1}$ in the following way

$$P_0 = \begin{pmatrix} -l_0 & l_1 & \dots & 0 \\ \vdots & \vdots & \ddots & \vdots \\ -l_0 & 0 & \dots & l_r \end{pmatrix}.$$

Now we associate to any such pair (A, P) a ring, graded by $K := \mathbb{Z}^{n+m}/\mathrm{im(P^*)}$, where P^* is the transpose of P. For every $0 \leq i \leq r$, define a monomial

$$T_i^{l_i} := T_{i1}^{l_{i1}} \cdots T_{in_i}^{l_{in_i}}.$$

Moreover, for any two indices $0 \leq i, j \leq r$, set $\alpha_{ij} := \det(a_i, a_j) = b_i c_j - b_j c_i$ and for any three indices $0 \leq i < j < k \leq r$ define a trinomial

$$g_{i,j,k} := \alpha_{jk} T_i^{l_i} \; + \; \alpha_{ki} T_j^{l_j} \; + \; \alpha_{ij} T_k^{l_k}.$$

Note that all trinomials $g_{i,j,k}$ are K-homogeneous of the same degree. Setting $g_i := g_{i,i+1,i+2}$, we obtain a K-graded factor algebra

$$R(A, P) := \mathbb{K}[T_{ij}, S_k;\ 0 \leq i \leq r,\ 1 \leq j \leq n_i, 1 \leq k \leq m] \;/\; \langle g_i;\ 0 \leq i \leq r - 2 \rangle.$$

Remark 1.98 The polynomials $g_{i,j,k}$ can be written as determinants in the following way:

$$g_{i,j,k} = \det \begin{pmatrix} b_i & b_j & b_k \\ c_i & c_j & c_k \\ T_i^{l_i} & T_j^{l_j} & T_k^{l_k} \end{pmatrix}.$$

Theorem 1.99 *Let (A, P) be data as in Construction 1.97. Then the algebra $R := R(A, P)$ is a normal factorially K-graded complete intersection; we have $R^* = \mathbb{K}^*$, the K-grading is almost free and $R_0 = \mathbb{K}$ holds. Moreover, the variables T_{ij}, S_k define a system of pairwise nonassociated K-prime generators for R.*

We say that the pair (A, P) is *sincere*, if $r \geq 2$ and $n_i l_{ij} > 1$ for all i, j; this ensures that there exist relations $g_{i,j,k}$ and none of these relations contains a linear term. The following statement tells us which of the $R(A, P)$ are factorial.

Theorem 1.100 *Let (A, P) be a sincere pair. Then the following statements are equivalent.*

(i) *The algebra $R(A, P)$ is a unique factorization domain.*
(ii) *The group $\mathbb{Z}^r/\mathrm{im}(P_0)$ is torsion free.*
(iii) *The numbers $\gcd(l_i)$ and $\gcd(l_j)$ are coprime for any $0 \leq i < j \leq r$.*

More generally, one can in a similar manner to Theorem 1.99 determine all affine algebras R with an effective factorial K-grading of complexity one such that $R_0 = \mathbb{K}$ holds, see [20]. The characterization of the factorial ones is the same as in Theorem 1.100. In dimensions two and three, the factorial algebras with a complexity one multigrading are described in early work of Mori [30] and Ishida [27].

Example 1.101 The algebra $\mathbb{K}[T_{01}, T_{11}, T_{21}]/\langle T_{01}^2 + T_{11}^2 + T_{21}^2 \rangle$ becomes graded by the group $K = \mathbb{Z} \oplus \mathbb{Z}/2\mathbb{Z} \oplus \mathbb{Z}/2\mathbb{Z}$ via

$$\deg(T_{01}) = (1, \bar{0}, \bar{0}), \qquad \deg(T_{11}) = (1, \bar{1}, \bar{0}), \qquad \deg(T_{21}) = (1, \bar{0}, \bar{1}).$$

This is an effective factorial grading of complexity one. However, the grading is not almost free. Thus, the algebra is not a Cox ring.

Let us turn to geometric aspects. We want to see the effective complexity one torus action on the varieties having a Cox ring $R(A, P)$. Existence of this action can be obtained by looking at the maximal possible grading of $R(A, P)$. We use here the canonical toric embedding which provides a little more geometric information.

Construction 1.102 Take a K-graded algebra $R = R(A, P)$ as constructed in Construction 1.97 and let $\mathfrak{F}$ be the system of pairwise nonassociated K-prime generators defined by the variables T_{ij} and S_k. Given any $\mathfrak{F}$-bunch Φ, we obtain a bunched ring $(R, \mathfrak{F}, \Phi)$ and an associated variety X. Consider the mutually dual sequences

$$0 \longrightarrow L \longrightarrow F \xrightarrow{P} N,$$

$$0 \longleftarrow K \underset{Q}{\longleftarrow} E \underset{P^*}{\longleftarrow} M \longleftarrow 0.$$

Recall that $Q\colon E \to K$ defines the K-degrees of the variables T_{ij}, S_k and note that P is indeed our defining matrix of the algebra $R(A, P)$. Now, let $\widehat{\Sigma}$ denote the fan in F generated by Φ and let Σ be its quotient fan in N; its rays have the columns v_{ij} and v_k of P as their primitive generators Moreover, consider

$$\overline{X} \ := \ V(g_0, \ldots, g_{r-2}) \ \subseteq \ \mathbb{K}^{n+m}.$$

The fan $\widehat{\Sigma}$ defines an open toric subvariety $\widehat{Z} \subseteq \mathbb{K}^{n+m}$ and the toric morphism $p\colon \widehat{Z} \to Z$ defined by $P\colon F \to N$ onto the toric variety Z associated to Σ is a characteristic space. The canonical toric embedding of X was obtained via the commutative diagram

$$\begin{array}{ccccccc} \overline{X} & \supseteq & \widehat{X} & \longrightarrow & \widehat{Z} & \subseteq & \mathbb{K}^{n+m}. \\ & & {\scriptstyle /\!\!/ H} \big\downarrow & & \big\downarrow {\scriptstyle /\!\!/ H} & & \\ & & X & \longrightarrow & Z & & \end{array}$$

Let $T_Z :=$ Spec $\mathbb{K}[M]$ be the acting torus of Z and let $T \subseteq T_Z$ be the subtorus corresponding to the inclusion $0 \times \mathbb{Z}^s \to N$. Then T acts on Z leaving $X \subseteq Z$ invariant and $T \times X \to X$ is an effective complexity one action. We also write $X = X(A, P, \Phi)$ for this T-variety.

Theorem 1.103 *Let X be a normal complete A_2-variety with an effective complexity one torus action. Then X is equivariantly isomorphic to a variety $X(A, P, \Phi)$ constructed in Construction 1.102.*

In particular, this enables us to apply the language of bunched rings to varieties with a complexity one torus action. The following is an application of Theorem 1.74 and Theorem 1.77.

Proposition 1.104 *Let X be a complete normal rational variety with an effective algebraic torus action $T \times X \to X$ of complexity one.*

(i) *The cone of divisor classes without fixed components is given by*

$$\bigcap_{1\le k\le m} \operatorname{cone}([E_s],[D_{ij}];\ s\neq k)\ \cap \bigcap_{\substack{0\le i\le r\\ 1\le j\le n_i}} \operatorname{cone}([E_k],[D_{st}];\ (s,t)\neq(i,j)).$$

(ii) *For any $0\le i\le r$, one obtains a canonical divisor for X by*

$$\max(0,r-1)\cdot\sum_{j=0}^{n_i} l_{ij}D_{ij} \ -\ \sum_{k=1}^{m} E_k \ -\ \sum_{i,j} D_{ij}.$$

The explicit description of rational varieties with a complexity one torus action by the data (A,P) can be used for classifications. We will present a result of [21] on Fano threefolds with free class group of rank one; we first note a general observation on varieties of this type made there.

Let X be an arbitrary complete d-dimensional variety with divisor class group $\mathrm{Cl}(X)\cong\mathbb{Z}$; thus, we do not require the presence of a torus action. The Cox ring $\mathcal{R}(X)$ is finitely generated and the total coordinate space $\overline{X}:=\operatorname{Spec}\mathcal{R}(X)$ is a factorial affine variety coming with an action of $\mathbb{K}^*$ defined by the $\mathrm{Cl}(X)$-grading of $\mathcal{R}(X)$. Choose a system $f_1,\ldots,f_\nu$ of homogeneous pairwise nonassociated prime generators for $\mathcal{R}(X)$. This provides a $\mathbb{K}^*$-equivariant embedding

$$\overline{X}\ \to\ \mathbb{K}^\nu, \qquad \overline{x}\ \mapsto\ (f_1(\overline{x}),\ldots,f_\nu(\overline{x})),$$

where $\mathbb{K}^*$ acts diagonally with the weights $w_i=\deg(f_i)\in\mathrm{Cl}(X)\cong\mathbb{Z}$ on $\mathbb{K}^\nu$. Moreover, X is the geometric $\mathbb{K}^*$-quotient of $\widehat{X}:=\overline{X}\setminus\{0\}$, and the quotient map $p\colon\widehat{X}\to X$ is a characteristic space.

Proposition 1.105 *For any $\overline{x}=(\overline{x}_1,\ldots,\overline{x}_\nu)\in\widehat{X}$ the local divisor class group $\mathrm{Cl}(X,x)$ of $x:=p(\overline{x})$ is finite of order $\gcd(w_i;\ \overline{x}_i\neq 0)$. The index of the Picard group $\mathrm{Pic}(X)$ in $\mathrm{Cl}(X)$ is given by*

$$[\mathrm{Cl}(X):\mathrm{Pic}(X)] = \operatorname{lcm}_{x\in X}(|\,\mathrm{Cl}(X,x)|).$$

Suppose that the ideal of $\overline{X}\subseteq\mathbb{K}^\nu$ is generated by $\mathrm{Cl}(X)$-homogeneous polynomials $g_1,\ldots,g_{\nu-d-1}$ of degree $\gamma_j:=\deg(g_j)$. Then one obtains

$$-\mathcal{K}_X=\sum_{i=1}^{\nu} w_i-\sum_{j=1}^{\nu-d-1}\gamma_j,$$

$$(-\mathcal{K}_X)^d=\left(\sum_{i=1}^{\nu} w_i-\sum_{j=1}^{\nu-d-1}\gamma_j\right)^d \frac{\gamma_1\cdots\gamma_{\nu-d-1}}{w_1\cdots w_\nu}$$

for the anticanonical class $-\mathcal{K}_X \in \mathrm{Cl}(X) \cong \mathbb{Z}$. *In particular,* X *is a Fano variety if and only if the following inequality holds*

$$\sum_{j=1}^{\nu-d-1} \gamma_j < \sum_{i=1}^{\nu} w_i.$$

Combining this with the explicit description in the presence of a complexity one torus action, one obtains bounds for the defining data. This leads to classification results. We present here the results for locally factorial threefolds; see [21] for details and more.

Theorem 1.106 *The following table lists the Cox rings* $\mathcal{R}(X)$ *of the three-dimensional locally factorial non-toric Fano varieties* X *with an effective two torus action and* $\mathrm{Cl}(X) = \mathbb{Z}$.

No.	$\mathcal{R}(X)$	$(w_1, \ldots, w_5)$	$(-K_X)^3$
1	$\mathbb{K}[T_1, \ldots, T_5] \;/\; \langle T_1T_2^5 + T_3^3 + T_4^2 \rangle$	$(1,1,2,3,1)$	8
2	$\mathbb{K}[T_1, \ldots, T_5] \;/\; \langle T_1T_2T_3^4 + T_4^3 + T_5^2 \rangle$	$(1,1,1,2,3)$	8
3	$\mathbb{K}[T_1, \ldots, T_5] \;/\; \langle T_1T_2^2T_3^3 + T_4^3 + T_5^2 \rangle$	$(1,1,1,2,3)$	8
4	$\mathbb{K}[T_1, \ldots, T_5] \;/\; \langle T_1T_2 + T_3T_4 + T_5^2 \rangle$	$(1,1,1,1,1)$	54
5	$\mathbb{K}[T_1, \ldots, T_5] \;/\; \langle T_1T_2^2 + T_3T_4^2 + T_5^3 \rangle$	$(1,1,1,1,1)$	24
6	$\mathbb{K}[T_1, \ldots, T_5] \;/\; \langle T_1T_2^3 + T_3T_4^3 + T_5^4 \rangle$	$(1,1,1,1,1)$	4
7	$\mathbb{K}[T_1, \ldots, T_5] \;/\; \langle T_1T_2^3 + T_3T_4^3 + T_5^2 \rangle$	$(1,1,1,1,2)$	16
8	$\mathbb{K}[T_1, \ldots, T_5] \;/\; \langle T_1T_2^5 + T_3T_4^5 + T_5^2 \rangle$	$(1,1,1,1,3)$	2
9	$\mathbb{K}[T_1, \ldots, T_5] \;/\; \langle T_1T_2^5 + T_3^3T_4^3 + T_5^2 \rangle$	$(1,1,1,1,3)$	2

1.3.3 $\mathbb{K}^*$-surfaces

Here we take a closer look at the first non-trivial examples of complexity one torus actions. $\mathbb{K}^*$-surfaces are studied by many authors; a classical reference is the work of Orlik and Wagreich [37]. Recall that a fixed point of a $\mathbb{K}^*$-surface is said to be

- *elliptic* if it is isolated and lies in the closure of infinitely many $\mathbb{K}^*$-orbits,

- *parabolic* if it belongs to a fixed point curve,
- *hyperbolic* if it is isolated and lies in the closure of two $\mathbb{K}^*$-orbits.

These are in fact the only possible types of fixed points for a normal $\mathbb{K}^*$-surface. For normal projective $\mathbb{K}^*$-surfaces X, there is always a *source* $F^+ \subseteq X$ and a *sink* $F^- \subseteq X$. They are characterized by the behaviour of general points: there is a non-empty open set $U \subseteq X$ with

$$\lim_{t \to 0} t{\cdot}x \in F^+, \qquad \lim_{t \to \infty} t{\cdot}x \in F^- \qquad \text{for all } x \in U.$$

The source can either consist of an elliptic fixed point or it is a curve of parabolic fixed points; the same holds for the sink. Any fixed point outside source or sink is hyperbolic. Note that in Example 1.90, we have two curves of parabolic fixed points and three hyperbolic fixed points. To every smooth projective $\mathbb{K}^*$-surface X having no elliptic fixed points Orlik and Wagreich associated a graph of the following shape:

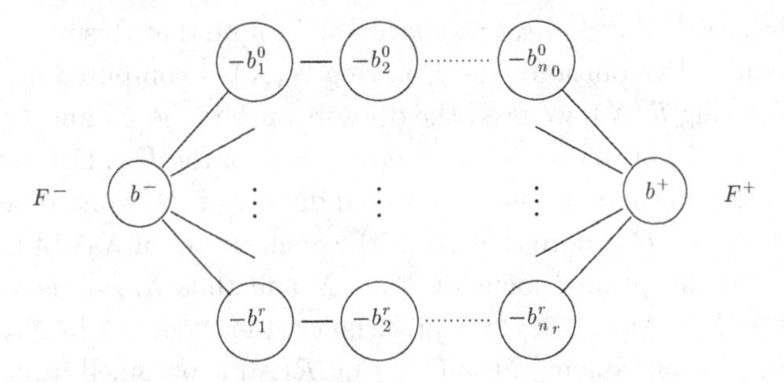

The two fixed point curves of X occur as F^- and F^+ in the graph. The other vertices represent the invariant irreducible contractible curves $D_{ij} \subseteq X$ different from F^- and F^+. The label $-b_j^i$ is the self intersection number of D_{ij}, and two of the D_{ij} are joined by an edge if and only if they have a common (fixed) point. Every D_{ij} is the closure of a non-trivial $\mathbb{K}^*$-orbit.

We show how to read off the Cox ring of a rational X from its Orlik-Wagreich graph. By [37, Sec. 3.5], the order l_{ij} of generic isotropy group of D_{ij} equals the numerator of the canceled continued fraction

$$b_1^i - \cfrac{1}{b_2^i - \cfrac{1}{\cdots - \cfrac{1}{b_{j-1}^i}}}.$$

Moreover, there is a canonical isomorphism $\mathbb{P}_1 = Y = F^-$ identifying $a_i \in Y$ with the point in $F^- \cap D_{i1}$. In the terminology introduced before, we thus obtain the following.

Theorem 1.107 *Let X be a smooth complete rational $\mathbb{K}^*$-surface without elliptic fixed points. Then the assignments $S^\pm \mapsto 1_{F^\pm}$ and $T_{ij} \mapsto 1_{D_{ij}}$ define an isomorphism*

$$\mathcal{R}(X) \cong \mathbb{K}[S^+, S^-, T_{ij};\ 0 \le i \le r,\ 1 \le j \le n_i] \ / \ \langle g_i;\ 0 \le i \le r-2 \rangle$$

of $\mathrm{Cl}(X)$-graded rings, where the $\mathrm{Cl}(X)$-grading on the right hand side is defined by associating to $S^\pm$ the class of $F^\pm$ and to T_{ij} the class of D_{ij}.

The Cox ring of a general rational $\mathbb{K}^*$-surface X is then computed as follows. Blowing up the (possible) elliptic fixed points suitably often gives a surface with two parabolic fixed point curves. Resolving the remaining singularities, we arrive at a smooth $\mathbb{K}^*$-surface $\widetilde{X}$ without parabolic fixed points, the so called *canonical resolution* of X; note that this is in general not the minimal resolution. The Cox ring $\mathcal{R}(\widetilde{X})$ is computed as above. For the Cox ring $\mathcal{R}(X)$, we need the divisors of the type E_k and D_{ij} in X and the orders l_{ij} of the generic isotropy groups of the D_{ij}. Each of these divisors is the image of a non-exceptional divisor of the same type in $\widetilde{X}$; to see this for the D_{ij}, note that X_0 is the open subset of $\widetilde{X}_0$ obtained by removing the exceptional locus of $\widetilde{X} \to X$ and thus $X_0/\mathbb{K}^*$ is an open subset of $\widetilde{X}_0/\mathbb{K}^*$. Moreover, by equivariance, the orders l_{ij} in X are the same as in $\widetilde{X}$. Consequently, the Cox ring $\mathcal{R}(X)$ is obtained from $\mathcal{R}(\widetilde{X})$ by removing those generators that correspond to the exceptional curves arising from the resolution.

Now let us look at rational normal complete $\mathbb{K}^*$-surfaces X with the methods presented in the preceding subsection. First recall that by finite generation of its Cox ring, X is $\mathbb{Q}$-factorial and projective. Moreover, X arises from a ring $R(A, P)$ as in Construction 1.102. As any surface, X is the only representative of its small birational class which means that no $\mathfrak{F}$-bunch Φ needs to be specified in the defining data.

Remark 1.108 Let (A, P) be data as in Construction 1.97. In order that $R(A, P)$ defines a surface X, we necessarily have $s = 1$ and for m we have the following three possibilities:

- $m = 0$. This holds if and only if X has two elliptic fixed points. In this case d' is empty.

- $m = 1$. This holds if and only if X has one fixed point curve and one elliptic fixed point. In this case, we may assume $d' = (-1)$.
- $m = 2$. This holds if and only if X has two elliptic fixed points. In this case, we may assume $d' = (1, -1)$.

In the fan Σ of the canonical ambient toric variety Z of X, the parabolic fixed point curves correspond to rays through $(0, \ldots, 0, \pm 1)$, the elliptic fixed points to full dimensional cones $\sigma^{\pm}$ and the hyperbolic fixed points to two-dimensional cones.

We demonstrate the concrete working with $\mathbb{K}^*$-surfaces by resolving singularities in two examples; for a detailed general treatment we refer to [24]. Similar to the canonical resolution of Orlik and Wagreich, the first step is resolving singular elliptic fixed points by inserting rays through $(0, \ldots, 0, \pm 1)$. The remaining singularities are then resolved by regular subdivision of two-dimensional singular cones of Σ. The first example is our surface from Example 1.68.

Example 1.109 In the setting of Construction 1.97, let $r = 2$, set $n_0 = 2$, $n_1 = 1$, $n_2 = 2$ and $m = 0$. For A choose the vectors $(-1, 0)$, $(1, -1)$ and $(0, 1)$. Consider the matrix

$$P = \begin{pmatrix} -1 & -1 & 2 & 0 & 0 \\ -1 & -1 & 0 & 1 & 1 \\ -1 & 0 & 1 & -1 & 0 \end{pmatrix}.$$

Then the defining equation of $\overline{X} \subseteq \mathbb{K}^5$ is $g = T_{01}T_{02} + T_{11}^2 + T_{21}T_{22}$. Observe that the Gale dual matrix is

$$Q = \begin{pmatrix} 1 & -1 & 0 & -1 & 1 \\ 1 & 1 & 1 & 0 & 2 \end{pmatrix}.$$

Thus, X is the del Pezzo surface we already encountered in Example 1.68. Consider the canonical toric ambient variety Z. Its fan Σ has the following appearance:

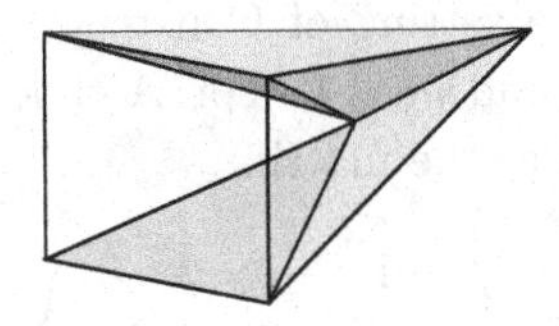

More precisely, in terms of the columns v_{01}, v_{02}, v_{11}, v_{21}, v_{22} of P, the maximal cones of Σ are

$$\begin{aligned} \sigma^- &:= \operatorname{cone}(v_{01}, v_{11}, v_{21}), & \sigma^+ &:= \operatorname{cone}(v_{02}, v_{12}, v_{22}), \\ \tau_{012} &:= \operatorname{cone}(v_{01}, v_{02}), & \tau_{212} &:= \operatorname{cone}(v_{21}, v_{22}). \end{aligned}$$

As we already observed, X comes with one singularity $x_0 \in X$; it is the toric fixed point corresponding to the cone σ^-. Set

$$v_1 := (0,0,-1), \qquad v_{10} := (1,0,0).$$

Inserting these vectors as columns at the right places of the matrix P gives the describing matrix for the resolution $\widetilde{X}$ of X:

$$\widetilde{P} = \begin{pmatrix} -1 & -1 & 1 & 2 & 0 & 0 & 0 \\ -1 & -1 & 0 & 0 & 1 & 1 & 0 \\ -1 & 0 & 0 & 1 & -1 & 0 & -1 \end{pmatrix}.$$

This allows us to read off the defining relation of the Cox ring; the degrees of the generators are the columns of the Gale dual matrix $\widetilde{Q}$. Concretely, we obtain:

$$\mathcal{R}(X) = \mathbb{K}[T_{ij}, S;\ i = 0,1,2\ j = 1,2] \ / \ \langle T_{01}T_{02} + T_{11}T_{12}^2 + T_{21}T_{22} \rangle,$$

$$\widetilde{Q} := \begin{pmatrix} 0 & 1 & -1 & 1 & 1 & 0 & 0 \\ 1 & 0 & -1 & 1 & 0 & 1 & 0 \\ 0 & -1 & -1 & 0 & 0 & -1 & 0 \\ 0 & 0 & 0 & 0 & -1 & 1 & 1 \end{pmatrix}.$$

The self intersection numbers of the curves corresponding to v_1 and v_{10} are computed as in Remark 1.78 and both equal -2. Thus, the singularity of X is of type A_2.

In the second example, we compute the Cox ring of the minimal resolution of the E_6 cubic surface from Example 1.96; the result was first obtained by Hassett and Tschinkel, without using the $\mathbb{K}^*$-action [23, Theorem 3.8].

Example 1.110 In the setting of Construction 1.97, let $r = 2$, set $n_0 = 2$, $n_1 = 1$, $n_2 = 1$ and $m = 0$. For A choose the vectors $(-1,0)$, $(1,-1)$ and $(0,1)$. Consider the matrix

$$P = \begin{pmatrix} -1 & -3 & 3 & 0 \\ -1 & -3 & 0 & 2 \\ -1 & -2 & 1 & 1 \end{pmatrix}.$$

The resulting surface X is the E_6 cubic surface of Example 1.96. Its Cox ring and degree matrix are given as

$$\mathcal{R}(X) = \mathbb{K}[T_{01}, T_{02}, T_{11}, T_{21}] \;/\; \langle T_{01}T_{02}^3 + T_{11}^3 + T_{21}^2 \rangle$$

$$Q := \begin{pmatrix} 3 & 1 & 2 & 3 \end{pmatrix}.$$

Let us look at the fan of the toric ambient variety. In terms of the columns $v_{01}, v_{02}, v_{11}, v_{21}$ of P, its maximal cones are

$$\sigma^- := \operatorname{cone}(v_{01}, v_{11}, v_{21}), \qquad \sigma^+ := \operatorname{cone}(v_{02}, v_{11}, v_{22}),$$
$$\tau_{012} := \operatorname{cone}(v_{01}, v_{02}).$$

The toric fixed point corresponding to σ^+ is the singularity $x_0 \in X$. It is singular for two reasons: firstly σ^+ is singular, secondly the total coordinate space

$$\overline{X} = V(T_{01}T_{02}^3 + T_{11}^3 + T_{21}^2)$$

is singular along the fiber $\mathbb{K}^* \times 0 \times 0 \times 0$ over x_0. Subdividing along $(0,0,1)$ and further resolving gives

$$\widetilde{P} = \begin{pmatrix} -1 & -3 & -2 & -1 & 3 & 2 & 1 & 0 & 0 & 0 \\ -1 & -3 & -2 & -1 & 0 & 0 & 0 & 2 & 1 & 0 \\ -1 & -2 & -1 & 0 & 1 & 1 & 1 & 1 & 1 & 1 \end{pmatrix}.$$

Note that inserting first $(0,0,1)$ is part of the first step of the canonical resolution. The Cox ring of the resolution and its degree matrix are thus given by

$$\mathcal{R}(\widetilde{X}) = \frac{\mathbb{K}[T_{01}, T_{02}, T_{03}, T_{04}, T_{11}, T_{12}, T_{13}, T_{21}, T_{21}, S]}{\langle T_{01}T_{02}^3T_{03}^2T_{04} + T_{11}^3T_{12}^2T_{13} + T_{21}^2T_{22} \rangle},$$

$$\widetilde{Q} = \begin{pmatrix} -1 & 1 & -1 & 0 & 0 & 0 & 0 & 0 & 0 & 0 \\ 1 & 0 & -1 & 1 & 0 & 0 & 0 & 0 & 0 & 0 \\ 0 & 1 & 0 & -1 & 0 & 1 & 0 & 1 & 0 & 0 \\ 0 & 0 & -1 & 0 & -1 & 0 & 1 & -1 & 0 & 0 \\ 1 & 0 & 0 & 0 & 0 & 1 & -1 & 0 & 1 & 0 \\ -1 & 0 & 0 & 0 & -1 & 1 & 0 & 0 & -1 & 0 \\ 1 & 0 & 0 & -1 & 0 & 0 & 0 & 0 & 0 & 1 \end{pmatrix}.$$

Remark 1.111 Inserting the rays through $(0,\ldots,0,\pm 1)$ as (partially) done in the preceding two examples is a "tropicalization step": the two rays arise when we intersect the fan of the toric ambient variety with the tropicalization $\operatorname{trop}(X)$ which, in the example case, is the union of the one codimensional cones of the normal fan of the Newton polytope of the defining equation.

Combining the description in terms of data (A, P) with the language of bunched rings makes $\mathbb{K}^*$-surfaces an easily accessible class of examples. This allows among other things explicit classifications. For example, the Gorenstein del Pezzo $\mathbb{K}^*$-surfaces are classified by these methods in [24]; part of the results has been obtained by other methods in [16] and [41, 22].

Theorem 1.112 *Tables 1.1 to 1.4 list the Cox rings $\mathcal{R}(X)$ and the singularity types $\mathrm{S}(X)$ of the non-toric Gorenstein Fano $\mathbb{K}^*$-surfaces X of Picard numbers one, two, three and four respectively.*

Table 1.1:

$\mathcal{R}(X)$	$\mathrm{Cl}(X)$	$(w_1, \ldots, w_r)$	$\mathrm{S}(X)$
$\mathbb{K}[T_1, T_2, T_3, S_1] \;/\; \langle T_1^2 + T_2^2 + T_3^2 \rangle$	$\mathbb{Z} \oplus \mathbb{Z}/2\mathbb{Z} \oplus \mathbb{Z}/2\mathbb{Z}$	$\begin{pmatrix} 1 & 1 & 1 & 1 \\ \bar{1} & \bar{0} & \bar{1} & \bar{0} \\ \bar{1} & \bar{1} & \bar{0} & \bar{0} \end{pmatrix}$	$D_4 3A_1$
$\mathbb{K}[T_1, \ldots, T_4] \;/\; \langle T_1T_2 + T_3^2 + T_4^2 \rangle$	$\mathbb{Z} \oplus \mathbb{Z}/4\mathbb{Z}$	$\begin{pmatrix} 1 & 1 & 1 & 1 \\ \bar{1} & \bar{3} & \bar{2} & \bar{0} \end{pmatrix}$	$2A_3 A_1$
$\mathbb{K}[T_1, \ldots, T_4] \;/\; \langle T_1T_2 + T_3^2 + T_4^3 \rangle$	$\mathbb{Z}$	$\begin{pmatrix} 1 & 5 & 3 & 2 \end{pmatrix}$	A_4
$\mathbb{K}[T_1, \ldots, T_4] \;/\; \langle T_1T_2 + T_3^2 + T_4^4 \rangle$	$\mathbb{Z} \oplus \mathbb{Z}/2\mathbb{Z}$	$\begin{pmatrix} 1 & 3 & 2 & 1 \\ \bar{1} & \bar{1} & \bar{1} & \bar{0} \end{pmatrix}$	$A_5 A_1$
$\mathbb{K}[T_1, \ldots, T_4] \;/\; \langle T_1T_2 + T_3^3 + T_4^3 \rangle$	$\mathbb{Z} \oplus \mathbb{Z}/3\mathbb{Z}$	$\begin{pmatrix} 1 & 2 & 1 & 1 \\ \bar{1} & \bar{2} & \bar{2} & \bar{0} \end{pmatrix}$	$A_5 A_2$
$\mathbb{K}[T_1, \ldots, T_4] \;/\; \langle T_1T_2^2 + T_3^3 + T_4^2 \rangle$	$\mathbb{Z}$	$\begin{pmatrix} 4 & 1 & 2 & 3 \end{pmatrix}$	D_5
$\mathbb{K}[T_1, \ldots, T_4] \;/\; \langle T_1^2T_2 + T_3^2 + T_4^4 \rangle$	$\mathbb{Z} \oplus \mathbb{Z}/2\mathbb{Z}$	$\begin{pmatrix} 1 & 2 & 2 & 1 \\ \bar{1} & \bar{0} & \bar{1} & \bar{0} \end{pmatrix}$	$D_6 A_1$
$\mathbb{K}[T_1, \ldots, T_4] \;/\; \langle T_1T_2^2 + T_3^3 + T_4^3 \rangle$	$\mathbb{Z} \oplus \mathbb{Z}/3\mathbb{Z}$	$\begin{pmatrix} 1 & 1 & 1 & 1 \\ \bar{1} & \bar{1} & \bar{2} & \bar{0} \end{pmatrix}$	$E_6 A_2$
$\mathbb{K}[T_1, \ldots, T_4] \;/\; \langle T_1T_2^3 + T_3^3 + T_4^2 \rangle$	$\mathbb{Z}$	$\begin{pmatrix} 3 & 1 & 2 & 3 \end{pmatrix}$	E_6
$\mathbb{K}[T_1, \ldots, T_4] \;/\; \langle T_1T_2^3 + T_3^4 + T_4^2 \rangle$	$\mathbb{Z} \oplus \mathbb{Z}/2\mathbb{Z}$	$\begin{pmatrix} 1 & 1 & 1 & 2 \\ \bar{0} & \bar{0} & \bar{1} & \bar{1} \end{pmatrix}$	$E_7 A_1$
$\mathbb{K}[T_1, \ldots, T_4] \;/\; \langle T_1T_2^4 + T_3^3 + T_4^2 \rangle$	$\mathbb{Z}$	$\begin{pmatrix} 2 & 1 & 2 & 3 \end{pmatrix}$	E_7
$\mathbb{K}[T_1, \ldots, T_4] \;/\; \langle T_1T_2^5 + T_3^3 + T_4^2 \rangle$	$\mathbb{Z}$	$\begin{pmatrix} 1 & 1 & 2 & 3 \end{pmatrix}$	E_8
$\mathbb{K}[T_1, \ldots, T_5] \;/\; \left\langle \begin{smallmatrix} T_1T_2 + T_3^2 + T_4^2, \\ \lambda T_3^2 + T_4^2 + T_5^2 \end{smallmatrix} \right\rangle$	$\mathbb{Z} \oplus \mathbb{Z}/2\mathbb{Z} \oplus \mathbb{Z}/2\mathbb{Z}$	$\begin{pmatrix} 1 & 1 & 1 & 1 & 1 \\ \bar{1} & \bar{1} & \bar{0} & \bar{1} & \bar{0} \\ \bar{0} & \bar{0} & \bar{1} & \bar{1} & \bar{0} \end{pmatrix}$	$2D_4$

Table 1.2:

$\mathcal{R}(X)$	$\mathrm{Cl}(X)$	$(w_1,\dots,w_r)$	$\mathrm{S}(X)$
$\mathbb{K}[T_1,\dots,T_4,S_1]\ /\ \langle T_1T_2+T_3^2+T_4^2\rangle$	$\mathbb{Z}^2\oplus\mathbb{Z}/2\mathbb{Z}$	$\begin{pmatrix}1&1&1&1&1\\1&-1&0&0&1\\\bar{1}&\bar{1}&\bar{1}&\bar{0}&\bar{0}\end{pmatrix}$	A_32A_1
$\mathbb{K}[T_1,\dots,T_5]\ /\ \langle T_1T_2+T_3T_4+T_5^2\rangle$	$\mathbb{Z}^2$	$\begin{pmatrix}1&1&1&1&1\\-1&1&2&-2&0\end{pmatrix}$	$2A_2A_1$
$\mathbb{K}[T_1,\dots,T_5]\ /\ \langle T_1T_2+T_3T_4+T_5^2\rangle$	$\mathbb{Z}^2$	$\begin{pmatrix}1&3&1&3&2\\1&1&0&2&1\end{pmatrix}$	A_2
$\mathbb{K}[T_1,\dots,T_5]\ /\ \langle T_1T_2+T_3T_4+T_5^3\rangle$	$\mathbb{Z}^2$	$\begin{pmatrix}1&2&1&2&1\\1&-1&-1&1&0\end{pmatrix}$	A_1A_3
$\mathbb{K}[T_1,\dots,T_5]\ /\ \langle T_1T_2+T_3^2T_4+T_4^2\rangle$	$\mathbb{Z}^2$	$\begin{pmatrix}1&3&1&2&2\\0&2&1&0&1\end{pmatrix}$	A_3
$\mathbb{K}[T_1,\dots,T_5]\ /\ \langle T_1T_2+T_3^2T_4+T_5^3\rangle$	$\mathbb{Z}^2$	$\begin{pmatrix}1&2&1&1&1\\-1&1&1&-2&0\end{pmatrix}$	A_4A_1
$\mathbb{K}[T_1,\dots,T_5]\ /\ \langle T_1T_2+T_3^3T_4+T_5^2\rangle$	$\mathbb{Z}^2$	$\begin{pmatrix}1&3&1&1&2\\-1&-1&0&-2&-1\end{pmatrix}$	A_4
$\mathbb{K}[T_1,\dots,T_5]\ /\ \langle T_1T_2^2+T_3T_4^2+T_5^2\rangle$	$\mathbb{Z}^2$	$\begin{pmatrix}2&1&2&1&2\\0&-1&-2&0&-1\end{pmatrix}$	D_4
$\mathbb{K}[T_1,\dots,T_5]\ /\ \langle T_1T_2^2+T_3T_4^2+T_5^3\rangle$	$\mathbb{Z}^2$	$\begin{pmatrix}1&1&1&1&1\\2&-1&-2&1&0\end{pmatrix}$	D_5A_1
$\mathbb{K}[T_1,\dots,T_5]\ /\ \langle T_1T_2^3+T_3T_4^2+T_5^2\rangle$	$\mathbb{Z}^2$	$\begin{pmatrix}1&1&2&1&2\\1&-1&-2&0&-1\end{pmatrix}$	D_5
$\mathbb{K}[T_1,\dots,T_5]\ /\ \langle T_1T_2^3+T_3T_4^3+T_5^2\rangle$	$\mathbb{Z}^2$	$\begin{pmatrix}1&1&1&1&2\\1&-1&-2&0&-1\end{pmatrix}$	E_6
$\mathbb{K}[T_1,\dots,T_6]\ /\ \left\langle {T_1T_2+T_3T_4+T_5^2, \atop \lambda T_3T_4+T_5^2+T_6^2} \right\rangle$	$\mathbb{Z}^2\oplus\mathbb{Z}/2\mathbb{Z}$	$\begin{pmatrix}1&1&1&1&1&1\\-1&1&1&-1&0&0\\\bar{0}&\bar{0}&\bar{1}&\bar{1}&\bar{1}&\bar{0}\end{pmatrix}$	$2A_3$

Table 1.3:

$\mathcal{R}(X)$	$\mathrm{Cl}(X)$	$(w_1, \dots, w_r)$	$\mathrm{S}(X)$
$\mathbb{K}[T_1, \dots, T_5, S_1] \ / \ \langle T_1T_2 + T_3T_4 + T_5^2 \rangle$	$\mathbb{Z}^3$	$\begin{pmatrix} 1 & 1 & 1 & 1 & 1 & 1 \\ 1 & -1 & 0 & 0 & 0 & 1 \\ 1 & -1 & -1 & 1 & 0 & 0 \end{pmatrix}$	A_1A_2
$\mathbb{K}[T_1, \dots, T_6] \ / \ \langle T_1T_2 + T_3T_4 + T_5T_6 \rangle$	$\mathbb{Z}^3$	$\begin{pmatrix} 1 & 1 & 1 & 1 & 1 & 1 \\ 1 & 0 & 0 & 1 & 0 & 1 \\ 0 & 0 & 1 & -1 & -1 & 1 \end{pmatrix}$	$3A_1$
$\mathbb{K}[T_1, \dots, T_6] \ / \ \langle T_1T_2 + T_3T_4 + T_5T_6 \rangle$	$\mathbb{Z}^3$	$\begin{pmatrix} 1 & 2 & 1 & 2 & 1 & 2 \\ 0 & 1 & 0 & 1 & 1 & 0 \\ 1 & -1 & -1 & 1 & 0 & 0 \end{pmatrix}$	A_1
$\mathbb{K}[T_1, \dots, T_6] \ / \ \langle T_1T_2 + T_3T_4 + T_5^2T_6 \rangle$	$\mathbb{Z}^3$	$\begin{pmatrix} 1 & 2 & 1 & 2 & 1 & 1 \\ 1 & 0 & 0 & 1 & 0 & 1 \\ 1 & -1 & -1 & 1 & 0 & 0 \end{pmatrix}$	A_2
$\mathbb{K}[T_1, \dots, T_6] \ / \ \langle T_1T_2 + T_3T_4^2 + T_5T_6^2 \rangle$	$\mathbb{Z}^3$	$\begin{pmatrix} 2 & 1 & 1 & 1 & 1 & 1 \\ 0 & 1 & 1 & 0 & 1 & 0 \\ 1 & 0 & -1 & 1 & 1 & 0 \end{pmatrix}$	A_3
$\mathbb{K}[T_1, \dots, T_6] \ / \ \langle T_1T_2^2 + T_3T_4^2 + T_5T_6^2 \rangle$	$\mathbb{Z}^3$	$\begin{pmatrix} 1 & 1 & 1 & 1 & 1 & 1 \\ 1 & -1 & -1 & 0 & -1 & 0 \\ 1 & 0 & -1 & 1 & 1 & 0 \end{pmatrix}$	D_4
$\mathbb{K}[T_1, \dots, T_7] \ / \ \left\langle \begin{smallmatrix} T_1T_2+T_3T_4+T_5T_6, \\ \lambda T_3T_4+T_5T_6+T_7^2 \end{smallmatrix} \right\rangle$	$\mathbb{Z}^3$	$\begin{pmatrix} 1 & 1 & 1 & 1 & 1 & 1 & 1 \\ 0 & 0 & 1 & -1 & -1 & 1 & 0 \\ -1 & 1 & 1 & -1 & 0 & 0 & 0 \end{pmatrix}$	$2A_2$

Table 1.4:

$\mathcal{R}(X)$	$\mathrm{Cl}(X)$	$(w_1, \dots, w_r)$	$\mathrm{S}(X)$
$\mathbb{K}[T_1, \dots, T_6, S_1] \ / \ \langle T_1T_2 + T_3T_4 + T_5T_6 \rangle$	$\mathbb{Z}^4$	$\begin{pmatrix} 1 & 1 & 1 & 1 & 1 & 1 & 1 \\ 1 & 0 & 0 & 1 & 0 & 1 & 0 \\ 0 & 1 & 0 & 1 & 1 & 0 & 0 \\ 1 & -1 & -1 & 1 & 0 & 0 & 0 \end{pmatrix}$	A_1
$\mathbb{K}[T_1, \dots, T_8] \ / \ \left\langle \begin{smallmatrix} T_1T_2+T_3T_4+T_5T_6, \\ \lambda T_3T_4+T_5T_6+T_7T_8 \end{smallmatrix} \right\rangle$	$\mathbb{Z}^4$	$\begin{pmatrix} 1 & 1 & 1 & 1 & 1 & 1 & 1 & 1 \\ 0 & 1 & 1 & 0 & 0 & 1 & 1 & 0 \\ 0 & 0 & 1 & -1 & -1 & 1 & 0 & 0 \\ -1 & 1 & 1 & -1 & 0 & 0 & 0 & 0 \end{pmatrix}$	$2A_1$

References

[1] A. A'Campo-Neuen, J. Hausen: Toric prevarieties and subtorus actions. Geom. Dedicata 87 (2001), no. 1-3, 35–64.

[2] D.F. Anderson: Graded Krull domains. Comm. Algebra 7 (1979), no. 1, 79–106.

[3] M. Artebani, J. Hausen, A. Laface: On Cox rings of K3 surfaces. Compos. Math. 146 (2010), no. 4, 964–998.

[4] I. Arzhantsev, U. Derenthal, J. Hausen, A. Laface: Cox rings. arXiv:1003.4229.

[5] I.V. Arzhantsev, J. Hausen: Geometric Invariant Theory via Cox rings. J. Pure Appl. Algebra 213 (2009), no. 1, 154–172.

[6] M. Audin: The topology of torus actions on symplectic manifolds. Prog. Math., 93. Birkhäuser Verlag, Basel, 1991.

[7] V.V. Batyrev: Quantum cohomology rings of toric manifolds. In: Journées de Géométrie Algébrique d'Orsay, Astérisque 218, 9–34 (1993).

[8] A. Borel: Linear algebraic groups. Second edition. Graduate Texts in Mathematics 126, Springer-Verlag, New York, 1991.

[9] F. Berchtold, J. Hausen: GIT-equivalence beyond the ample cone. Michigan Math. J. 54 (2006), no. 3, 483–515.

[10] F. Berchtold, J. Hausen: Cox rings and combinatorics. Trans. Amer. Math. Soc. 359 (2007), no. 3, 1205–1252.

[11] C. Birkar, P. Cascini, C. Hacon, J. McKernan: Existence of minimal models for varieties of log general type. J. Amer. Math. Soc. 23 (2010), no. 2, 405–468.

[12] J.L. Colliot-Thélène, J.-J. Sansuc: Torseurs sous des groupes de type multiplicatif; applications à l'étude des points rationnels de certaines variétés algébriques. C. R. Acad. Sci. Paris Sér. A-B 282 (1976), no. 18, Aii, A1113–A1116.

[13] D.A. Cox: The homogeneous coordinate ring of a toric variety. J. Alg. Geom. 4 (1995), no. 1, 17–50.

[14] D.A. Cox, J.B. Little, H.K. Schenck: Toric Varieties. Graduate Studies in Mathematics 124, American Mathematical Society (2011).

[15] V.I. Danilov: The geometry of toric varieties. Uspekhi Mat. Nauk 33 (1978), no. 2(200), 85–134 (Russian); English transl: Russian Math. Surveys 33 (1978), no. 2, 97–154.

[16] U. Derenthal: Singular Del Pezzo surfaces whose universal torsors are hypersurfaces. arXiv:math/0604194, 2006.

[17] I. Dolgachev: Weighted projective varieties. Group actions and vector fields (Vancouver, B.C., 1981), 34–71, Lecture Notes in Math., 956, Springer, Berlin, 1982.

[18] W. Fulton: Introduction to Toric Varieties. Princeton Univ. Press, Princeton, 1993.

[19] J. Hausen: Cox rings and combinatorics II. Mosc. Math. J. 8 (2008), no. 4, 711–757.

[20] J. Hausen, E. Herppich: Factorially graded rings. This volume.

[21] J. Hausen, E. Herppich, H. Süß: Multigraded factorial rings and Fano varieties with torus action. Documenta Math. 16 (2011) 71-109.

[22] J. Hausen, H. Süß: The Cox ring of an algebraic variety with torus action. Advances Math. 225 (2010), 977–1012.

[23] B. Hassett, Yu. Tschinkel: Universal torsors and Cox rings. In: Arithmetic of higher-dimensional algebraic varieties, Progr. Math., 226, Birkhäuser Boston, Boston, MA, 149–173 (2004).

[24] E. Herppich: Rational $\mathbb{K}^*$-surfaces. In preparation.

[25] Y. Hu, S. Keel: Mori dream spaces and GIT. Michigan Math. J. 48 (2000) 331–348.

[26] A.R. Iano-Fletcher: Working with weighted complete intersections. Explicit birational geometry of 3-folds, 101–173, London Math. Soc. Lecture Note Ser., 281, Cambridge Univ. Press, Cambridge, 2000.

[27] M.-N. Ishida: Graded factorial rings of dimension 3 of a restricted type. J. Math. Kyoto Univ. 17 (1977), no. 3, 441–456.

[28] H. Kraft: Geometrische Methoden in der Invariantentheorie. Vieweg Verlag, Braunschweig, 1984.

[29] J. McKernan: Mori dream spaces. Jpn. J. Math. 5 (2010), no. 1, 127–151.

[30] S. Mori: Graded factorial domains. Japan J. Math. 3 (1977), no. 2, 223–238.

[31] D. Mumford, J. Fogarty, F. Kirwan: Geometric invariant theory. Third edition. Ergebnisse der Mathematik und ihrer Grenzgebiete 34. Springer-Verlag, Berlin, 1994.

[32] I.M. Musson: Differential operators on toric varieties. J. Pure Appl. Algebra 95 (1994), no. 3, 303–315.

[33] T. Oda: Convex bodies and algebraic geometry. An introduction to the theory of toric varieties. Ergebnisse der Mathematik und ihrer Grenzgebiete (3), Springer-Verlag, Berlin, 1988.

[34] T. Oda, H.S. Park: Linear Gale transforms and Gelfand-Kapranov-Zelevinskij decompositions. Tohoku Math. J. (2) 43 (1991), no. 3, 375–399.

[35] A.L. Onishchik, E.B. Vinberg: Lie Groups and Algebraic Groups. Springer-Verlag, Berlin Heidelberg, 1990.

[36] J.C. Ottem: On the Cox ring of $\mathbb{P}^2$ blown up in points on a line. Math. Scand. 109 (2011), 22–30.

[37] P. Orlik, P. Wagreich: Algebraic surfaces with k^*-action. Acta Math. 138 (1977), no. 1-2, 43–81.

[38] A.N. Skorobogatov: Torsors and rational points. Cambridge Tracts in Math. 144, Cambridge University Press, 2001.

[39] T.A. Springer: Linear algebraic groups. Reprint of the 1998 second edition. Modern Birkhäuser Classics. Birkhäuser Boston, Inc., Boston, MA, 2009.

[40] H. Sumihiro: Equivariant completion. J. Math. Kyoto Univ. 14 (1974), 1–28.

[41] H. Süß: Canonical divisors on T-varieties. Preprint, arXiv:0811.0626.

[42] J. Tevelev: Compactifications of subvarieties of tori. Amer. J. Math. 129 (2007), 1084–1104.

[43] B. Totaro: The cone conjecture for Calabi-Yau pairs in dimension 2. Duke Math. J. 154 (2010), no. 2, 241–263.

[44] J. Wlodarczyk: Embeddings in toric varieties and prevarieties. J. Alg. Geom. 2 (1993), no. 4, 705–726.

2

A very brief introduction to étale homotopy

Tomer M. Schlank
Einstein Institute of Mathematics, The Hebrew University of Jerusalem

Alexei N. Skorobogatov
Department of Mathematics, Imperial College London

Abstract The task of these notes is to supply the reader who has little or no experience of simplicial topology with a phrase-book on étale homotopy, enabling them to proceed directly to [5] and [10]. This text contains no proofs, for which we refer to the foundational book by Artin and Mazur [1] in the hope that our modest introduction will make it more accessible. This is only a rough guide and is no substitute for a rigorous and detailed exposition of simplicial homotopy for which we recommend [4] and [8].

Let X be a Noetherian scheme which is locally unibranch (this means that the integral closure of every local ring of X is again a local ring), e.g., a Noetherian normal scheme (all local rings are integrally closed). All smooth schemes over a field fall into this category. The aim of the Artin–Mazur theory is to attach to X its étale homotopy type $\acute{E}t(X)$. This is an object of a certain category pro $-\ \mathcal{H}$, the pro-category of the homotopy category of CW-complexes. The aim of these notes is to explain this construction.

2.1 Simplicial objects

2.1.1 Simplicial objects and sisets

The ordinal category Δ is the category whose objects are finite ordered sets $[n] = \{0, \ldots, n\}$, one for each non-negative integer, where the morphisms are the order-preserving maps $[n] \to [m]$. It is easy to see that all these maps are compositions of the *face* maps $\delta_i : [n] \to [n+1]$ (the image misses i) and the *degeneracy* maps $\sigma_i : [n+1] \to [n]$ (hitting i twice).

A *simplicial object* with values in a category $\mathcal{C}$ is a contravariant functor $\Delta \to \mathcal{C}$. Simplicial objects in $\mathcal{C}$ form a category $\mathcal{SC}$, where the morphisms are the natural transformations of such functors.

For example, a simplicial set (or a *siset*) is a simplicial object $S = (S_n)$ with values in $\mathcal{S}ets$, the category of sets. Let $\mathcal{SS}ets$ be the category of sisets. The elements of S_n are called n-simplices of S. A simplex x is non-degenerate if it is not of the form $\sigma_i(y)$ for some simplex $y \in S$.

We denote by $\Delta[n]$ the siset given by the contravariant functor $\Delta \to \mathcal{S}ets$ sending $[m]$ to $\mathrm{Hom}_\Delta([m],[n])$. For example, $\Delta[0]$ has the one-element set in every degree m, that is, the zero function $[m] \to [0]$. Next, $\Delta[1]_m$ can be identified with the set of non-decreasing sequences of 0 and 1 of length $m+1$. The degeneracy operators repeat the i-th coordinate; the face operators erase the i-th coordinate. The siset $\Delta[1]$ has only three non-degenerate simplices: the constant functions $[0] \to 0$ and $[0] \to 1$ in degree 0, and the identity function $[1] \to [1]$ in degree 1. In general, $\Delta[n]_m$ is the set of increasing sequences of $0, 1, \ldots, n$ of length $m+1$, so that the identity $[n] \to [n]$ is the unique non-degenerate n-simplex of $\Delta[n]$. For any n-simplex of a siset S there is a unique map $\Delta[n] \to S$ that sends the identity function $[n] \to [n]$ to this simplex, hence $S_n = \mathrm{Hom}_{\mathcal{SS}ets}(\Delta[n], S)$.

The product of two sisets $R \times S$ is defined levelwise: $(R \times S)_n = R_n \times S_n$ with face and degeneracy operators acting simultaneously on both factors.

The maps of sisets $f : R \to S$ and $g : R \to S$ are strictly homotopy equivalent if there is a map of sisets $R \times \Delta[1] \to S$ whose restrictions to $R \times (0)$ and $R \times (1)$ give f and g, respectively. Two maps of sisets are homotopy equivalent if they can be connected by a chain of strict homotopies.

If S is a siset and A is an abelian group, then the *simplicial cohomology groups* $\mathrm{H}^n(S, A)$ are defined as the cohomology groups of the complex

$$\ldots \to \mathrm{Maps}(S_1, A) \to \mathrm{Maps}(S_0, A) \to 0,$$

where the differentials are alternating sums of the maps induced by the face maps.

2.1.2 Topological realisation

Let Top be the category of topological spaces, where the morphisms are continuous maps. Recall that a continuous map of topological spaces is a *weak equivalence* if it induces a bijection of the sets of connected components, and it induces isomorphisms on all the homotopy groups.

Consider the standard n-dimensional simplex:

$$\Delta^n = \{(x_0, \ldots, x_n) \in \mathbb{R}^{n+1} | x_i \geq 0, \sum x_i = 1\}.$$

Any siset S has the *topological realisation* $|S|$ which is a topological space defined as the quotient of the disjoint union of $S_n \times \Delta^n$ by the equivalence relation generated by the following relations: for any order preserving map $\alpha : [m] \to [n]$ we identify $(\alpha^*(x), t)$ with $(x, \alpha(t))$ (it is enough to do this for the face maps and the degeneracy maps). The natural map $|\Delta[n]| \to \Delta^n$ is a homeomorphism. The topological realisation $|S|$ of a siset S can be given the structure of a CW-complex: the n-cells are all non-degenerate n-simplices of S, with the face maps as the gluing maps, see [8, Thm. 14.1].

Any map of sisets induces a continuous map of topological realisations, hence the topological realisation is a functor $\mathcal{SS}ets \to Top$. Homotopic maps of sisets give rise to topologically homotopic maps. One can define the homotopy groups of a pointed siset directly, see [4, I.7], or, equivalently, as the homotopy groups of its topological realisation. The topological realisation commutes with products: we have $|R \times S| = |R| \times |S|$ for any sisets R and S such that $|R| \times |S|$ is a CW-complex (e.g. when R and S are both countable). See [8], Thm. 14.3 and Remark 14.4.

The *singular functor* $Top \to \mathcal{SS}ets$ attaches to a topological space T the siset ST, where $(ST)_n$ is the set of continuous maps $\Delta_n \to T$. It is a crucial property that the singular functor is right adjoint to the functor of topological realisation.

Let $\mathcal{H}$ be the homotopy category of CW-complexes. If X and Y are CW-complexes, then the set of morphisms from X to Y in $\mathcal{H}$ is the set of homotopy classes of maps $X \to Y$. It is denoted by $[X, Y]$.

Let $\mathcal{H}_0$ be the pointed homotopy category of connected pointed CW-complexes (the homotopy is assumed to preserve the base point).

2.1.3 Simplicial mapping space

In the category of sisets there is a natural notion of internal Hom.

Definition 2.1 Let A and B be sisets. The internal mapping space $\underline{\mathrm{Maps}}(A, B)$ is an object of $\mathcal{S}Sets$ defined by

$$\underline{\mathrm{Maps}}(A, B)_n := \mathrm{Hom}_{\mathcal{S}Sets}(A \times \Delta[n], B),$$

with the natural face and degeneracy maps.

There is a natural map

$$\phi : |\underline{\mathrm{Maps}}(A, B)| \to \mathrm{Maps}_{Top}(|A|, |B|).$$

However, sisets are too rigid in the sense that some maps of their realisations cannot be represented up to homotopy by maps of sisets, so that ϕ is not always a weak equivalence. For example, let C be the siset obtained by identifying the two 0-simplices of $\Delta[1]$. It is clear that $|C|$ is homotopy equivalent to S^1, however one can check that the only maps $C \to C$ are the constant map and the identity. Thus all the surjective maps $S^1 \to S^1$ of degrees more than 1 are not in the image of $|\underline{\mathrm{Maps}}(C, C)| \to \mathrm{Maps}_{Top}(|C|, |C|)$.

Kan noticed that ϕ is a weak equivalence if B (but not necessarily A) belongs to a special class of sisets that are now called Kan sisets.

2.1.4 Kan sisets and Kan fibrations

Let $m \in \{0, \ldots, n\}$. The *horn* $\Lambda^m[n]$ is the smallest sub-siset of $\Delta[n]$ containing all the non-degenerate $(n-1)$-simplices of which m is an element (that is, all except one). A map of sisets $p : X \to B$ is a *Kan fibration* if for any $n \geq 1$ and any commutative diagram

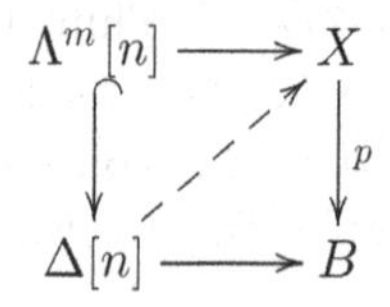

there exists a dotted map of sisets such that the two resulting triangles are commutative ("every horn has a filler"). If (B, b) is a pointed siset, then $p^{-1}(b)$ is called the fibre of f. A siset Y is called *fibrant* or a *Kan siset* if the map sending Y to a point is a Kan fibration. For example, if T is a topological space, then ST is a Kan siset. Note, however, that $\Delta[n]$ is a not a Kan siset for $n \geq 1$ (exercise).

Recall that a continuous map of topological spaces $f : V \to U$ is a *Serre fibration* if the dotted arrow exists in every commutative diagram of commutative maps

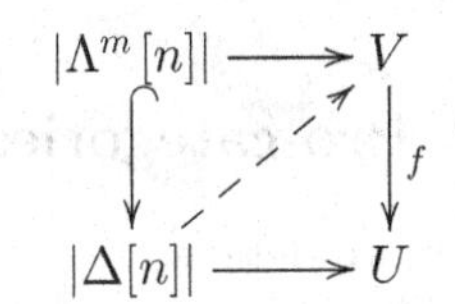

turning it into two commutative triangles. Kan fibrations of simplicial sets behave similarly to Serre fibrations of topological spaces. For example, Kan fibrations have the homotopy lifting property: if $E \to B$ is a Kan fibration and X is any siset, then for every diagram

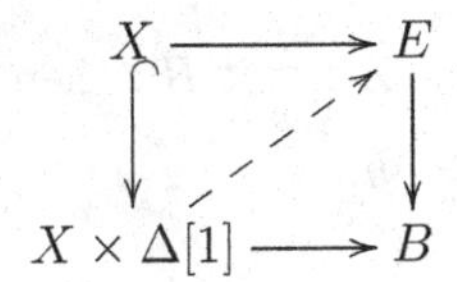

there exists a dotted map of sisets such that the two resulting triangles are commutative. The adjointness of S and $| \; |$ implies that $f : V \to U$ is a Serre fibration if and only if $f : S(V) \to S(U)$ is a Kan fibration. Quillen proved that if $X \to B$ is a Kan fibration, then $|X| \to |B|$ is a Serre fibration, see [4, Thm. I.10.10].

As mentioned above, if A is a siset and B is a Kan siset, then the natural map ϕ from the previous section is a weak equivalence[1]. This fact makes Kan simplicial sets a good model for topological spaces. Indeed, the homotopy category of Kan sisets (where the morphisms are the simplicial homotopy classes of maps) is equivalent to the homotopy category of CW-complexes, via the topological realisation functor and the singular functor. Thus in applications is it possible to use constructions from either category.

For an arbitrary siset A the adjunction between realisation and the singular functors gives rise to a natural map

$$A \to S|A|.$$

[1] This is a consequence of the fact that sisets and topological spaces are Quillen equivalent as simplicial model categories, and the Kan sisets are fibrant in the model category of sisets. See [6] for more on this subject.

This map is a weak equivalence, and $S|A|$ is a Kan siset (cf. [4], the proof of Thm. 11.4). Therefore, every siset can be functorially replaced with a weakly equivalent Kan siset. It is common to use for this purpose a different functor with a better combinatorial description, namely the Kan replacement functor Ex^∞, see [4, Thm. III.4.8].

2.2 Pro-categories

See the appendix to [1] for more details.

Definition 2.2 A category is cofiltering if it satisfies the following conditions:

(1) for any objects A and B there is an object C with morphisms to both A and B,

(2) if there are morphisms $A \underset{g}{\overset{f}{\rightrightarrows}} B$, then there is a morphism $h : C \to A$ such that $fh = gh$.

In other words, there is a diagram $C \xrightarrow{h} A \underset{g}{\overset{f}{\rightrightarrows}} B$.

Examples

(1) The category whose objects are natural numbers, and the morphisms are $n \to m$ whenever $n \geq m$.
(2) The category whose objects are positive natural numbers, and the morphisms are $n \to m$ whenever $m|n$.
(3) Connected pointed étale coverings of a pointed scheme.

Definition 2.3 Let $\mathcal{C}$ be a category. The objects of the **pro-category** $\mathrm{pro}-\mathcal{C}$ are functors $F : \mathcal{I} \to \mathcal{C}$, where the category $\mathcal{I}$ is cofiltering. The morphisms are defined as follows:

$$\mathrm{Hom}_{\mathrm{pro}-\mathcal{C}}(\{C_i\}_{i\in\mathcal{I}}, \{D_j\}_{j\in\mathcal{J}}) = \varprojlim_{j\in\mathcal{J}} \varinjlim_{i\in\mathcal{I}} \mathrm{Hom}(C_i, D_j).$$

This definition of morphisms between pro-objects should be familiar: homomorphisms of pro-finite groups are defined in the same way.

Define $\mathrm{pro}-\mathcal{H}$ as the pro-category of the homotopy category of CW-complexes $\mathcal{H}$. For the objects of $\mathrm{pro}-\mathcal{H}$ one defines the analogues of homology groups, but note that these are defined as pro-groups, not groups. Let pro $-\mathcal{H}_0$ be the pro-category of the pointed homotopy category of connected pointed CW-complexes $\mathcal{H}_0$. For the objects of pro $-\mathcal{H}$ one defines the homotopy pro-groups.

2.3 Coverings and hypercoverings

2.3.1 Coverings of CW-complexes and étale coverings of schemes

Let X be a CW-complex. An open covering $X = \cup_{\alpha \in J} U_\alpha$ is called *excellent* if the intersection of any number of open sets U_α is contractible or empty. The homotopy type of X can be recovered from the siset attached to such a covering. Indeed, let $\mathbf{U}_n$ be the set of functions $f : [n] \to J$ such that $\cap_{i=1}^n U_{f(i)} \neq \emptyset$. The face and degeneracy maps in $\mathbf{U}_\bullet$ are defined in the obvious way. Then the topological realisation $|\mathbf{U}_\bullet|$ is weakly equivalent to X. This means that X and $|\mathbf{U}_\bullet|$ can be connected by a zigzag of morphisms that induce isomorphisms on all homotopy groups.

We can think of a covering $X = \cup_{\alpha \in J} U_\alpha$ as a map $\mathcal{U} \to X$, where $\mathcal{U}$ is the disjoint union of open sets U_α. Note that $U_\alpha \cap U_\beta = U_\alpha \times_X U_\beta$, so that the fibred product $\mathcal{U} \times_X \mathcal{U}$ is just the disjoint union of pairwise intersections $U_\alpha \cap U_\beta$, the fibred product $\mathcal{U} \times_X \mathcal{U} \times_X \mathcal{U}$ is the disjoint union of triple intersections, and so on.

Definition 2.4 Let $\mathcal{U} \to X$ be a covering of a topological space X. The Čech nerve $\pi_0(\mathcal{U})$ of $\mathcal{U} \to X$ is the siset such that

$$\pi_0(\mathcal{U})_n = \pi_0(\mathcal{U} \times_X \ldots \times_X \mathcal{U}) \qquad (n+1 \text{ times}).$$

The face maps in $\pi_0(\mathcal{U})$ are obvious projections, and the degeneracy maps are various diagonal embeddings.

If we allow all the n-fold intersections to be disjoint unions of contractible sets, we arrive at the notion of a *good* covering. If $\mathcal{U} \to X$ is a good covering, then the topological realisation $|\pi_0(\mathcal{U})|$ of its Čech nerve is weakly equivalent to X.

Representing a covering by a morphism $\mathcal{U} \to X$ is convenient because this carries over to étale coverings. One would like to apply this construction to a Grothendieck topology on a scheme X, for example, to the small étale site $X_{\text{ét}}$ of X. The obvious problem is that open étale "subsets" are rarely contractible. However, it is clear that to recover the space we started with, the connected components must be contractible. Instead, we can think of the Čech nerve of an étale covering $\mathcal{U} \to X$ as an "approximation" to the homotopy type of the scheme X. This approximation becomes better when we take finer and finer coverings. On passing to a limit in $\mathcal{U}$ this system of approximations computes the Čech cohomology of X. Pro-objects in étale homotopy theory are used exactly in order to formalise this notion of "a hierarchic system of approximations". Naively, one might want to define the étale homotopy type as the pro-space $\acute{E}t : Cov(X) \to Top$, where $Cov(X)$ is the category of étale coverings of X, and $\acute{E}t(\mathcal{U}) = |\pi_0(\mathcal{U})|$ is the topological realisation of the Čech nerve of $\mathcal{U}$. There are two problems with this definition. The first problem is that the system of étale coverings does not always compute the correct cohomology groups. The second problem is that the category of étale coverings $Cov(X)$ is not cofiltering (cf. [9], III, Remark 2.2 (a)).

To understand the first problem better, we note that for an étale covering $\mathcal{U} \to X$ the cohomology groups $\mathrm{H}^n(|\pi_0(\mathcal{U})|, A)$ coincide with the Čech étale cohomology groups $\check{\mathrm{H}}^n(\mathcal{U}, A)$ (by definition, see [9], III.2). Passing to the limit at the level of cohomology groups is well defined (see [9], *loc. cit.*). Thus we obtain the groups

$$\check{\mathrm{H}}^n_{\text{ét}}(X, A) = \varinjlim \check{\mathrm{H}}^n(\mathcal{U}, A).$$

Unfortunately, $\check{\mathrm{H}}^n_{\text{ét}}(X, A)$ is not always equal to the étale cohomology group $\mathrm{H}^n_{\text{ét}}(X, A)$. For a quasicompact scheme X the canonical morphism $\check{\mathrm{H}}^n_{\text{ét}}(X, A) \to \mathrm{H}^n_{\text{ét}}(X, A)$ is an isomorphism when any finite subscheme of X is contained in an affine subset, e.g., X is quasi-projective over an affine scheme (for example, X is quasi-projective over a field). This is Artin's theorem, see [9, III.2.17].

A theorem of Verdier points at a way to remedy this situation by replacing coverings by a more general notion of hypercoverings. To define hypercoverings we first need to introduce the notions of skeleton and coskeleton which we do in the next section.

Unfortunately, even replacing the category $Cov(X)$ by the category of hypercoverings $Hyp(X)$ is not enough to solve the second problem. The reason is that $Hyp(X)$ is not cofiltering, either. To circumvent this problem one can work with the notion of the homotopy category of hypercoverings $HC(X)$ which is cofiltering. But passing to the homotopy classes of hypercoverings comes with a price: we obtain an object in $Pro - \mathcal{H}$ rather than in $Pro - Top$. This suffices for our needs, but in some cases one would like to get the actual "étale topological type". This can be done; we refer the interested reader to Friedlander's book [3] and to the recent paper [2].

2.3.2 Skeleton and coskeleton

Let $\mathcal{C}$ be a category closed under finite limits and colimits (e.g. the category $\mathcal{S}ets$ of sets, the category of pointed sets, or the étale site of a scheme). For $n \geq 0$ one defines functors sk_n and cosk_n from $\mathcal{SC}$ to itself, as follows. Let Δ/n be the ordinal category Δ truncated at level n, i.e. the full subcategory of Δ whose objects are $[m]$ for $m \leq n$. Let $\mathcal{S}_n\mathcal{C}$ be the category of functors $\Delta/n \to \mathcal{C}$. Since Δ/n is a finite category and $\mathcal{C}$ is closed under finite colimits, the obvious truncation functor $\tau_n : \mathcal{SC} \to \mathcal{S}_n\mathcal{C}$ has a left adjoint (the left Kan extension, cf. [7, X.4]). The composition of the truncation functor and its left adjoint

$$\mathrm{sk}_n : \mathcal{SC} \to \mathcal{S}_n\mathcal{C} \to \mathcal{SC}$$

is called the *skeleton* functor. For example, if $\mathcal{C} = \mathcal{S}ets$, then $\mathrm{sk}_n(X)$ is the simplicial subset of X that agrees with X up to the level n, and which has no non-degenerate simplices in dimensions greater than n.

Since Δ/n is a finite category and $\mathcal{C}$ is closed under finite limits, the functor $\tau_n : \mathcal{SC} \to \mathcal{S}_n\mathcal{C}$ also has a right adjoint (the right Kan extension, cf. [7, X.3]). The composed functor

$$\mathrm{cosk}_n : \mathcal{SC} \to \mathcal{S}_n\mathcal{C} \to \mathcal{SC}$$

is called the *coskeleton* functor. By definition we have

$$\mathrm{Hom}_{\mathcal{SC}}(\mathrm{sk}_n(A), B) = \mathrm{Hom}_{\mathcal{S}_n\mathcal{C}}(\tau_n(A), \tau_n(B)) = \mathrm{Hom}_{\mathcal{SC}}(A, \mathrm{cosk}_n(B)),$$

and taking $A = B$ we obtain the canonical morphisms

$$\mathrm{sk}_n(A) \to A \quad \text{and} \quad A \to \mathrm{cosk}_n(A).$$

Now let $\mathcal{C}$ be the category of sets or pointed sets. If A is a (pointed) siset, then it is clear that $\mathrm{cosk}_n(A)$ is the siset such that

$$\mathrm{cosk}_n(A)_m = \mathrm{Hom}_{\mathcal{SS}ets}(\Delta[m], \mathrm{cosk}_n(A)) = \mathrm{Hom}_{\mathcal{SS}ets}(\mathrm{sk}_n(\Delta[m]), A).$$

In particular, $\mathrm{cosk}_0(A)$ is the siset such that

$$\mathrm{cosk}_0(A)_m = \mathrm{Hom}_{\mathcal{S}ets}(\mathrm{sk}_0(\Delta[m]), A) = A_0^{m+1}.$$

Also, $\mathrm{cosk}_n(\Delta[r]) = \Delta[r]$ if $n \geq r$.

The coskeleton functor preserves Kan sisets (but not Kan fibrations[2]). If X is an object of $\mathcal{H}_0$, then the coskeleton $\mathrm{cosk}_n(X)$ is characterised by the following universal property: $\pi_m(\mathrm{cosk}_n(X)) = 0$ for $m \geq n$, and the canonical map $X \to \mathrm{cosk}_n(X)$ is universal in the homotopy category among the maps to objects with vanishing π_m for $m \geq n$, cf. [1], (2.4). For $m < n$ the map $X \to \mathrm{cosk}_n(X)$ induces an isomorphism $\pi_m(X) \xrightarrow{\sim} \pi_m(\mathrm{cosk}_n(X))$. In other words,

$$\ldots \to \mathrm{cosk}_{n+1}(X) \to \mathrm{cosk}_n(X) \to \ldots \to \mathrm{cosk}_0(X)$$

is a Postnikov tower of X.

Next, the homotopy fibre of $\mathrm{cosk}_{n+1}(X) \to \mathrm{cosk}_n(X)$ is the Eilenberg–MacLane space $K(\pi_n(X), n)$. Recall that for a group G a connected CW-complex X is called an Eilenberg–MacLane space $K(G, n)$ if the only non-trivial homotopy group of X is $\pi_n(X) = G$; all such CW-complexes are homotopy equivalent.

An important fact in homotopy theory is Whitehead's theorem that states that a map of CW-complexes is a homotopy equivalence if and only if it induces isomorphism on all homotopy groups. Whitehead's theorem allows us to study homotopy types "one homotopy group at a time" which makes it an essential tool in homotopy theory. However the analog of Whitehead's theorem is false for pro-CW-complexes. To remedy this situation Artin and Mazur [1, §3] present the following construction.

For an object $(X_\bullet) = (X_i)$ of $\mathrm{pro} - \mathcal{H}_0$ the universal property of coskeleton allows us to define the following object of $\mathrm{pro} - \mathcal{H}_0$:

$$X^\natural = \{\mathrm{cosk}_n(X_i)\}.$$

It is indexed by pairs (n, i) with the obvious canonical maps $\mathrm{cosk}_m(X_i) \to \mathrm{cosk}_n(X_j)$, $m \geq n$, where $i \to j$ is a morphism in the indexing category $\mathcal{I}$ of X. The canonical map $X \to X^\natural$ is a weak homotopy equivalence, but not necessarily an isomorphism in $\mathrm{pro} - \mathcal{H}_0$. Indeed, if X is a CW-complex, then

$$[X^\natural, X] = \varinjlim\, [\mathrm{cosk}_n(X), X],$$

so for the canonical map $X \to X^\natural$ to be invertible in $\mathcal{H}_0$, X must be *bounded*, i.e., $\pi_n(X) = 0$ for large n. In such a case X is homotopy equivalent to $\mathrm{cosk}_n(X)$ for some n.

By functoriality, any map $f : X \to Y$ in $\mathrm{pro}-\mathcal{H}_0$ induces a map $X^\natural \to Y^\natural$. The latter is an isomorphism if and only if f induces isomorphisms of

[2] As it is incorrectly stated on page 8 of [1].

all coskeletons of X and Y, and, equivalently, of all homotopy pro-groups of X and Y (see [1, Cor. 4.4]).

The functor $X \to X^\natural$ can also be defined in pro$-\mathcal{H}$. Informally speaking, a base point is not needed to "Postnikov-filter" the homotopy information by dimension, but is required to define the "associated graded filtration" in terms of groups.

2.3.3 Hypercoverings

Let X be a scheme, and let $X_{\text{ét}}$ be the small étale site of X. This is the category of all schemes Y étale over X; the coverings in $X_{\text{ét}}$ are surjective families of étale morphisms. To a covering $Y \to X$ we can associate the simplicial étale X-scheme $(Y_\bullet)$, where

$$Y_n = Y \times_X \ldots \times_X Y \quad (n \text{ times}).$$

The simplicial scheme $(Y_\bullet)$ is an example of a hypercovering.

Definition 2.5 A simplicial étale X-scheme $\mathcal{U}_\bullet$ is called a **hypercovering** if

(1) $\mathcal{U}_0 \to X$ is a covering;
(2) for every n the canonical morphism $\mathcal{U}_{n+1} \to \operatorname{cosk}_n(\mathcal{U}_\bullet)_{n+1}$ is a covering.

Note that $\mathcal{U}_0 \to X$ is a covering, by (1), and $\operatorname{cosk}_0(\mathcal{U}_\bullet)$ is the corresponding simplicial étale X-scheme

$$\operatorname{cosk}_0(\mathcal{U}_\bullet)_n = \mathcal{U}_0 \times_X \ldots \times_X \mathcal{U}_0.$$

The definition of a hypercovering implies that $\mathcal{U}_1 \to \mathcal{U}_0 \times_X \mathcal{U}_0$ is a covering, so it refines the notion of covering by allowing greater freedom at every level. This very general categorical construction is due to Verdier (SGA 4, Exp. V). It can be used with any site on X, see [1, Ch. 8].

A very important example is when X is a point and the site is the category $\mathcal{S}ets$, where a covering is just a surjective family of maps. Then a hypercovering is the same thing as a *contractible* Kan siset, see Enlightenment 8.5 (a) in [1].

2.3.4 Homotopy category of hypercoverings

Assume that $\mathcal{C}$ is a category with finite direct sums. If K is an object of $\mathcal{C}$, and S is a finite set, then $K \otimes S$ denotes the direct sum of copies of K indexed by the elements of S. Now let $(K_\bullet)$ be a simplicial object with values in $\mathcal{C}$. Define the simplicial object $(K_\bullet) \otimes \Delta[1]$ in $\mathcal{SC}$ by the formula

$$((K_\bullet) \otimes \Delta[1])_n = K_n \otimes \Delta[1]_n,$$

with the simultaneous action of face and degeneracy operators on both factors. There are two obvious inclusions $\Delta[0] \to \Delta[1]$, indexed by 0 and 1. Let e_0 and e_1 be the corresponding inclusions $(K_\bullet) \to (K_\bullet) \otimes \Delta[1]$.

Definition 2.6 The maps $f_0 : (K_\bullet) \to (L_\bullet)$ and $f_1 : (K_\bullet) \to (L_\bullet)$ are strictly homotopic if there is a map $(K_\bullet) \otimes \Delta[1] \to (L_\bullet)$ such that $f_0 = fe_0$ and $f_1 = fe_1$. Two maps are homotopic if they are related by a chain of strict homotopies.

We apply this to the case when $\mathcal{C}$ is the small étale site $X_{\text{ét}}$.

Definition 2.7 The homotopy category of hypercoverings $HC(X_{\text{ét}})$ is the category whose objects are étale hypercoverings of X, and whose maps are homotopy classes of morphisms of simplicial étale X-schemes.

An important result is that $HC(X_{\text{ét}})$ is a cofiltering category [1, Cor. 8.13 (i)]. Passing to the limit over $HC(X_{\text{ét}})$ one establishes a canonical isomorphism [1, Thm. 8.16]

$$\mathrm{H}^n_{\text{ét}}(X, F) = \varinjlim \mathrm{H}^n(\mathcal{U}_\bullet, A).$$

We will also need the pointed versions of the above constructions: the pointed étale site on X (with the choice of a geometric point on every étale X-scheme), pointed sisets, pointed hypercoverings, homotopy classes of pointed morphism (the strict homotopies are assumed to preserve the base point).

2.3.5 Étale homotopy type

Now we are ready to implement the strategy outlined earlier.

Let X be a locally Noetherian scheme. Then every scheme Y étale over X is a finite disjoint union of connected schemes. Write $\pi_0(Y)$ for the set of connected components of Y. Let $\pi_0(\mathcal{U}_\bullet)$ be the siset obtained by applying the functor π_0 to a simplicial étale hypercovering $\mathcal{U}_\bullet$. Since $HC(X_{\text{ét}})$ is cofiltering we can consider the pro-object

$$\pi_0(X_{\text{ét}}) = \{\pi_0(\mathcal{U}_\bullet)\},$$

which is an object in the pro-category of the homotopy category of sisets.

Applying the topological realisation functor we obtain the *étale homotopy type* $\acute{E}t(X)$ of X as an object in pro $-\mathcal{H}$.

One can now define the homology pro-groups of $\acute{E}t(X)$. The situation with cohomology is much better because cohomology is contravariant, the direct limit is an exact functor and the direct limit of abelian groups is also an abelian group. For an abelian group A this leads to the definition of the cohomology groups $\mathrm{H}^n(\acute{E}t(X), A)$. Using the previous theory one obtains a canonical isomorphism

$$\mathrm{H}^n(\acute{E}t(X), A) = \mathrm{H}^n_{\text{ét}}(X, A),$$

see [1, Cor. 9.3].

If X is equipped with a geometric base point, then we can consider the pointed étale site of X and define $\acute{E}t(X)$ as an object in pro $-\mathcal{H}_0$. A base point allows us to define homotopy pro-groups. By [1, Cor. 10.7] for every pointed scheme we have

$$\pi_1(\acute{E}t(X)) = \pi_1(X_{\text{ét}}).$$

This makes it possible to define higher étale homotopy pro-groups for all $n \geq 0$

$$\pi_n(X_{\text{ét}}) := \pi_n(\acute{E}t(X)).$$

2.3.6 Profinite completion and the comparison theorem

Let X be a pointed connected geometrically unibranch scheme over $\mathbb{C}$. Let us consider $X(\mathbb{C})$ as a topological space with the classical topology on $\mathbb{C}$. By the Riemann existence theorem we have

$$\pi_1(X_{\text{ét}}) = \pi_1\widehat{(X(\mathbb{C}))},$$

where $\widehat{G}$ is the profinite completion of the group G, see [9, §5]. It is natural to ask if for $n > 1$ there is an isomorphism between $\pi_n(X_{\text{ét}})$ and the profinite completion of $\pi_n(X(\mathbb{C}))$. In general this is not the case. Here one could recall a guiding principle of homotopy theory that says that functors should always be applied to homotopy types rather than to homotopy groups.

Let $\mathcal{H}_0^{\text{fin}}$ be the full subcategory of $\mathcal{H}_0$ consisting of pointed connected CW-complexes all of whose homotopy groups are finite. The following result is Theorem 3.4 of [1].

Theorem 2.8 *For any X in* pro$-\mathcal{H}_0$ *there are an object $\widehat{X}$ of* pro$-\mathcal{H}_0^{\text{fin}}$

and a map $X \to \widehat{X}$, *which are universal with respect to maps from* X *to objects of* pro $- \mathcal{H}_0^{\text{fin}}$.

Let us call $\widehat{X}$ the *profinite completion* of X. Associating to X its profinite completion defines a left adjoint functor to the inclusion of pro $- \mathcal{H}_0^{\text{fin}}$ into pro $- \mathcal{H}_0$.

Now we have the following generalisation of the isomorphism $\pi_1(X_{\text{ét}}) = \pi_1(\widehat{X(\mathbb{C})})$, see [1], Thm. 12.9 and Cor. 12.10.

Theorem 2.9 *Let* X *be a pointed connected geometrically unibranch scheme over* $\mathbb{C}$. *Let* $Cl(X)$ *be the object of* $\mathcal{H}_0$ *given by the homotopy type of the topological space* $X(\mathbb{C})$. *Consider* $\mathcal{H}_0$ *as a subcategory of* pro $- \mathcal{H}_0$. *There is a natural map*

$$Cl(X) \to \acute{E}t(X)$$

which makes $\acute{E}t(X)$ *the profinite completion of* $Cl(X)$.

Note that in the light of this theorem $\acute{E}t(X)$ can be computed from the homotopy type of $X(\mathbb{C})$, which makes $\acute{E}t(X)$ a tractable object.

References

[1] M. Artin and B. Mazur. *Etale homotopy.* Lecture Notes in Mathematics **100**, Springer-Verlag, 1969.

[2] I. Barnea and T.M. Schlank. A projective model structure on pro simplicial sheaves, and the relative étale homotopy type. arXiv:1109.5477

[3] E.M. Friedlander. *Étale homotopy of simplicial schemes.* Annals of Mathematical Studies **104**, Princeton University Press, 1982.

[4] P. Goerss and J. Jardine. *Simplicial homotopy theory.* Birkhäuser, 1999.

[5] Y. Harpaz and T.M. Schlank. Homotopy obstructions to rational points, *this volume.*

[6] M. Hovey. *Model categories.* Mathematical Surveys and Monographs **63**, American Mathematical Society, 1998.

[7] S. Mac Lane. *Categories for the working mathematician.* Springer-Verlag, 1998.

[8] J. Peter May. *Simplicial objects in algebraic topology.* D. Van Nostrand, 1967.

[9] J.S. Milne. *Étale cohomology.* Princeton University Press, 1980.

[10] A. Pál. Homotopy sections and rational points on algebraic varieties. arXiv:1002.1731

3

Torsors and representation theory of reductive groups

Vera Serganova
Department of Mathematics, University of California, Berkeley

3.1 Introduction

It is known from classical algebraic geometry that the Picard group of a del Pezzo surface has a remarkable group of symmetries preserving the intersection form. Recently the Cox rings of del Pezzo surfaces were studied extensively by different methods in [1], [5], [6], [14], [20], [26], [29]. The result is a presentation of the Cox ring of a del Pezzo surface as the coordinate ring of an intersection of quadrics. In [28] a similar result is obtained for $\mathbb{P}^n$ blown up at $n+3$ points.

All these examples fall into a general pattern conjectured by Victor Batyrev some time ago. Following his idea, Alexei Skorobogatov and the author developed a general approach for dealing with all the above examples, see [21], [22] and [23]. The aim of these lectures is to explain this approach.

Let k be an algebraically closed field of characteristic zero, and let X be a smooth projective variety over k whose Picard group $\mathbf{Pic}\,X$ is torsion free. Let $\mathcal{T} \to X$ be a universal torsor. The structure group of this torsor is the torus $\mathbb{T}$ whose group of characters $\hat{\mathbb{T}}$ is $\mathbf{Pic}\,X$. We identify the Cox ring of X with the ring $k[\mathcal{T}]$ of regular functions on $\mathcal{T}$. (Note that in general $k[\mathcal{T}]$ is not a finitely generated k-algebra.)

The main idea is as follows. Let X be a del Pezzo surface over k, or the projective space $\mathbb{P}^n_k$ blown up at $n+3$ points such that no $n+1$

of them are contained in a hyperplane. It is possible to construct a $\mathbb{T}$-equivariant embedding of $\mathcal{T}$ into the affine cone $(G/P)_a$ over a projective homogeneous space G/P of a reductive algebraic group G, where $\mathbb{T}$ is identified with a maximal torus of G extended by the group of scalars, with its natural left action on $(G/P)_a$. Now one can use the realization of $(G/P)_a$ as the minimal G-orbit in an irreducible representation V of G, together with some other classical results of representation theory, to find generators and relations of $k[\mathcal{T}]$.

In Sections 2 and 3 we briefly recall the well known theory of reductive groups, their representations and homogeneous projective spaces. Here we only give proofs which are not too long or technical. In Section 4 we review basic facts about GIT quotients and universal torsors. Section 5 contains a key result which we call the blow-up theorem. It says that in certain cases the GIT quotient $(G/P)/\!/\mathbb{G}_m$ is isomorphic to the blow-up of a projective space at some smooth closed subvariety. In Section 6 we show how Batyrev's conjecture for del Pezzo surfaces of degree at least 2 can be proved inductively using the blow-up theorem. In Section 7 we give a description of relations in the Cox ring. Finally, Section 8 deals with the last remaining case of del Pezzo surfaces of degree 1.

I would like to thank Alexei Skorobogatov for advice and help in preparing and editing these lectures, and the ICMS in Edinburgh for hospitality.

3.2 Reductive algebraic groups and their representations

In this section we present basic facts about reductive groups and their Lie algebras. They can be found in [2], [13] or [18]. We always assume that k is an algebraically closed field of characteristic zero, except in Section 3.7.1.

3.2.1 Structure theory of reductive groups

An affine algebraic group G is called *reductive* if every finite-dimensional representation of G is completely reducible, i.e. it splits into a direct sum of irreducible subrepresentations.

Let G be a connected reductive group.

Example 3.1 $G = T \simeq (k^*)^n$. The irreducible representations of T

are parametrized by elements of the character lattice $\hat{T}$. Every finite-dimensional representation V of T has a weight decomposition $V = \bigoplus_{\mu \in \hat{T}} V_\mu$, where

$$V_\mu = \{v \in V | tv = \mu(t)v, \quad \forall\, t \in T\}.$$

The Lie algebra $\mathfrak{g} = T_e G$ is the algebra of all left-invariant derivations of $k[G]$. If G is reductive, then $\mathfrak{g}$ is a direct sum of its ideals

$$\mathfrak{g} = \mathfrak{g}_1 \oplus \cdots \oplus \mathfrak{g}_m \oplus \mathfrak{z},$$

where the Lie algebras $\mathfrak{g}_i$ are simple (that is, non-abelian algebras that do not contain proper non-trivial ideals), and the Lie algebra $\mathfrak{z}$ is abelian. When $\mathfrak{z} = 0$ the Lie algebra $\mathfrak{g}$ is *semisimple.*

By T we denote a maximal torus in G. All maximal tori in G are conjugate. The adjoint representation $Ad : T \to GL(\mathfrak{g})$ has a weight decomposition:

$$\mathfrak{g} = \mathfrak{t} \oplus \bigoplus_{\alpha \in R} \mathfrak{g}_\alpha,$$

here $\mathfrak{t} = \mathfrak{g}_0$ is the Lie algebra of T. The finite set $R \subset \hat{T} \setminus \{0\}$ is called the *root system* of $\mathfrak{g}$ and G, elements of R are called *roots.*

Note that $\mu \in \hat{T}$ induces a one-dimensional representation of T and hence of $\mathfrak{t}$. This defines a canonical embedding $\hat{T} \to \mathfrak{t}^*$. Thus we may consider $\hat{T}$ as a canonical sublattice of $\mathfrak{t}^*$.

By Q we denote the sublattice of $\hat{T}$ generated by roots. The quotient group $\hat{T}/Q$ is isomorphic to the group of characters of the center of G.

Let us list some important properties of roots:

(1) $\dim \mathfrak{g}_\alpha = 1$;
(2) $[\mathfrak{g}_\alpha, \mathfrak{g}_\beta] \subset \mathfrak{g}_{\alpha+\beta}$;
(3) If $\alpha \in R$, then $-\alpha \in R$.

Example 3.2 Let $G = SL(2)$. Then T is the subgroup of diagonal matrices

$$\begin{pmatrix} a & 0 \\ 0 & a^{-1} \end{pmatrix}, \quad a \in k^*.$$

If $\omega \in \hat{T}$ is a generator, then

$$R = \{\alpha, -\alpha\}$$

with $\alpha = 2\omega$, $\mathfrak{g}_\alpha$ is the set of strictly upper-triangular matrices, and $\mathfrak{g}_{-\alpha}$ is the set of strictly lower-triangular matrices.

Exercise 3.3 Let V be a finite-dimensional representation of G. If $V = \oplus_{\mu \in \hat{T}} V_\mu$, then $\mathfrak{g}_\alpha V_\mu \subset V_{\mu+\alpha}$ for any $\alpha \in R$.

Let $\hat{T}_\mathbb{R} := \hat{T} \otimes_\mathbb{Z} \mathbb{R}$. Consider a hyperplane in $\hat{T}_\mathbb{R}$ through the origin which does not contain any root $\alpha \in R$. It divides $\hat{T}_\mathbb{R}$ in two half-spaces and thus the set of roots R is divided accordingly: $R = R^+ \cup R^-$.

Proposition 3.4 *There exists a unique linearly independent set $\alpha_1, \dots, \alpha_r$ such that every $\alpha \in R^+$ can be written uniquely as*

$$\alpha = \sum_{i=1}^{n} m_i \alpha_i$$

with $m_i \in \mathbb{Z}_{\geq 0}$.

The set $\alpha_1, \dots, \alpha_r$ is called *a base*, and the α_i are called the *simple roots* of $\mathfrak{g}$.

By property (2)

$$\mathfrak{b} = \mathfrak{t} \oplus \bigoplus_{\alpha \in R^+} \mathfrak{g}_\alpha$$

is a subalgebra. The corresponding subgroup B is called a *Borel subgroup.* One can show that it is a maximal solvable subgroup in G.

The Weyl group W is the quotient by T of the normalizer $N_G(T)$ of T in G. The action of W on T induces an action of W on $\hat{T}$. If $w \in W$ and $\alpha \in R$, then

$$\mathrm{Ad}_w(\mathfrak{g}_\alpha) = \mathfrak{g}_{w(\alpha)}.$$

Hence W permutes the roots.

Example 3.5 Let $G = GL(n)$, let T be the subgroup of diagonal matrices. Let $\varepsilon_1, \dots, \varepsilon_n$ be the natural basis in $\hat{T}$. Then

$$R = \{\varepsilon_i - \varepsilon_j | 1 \leq i, j \leq n\}.$$

One can choose $\varepsilon_1 - \varepsilon_2, \dots, \varepsilon_{n-1} - \varepsilon_n$ for a base. The corresponding Borel subgroup is the subgroup of upper triangular matrices. The Weyl group W is isomorphic to the symmetric group S_n.

The Killing form. Let us assume now that $\mathfrak{g}$ is semisimple ($\mathfrak{z} = 0$). Then the bilinear symmetric form

$$(x, y) = \mathrm{tr}(\mathrm{ad}_x\, \mathrm{ad}_y)$$

is invariant under the adjoint action of G. It is non-degenerate (Cartan's criterion).

The restriction of the Killing form to $\mathfrak{t}$ is non-degenerate, and so one can identify $\mathfrak{t}$ and $\mathfrak{t}^*$ via this form. Recall that there is a natural embedding $\hat{T} \to \mathfrak{t}^*$. The restriction of the Killing form to $\hat{T}$ is a positive definite integral quadratic form.

Theorem 3.6 (1) *The Weyl group W is generated by the simple reflections $s_1, \ldots, s_r$*

$$s_i(\xi) = \xi - 2\frac{(\xi, \alpha_i)}{(\alpha_i, \alpha_i)}\alpha_i,$$

where $\xi \in \hat{T}$;

(2) *W acts simply transitively on the set of bases.*

The Casimir element. Let us choose a basis $\{x_i\}$ in $\mathfrak{g}$, and let $\{y_i\}$ be the dual basis with respect to the Killing form. In the universal enveloping algebra of $\mathfrak{g}$ we can form the element

$$\Omega = \sum_i x_i y_i.$$

It is called the Casimir element.

Exercise 3.7 Exercise 2 Show that Ω does not depend on the choice of a basis $\{x_i\}$. Show that for any finite-dimensional representation V of $\mathfrak{g}$ we have $x\Omega v = \Omega x v$ for all $x \in \mathfrak{g}$ and $v \in V$, see [2, I.3.7].

Coroots. For each simple root α_i we can choose $x_i \in \mathfrak{g}_{\alpha_i}$, $h_i \in \mathfrak{t}$ and $y_i \in \mathfrak{g}_{-\alpha_i}$ satisfying the relation $[x_i, y_i] = h_i$ and $\alpha_i(h_i) = 2$. These three elements span a Lie subalgebra isomorphic to $sl(2)$. The following identity is very important:

$$\xi(h_i) = 2\frac{(\xi, \alpha_i)}{(\alpha_i, \alpha_i)}$$

for any $\xi \in \mathfrak{t}^*$.

The matrix $a_{ij} := \alpha_i(h_j)$ is called the *Cartan matrix* of $\mathfrak{g}$. Using the fact that the Killing form is positive definite it is not hard to show that

$$a_{ii} = 2, \text{ and } a_{ij} = 0, -1, -2, -3 \text{ if } i \neq j.$$

Dynkin diagram. This is a directed graph whose nodes correspond to the simple roots. The number of edges joining i and j equals $\max\{\alpha_i(h_j), \alpha_j(h_i)\}$. The arrow goes from the longer of the two roots to the shorter one.

Example 3.8 Let $G = SL(n)$. The attached Dynkin diagram is A_{n-1}:

$$\bullet\!\!-\!\!\!-\!\!\bullet\!\!-\!\!\!-\quad \cdots \quad -\!\!\!-\!\!\bullet$$

Let $G = SO(5)$. Then the Lie algebra $\mathfrak{g} = so(5)$ is isomorphic to $sp(4)$. The Dynkin diagram is B_2:

$$\bullet\!\!\Rightarrow\!\!\bullet$$

Connected Dynkin diagrams are in bijection with simple Lie algebras, and also with simply connected simple algebraic groups (for the definition of a simply connected algebraic group see the next section). They are of two kinds, as follows.

- Classical: A_r $(sl(r+1))$, $r \geq 1$; B_r $(so(2r+1))$, $r \geq 2$; C_r $(sp(2r))$, $r \geq 2$ and D_r $(so(2r))$, $r \geq 4$;
- Exceptional: G_2, F_4, E_6, E_7, E_8.

3.2.2 Representations of reductive groups

If G is an affine algebraic group, then the commutator $[G, G]$ is a closed subgroup, see [13], VII.17.1. Set $\mathcal{D}^0 G = G$, $\mathcal{D}^{i+1} G = [\mathcal{D}^i G, \mathcal{D}^i G]$. Recall that G is called *solvable* if $\mathcal{D}^i G = \{e\}$ for some i.

Theorem 3.9 (Lie–Kolchin) *If V is a representation of a connected solvable group B, then there exists a full flag $0 \subset V_1 \subset \cdots \subset V_m = V$ invariant under the action of B.*

For the proof see, for example, [13], VII.17.3.

Now let V be an irreducible representation of a reductive connected algebraic group G. Then V has the induced structure of a simple $\mathfrak{g}$-module. By the Lie–Kolchin theorem V contains a B-invariant one-dimensional subspace kv. Then $\mathfrak{b}v \subset kv$ and $[\mathfrak{b}, \mathfrak{b}]v = 0$, in particular, $\mathfrak{g}_\alpha v = 0$ for all $\alpha \in R^+$. Let λ be the weight of this subspace with respect to the torus T. Since V is irreducible, for any weight μ of V we have $\lambda = \mu + \sum n_i \alpha_i$ for $n_i \geq 0$. That implies the uniqueness of v (up to proportionality). Such a vector v is called a *highest vector* of V, and λ is called the *highest weight* of V.

Define

$$\hat{T}^+ = \{\lambda \in \hat{T} | \lambda(h_i) = 2\frac{(\lambda, \alpha_i)}{(\alpha_i, \alpha_i)} \in \mathbb{Z}_{\geq 0},\ i = 1, \ldots, r\}.$$

The weights in $\hat{T}^+$ are called *dominant.*

Exercise 3.10 Check that any W-orbit in $\hat{T}$ meets $\hat{T}^+$ in exactly one point.

Theorem 3.11 *For any $\lambda \in \hat{T}^+$ there exists a unique up to isomorphism irreducible representation $V(\lambda)$ with highest weight λ. Every finite-dimensional irreducible representation of G is isomorphic to $V(\lambda)$ for some $\lambda \in \hat{T}^+$. Finally, $V(\lambda)$ is isomorphic to $V(\mu)$ if and only if $\mu = \lambda$.*

Example 3.12 Let $G = SL(2)$. Then $\hat{T}^+ = \{m\omega | m = 0, 1, 2, \dots\}$. One can check that $V(m\omega) \simeq S^m(V)$, where V is the natural two-dimensional representation of $SL(2)$.

It is in fact a key example. Indeed, the condition of dominance for the highest weight follows immediately from that for the $sl(2)$-triples associated with simple roots.

Fundamental weights. Let G be a semisimple connected algebraic group. Define $\omega_i \in \mathfrak{t}^*$ by the condition $\omega_i(h_j) = \delta_{ij}$. The weights $\omega_1, \dots, \omega_r$ are called the *fundamental* weights. Every dominant weight is a linear combination of fundamental weights with non-negative integral coefficients. A group G is called *simply connected* if the lattice $\hat{T}$ is generated by the fundamental weights. In the case $k = \mathbb{C}$ this condition is equivalent to G being simply connected as a complex manifold.

If v is a highest vector of $V(\lambda)$ and w is a highest vector of $V(\mu)$, then $v \otimes w$ is a highest vector of $V(\lambda) \otimes V(\mu)$ with weight $\lambda + \mu$. Note that the tensor product is usually reducible. Any irreducible representation is a subrepresentation in some tensor product of fundamental representations.

Example 3.13 If $G = SL(r+1)$, then the highest weight of the natural representation V is ω_1, the highest weight of the exterior power $\Lambda^p(V)$ is ω_p, and the highest weight of the adjoint representation is $\omega_1 + \omega_r$.

Exercise 3.14 Let $G = G_1 \times G_2$ be the direct product of two reductive groups. Then any irreducible representation V is isomorphic to the tensor product $V_1 \otimes V_2$ of irreducible representations of G_1 and G_2.

Exercise 3.15 An irreducible representation of a semisimple group G is called minuscule if all weights of V form one orbit under the action of W. Prove that the highest weight of a minuscule representation is fundamental, and that the number of minuscule representations equals the order of $\hat{T}/Q$. Find all minuscule representations for all simple G, see [2, VIII.7.3].

By Schur's lemma the Casimir element Ω acts by a scalar on any irreducible representation $V(\lambda)$. It is not difficult to compute this scalar. For any $\alpha \in R^+$ choose $x_\alpha \in \mathfrak{g}_\alpha$ and $y_\alpha \in \mathfrak{g}_{-\alpha}$ so that $(x_\alpha, y_\alpha) = 1$. Let u_i be an orthonormal basis in $\mathfrak{t}$. Then

$$\Omega = \sum_{\alpha\in R^+} x_\alpha y_\alpha + y_\alpha x_\alpha + \sum_i u_i^2 = 2\sum_{\alpha\in R^+} y_\alpha x_\alpha + \sum_{\alpha\in R^+} h_\alpha + \sum_i u_i^2,$$

where $h_\alpha = [x_\alpha, y_\alpha]$. If v is a highest vector of $V(\lambda)$, then

$$\Omega v = \sum_{\alpha\in R^+} h_\alpha v + \sum_i u_i^2 v = (\lambda + 2\rho, \lambda)v, \tag{3.1}$$

where

$$\rho = \frac{1}{2}\sum_{\alpha\in R^+} \alpha.$$

Since Ω acts as a scalar operator on $V(\lambda)$, we have $\Omega u = (\lambda + 2\rho, \lambda)u$ for any $u \in V$.

Exercise 3.16 Show that $(\rho, \alpha_i) = \frac{(\alpha_i, \alpha_i)}{2}$ for every simple root α_i.

3.3 Projective homogeneous spaces

3.3.1 General facts about homogeneous spaces

Let K be a closed subgroup of an affine algebraic group G, then one can equip the space G/K of left cosets with the structure of a quasiprojective algebraic variety. Most of the results of this section can be found in [13].

Lemma 3.17 *(Chevalley trick) There exists a finite-dimensional representation V of G, and a line l in V such that K is the stabilizer of l.*

Proof Let I_K be the ideal of K in $k[G]$. The right multiplication action of G on itself induces the left linear action of G on $k[G]$ given by

$$R_g f(x) = f(xg).$$

Observe that K coincides with the stabilizer of I_K. Indeed, the inclusion $K \subset \operatorname{Stab} I_K$ is obvious. On the other hand, if $g \in \operatorname{Stab} I_K$, then $f(g) = R_g f(e) = 0$ for any $f \in I_K$. Hence $\operatorname{Stab} I_K \subset K$.

Let $\Delta : k[G] \to k[G] \otimes k[G]$ denote the coalgebra map dual to the multiplication $G \times G \to G$. Let $f \in k[G]$ and $\Delta(f) = \sum_{i=1}^s f^i \otimes f_i$. Then for any $g \in G$ we have $R_g f = \sum_{i=1}^s f_i(g) f^i$. Therefore f is contained in some finite dimensional G-invariant subspace spanned by $R_g f$ for all $g \in G$. Similarly any finite subset of $k[G]$ is contained in a finite-dimensional G-invariant subspace of $k[G]$.

Let $f_1, \ldots, f_n$ be generators of I_K, and let $W \subset k[G]$ be the minimal G-invariant subspace containing $f_1, \ldots, f_n$. By the above we see that W is finite-dimensional. Then K coincides with the stabilizer of $W \cap I_K$. Set $V = \Lambda^d(W)$, $l = \Lambda^d(W \cap I_K)$ where $d = \dim(W \cap I_K)$. □

3.3.2 Parabolic subgroups and projective homogeneous spaces

Theorem 3.18 *Let B be a connected solvable group acting on a projective variety X. Then B has a fixed point on X. Moreover, any closed B-orbit on X is a point.*

Proof Since any projective variety X contains a closed B-orbit, it is sufficient to prove that any homogeneous projective variety X of B is a point. By Lemma 3.17 we may assume that X is a closed B-orbit in $\mathbb{P}(V)$ for some representation V of B. Since B is connected, X is irreducible. We may also assume without loss of generality that the affine cone X_a spans V. By Lie's theorem there exists a non-zero $\varphi \in V^*$ such that $B\varphi \subset k\varphi$. Then $X_a \cap \operatorname{Ker} \varphi$ is B-invariant. Since X_a spans V, we have $X_a \cap \operatorname{Ker} \varphi = \{0\}$. Therefore $\dim V = 1$ and X is a point. □

A subgroup P of G is called *parabolic* if G/P is a projective variety.

Theorem 3.19 *Let G be a connected reductive group, and let B be a Borel subgroup.*

(1) *Any subgroup $P \subset G$ which contains B is parabolic.*
(2) *Any parabolic subgroup of G is conjugate to one in* (1).

Proof (1) Let $\lambda = \sum_{i=1}^r \omega_i$ and $V = V(\lambda)$. Then $\mathbb{P}(V)$ has a unique closed orbit $G(kv)$, where v is a highest vector. Moreover, $\operatorname{Stab}_G(kv) = B$. Hence G/B is projective. Since the natural map $G/B \to G/P$ is surjective, the image G/P is complete and quasiprojective. Hence G/P is projective.

(2) Let $X = G/K$ be homogeneous and projective. By Theorem 3.18 B has a fixed point on X. So one can find a point $x \in X$ whose stabilizer contains B. Let $P = \operatorname{Stab}_G x$. Then K is conjugate to P. Note that P is parabolic by (1). □

Corollary 3.20 *Any projective homogeneous variety is isomorphic to the orbit of a highest vector in the projectivization of some irreducible representation of G.*

A parabolic subgroup containing a fixed B as in (1) of Theorem 3.19 is called *standard.* All standard parabolic subgroups are in bijection with subsets $I \subset \{1, \ldots, r\}$ of nodes of the Dynkin diagram. If $\omega_I = \sum_{i \in I} \omega_i$, then the corresponding standard parabolic P is the stabilizer of the highest vector line in $\mathbb{P}(V(\omega_I))$.

Let us describe the structure of $\mathfrak{p}$. Let $\gamma \in \mathrm{Hom}(\hat{T}, \mathbb{Q})$ be such that $\langle \gamma, \alpha_i \rangle = 0$ if $i \notin I$, and $\langle \gamma, \alpha_i \rangle = 1$ if $i \in I$. Define a $\mathbb{Z}$-grading $\mathfrak{g} = \bigoplus_{i \in \mathbb{Z}} \mathfrak{g}_i$ by setting

$$\mathfrak{g}_p = \bigoplus_{\langle \gamma, \alpha \rangle = p} \mathfrak{g}_\alpha .$$

Then $\mathfrak{p} = \bigoplus_{p \geq 0} \mathfrak{g}_p$. Note that $\mathfrak{g}_0$ is a reductive subalgebra of $\mathfrak{g}$.

Let $G_0 \subset G$ be the reductive subgroup with the Lie algebra $\mathfrak{g}_0$. The dimension of the center of G_0 equals $|I|$. We denote by G' and $\mathfrak{g}'$ the commutators of G_0 and $\mathfrak{g}_0$, respectively. Clearly, G' is semisimple and its Dynkin diagram is obtained from that of G by removing all nodes in I. The unipotent subgroup G_+ with the nilpotent Lie algebra $\mathfrak{g}_+ = \bigoplus_{p>0} \mathfrak{g}_p$ is the maximal normal unipotent subgroup of P, and P is a semidirect product of G_0 and G_+. The connected unipotent subgroup G_- with the Lie algebra $\mathfrak{g}_- = \bigoplus_{p<0} \mathfrak{g}_p$ is transversal to P in G. The tangent space $T_e(G/P)$ can be identified with $\mathfrak{g}_-$. Since G/P is irreducible, G_-P is a dense open subset in G/P isomorphic to the affine space G_-.

The Picard group of G/P is a free abelian group of rank $|I|$. It can be identified with the lattice of characters of P, which is the same as the lattice of characters of G_0 (see [11, 1.3], or [19]). More precisely, let $\hat{T}_I$ denote the sublattice of $\hat{T}$ generated by ω_i for all $i \in I$. Every $\chi \in \hat{T}_I$ induces a one-dimensional representation C_χ of G_0 and therefore of P. It is not hard to see that all characters of P can be obtained in this way. Define $G \times^P C_\chi$ as the quotient of $G \times k$ by the action of P, which is its natural right action on G, and the action by the character χ on k. Then $G \times^P C_\chi$ is a line bundle on G/P representing the element of $\mathbf{Pic}\, G/P$ corresponding to χ. This gives a canonical isomorphism $\mathbf{Pic}\, G/P = \hat{T}_I$.

3.3.3 The affine cone $(G/P)_a$

Theorem 3.21 *[15] Let $V(\lambda)$ be an irreducible representation of G, and let $G/P \subset \mathbb{P}(V(\lambda))$ be the orbit of a highest vector. Let $S^2(V(\lambda)) = V(2\lambda) \oplus N$, where N is a subrepresentation of G, and let $\psi : S^2(V(\lambda)) \to N$ be the G-invariant projection. If $\sigma : V(\lambda) \to S^2(V(\lambda))$ denotes the map $\sigma(v) = v \otimes v$, then the affine cone $(G/P)_a$ is defined by the equations*

$$\psi\sigma(v) = 0. \tag{3.2}$$

Moreover, the ideal of $(G/P)_a$ in V is generated by these equations. In particular, $(G/P)_a$ is an intersection of quadrics in $V(\lambda)$.

Proof If v is a highest vector of $V(\lambda)$, then $\sigma(v)$ is a highest vector of $V(2\lambda)$. Therefore, by G-invariance, $\sigma(v) \in V(2\lambda)$ for any $v \in (G/P)_a$.

Next, note that if $V(\mu)$ is an irreducible component of $S^2(V(\lambda))$, then $\mu = 2\lambda - \sum_i n_i\alpha_i$ for some $n_i \geq 0$. If we have

$$(2\lambda + 2\rho, 2\lambda) = (\mu + 2\rho, \mu),$$

then

$$(2\lambda + \mu + 2\rho, 2\lambda - \mu) = (2\lambda + \mu + 2\rho, \sum_i n_i\alpha_i) = 0.$$

Since λ and μ are dominant, from Exercise 3.16 we obtain $(2\lambda + \mu + 2\rho, \sum_i n_i\alpha_i) > 0$. Therefore, all $n_i = 0$ and $\mu = 2\lambda$. Hence $V(2\lambda)$ is the subspace of $S^2(V(\lambda))$ given by $\Omega x = (2\lambda + 2\rho, 2\lambda)x$.

Therefore, (3.2) can be rewritten in the form

$$\Omega(v^2) = (2\lambda + 2\rho, 2\lambda)v^2, \tag{3.3}$$

where $v^2 = v \otimes v$. Using the expression

$$\Omega = \sum_{\alpha \in R^+} x_\alpha y_\alpha + y_\alpha x_\alpha + \sum_i u_i^2$$

and (3.1), one can simplify the system of equations (3.3) to the form

$$2 \sum_{\alpha \in R^+} x_\alpha v \cdot y_\alpha v + \sum_i (u_i v)^2 = (\lambda, \lambda)v^2.$$

Using a similar argument one can show that the condition $v^n \in V(n\lambda)$ is equivalent to

$$\Omega(v^n) = (n\lambda + 2\rho, n\lambda)v^n. \tag{3.4}$$

We claim that (3.3) implies (3.4). Indeed, since Ω is quadratic we have

$$\Omega(v^n) = \frac{n(n-1)}{2}\Omega(v^2)v^{n-2} - n(n-2)\Omega(v)v^{n-1} =$$

$$(\frac{n(n-1)}{2}(2\lambda+2\rho,2\lambda) - n(n-2)(\lambda+2\rho,\lambda))v^n = (n\lambda+2\rho,n\lambda)v^n.$$

If we denote by J the ideal in $S(V(\lambda)^*)$ generated by (3.3), then, by the above, we have

$$S(V(\lambda)^*)/J = \oplus_{n\geq 0}V(n\lambda)^*.$$

We see that every graded component in $S(V(\lambda)^*)/J$ is isomorphic to $V(n\lambda)^*$ and hence is an irreducible representation of G. Let I denote the ideal of $(G/P)_a$. Then $J \subset I$. Note that I is G-invariant. If $J \neq I$, then $S^n(V(\lambda)^*) \subset I$ for some n and hence the zero set of I is 0. That implies that $J = I$. □

Remark Note that $(G/P)_a = Gv \cup \{0\}$ since $Tv = k^*v$.

Example 3.22 Let $G = SL(n)$, $I = \{p\}$. Then G/P is isomorphic to the Grassmannian $Gr(p,n)$, $V(\lambda) = \Lambda^p(V)$, and the embedding $G/P \to \mathbb{P}(\Lambda^p(V))$ is the Plücker embedding.

3.3.4 Schubert cells

Let us fix a standard parabolic subgroup $P \subset G$ and an irreducible representation V of G with highest weight λ such that G/P is the G-orbit of the highest vector line V_λ. For any $\mu \in W\lambda$ we have $\dim V_\mu = 1$. Let

$$\mathfrak{n}_- = \bigoplus_{\alpha\in R^-} \mathfrak{g}_\alpha,$$

and let N_- be the connected unipotent subgroup of G with Lie algebra $\mathfrak{n}_-$. The N_--orbit $X_\mu = N_-V_\mu \subset G/P$ is called a *Schubert cell.* In the case $\mu = \lambda$, $X_\lambda = G_-V_\lambda$ is an open dense subset of G/P isomorphic to G_-. It is called the big Schubert cell. In fact, $\bigcup_{\mu\in W\lambda} X_\mu$ gives a cell decomposition of G/P as follows from the following theorem.

Theorem 3.23 *G/P is the disjoint union of Schubert cells X_μ. If $\mu = w\lambda$, then the Schubert cell X_μ is isomorphic to the affine space $\mathfrak{n}_- \cap \mathrm{Ad}_w(\mathfrak{g}_-)$.*

Proof Fix $\epsilon \in \mathrm{Hom}(\hat{T},\mathbb{Z})$ such that $\langle\epsilon,\alpha_i\rangle < 0$ for all simple roots α_i and $\langle\epsilon,\mu\rangle \neq \langle\epsilon,\nu\rangle$ for two distinct weights μ and ν of V. Set $\mu < \nu$ if $\langle\epsilon,\mu\rangle < \langle\epsilon,\nu\rangle$.

Let μ be a weight of V. Define

$$(Y_\mu)_a = \{u \in (G/P)_a | u = u_\mu + \sum_{\nu > \mu} u_\nu,\ u_\mu \neq 0,\ u_\nu \in V_\nu\},$$

$$(Z_\mu)_a = \{u \in (G/P)_a | u = u_\mu + \sum_{\nu \neq \mu} u_\nu,\ u_\mu \neq 0,\ u_\nu \in V_\nu\}.$$

Let Y_μ and Z_μ be the corresponding subsets of G/P. Clearly, Y_μ are disjoint and $Y_\mu \subset Z_\mu$. A straightforward calculation shows that Y_μ are N_--invariant.

First we will show that Y_μ is not empty if and only if $\mu \in W\lambda$. Indeed, let $u \in (Y_\mu)_a$ and $\epsilon(t)$ be the one-parameter subgroup associated with ϵ. Then for any $t \in k^*$ we have

$$t^{-\langle \epsilon, \mu \rangle} \epsilon(t) u = u_\mu + \sum_{\nu > \mu} t^{\langle \epsilon, \nu - \mu \rangle} u_\nu.$$

By taking limit at $t = 0$ we get $u_\mu \in (G/P)_a$. That implies $\mu \in W\lambda$.

Next we claim that $Y_\mu = X_\mu$. In the case $\mu = \lambda$ we have $Z_\lambda = Y_\lambda$. Note that the action of G_- on Y_λ is free. Since G_- is unipotent, this can be done by showing that the annihilator of any $x \in (Y_\lambda)_a$ in the Lie algebra $\mathfrak{g}_-$ is trivial. Thus, all G_--orbits in Y_λ have the same dimension. The irreducibility of G/P and the density of X_λ in G/P imply that Y_λ consists of one G_--orbit X_λ. Furthermore, $Z_\lambda = Y_\lambda = X_\lambda$.

Now let $\mu = w\lambda$. Then

$$Z_\mu = w(Z_\lambda) = w(X_\lambda) = wG_- w^{-1} V_\mu. \tag{3.5}$$

The Lie algebra of wG_-w^{-1} is $\mathrm{Ad}_w(\mathfrak{g}_-)$. Consider the decomposition

$$\mathrm{Ad}_w(\mathfrak{g}_-) = \mathfrak{m}_+ \oplus \mathfrak{m}_-,$$

where $\mathfrak{m}_- := \mathrm{Ad}_w(\mathfrak{g}_-) \cap \mathfrak{n}_-$ and $\mathfrak{m}_+ := \mathrm{Ad}_w(\mathfrak{g}_-) \cap \mathfrak{b}$. Since wG_-w^{-1} is unipotent, the exponential map is an isomorphism of algebraic varieties $\mathrm{Ad}_w(\mathfrak{g}_-)$ and wG_-w^{-1}. Moreover, we have

$$wG_-w^{-1} = \exp(\mathfrak{m}_-)\exp(\mathfrak{m}_+).$$

Let $y \in Y_\mu$. Then $y \in Z_\mu$ and by (3.5) we have $y = \exp(\xi_-)\exp(\xi_+)V_\mu$ for some $\xi_\pm \in \mathfrak{m}_\pm$. However, if $\xi_+ \neq 0$, then $\exp(\xi_+)V_\mu \in Y_\nu$ for some $\nu < \mu$. Since Y_ν is N_--invariant and $\exp(\xi_-) \in N_-$, we have $y \in Y_\nu$. Therefore $y \in Y_\mu$ implies $y = \exp(\xi_-)V_\mu$ for some $\xi_- \in \mathfrak{m}_-$. Hence $y \in N_-V_\mu = X_\mu$. Moreover, by the above we have

$$Y_\mu = X_\mu = \exp(\mathfrak{m}_-)V_\mu.$$

The stabilizer of V_μ in $\exp(\mathfrak{m}_-)$ is trivial. That gives an isomorphism $\mathfrak{m}_- \simeq X_\mu$. □

3.3.5 The case of a maximal proper parabolic subgroup

We are especially interested in the case when P is a maximal proper parabolic subgroup, which is equivalent to the condition $I = \{i\}$. In this case we have the following properties:

(1) **Pic** G/P has rank 1;
(2) $\mathfrak{g}_{-1}$ is an irreducible representation of G_0 with highest weight $-\alpha_i$;
(3) If $\mathfrak{g}$ is simple and θ is the highest root of $\mathfrak{g}$, then the length of the $\mathbb{Z}$-grading equals the coefficient m_i in the decomposition $\theta = \sum_{j=1}^r m_j\alpha_j$;
(4) The subgroup G_+ is abelian iff $m_i = 1$.

Set $\alpha = \alpha_i$, $\omega = \omega_i$. Let $V = V(\omega)$ be the irreducible representation of G with highest weight ω, and let v denote a highest vector of V, i.e. a vector of weight ω. Let $\gamma \in \mathrm{Hom}(\hat{T}, \mathbb{Q})$ be as in Section 3.2. Recall that it gives rise to a $\mathbb{Z}$-grading on the Lie algebra $\mathfrak{g} = \bigoplus_{i\in\mathbb{Z}} \mathfrak{g}_i$. We define a $\mathbb{Z}$-grading

$$V = \bigoplus_{i=0}^{l} V_i$$

by letting V_i be the direct sum of the weight subspaces V_μ such that $\langle\mu,\omega\rangle = \langle\gamma,\omega\rangle - i$. Then $V_0 = kv$ and $V_1 = \mathfrak{g}_{-1}v$. Therefore we have a G'-equivariant isomorphism $\phi : \mathfrak{g}_{-1} \to V_1$ defined by $\phi(x) = xv$.

Consider the exponential map

$$\exp\ :\ \mathfrak{g}_{-1} \to G/P, \qquad \exp(x) = v + xv + \frac{1}{2}x^2v + \cdots + \frac{1}{l!}x^l v,$$

where $x^i v \in V_i$. Thus we have a G'-invariant polynomial map of degree i:

$$p_i = \pi_i \circ \exp \circ \phi^{-1} : V_1 \to V_i,$$

where $\pi_i : (G/P)_a \to V_i$ denotes the restriction of the natural projection $V \to V_i$ to $(G/P)_a$. We denote by the same letter p_i the polarization map $p_i : S^i(V_1) \to V_i$.

Lemma 3.24 (i) *We have $V_2 = V_2^+ \oplus V_2^-$, where $V_2^- = \mathfrak{g}_{-2}v$ and V_2^+ is the image of the polarization map $p_2 : S^2(V_1) \to V_2$.*

(ii) *We have the following isomorphisms of G'-modules*

$$S^2(\mathfrak{g}_{-1}) \simeq S^2(V_1) \simeq V_2^+ \oplus V'(-2\alpha),$$

where $V'(-2\alpha)$ is the irreducible representation of G' with the highest weight -2α.

Proof Let $U(\mathfrak{l})$ denote the universal enveloping algebra of a Lie algebra $\mathfrak{l}$. Consider the induced $\mathfrak{g}$-module $M = U(\mathfrak{g}) \otimes_{U(\mathfrak{p})} kv$. Note that M is infinite-dimensional, and there are isomorphisms of $\mathfrak{g}'$-modules

$$M \simeq U(\mathfrak{g}_-) \otimes kv \simeq U(\mathfrak{g}_-).$$

We have a surjection $s : M \to V$ defined by $s(X \otimes v) = Xv$. Denote by N the kernel of s. It is clear that one can define a $\mathbb{Z}$-grading $M = \bigoplus_{i \geq 0} M_i$ compatible with s such that $M_0 = 1 \otimes kv$. Then M_2 is isomorphic to $(S^2(\mathfrak{g}_{-1}) \oplus \mathfrak{g}_{-2}) \otimes v$ as a module over $\mathfrak{g}'$, and, furthermore, we have $V_2^+ = s(S^2(\mathfrak{g}_{-1}) \otimes v)$ and $V_2^- = s(\mathfrak{g}_{-2} \otimes v)$. Since $s(S^2(\mathfrak{g}_{-1}) \otimes v)$ is by definition the image of the polarization map p_2, we obtain the first assertion.

To prove the second assertion we have to show that $N_2 = N \cap M_2$ is isomorphic to $V'(-2\alpha)$. Let $y_\alpha \in \mathfrak{g}_{-\alpha}$ be a non-zero element. A simple calculation shows that $y_\alpha^2 v = 0$. Hence $y_\alpha^2 \otimes v \in N_2$. Moreover, it generates a $\mathfrak{g}'$-submodule $V'(-2\alpha)$ in N_2. Let N' be the $\mathfrak{g}$-submodule in M generated by $y_\alpha^2 \otimes v$. It suffices to show that $N' = N$.

Recall that x acts locally nilpotently on U if for any $u \in U$ there exists $n_u > 0$ such that $x^{n_u} u = 0$. We note that for any simple root β, the elements x_β and y_β act nilpotently on the image of $1 \otimes v$ in M/N'. Since the adjoint action of $\mathfrak{n}_\pm$ on $U(\mathfrak{g})$ is locally nilpotent, it is easy to show that every element in the Lie algebras $\mathfrak{n}_\pm$ acts locally nilpotently on M/N'. Hence the action of $\mathfrak{g}$ on M/N' can be lifted to an action of the group G using the exponential map for locally nilpotent elements and the already defined action of the maximal torus T. Thus, M/N' is a representation of the algebraic group G. It is completely reducible and every invariant subspace containing a vector of weight ω coincides with M/N'. The complete reducibility of M/N' implies the irreducibility of M/N'. Hence $M/N' \simeq V$ and $N = N'$. □

Lemma 3.25 *Let $V' = V_1$, let v' be a highest vector with respect to $G' \cap B$, and let P' be the stabilizer of kv'. Then*

$$(G'/P')_a = (G/P)_a \cap V_1 = p_2^{-1}(0),$$

and the equations $p_2(x) = 0$ generate the ideal of $(G'/P')_a$ in $S(V_1^*)$.

Proof Let us prove the first equality. For any $u \in V$ the tangent space at u to the orbit Gu coincides with $\mathfrak{g}u$. Since $(G/P)_a \setminus \{0\} = Gu$, we have $T_{u,(G/P)_a} = \mathfrak{g}u$ for any $u \in (G/P)_a \setminus \{0\}$. If $u \in (G'/P')_a \subset V_1$, then

$$T_{u,(G/P)_a} \cap V_1 = \mathfrak{g}u \cap V_1 = \mathfrak{g}_0 u = \mathfrak{g}'u = T_{u,(G'/P')_a}.$$

Hence $(G'/P')_a$ is an irreducible component of $(G/P)_a \cap V_1$. On the other hand, the closed set $(G/P)_a \cap V_1$ is a union of G'-orbits, but the closure of any non-zero orbit contains $(G'/P')_a$. Hence $(G'/P')_a = (G/P)_a \cap V_1$.

The scheme-theoretic equality $(G'/P')_a = p_2^{-1}(0)$ follows from Lemma 3.24 and Theorem 3.21 (with $G = G'$). □

3.4 GIT quotients and universal torsors

3.4.1 GIT quotients

We start with recalling definitions and basic results of the geometric invariant theory, see [7] and [17].

Let X be a projective variety, and let a reductive group T act on X. By a linearization we understand a T-equivariant embedding $X \to \mathbb{P}(V)$ for some representation V of T. Recall that a point $x = ku \in X$ is called *stable* if the stabilizer of u in T is finite, and the orbit Tu is closed in V. A point is *semistable* if the closure of Tu does not contain 0. Finally x is *unstable* if the closure of Tu contains 0. Let X^s and X^{ss} denote the set of stable and semistable points in X, respectively. Both sets are open in X.

The GIT quotient $X/\!\!/T$ is defined as $\operatorname{Proj} k[X_a]^T$, where $k[X_a]^T$ is the ring of T-invariant functions on the affine cone X_a over X. There is a morphism of varieties $X^{ss} \to X/\!\!/T$ whose fibers are unions of T-orbits of semistable points of X. The GIT quotient $X/\!\!/T$ contains an open subset X^s/T whose points bijectively correspond to the T-orbits of stable points of X.

Assume that T is a torus. Then V has a weight decomposition $V = \oplus_{\mu \in \hat{T}} V_\mu$, and so every $u \in V$ can be written $u = \sum u_\mu$ for some $u_\mu \in V_\mu$. Let $\mathrm{wt}(u)$ be the set of all μ such that $u_\mu \neq 0$.

By the Hilbert–Mumford criterion a point $x = ku \in \mathbb{P}(V)$ is semistable if and only if the convex hull of $\mathrm{wt}(u)$ in $\hat{T}_{\mathbb{R}}$ contains 0, and it is stable if and only if 0 is an interior point of the convex hull of $\mathrm{wt}(0)$.

Theorem 3.26 *Let $G = GL(2) \times SL(n)$, $n \geq 3$. Let V' be the natural 2-dimensional representation of $GL(2)$, let V'' be the conatural n-dimensional representation of $SL(n)$, and let $\mathbb{M}_{2,n} = V' \otimes V''$. Let $T' \subset GL(2)$ and $T'' \subset SL(n)$ be maximal tori. Let $\mathbb{T}$ be the torus in $GL(\mathbb{M}_{2,n})$ generated by T' and T''. Let $S \subset \mathbb{T}$ be the torus of codimension 1 generated by T'' and the torus $\{(\begin{smallmatrix} t^m & 0 \\ 0 & t^{-1} \end{smallmatrix}) | t \in k^*\} \subset GL(2)$. Assume that $\frac{n-2}{2} < m < n-1$, for example, $m = n-2$. Then $\mathbb{M}_{2,n}^{ss} = \mathbb{M}_{2,n}^{s}$, and $\mathbb{P}(\mathbb{M}_{2,n})/\!\!/S$ is isomorphic to $\mathbb{P}^{n-1}$ blown up at n points in general position.* [1]

Proof One can identify $\mathbb{M}_{2,n}$ with the space of $2 \times n$ matrices (x_{ij}) with the action of G given by multiplication on the right by elements of $SL(n)$ and on the left by elements of $GL(2)$. First, we prove that $\mathbb{M}_{2,n}^{ss} = \mathbb{M}_{2,n}^{s}$ by showing that this set consists of matrices which have no zero columns, at most one zero in the second row and at most $n-2$ zeros in the first row. We use the Hilbert–Mumford criterion. The weights of $\mathbb{M}_{2,n}$ are $\{\gamma_i + m\delta, \gamma_i - \delta | i = 1, \ldots n\}$, where $\gamma_1, \ldots, \gamma_{n-1}, \delta$ are linearly independent and $\gamma_n = -\gamma_1 - \cdots - \gamma_{n-1}$. Due to the symmetry with respect to S_n-action on $\gamma_1, \ldots, \gamma_n$ it suffices to show that

- If $\mathrm{wt}(x) \subset \{\gamma_1 - \delta, \ldots, \gamma_{n-2} - \delta, \gamma_1 + m\delta, \ldots, \gamma_n + m\delta\}$, then x is unstable;
- If $\mathrm{wt}(x) \subset \{\gamma_1 - \delta, \ldots, \gamma_n - \delta, \gamma_1 + m\delta\}$, then x is unstable;
- If $\mathrm{wt}(x) \supset \{\gamma_1 - \delta, \ldots, \gamma_{n-1} - \delta, \gamma_1 + m\delta, \gamma_n + m\delta\}$, then x is stable.

Let us start with the first statement. Assume that x is semistable. Then there exist non-negative $a_1, \ldots, a_n, b_1, \ldots, b_{n-2} \in \mathbb{R}$ such that

$$a_1(\gamma_1 + m\delta) + \cdots + a_n(\gamma_n + m\delta) + b_1(\gamma_1 - \delta) + \cdots + b_{n-2}(\gamma_{n-2} - \delta) = 0.$$

The above equation is equivalent to the following conditions

$$a_1 + b_1 = \cdots = a_{n-2} + b_{n-2} = a_{n-1} = a_n,$$

$$m(a_1 + \cdots + a_n) - (b_1 + \cdots + b_{n-2}) = 0.$$

Without loss of generality we may assume that $a_{n-1} = a_n = 1$. After substituting $a_i = 1 - b_i$ for $i = 1, \ldots, n-2$ in the second equation we obtain

$$nm = (m+1)(b_1 + \cdots + b_{n-2}).$$

[1] N points in $\mathbb{P}^{n-1}$ are in general position if no n-point subset is contained in a hyperplane. Given two sets of $n+1$ points in general position, there exists an element in $PGL(n)$ which moves one set to another.

Recall that $b_i \le 1$ and hence $nm \le (m+1)(n-2)$. That implies $2m \le n-2$ and we obtain a contradiction.

Similarly we can deal with the second statement. The condition

$$a_1(\gamma_1 + m\delta) + b_1(\gamma_1 - \delta) + \cdots + b_n(\gamma_n - \delta) = 0$$

implies

$$a_1 + b_1 = b_2 = \cdots = b_n,$$

$$ma_1 - b_1 - \cdots - b_n = 0.$$

As in the previous case we may assume $a_1 + b_1 = b_2 = \cdots = b_n = 1$. Now solving the equations we obtain $a_1 = \frac{n}{m+1}, b_1 = \frac{m+1-n}{m+1}$. Since $b_1 \ge 0$ we have $m \ge n-1$. That provides a contradiction.

Finally, for the last statement we observe that

$$0 = \frac{n-m-1}{m+1}(\gamma_1 + m\delta) + \frac{2m+2-n}{m+1}(\gamma_1 - \delta) + (\gamma_2 - \delta) + \cdots \\ + (\gamma_{n-1} - \delta) + (\gamma_n + m\delta).$$

Observe also that $\gamma_1 + m\delta, \gamma_n + m\delta, \gamma_1 - \delta, \ldots, \gamma_{n-1} - \delta$ span $\hat{S}_{\mathbb{R}}$. That implies that 0 as an interior point of the convex hull of $\mathrm{wt}(x)$. Hence x is stable.

Now we can determine the GIT quotient $\mathbb{P}(\mathrm{M}_{2,n})/\!\!/S$. We decompose $\mathrm{M}_{2,n}^s$ as the disjoint union of $\mathrm{M}_{2,n}^0$ and $D_1, \ldots, D_n$, where $\mathrm{M}_{2,n}^0 = \{x \in \mathrm{M}_{2,n}^s | x_{2i} \neq 0, i = 1, \ldots n\}$ and $D_i = \{x \in \mathrm{M}_{2,n}^s | x_{2i} = 0\}$. Let $e_1, \ldots, e_n$ be a basis of eigenvectors of T'' in V'', and let $p_i \in \mathbb{P}(V'')$ be the image of e_i, $i = 1, \ldots, n$. It is not hard to check that $\mathrm{M}_{2,n}^0/\mathbb{T} = \mathbb{P}(\mathrm{M}_{2,n}^0)/S$ is isomorphic to $\mathbb{P}(V'') \setminus \{p_1, \ldots, p_n\}$, and $D_i/\mathbb{T} = \mathbb{P}(D_i)/S$ are isomorphic to $\mathbb{P}^{n-2}$. Define the map $\varphi : \mathbb{P}(\mathrm{M}_{2,n}^s) \to \mathbb{P}^{n-1}$ in the homogeneous coordinates by

$$\varphi(x) = (\frac{x_{11}}{x_{21}} : \ldots : \frac{x_{1n}}{x_{2n}}) = (\frac{x_{11}x_{2i}}{x_{21}} : \ldots : x_{1i} : \ldots : \frac{x_{1n}x_{2i}}{x_{2n}}),$$

where $i = 1, \ldots, n$. Clearly, φ is a morphism of projective varieties since it is well defined on $\mathrm{M}_{2,n}^0$ and on all D_i. Moreover, φ is S-invariant and its fibers outside the points $p_1, \ldots, p_n$ are single S-orbits. The induced morphism $\bar{\varphi} : \mathbb{P}(\mathrm{M}_{2,n}^s)/S \to \mathbb{P}^{n-1}$ is inverse to the blowing up at these n points. Therefore $\mathbb{P}(\mathrm{M}_{2,n})/\!\!/S$ is isomorphic to $\mathbb{P}^{n-1}$ blown up at n points in general position. $\square$

3.4.2 Universal torsors

Let T be a torus, and let X be an irreducible variety. Recall that an X-*torsor* is a variety $\mathcal{T}$ with an action of T and a T-equivariant morphism $f : \mathcal{T} \to X$ such that locally in étale topology $\mathcal{T}$ is equivariantly isomorphic to $T \times X$. Since k is assumed to be algebraically closed, T is a split torus, and therefore $f : \mathcal{T} \to X$ is locally trivial in Zariski topology.

Colliot-Thélène and Sansuc associated to $f : \mathcal{T} \to X$ the exact sequence

$$1 \to k[X]^*/k^* \xrightarrow{f^*} k[\mathcal{T}]^*/k^* \xrightarrow{j} \hat{T} \xrightarrow{\beta} \mathbf{Pic}\, X \xrightarrow{f^*} \mathbf{Pic}\, \mathcal{T} \to 0, \quad (3.6)$$

see [4], formula (2.1.1). Here j is the restriction to a fiber. To define β consider the natural pairing:

$$H^1(X,T) \times \hat{T} \to H^1(X,\mathbb{G}_m) = \mathbf{Pic}\, X,$$

where the cohomology groups are in étale topology. Then β sends $\chi \in \hat{T}$ to $\langle[\mathcal{T}],\chi\rangle$, where $[\mathcal{T}] \in H^1(X,T)$ is the class of the torsor $\mathcal{T} \to X$. The map β is called the *type* of the torsor $\mathcal{T} \to X$. A torsor $\mathcal{T} \to X$ is called *universal* if its type is an isomorphism.

If the variety X is projective, the exact sequence (3.6) implies the following criterion: an X-torsor under T is universal if and only if $\mathbf{Pic}\, \mathcal{T} = 0$ and $k[\mathcal{T}]^* = k^*$, that is, $\mathcal{T}$ has no non-constant invertible regular functions.

Let $\mathcal{T} \to X$ be a universal torsor. Consider the weight decomposition

$$k[\mathcal{T}] = \bigoplus_{\chi \in \hat{T}} k[\mathcal{T}]_\chi,$$

where

$$k[\mathcal{T}]_\chi = \{f \in k[\mathcal{T}] | f(tx) = \chi(t)f(x), \ \ \forall x \in \mathcal{T}, \ \forall t \in T\}.$$

Then

$$k[\mathcal{T}]_\chi \simeq \Gamma(X, L_{\beta(\chi)}),$$

where $L_{\beta(\chi)}$ is the line bundle associated with $\beta(\chi) \in \mathbf{Pic}\, X$. Therefore $k[\mathcal{T}]$ is isomorphic to the Cox ring of X (see [1]).

Now let G be a simple connected algebraic group, let T be a maximal torus and let $P \subset G$ be a maximal parabolic subgroup. Then $I = \{i\}$. We will use the following notation: $\alpha = \alpha_i$, $\omega = \omega_i$, $V = V(\omega)$, and denote by v a highest vector of V. Recall that in this case $\mathbf{Pic}\, G/P = \hat{T}_{\{i\}}$

has rank 1 and is generated by the class of a hyperplane section. We can assume without loss of generality that V is a faithful representation of G by taking the quotient of G by the kernel of the representation if necessary.

By $\mathbb{T}$ we denote the subgroup in $GL(V)$ generated by T and the scalar operators.

Let $(G/P)_a^f$ be the set of points of $(G/P)_a$ with trivial stabilizer in $\mathbb{T}$, and let $(G/P)_a^{sf}$ be the intersection of $(G/P)_a^f$ with the set $(G/P)_a^s$ of stable points. Note that a non-zero vector in V has the trivial stabilizer in $\mathbb{T}$ if and only if the stabilizer in T of the corresponding point in $\mathbb{P}(V)$ coincides with $Z(G)$.

Proposition 3.27 *[21] Assume that the Dynkin diagram of G is simply laced (i.e. of types A, D or E), and (G, ω) is not one of the following list:*

$$(G, \omega_1), \text{ where } G \text{ is classical}, (A_r, \omega_r), (A_3, \omega_2), (D_4, \omega_3), (D_4, \omega_4).$$

Then the codimension of $(G/P)_a \setminus (G/P)_a^{sf}$ in $(G/P)_a$ is at least 2.

The proof is based on the following arguments. Let $u \in V$. Using the Hilbert–Mumford criterion one proves that if $W\omega \backslash \mathrm{wt}(u)$ has at most one point, then u is stable. Since the intersection of two hyperplane sections in $(G/P)_a$ has codimension 2, the closed subset $(G/P)_a \setminus (G/P)_a^s$ has codimension at least 2.

To prove that $(G/P)_a \setminus (G/P)_a^f$ has codimension at least 2 we use the following statements.

Proposition 3.28 *Let $x \in G/P$, $\mathrm{St}_T(x)$ denote the stabilizer of x in T, and let K_x be the connected component of the centralizer of $\mathrm{St}_T(x)$ in G. Then*

(i) *K_x is a reductive subgroup of G, $T \subset K_x$;*
(ii) *$x \in K_x wv = K_x/(wPw^{-1} \cap K_x)$ for some $w \in W$;*
(iii) *$Z(K_x) = \mathrm{St}_T(x)$;*
(iv) *$\mathrm{St}_T(x)$ is finite if and only if K_x is semisimple, in which case the ranks of K_x and G are equal.*

For the proof see Proposition 2.2 in [21].

Corollary 3.29 *Assume that $\mathfrak{g}$ is simply laced. Then the codimension of the set of points $x \in G/P$ such that $\mathrm{St}_T(x) \neq Z(G)$, is at least 2.*

For the proof see Corollary 2.3 in [21]. This implies that the codimension of $(G/P)_a \setminus (G/P)^f_a$ in $(G/P)_a$ is at least 2, and so concludes the proof of Proposition 3.27.

In the next section we will use the following

Corollary 3.30 *Let $G = SL(n)$, and let $x \in G/P$ be such that* $\mathrm{St}_T(x)$ *is finite. Then* $\mathrm{St}_T(x) = Z(G)$.

Proof By Proposition 3.28, $\mathrm{St}_T(x) = Z(K)$ for a semisimple subgroup of $K \subset G$ containing T. The roots of K form a root subsystem of A_{n-1} of rank $n-1$. It follows from a classical result of Dynkin (or can be checked directly), that the root system of K coincides with A_{n-1}. Hence $K = G$. □

Theorem 3.31 *[21] Assume that G/P satisfies the conditions of Proposition* 3.27. *Let $Y = (G/P)^{sf}_a/\mathbb{T}$. Then $f : (G/P)^{sf}_a \to Y$ is a universal torsor.*

Proof By construction $\mathbb{T}$ acts freely on $(G/P)^{sf}_a$. By GIT we see that $f : (G/P)^{sf}_a \to Y$ is an affine morphism whose fibers are $\mathbb{T}$-orbits. Hence $f : (G/P)^{sf}_a \to Y$ is a torsor (see Lemma 1.1 in [21]).

The complement to $(G/P)^{sf}_a$ in $(G/P)_a$ has codimension at least 2. Since all invertible functions on $(G/P)_a$ are constant, and $\mathbf{Pic}\,(G/P)_a$ is trivial, the same is true for $(G/P)^{sf}_a$, and the statement follows. □

Remark Recall Theorem 3.26. By a similar argument $f : \mathbb{M}^s_{2,n} \to Y$, where Y is $\mathbb{P}^{n-1}$ blown up at n points in general position, is a universal torsor with the structure torus $\mathbb{T}$.

3.5 The blow-up theorem

In this section, for certain projective homogeneous spaces G/P and certain tori S, we describe $(G/P)/\!\!/S$ as a blow-up of some toric variety.

Let us assume that the parabolic subgroup $P \subset G$ has abelian unipotent radical. Then the corresponding grading is of the form

$$\mathfrak{g} = \mathfrak{g}_{-1} \oplus \mathfrak{g}_0 \oplus \mathfrak{g}_1,$$

and $\mathfrak{g}_0 = \mathfrak{g}' \oplus k$, where $\mathfrak{g}'$ is a semisimple Lie algebra whose Dynkin diagram is obtained from that of $\mathfrak{g}$ by removing the node α.

Here is the list of all such pairs (G, G'):

$$(A_r, A_{r-1}),\ (A_r, A_i \times A_{r-i-1}),\ (B_r, B_{r-1}),\ (D_r, D_{r-1}),\ (D_r, A_{r-1}),$$
$$(C_r, A_{r-1}),\quad (E_6, D_5),\quad (E_7, E_6).$$

This list can be obtained just by direct inspection of Dynkin diagrams. Indeed, the Dynkin diagram of G' is obtained from that of G by removing the node corresponding to a simple root α_i such that $m_i = 1$ in the decomposition $\theta = \sum m_j \alpha_j$ (see Section 3.5). The coefficients m_j for all Dynkin diagrams are listed, for instance, in [2] and [18].

Recall the grading $V = \bigoplus_{i=1}^{l} V_i$ defined in Section 3.5. Recall also that the G'-equivariant isomorphism $\phi : \mathfrak{g}_{-1} \to V_1$ is defined by $\phi(x) = xv$. In what follows we assume that $V_2 \neq 0$; this excludes the pair (A_r, A_{r-1}).

In our situation $\mathfrak{g}_{-2} = 0$, hence $V_2 = V_2^+$. The image of the exponential map $\exp\ :\ \mathfrak{g}_{-1} \to G/P$ is the big Schubert cell.

Example 3.32 For the pair $(A_r, A_1 \times A_{r-2})$ we have $\omega = \omega_2$, $V = V_0 \oplus V_1 \oplus V_2$, $V_1 = k^2 \otimes k^{r-1}$, $V_2 = \Lambda^2(k^{r-1})$.

Example 3.33 For the pair (E_7, E_6) we have $\omega = \omega_1$, $V = V_0 \oplus V_1 \oplus V_2 \oplus V_3$, where $V_1 = V(\omega_1)$ and $V_2 = V(\omega_5)$ are fundamental representation of dimension 27 dual to each other. Finally, V_3 is the trivial one-dimensional representation of $G' = E_6$. The map $p_3 : V_1 \to V_3$ defines a G'-invariant cubic form on V_1. One can check that

$$p_3(u) = \langle u, p_2(u) \rangle.$$

Let $D \subset \mathbb{T}$ be the one-parameter subgroup consisting of the elements which act on V_i as multiplication by t^{1-i}, $t \in k^*$.

Let $(G/P)_a^+$ be the open subset of $(G/P)_a$ which is the complement to the union of closed subsets $(G/P)_a \cap V_{>1}$ and $(G/P)_a \cap V_{\leq 1}$. The action of D on $(G/P)_a^+$ is free. The projection $\pi_1 : (G/P)_a \to V_1$ gives rise to the D-equivariant morphism $\pi_1 : (G/P)_a^+ \to V_1 \backslash \{0\}$. Let $(G/P)^+$ be the image of $(G/P)_a^+$ in $\mathbb{P}(V)$, and let $\bar{\pi}_1$ be the morphism $(G/P)^+ \to \mathbb{P}(V_1)$.

Theorem 3.34 *[21] There exist a D-torsor $\pi : (G/P)_a^+ \to \mathcal{B}$ and the morphism $\tau : \mathcal{B} \to V_1 \setminus \{0\}$ such that $\pi_1 = \tau \circ \pi$, and τ is the inverse to the blowing-up of $V_1 \setminus \{0\}$ at $(G'/P')_a \setminus \{0\}$. The morphism $\bar{\pi}_1 : (G/P)^+ \to \mathbb{P}(V_1)$ is the composition of a D-torsor and the inverse to the blowing-up of G'/P'.*

Remark If $\dim V_2 = 1$, then by Lemma 3.25 the closed subset $(G'/P')_a$ of V_1 has codimension 1, so that τ is an isomorphism. Let $\dim V_2 > 1$. Then $(G/P)_a \cap V_{\leq 1}$ has codimension at least 2, $(G/P)_a \cap V_{>1}$ is the union of Schubert cells X_μ for all weights μ of $V_2 \oplus \cdots \oplus V_l$. Those cells have codimension at least 2. Hence the codimension of $(G/P)_a \cap (V_{>1} \cup V_{\leq 1})$ is at least 2.

Proof Let

$$\mathcal{B}' = \{(x, kp_2(x)) \in V_1 \times \mathbb{P}(V_2) | x \in V_1 \setminus (G'/P')_a\},$$

and let $\mathcal{B}$ be the Zariski closure of $\mathcal{B}'$ in $(V_1 \setminus \{0\}) \times \mathbb{P}(V_2)$. By [12, Prop. II.7.14] and Lemma 3.25, $\mathcal{B}$ is the blow-up of $V_1 \setminus \{0\}$ at $(G'/P')_a \setminus \{0\}$, and the projection τ onto $V_1 \setminus \{0\}$ is the inverse to the blowing-up of $V_1 \setminus \{0\}$ at $(G'/P')_a \setminus \{0\}$.

Let $\bar{\pi}_2$ be the projection $\pi_2 : V \to V_2$ followed by the natural map $V_2 \setminus \{0\} \to \mathbb{P}(V_2)$. Let $\pi := (\pi_1, \bar{\pi}_2)$. It is easy to see that π is a well defined map on $(G/P)_a^+$, and that π sends the points in the big Schubert cell to $\mathcal{B}'$. Thus we can view it as a map

$$\pi : (G/P)_a^+ \to V_1 \setminus \{0\} \times \mathbb{P}(V_2).$$

By construction $\pi_1 = \tau \circ \pi$. It remains to prove that π is a D-torsor. We start by calculating the fiber $\pi^{-1}(y)$ for all $y \in \mathcal{B}$.

Let $y = (x, kp_2(x)) \in \mathcal{B}'$. Then $\pi^{-1}(y)$ belongs to the big Schubert cell X_λ and $\pi^{-1}(y) = D(\exp \phi^{-1}(x))v$ is a single D-orbit.

Now let $y \in \mathcal{B} \setminus \mathcal{B}'$. Denote by $p_2(\xi_1, \xi_2)$ the polarization of the quadratic map p_2. From the definition of the blow-up we deduce that $y = (x, kp_2(x, u))$ for some $x \in (G'/P')_a$ and $u \in V_1$. Note that $\pi^{-1}(y)$ lies in the union of the Schubert cells X_μ for all weights μ of V_1. That implies $\pi^{-1}(y) \subset G_-(G'/P')_a$. On the other hand, we have $\pi_1(G_- u) = u$ for any $u \in V_1$, and $\pi_1(\pi^{-1}(y)) = x$. Hence $\pi^{-1}(y) \subset G_- x$. This implies $\pi^{-1}(y) = D\exp(\zeta)x$, where $\zeta \in \mathfrak{g}_{-1}$ is such that $\zeta x = u$. Observe that $\zeta x = p_2(x, \phi(\zeta))$ for any $\zeta \in \mathfrak{g}_{-1}$ and any $x \in V_1$. Indeed, $[\phi^{-1}(x), \zeta] = 0$ implies

$$p_2(x, \phi(\zeta)) = \frac{1}{2}(\zeta\phi^{-1}(x)v + \phi^{-1}(x)\zeta v) = \zeta(\phi^{-1}(x)v) = \zeta x,$$

since $\phi^{-1}(x)v = x$. Therefore, $\pi^{-1}(y) = D\exp(\phi^{-1}(u))x$ is a single D-orbit.

We have proved that $\pi^{-1}(y)$ is a single D-orbit for any $y \in \mathcal{B}$. The local triviality of π in Zariski topology follows from the local triviality of the tautological bundle $V_2 \setminus \{0\} \to \mathbb{P}(V_2)$. □

Remark This proof shows more, namely, that τ is an isomorphism on $\mathcal{B}'$, and that the exceptional divisor of τ is $\mathcal{B} \setminus \mathcal{B}'$ which τ contracts onto

$(G'/P')_a \setminus \{0\}$. Let $H_\omega = \{u \in V | u_\omega = 0\}$ be the weight coordinate hyperplane defined by the highest weight ω. The proof also shows that the inverse image of the exceptional divisor is

$$\pi^{-1}(\mathcal{B} \setminus \mathcal{B}') = \pi_1^{-1}((G'/P')_a \setminus \{0\}) = (G/P)_a^+ \cap H_\omega. \tag{3.7}$$

For future reference we point out that this implies

$$\pi_1((G/P)_a \cap H_\omega) = (G'/P')_a. \tag{3.8}$$

Let $a, b \in \mathbb{Z}$. By $D_{a,b}$ we denote the one-parameter subgroup in $\mathbb{T}$ consisting of the elements which act on V_i as multiplication by t^{a-bi}. Note that $D = D_{1,1}$, and that $D_{1,0}$ is the subgroup of scalars in $GL(V)$.

Lemma 3.35 *If $2b > a > b > 0$, then the set of $D_{a,b}$-stable points in $(G/P)_a$ equals the set of $D_{a,b}$-semistable points, and also coincides with $(G/P)_a^+$. Moreover, $(G/P)/\!\!/D_{a,b} = (G/P)^+/D_{a,b}$ is isomorphic to $\mathbb{P}(V_1)$ blown up at G'/P'.*

Proof Since $D_{a,b}$ is one-dimensional, the set of semistable but not stable points in V is non-empty if and only if the space of $D_{a,b}$-invariant vectors in V is non-zero. But the conditions on a and b imply that this space is zero. Hence the first assertion.

Let $u \in V$. Write $u = u_+ + u_-$ where $u_- \in V_{\leq 1}$, $u_+ \in V_{>1}$. By the Hilbert–Mumford criterion the stability of u is equivalent to the condition $u_+ \neq 0, u_- \neq 0$. Hence the second statement.

Since $b \neq 0$, the tori D and $D_{1,0}$ generate the same two-dimensional subtorus in $\mathbb{T}$ as $D_{a,b}$ and $D_{1,0}$. Hence, by Theorem 3.34, the induced morphism

$$(G/P)^+/D_{a,b} = (G/P)^+/D \to \mathbb{P}(V_1)$$

is the inverse to the blowing-up of G'/P'. That implies the last statement. □

Let $\mathbb{T}'$ be the torus in $GL(V_1)$ generated by T' and the subgroup of scalar operators. The representation of $\mathbb{T}$ in V_1 induces a surjective homomorphism $r : \mathbb{T} \to \mathbb{T}'$ with kernel $D = D_{1,1}$. Denote by $\theta : \mathbb{T}' \to \mathbb{T}$ the unique homomorphism such that $r \circ \theta = \text{id}$ and $\omega(\theta(\mathbb{T}')) = \{1\}$. Clearly, $\theta : T' \to T$ is induced by the embedding $G' \subset G$, and θ maps the subgroup of scalar operators to $D_{0,1}$. Furthermore, $\mathbb{T} = \theta(\mathbb{T}') \times D$.

Theorem 3.36 *Let $S' \subset \mathbb{T}'$ be a torus such that the set of S'-semistable points in V_1 coincides with the set of S'-stable points. Let $S \subset \mathbb{T}$ be the* *torus generated by S' and the one-dimensional torus $D_{b+1,b}$ for some integer b. If b is sufficiently large, then the sets of S-stable and S-semistable points in G/P coincide, and $(G/P)/\!\!/S$ is isomorphic to $\mathbb{P}(V_1)/\!\!/S'$ blown up at $(G'/P')/\!\!/S'$.*

Proof Write $a = b+1$, and assume that $b > 1$. Note that $\hat{\mathbb{T}} \simeq \hat{\mathbb{T}}' \oplus \mathbb{Z}\omega$, and S is a quotient of $S' \times D_{a,b}$ by some finite subgroup. Hence we can use the identification

$$\hat{S}_{\mathbb{R}} \simeq \hat{S}'_{\mathbb{R}} \oplus \mathbb{R}\varepsilon,$$

where $\varepsilon = \frac{\omega}{a}$. Introduce a positive definite scalar product $\langle\cdot,\cdot\rangle$ such that $\langle\varepsilon, \hat{S}'_{\mathbb{R}}\rangle = 0$ and $\langle\varepsilon,\varepsilon\rangle = 1$.

If $\mathcal{M}$ denotes the set of weights of S' in V_1, and $\mathcal{M}_j$ denotes the set of weights of S in V_i, then $\mathcal{M}_0 = \{a\varepsilon\}$, $\mathcal{M}_1 = \{\mu+\varepsilon | \mu \in \mathcal{M}\}$ and for $i \geq 2$ all elements of $\mathcal{M}_j$ are of the form $\mu_1 + \cdots + \mu_j + (a-bj)\varepsilon$ for some $\mu_1, \ldots, \mu_j \in \mathcal{M}$. The last assertion is the consequence of the fact that V_j is an S'-submodule in $S^j(V_1)$.

First we note that S-semistability implies $D_{a,b}$-semistability. Therefore by Lemma 3.35 we have $(G/P)^{ss}_a \subset (G/P)^+_a$. Let $u \in (G/P)^+_a$. Without loss of generality we may assume that either

(1) $u = v + x + z_2 + \cdots + z_l$, or
(2) $u = x + z_2 + \cdots + z_l$

for some $x \in V_1$ and $z_i \in V_i$. In both cases $x \neq 0$, $z_2 \neq 0$. In the first case $x \in V_1 \setminus (G'/P')_a$ and $z_i = p_i(x)$. In the second case $x \in (G'/P')_a$.

We claim that if x is S'-stable, then u is S-stable. Indeed, let x be S'-stable and $\mathrm{wt}(x) = \{\mu_1+\varepsilon, \ldots, \mu_k+\varepsilon\}$. Let $\mu - (b-1)\varepsilon \in \mathrm{wt}(z_2)$. Since 0 is an interior point of the convex hull of $\mu_1, \ldots, \mu_k$ in $\hat{S}'_{\mathbb{R}}$, the convex cone generated by $\mu_1, \ldots, \mu_k$ coincides with $\hat{S}'_{\mathbb{R}}$. In particular, we can write $-\mu = a_1\mu_1 + \cdots + a_k\mu_k$ for some positive $a_1, \ldots, a_k$, and $0 = b_1\mu_1 + \cdots + b_k\mu_k$ for some positive $b_1, \ldots, b_k$. Assume that b is large enough so that $r = b - 1 - (a_1 + \cdots + a_k) > 0$. Set $c_i = a_i + sb_i$, where $s(b_1 + \cdots + b_k) = r$. Then $\sum_{i=1}^k c_i(\mu_i + \varepsilon) + (\mu - (b-1)\varepsilon) = 0$. Since the set

$$\{\mu_1+\varepsilon, \ldots, \mu_k+\varepsilon, \mu-(b-1)\varepsilon\}$$

is not contained in a hyperplane in $\hat{S}_{\mathbb{R}}$, the point 0 is in the interior of the convex hull of $\mu_1+\varepsilon, \ldots, \mu_k+\varepsilon$ and $\mu-(b-1)\varepsilon$. Therefore 0 is an interior point of the convex hull of $\mathrm{wt}(u)$. Hence u is S-stable.

Now we will prove that if x is S'-unstable, then u is S-unstable.

Consider case (1). Let $\mathrm{wt}(x) = \{\mu_1+\varepsilon, \ldots \mu_k+\varepsilon\}$. Since $z_j = p_j(x)$, any $\nu \in \mathrm{wt}(z_j)$ has the form $\nu = \mu_{i_1} + \cdots + \mu_{i_j} + (a - bj)\varepsilon$ for some $i_1, \ldots, i_j$ in the set $1, \ldots, k$. If x is S'-unstable there exists $\gamma \in \hat{S}'_{\mathbb{R}}$ such that $\langle \gamma, \mu_i \rangle > 1$ for all $i \leq k$. Let $\gamma' = \gamma + \frac{1}{b}\varepsilon$. Then

$$\langle \gamma', \nu \rangle = \langle \gamma, \mu_{i_1} \rangle + \cdots + \langle \gamma, \mu_{i_j} \rangle + \frac{1}{b}(a - bj) > \frac{a}{b} > 0$$

for all $\nu \in \mathrm{wt}(z_j)$. Furthermore, $\langle \gamma', a\varepsilon \rangle > 0$ and $\langle \gamma', \mu_i + \varepsilon \rangle > 0$ for all $i = 1, \ldots k$. Since $\mathrm{wt}(u) = \mathrm{wt}(v) \cup \mathrm{wt}(x) \cup \mathrm{wt}(z_2) \cup \cdots \cup \mathrm{wt}(z_l)$, we obtain $\langle \gamma', \nu \rangle > 0$ for all $\nu \in \mathrm{wt}(u)$. Hence u is unstable.

Next we consider case (2). Let again $\mathrm{wt}(x) = \{\mu_1 + \varepsilon, \ldots, \mu_k + \varepsilon\}$, and let γ be as in case (1). There exists a positive c such that $\langle \gamma, \mu \rangle > -c$ for all $\mu \in \mathcal{M}_2 \cup \cdots \cup \mathcal{M}_l$. Assume that $b > c + 1$. Set $\gamma' = \gamma - \varepsilon$. Then $\langle \gamma', \mu \rangle > 0$ for all $\mu \in \mathcal{M}_2 \cup \cdots \cup \mathcal{M}_l$ and $\langle \gamma', \mu_i + \varepsilon \rangle > 0$ for all $i = 1, \ldots, k$. Since $\mathrm{wt}(u) \subset \mathrm{wt}(x) \cup \mathcal{M}_2 \cup \cdots \cup \mathcal{M}_l$, we conclude that $\langle \gamma, \nu \rangle > 0$ for all $\nu \in \mathrm{wt}(u)$, so that u is unstable.

Let $(G/P)^s$, $(G/P)^{ss}$ be the sets of S-stable and S-semistable points, respectively, and let $\mathbb{P}(V_1)^s$, $(G'/P')^s$ be the sets of S'-stable points. The above implies that $(G/P)^s = (G/P)^{ss}$. Moreover, the map $\bar{\varphi} : (G/P)^+/D_{a,b} \to \mathbb{P}(V_1)$ constructed in the proof of Lemma 3.35 restricts to $\bar{\varphi} : (G/P)^s/D_{a,b} \to \mathbb{P}(V_1)^s$. If $Z = (G/P)^s/D_{a,b}$, then Z is the blow-up of $\mathbb{P}(V_1)^s$ at $(G'/P')^s$. Therefore Z/S' is the blow-up of $\mathbb{P}(V_1)^s/S'$ at $(G'/P')^s/S'$. But $Z/S' = (G/P)^s/S = (G/P)/\!\!/S, \mathbb{P}(V_1)^s/S' = \mathbb{P}(V_1)/\!\!/S'$ and $(G'/P')^s/S' = (G'/P')/\!\!/S'$. That completes the proof of the theorem. □

Corollary 3.37 *In the setting of Theorem* 3.36 *assume further that* $(G'/P')_a \setminus (G'/P')^s_a$ *has codimension at least* 2 *and* $\dim V_2 \geq 2$. *Then* $(G/P)_a \setminus (G/P)^s_a$ *has codimension at least* 2.

Proof As in the proof of Proposition 3.27 it suffices to check that any $u \in (G/P)^+_a$ such that $\mathrm{wt}(u) \supset W\omega \setminus \{\mu\}$ is stable. That easily follows from the fact that S-stability of u is equivalent to S'-stability of $x = \pi_1(u)$, as explained in the proof of Theorem 3.36. Recall the notation of that proof. If $\mu \in \mathcal{M}_i$ for $i \neq 1$, then $\mathrm{wt}(x) = \mathcal{M}$, x is S'-stable and hence u is S-stable. If $\mu \in \mathcal{M}_1$, then $\mathrm{wt}(x) = \mathcal{M} \setminus \{\mu\}$. Then x is S'-stable by our assumption on $(G'/P')^s_a$. Hence u is S-stable. □

Now we concentrate on the case when $G = SL(n+2)$, $G' = SL(2) \times SL(n)$ and $G/P = Gr(2, n+2)$. Then $V_1 = \mathbb{M}_{2,n}$, and $S' = S$ in the notation of Theorem 3.26, and $(G'/P')_a$ is the locus of matrices of rank ≤ 1. It is easy to see that G'/P' contains one open S'-orbit which corresponds to a single point in $P(V_1)/\!\!/S'$. Corollary 3.30 implies $Gr(2,n+2)^{sf}_a = Gr(2,n+2)^s_a$. Hence we have the following

Corollary 3.38 *Let $n \geq 3$. Let $\mathbb{T}$ be a maximal torus in $GL(n+2)$, and let $C \subset \mathbb{T}$ be the subgroup of scalar operators. There exists a torus $S \subset \mathbb{T}$ that maps surjectively onto $\mathbb{T}/C$ such that the GIT quotient $Gr(2,n+2)/\!\!/S$ is isomorphic to $\mathbb{P}^{n-1}$ blown up at $n+1$ points in general position. The map $Gr(2,n+2)^s_a \to Gr(2,n+2)/\!\!/S$ is a universal torsor with structure torus $\mathbb{T}$.*

Exercise 3.39 In the situation of Corollary 3.38, the Grassmannian $Gr(2,n+2)^s$ with its Plücker embedding has $\binom{n+2}{2}$ $\mathbb{T}$-invariant hyperplane sections $u_\mu = 0$. These hyperplane sections cut certain divisors in $\mathbb{P}^{n-1}$ blown up at $n+1$ points. Prove that these divisors are $n+1$ exceptional divisors corresponding to the blown up points, and the strict transforms of the hyperplanes passing through $n-1$ out of these $n+1$ points. It is sufficient to find the images of all $\mathbb{T}$-invariant hyperplane sections under the composition map $\varphi \circ \pi_1 : Gr(2,n+2)^s \to \mathbb{P}^{n-1}$, where φ is as in the proof of Theorem 3.26.

If $n = 3$, then $S' = T'$. We leave to the reader to check that $b = 4$ satisfies the condition of Theorem 3.36. In this case $S = T$. Recall that a del Pezzo surface of degree 5 (see the next section for the definition) is isomorphic to $\mathbb{P}^2$ blown up at 4 points in general position. Therefore we have the following corollary.

Corollary 3.40 *$Gr(2,5)/\!\!/T$ is a del Pezzo surface of degree 5.*

Exercise 3.41 $Gr(2,4)/\!\!/T \simeq \mathbb{P}^1$.

3.6 Batyrev's conjecture

A del Pezzo surface is a smooth birationally trivial projective surface with ample anticanonical sheaf. Over an algebraically closed field such a surface is the quadric $\mathbb{P}^1 \times \mathbb{P}^1$ or the projective plane $\mathbb{P}^2$ with $r = 9 - d$ points in general position blown up; the degree d of a del Pezzo surface can be any integer between 1 and 9. (Here we say that r points are in general position if there are no three collinear points, no six points lying on a conic, for $r \geq 7$ there are no 7 points lying on a cubic curve such that one of the points is a double point, for $r = 8$ all the points do not belong to a quartic such that 3 points are double points nor to a quintic such that 6 points are double points.)

It was noticed by Schoutte, Coxeter and Du Val that the intersection graph of exceptional curves on del Pezzo surfaces of degree $d \leq 6$ has a huge group of symmetries. In the case $d = 3$ the 27 lines on a smooth cubic surface were discovered by Cayley and Salmon, the corresponding group is isomorphic to the Weyl group of the roots system E_6.

In the late 1960s Manin ([16], IV) discovered that to a del Pezzo surface X of degree $d = 9 - r$, $d \leq 6$, one can attach a root system R_r in such a way that the automorphism group of the incidence graph of the exceptional curves on X is the Weyl group $W(R_r)$. The sequence R_r is

$$A_1 \times A_2 \subset A_4 \subset D_5 \subset E_6 \subset E_7 \subset E_8 \tag{3.9}$$

If $R_8 = E_8$ with the enumeration of simple roots given below, then $R_{r-1} = R_r \setminus \alpha_r$ for $4 \leq r \leq 8$.

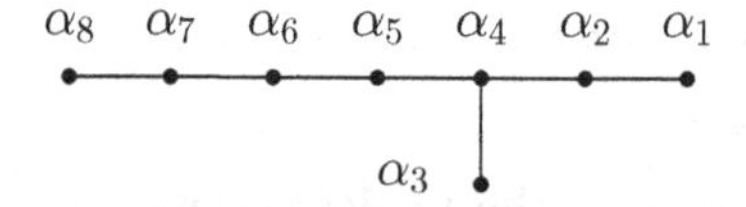

The Picard group $\mathbf{Pic}\, X = \oplus_{i=0}^r \mathbb{Z} l_i$ is equipped with an integral bilinear form given by intersection index. It can be described as follows:

$$(l_0^2) = 1, \quad (l_i^2) = -1, \ i \geq 0, \quad (l_i \cdot l_j) = 0, \ i \neq j.$$

The canonical class is $K_X = -3l_0 + l_1 + \cdots + l_r$. The roots are all $\alpha \in K_X^\perp$ such that $(\alpha \cdot \alpha) = -2$. The simple roots can be chosen as follows:

$$\alpha_1 = l_2 - l_1, \quad \alpha_2 = l_3 - l_2, \quad \alpha_3 = -l_0 + l_1 + l_2 + l_3,$$
$$\alpha_4 = l_4 - l_3, \quad \ldots, \quad \alpha_r = l_r - l_{r-1}.$$

The exceptional divisors correspond to the vectors $\mu \in \mathbf{Pic}\, X$ satisfying the condition $(\mu \cdot K_Z) = (\mu^2) = -1$. The Weyl group W acts transitively on the set of exceptional divisors.

The number of exceptional divisors on a del Pezzo surface of degree $d > 1$ equals the dimension of the fundamental representation $V = V(\omega)$ associated to a pair (R_r, R_{r-1}). These numbers are $6, 10, 16, 27, 56$. If $d = 1$, the number of exceptional divisors equals 240. That coincides with the number of roots of E_8. In this case $V = \mathfrak{g}$ is the adjoint representation and this is the only case when V is not minuscule.

The following result was conjectured by Batyrev and proved by Popov ([1], [20]) and Derenthal ([5], [6]) for $d \geq 2$. It was reproved in [21] by use of the blow-up Theorem, and recently generalized to the case $d = 1$ in [23].

Theorem 3.42 *Let X be a del Pezzo surface of degree $d \leq 5$, let R_r be the corresponding root system, let G be the corresponding simple simply connected algebraic group, and let P be the parabolic subgroup with semisimple part G' whose root system is R_{r-1}. Finally, let $f : \mathcal{T} \to X$ be a universal X-torsor. We use the notion of stability defined by a maximal torus $T \subset G$. There exists an equivariant embedding of $\mathcal{T}$ to the universal torsor $(G/P)_a^{sf} \to Y := (G/P)_a^{sf}/\mathbb{T}$. The images under f of the T-invariant hyperplane sections of $\mathcal{T}$ corresponding to the weights $W\omega$ are the exceptional divisors of X.*

We will give a proof for $d > 1$ by induction on r with the base case $r = 4$, $d = 5$ already proven, see Corollary 3.38, Corollary 3.40 and Exercise 3.39.

If $G' = SL(5)$, then $G = \mathrm{Spin}(10)$ (a double cover of $SO(10)$), V is a 16-dimensional spinor representation of G. When restricted to G', the representation V splits into the direct sum $V = V_0 \oplus V_1 \oplus V_2$, where V_1 and V_2 are the second and the fourth exterior powers of the natural representation of $SL(5)$, respectively.

If $G' = \mathrm{Spin}(10)$ and G is the simply connected group of type E_6, V is a 27-dimensional representation of G, which has a decomposition $V = V_0 \oplus V_1 \oplus V_2$, where V_1 and V_2 are the spinor representation and the natural representation of $\mathrm{Spin}(10)$, respectively.

The case $E_6 \subset E_7$ was already discussed in Example 3.33.

If G is of type E_8, G' is of type E_7, then $V = \mathfrak{g}$ is the adjoint representation,

$$\mathfrak{g} = \mathfrak{g}_{-2} \oplus \mathfrak{g}_{-1} \oplus \mathfrak{g}_0 \oplus \mathfrak{g}_1 \oplus \mathfrak{g}_2,$$

with $\mathfrak{g}_0 = \mathfrak{g}' \oplus k$, $\mathfrak{g}_{\pm 1}$ being the 56-dimensional representation of G', $\mathfrak{g}_{\pm 2}$ being the trivial representation of G'. This case requires a generalization of the blow-up theorem and some additional tricks, see Section 8.

Let X' be a del Pezzo surface of degree $d + 1$ such that X is obtained from X' by blowing up one point M. We assume that there is an equivariant embedding of the universal torsor $\mathcal{T}' \to X'$ into

$$f' : (G'/P')_a^{sf} \to (G'/P')_a^{sf}/\mathbb{T}'$$

satisfying all the requirements of the theorem. Observe that $V_1 = V'$ is invariant under the action of the reductive subgroup $G_0 \subset G$. The representation $r : G_0 \to GL(V')$ is faithful and irreducible, and $G_0 \simeq r(G_0) \subset GL(V')$ is generated by $r(G')$ and the subgroup of scalar operators in V'. In particular, $r : T \to \mathbb{T}' \subset r(G_0)$ is an isomorphism.

Recall that V is a minuscule representation, in particular, all weight spaces are one-dimensional. This implies that the centralizer $\mathbb{V}$ of T in $GL(V)$ is a torus. If $\{e_\mu\}$ is a T-eigenbasis in V, then for any $g \in \mathbb{V}$ we have $ge_\mu = x_\mu e_\mu$ for some $x_\mu \neq 0$. Thus, $\mathbb{V}$ is isomorphic to V with the coordinate hyperplanes removed. In what follows we always use this identification. By $\mathbb{V}'$ we denote the corresponding torus for G'.

Recall also the maps $p_i : V' \to V_i$. For a weight μ of V we denote by $p_i^\mu : V' \to (V_i)_\mu$ the polynomial function on V' such that

$$p_i(u) = \sum_{e_\mu \in V_i} p_i^\mu(u) e_\mu.$$

Since M does not belong to any of the exceptional divisors in X', we have $p_1^\mu(x) \neq 0$ for all weights μ of V_1, i. e. none of coordinates of x is zero.

Lemma 3.43 *There exists $s \in \mathbb{V}'$ such that $sT' \cap (G'/P')_a = Tsx$ is a single orbit, and the restriction of p_2^μ on sT' is not identically zero for all weights μ of V_2.*

Proof We are looking for s in the form yx^{-1} for some $y \in (G'/P')_a$. First we claim that for each weight μ of V_2 there is y such that p_2^μ is not identically zero on sT'. Assume the contrary. Then $p_2^\mu(yx^{-1}u) = 0$ for all $u \in T'$ and $y \in (G'/P')_a$. Consider $p_2^\mu(yx^{-1}u)$ as a function of y. By Lemma 3.25 all quadratic forms in $I((G'/P')_a) \subset S^2(V'^*)$ of weight μ are proportional to $p_2^\mu(y)$. Write

$$p_2^\mu(y) = \sum_{\mu_1+\mu_2=\mu} c_{\mu_1,\mu_2} y^{\mu_1} y^{\mu_2}.$$

We claim that $c_{\mu_1,\mu_2} \neq 0$ whenever $\mu_1+\mu_2 = \mu$. Indeed, V_2 is a minuscule representation of G', hence by W'-symmetry $p_2^\kappa(y)$ have the same number of non-zero coefficients for all weights κ of V_2. For any weight ν of V_1, let $V'(\nu) \subset V'$ be the subspace spanned by all vectors of weight $\nu - \beta$ for all roots β of G' such that $(\beta, \nu) = 1$. By Lemma 3.25 $(G'/P')_a \cap V'(\nu)$ is isomorphic to $(G''/P'')_a$ of the previous step. In particular, all T-homogeneous quadratic forms in the ideal of $(G'/P')_a \cap V'(\nu)$ have the same number of non-zero coefficients, and this number does not depend on ν. Assume that $\mu_1 + \mu_2 = \nu_1 + \nu_2 = \mu$, $c_{\mu_1,\mu_2} \neq 0$, $c_{\nu_1,\nu_2} = 0$. Then the restriction of the form $p_2^\mu(y)$ on $V'(\nu_1)$ has more non-zero coefficients than the restriction of $p_2^\mu(y)$ on $V'(\mu_1)$. This is a contradiction.

We can choose a point $u \in \mathcal{T}'$ such that $f'(u)$ belongs to exactly one exceptional curve of X'. Let ν be its class in $\mathbf{Pic}\, X'$. Then $u_\nu = 0$ and $u_\kappa \neq 0$ for any $\nu \neq \kappa$. Then

$$p_2^\mu(yx^{-1}u) = \sum_{\mu_1+\mu_2=\mu} d_{\mu_1,\mu_2} y^{\mu_1} y^{\mu_2},$$

where $d_{\mu_1,\mu_2} \neq 0$ if and only if $\mu_1 + \mu_2 = \mu$ and $\mu_1, \mu_2 \neq \nu$. Our assumption implies that $p_2^\mu(yx^{-1}u) = cp_2^\mu(y)$. Comparing the coefficients we see that $c = 0$. This is a contradiction.

Now since the set of such y is open for each μ we can find $s = yx^{-1}$ such that the restriction of p_2^μ on $s(\mathcal{T}')$ is not identically zero for all weights μ of V_2.

To each weight μ of V we can now associate the divisor in X' defined by

$$L'_\mu = f'(s^{-1}\{u \in s(\mathcal{T}') | p_i^\mu(u) = 0\}),$$

where i is such that μ is a weight of V_i. If μ is a weight of V_1, then by the induction assumption L'_μ is an exceptional divisor. If μ is a weight of V_2, then L'_μ is a conic in X' (i.e., a curve with self-intersection index zero). In fact, since $\mathbf{Pic}\, X'$ is generated by the exceptional divisors we can compute the intersection indices for all $[L'_\mu]$ and $[L'_\nu]$. It is easy to see that one can choose weights μ and ν of V_2 so that $[L'_\mu]$ and $[L'_\nu]$ have intersection index 1. To check that $s\mathcal{T}' \cap (G'/P')_a = Tsx = Ty$ observe that $u \in s\mathcal{T}' \cap (G'/P')_a$ implies $p_2^\mu(u) = p_2^\nu(u) = 0$, hence $f'(s^{-1}u) \in L'_\mu \cap L'_\nu$. Hence $f'(s^{-1}u) = M$. □

Now we are ready to prove Theorem 3.42. It is not hard to see from our construction that $\pi_1^{-1}(s\mathcal{T}' \setminus Tsx) \subset (G/P)_a^{sf}$. Let $\mathcal{T}$ be the Zariski closure of $\pi_1^{-1}(s\mathcal{T}'\setminus Tsx)$ in $(G/P)_a^{sf}$. Using the blow-up theorem and the functoriality of blowing-up we obtain the following commutative diagram

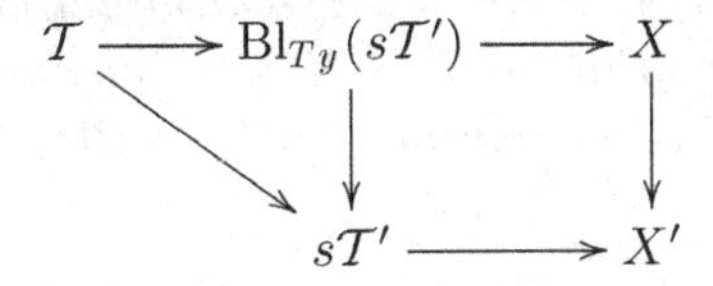

$$\begin{array}{ccccc} \mathcal{T} & \longrightarrow & \mathrm{Bl}_{Ty}(s\mathcal{T}') & \longrightarrow & X \\ & \searrow & \downarrow & & \downarrow \\ & & s\mathcal{T}' & \longrightarrow & X' \end{array}$$

where the horizontal arrows are torsors under tori, and the vertical arrows are smooth contractions. The map $\mathcal{T} \to \mathrm{Bl}_{Ty}(s\mathcal{T}')$ is a torsor with structure torus D. Therefore, the composition $f : \mathcal{T} \to X$ of two torsors is a torsor with structure torus $\mathbb{T}$.

Next we show that $f : \mathcal{T} \to X$ is a universal torsor. For each weight μ of V let $H_\mu = \{u \in V | u_\mu = 0\}$ be the corresponding hyperplane. Let $L_\mu = f(\mathcal{T} \cap H_\mu)$. By Lemma 3.43 and formula (3.7) the curve L_ω is the exceptional divisor corresponding to the blown-up point M. Furthermore, the classes $[L_\mu]$ for all weights μ of V_1, together with $[L_\omega]$, generate $\mathbf{Pic}\, X$.

Recall that $(G/P)_a^{sf} \to Y$ is a universal torsor, thus its type $\hat{\mathbb{T}} \to \mathbf{Pic}\, Y$ is an isomorphism. The restriction map $\mathbf{Pic}\, Y \to \mathbf{Pic}\, X$ is surjective, since $[L_\omega]$, and $[L_\mu]$ for all weights μ of V_1, are in the image. Since $\mathbf{Pic}\, Y$ and $\mathbf{Pic}\, X$ are free abelian groups of the same rank this restriction map is an isomorphism. But the type of f is the composition $\hat{\mathbb{T}} \to \mathbf{Pic}\, Y \to \mathbf{Pic}\, X$, so this type is also an isomorphism.

Finally, we need to show that the set of exceptional divisors of X coincides with the set of L_μ for all weights μ of V. This follows from the fact that the surjective homomorphism

$$\sigma : k[(G/P)_a^{sf}] = k[(G/P)_a] \to k[\mathcal{T}]$$

is $\mathbb{T}$-equivariant. Consider the weight decomposition

$$k[(G/P)_a] = \bigoplus_{\chi \in \hat{\mathbb{T}}} k[(G/P)_a]_\chi, \quad k[\mathcal{T}] = \bigoplus_{\chi \in \hat{\mathbb{T}}} k[\mathcal{T}]_\chi.$$

If μ is a weight of V, then $k[(G/P)_a]_\mu$ is one-dimensional and spanned by the coordinate function u^μ. By surjectivity $\sigma(u^\mu) \neq 0$, and hence the image of the hyperplane section $u^\mu = 0$ under f is an exceptional curve.

Remark Our proof depends on a choice of an exceptional divisor L_ω which we blow down to get a del Pezzo surface of higher degree. The weights of $V_i \subset V$ correspond to the exceptional divisors on X whose intersection index with L_ω equals $i - 1$.

The proof of Lemma 3.43 and Theorem 3.42 imply

Corollary 3.44 *There exists an open set $(G'/P')_a^\circ \subset (G'/P')_a$ such that for any $y \in (G'/P')_a^\circ$ there is a $\mathbb{T}$-equivariant embedding $\varphi : \mathcal{T} \to (G/P)_a^{sf}$ of the universal torsor $\mathcal{T} \to X$ satisfying $\pi_1(\varphi(\mathcal{T})) = yx^{-1}\mathcal{T}'$. More precisely, $\pi_1 \circ \varphi$ is the natural map $\mathcal{T} \to \mathrm{Bl}_{\mathcal{T}x}(\mathcal{T}') \to \mathcal{T}'$ followed by the action of $yx^{-1} \in \mathbb{V}'$.*

3.6.1 $\mathbb{P}^{n-3}$ blown up at n points

This example was previously considered in [3], [28]. Consider the inclusion of root systems $A_{n-1} \subset D_n$, where $n \geq 5$:

Then $\omega = \omega_n$ and $V = V_\omega$ is one of the two spinor representations; the group $G = Spin(2n)$ is a double cover of $SO(2n)$; the homogeneous space G/P is a connected component of the Grassmannian of maximal isotropic subspaces in the natural representation space of $SO(2n)$.

Recall the construction of the spinor representation (see [9] III.20 for details). The natural representation F of $SO(2n)$ is isomorphic to k^{2n} equipped with a $SO(2n)$-invariant quadratic form b. Consider the Clifford algebra

$$\mathrm{Cliff}(F) = T(F)/(xy + yx - b(x, y)),$$

where $T(F)$ is the tensor algebra of F.

Write $F = L \oplus L'$ as a direct sum of two isotropic subspaces. We now define the structure of a $\mathrm{Cliff}(F)$-module on the exterior algebra $\Lambda(L)$ as follows. For any $u \in L'$ define $\partial_u \in \mathrm{End}_k(\Lambda(L))$ by setting $\partial_u(w) := b(u, w)$ for all $w \in L$, and then extend ∂_u to a linear operator on $\Lambda(L)$ by the $\mathbb{Z}_2$-graded Leibniz identity

$$\partial_u(x \wedge y) = \partial_u(x) \wedge y + (-1)^{p(x)} x \wedge \partial_u(y),$$

where $p(x)$ is the parity of $x \in \Lambda(L)$. For any vector $w \in L$ define $j_w \in \mathrm{End}_k(\Lambda(L))$ by $j_w(x) := w \wedge x$. It is easy to check the relations

$$\partial_u \partial_{u'} + \partial_{u'} \partial_u = 0, \quad j_w j_{w'} + j_{w'} j_w = 0, \quad \partial_u j_w + j_w \partial_u = b(u, w),$$

for all $u, u' \in L'$ and all $w, w' \in L$. This defines a $\mathrm{Cliff}(F)$-module structure on $\Lambda(L)$.

By a direct calculation one can check that the span of $\partial_u \partial_{u'}$ for all $u, u' \in L'$, $j_w j_{w'}$ for all $w, w' \in L$ and $j_w \partial_u - \frac{1}{2} b(w, u)$ for all $w \in L$, $u \in L'$, is closed under the commutator and is isomorphic to the Lie algebra $\mathfrak{g} = so(2n)$. Moreover, $\mathfrak{g}$ has a grading $\mathfrak{g} = \mathfrak{g}_{-1} \oplus \mathfrak{g}_0 \oplus \mathfrak{g}_1$, where $\mathfrak{g}_1 = \mathrm{span}\{\partial_u \partial_{u'}\}$, $\mathfrak{g}_{-1} = \mathrm{span}\{j_w j_{w'}\}$, and $\mathfrak{g}_0 = \mathrm{span}\{j_w \partial_u - \frac{1}{2} b(w, u)\}$. Then $\mathfrak{p} = \mathfrak{g}_0 \oplus \mathfrak{g}_1$ is a maximal parabolic subalgebra of $\mathfrak{g}$. Define a representation of $\mathfrak{g}$ in $\Lambda(L)$ via the embedding $\mathfrak{g} \to \mathrm{Cliff}(F)$. This representation can be lifted to a representation of the algebraic group $G = Spin(2n)$ since $Spin(2n)$ is simply connected. Furthermore, $\Lambda(L)$ is a direct sum of two irreducible representations of $Spin(2n)$:

$$\Lambda(L) = \Lambda_{ev}(L) \oplus \Lambda_{odd}(L).$$

We define $V = \Lambda_{ev}(L)$, and set $v = 1 \in \Lambda$. Then $\mathfrak{p}$ coincides with the stabilizer of kv in $\mathfrak{g}$. Thus, v is a highest vector in V for a suitable choice of a Borel subalgebra. The $\mathbb{Z}$-grading on V is given by $V_i = \Lambda^{2i}(L)$. In particular, $V_1 = V' = \Lambda^2(L)$ and the maps $p_i : V_1 \to V_i$ are given by the formula

$$p_i(u) = u \wedge \cdots \wedge u \quad (i \text{ times}).$$

The affine cone in $(G'/P')_a = Gr(2,n)_a$ coincides with the set of decomposable bivectors in $\Lambda^2(L)$, i.e. bivectors of the form $u \wedge w$ for some $u, w \in L$. Indeed, this is the image of $Gr(2,n)_a$ under the Plücker embedding.

Let $S' \subset \mathbb{T}'$ be such that $Gr(2,n)/\!\!/S'$ is isomorphic to the projective space $\mathbb{P}^{n-3}$ blown up at $n-1$ points in general position. (Note that S' is S in the notation of Corollary 3.38.)

Lemma 3.45 *Let $S \subset \mathbb{T}$ be as in Theorem 3.36, and let $(G/P)_a^{sf}$ denote the set of S-stable points with trivial stabilizer in $\mathbb{T}$. Then the natural map*

$$f : (G/P)_a^{sf} \to Y := (G/P)_a^{sf}/\mathbb{T}$$

is a universal torsor.

Remark If C is the 1-dimensional group of scalar operators, then the natural projection $\mathbb{T} \to \mathbb{T}/C$ induces a surjective map $S \to \mathbb{T}/C$. Note that S is not invariant with respect to the whole Weyl group $W(D_n)$ but only with respect to the subgroup $S_n \simeq W(A_{n-1}) \subset W(D_n)$.

Proof Following the strategy of the proof of Theorem 3.31 it suffices to show that the complement of $(G/P)_a^{sf}$ in $(G/P)_a$ has codimension at least 2. This is an immediate consequence of Corollary 3.29 and Corollary 3.37. □

The following theorem is equivalent to Theorem 1.1 in [28].

Theorem 3.46 *Let X be the projective space $\mathbb{P}^{n-3}$ blown up at n points in general position, and let $f : \mathcal{T} \to X$ be a universal torsor. There exists a closed equivariant embedding of $\mathcal{T}$ into the universal torsor $f : (G/P)_a^{sf} \to (G/P)_a^{sf}/\mathbb{T}$. The images under f of the T-invariant hyperplane sections of $\mathcal{T}$ generate the ample cone in* **Pic** *X.*

Proof The proof is very similar to the proof of Theorem 3.42.

Let $\{w_1, \ldots, w_n\}$ be a T'-eigenbasis in L. Since $V_i = \Lambda^{2i}(L)$, every $u \in V_1$ can be written in the form

$$u = \sum_{i<j} u_{ij} w_i \wedge w_j,$$

and every $z \in V_2$ can be written in the form

$$z = \sum_{i<j<p<q} z_{ijpq} w_i \wedge w_j \wedge w_p \wedge w_q.$$

Therefore

$$p_2(u) = \sum_{i<j<p<q} p_2^{ijpq}(u) w_i \wedge w_j \wedge w_p \wedge w_q,$$

where

$$p_2^{ijpq}(u) = u_{ij}u_{pq} - u_{ip}u_{jq} + u_{iq}u_{jp}.$$

Set X' be the projective space $\mathbb{P}^{n-3}$ blown up at $n-1$ points $M_1, \ldots, M_{n-1}$ in general position. By $Gr(2,n)_a^s$ we denote the set of S'-stable points in $Gr(2,n)_a$. We recall from Corollary 3.38 that $f' : Gr(2,n)_a^s \to X'$ is an universal torsor. Assume that $M \in X'$ be such that its image in $\mathbb{P}^{n-3}$ under the blow-down is in general position with respect to $M_1, \ldots, M_{n-1}$. Let $x \in Gr(2,n)_a^s$ be such that $f'(x) = M$. The condition on M is equivalent to the condition $x_{ij} \neq 0$ for all $i < j$. The following lemma is an analogue of Lemma 3.43.

Lemma 3.47 *There exists $s \in \mathbb{V}'$ such that $s(Gr(2,n)_a^s) \cap Gr(2,n)_a^s = Tsx$ is a single orbit, and the restriction of p_2^{ijpq} to $s(Gr(2,n)_a^s)$ is not identically zero for all $i < j < p < q$.*

Proof As in the proof of Lemma 3.43 we set $s = yx^{-1}$ for some $y \in Gr(2,n)_a$. The proof of the second assertion is identical to the proof of the same assertion in Lemma 3.43. So it remains to prove the first assertion. It is possible to choose s so that $s_{ij}s_{pq}, s_{ip}s_{jq}, s_{iq}s_{jp}$ are distinct. By a straightforward calculation one can check that $p_2^{ijpq}(u) = p_2^{ijpq}(su) = p_2^{ijpq}(x) = p_2^{ijpq}(sx) = 0$ implies

$$\frac{u_{ij}u_{pq}}{x_{ij}x_{pq}} = \frac{u_{ip}u_{jq}}{x_{ip}x_{jq}} = \frac{u_{iq}u_{jp}}{x_{iq}x_{jp}}. \tag{3.10}$$

If $p_2(u) = 0$, the equalities in (3.10) hold for any $i < j < p < q$. On the other hand, (3.10) for all $i < j < p < q$ imply that u belongs to the Zariski closure of Tx. Thus, if $u \in s^{-1}(Gr(2,n)_a^s) \cap (G'/P')_a^s$, then u

belongs to the Zariski closure of Tx. Since u is S-stable, $u \in Tx$. Thus, $s(Gr(2,n)^s_a) \cap (G'/P')^s_a = s(Tx) = Tsx$. □

We proceed now as in the proof of Theorem 3.42. Set $\mathcal{T}$ to be the Zariski closure of $\pi_1^{-1}(s(Gr(2,n)^s_a) \setminus Tsx)$ in $(G/P)^{sf}_a$. Let $f : \mathcal{T} \to X$ be the restriction of the torsor $(G/P)^{sf}_a \to Y$. Then X is the blow-up of X' at M. Moreover, if we consider the surjective homomorphism $\sigma : k[(G/P)^{sf}_a] = k[(G/P)_a] \to k[\mathcal{T}]$, then $\sigma(u_\mu)$ for all weights μ of V generate $\mathrm{Cox}(X)$. Recall that $\mathrm{Cox}\, X = k[\mathcal{T}]$, and the ample cone in $\hat{\mathbb{T}} = \mathbf{Pic}\, X$ coincides with the set of weights of $k[\mathcal{T}]$. Every weight of $k[\mathcal{T}]$ is a non-negative linear combination of the weights of V. The proof of the theorem is complete. □

3.7 On the Cox ring of a del Pezzo surface

In the previous section we constructed an equivariant embedding of a universal torsor $\mathcal{T}$ of a del Pezzo surface X of degree $d \geq 2$ into $(G/P)^{sf}_a$. The construction is based on an inductive procedure. Here we are going to give an alternative description of such an embedding. We start with the following obvious observation.

Lemma 3.48 *Denote by V^{sf} the subset of T-stable points with trivial stabilizer in $\mathbb{T}$. Let $\varphi, \psi : \mathcal{T} \to V^{sf}$ be $\mathbb{T}$-equivariant embeddings of the universal torsor $\mathcal{T}$ which induce the same map from the set of weights of V to the set of exceptional divisors in* **Pic** *X. Then there exists $s \in \mathbb{V}$ such that $\varphi = s \circ \psi$.*

Proof A $\mathbb{T}$-homogeneous function $y^\mu \in V^*$ defines two functions $y^\mu \circ \varphi$ and $y^\mu \circ \psi$ on $\mathcal{T}$ with the same divisor. Since $k[\mathcal{T}]^* = k^*$, we obtain $y^\mu \circ \varphi = s_\mu y^\mu \circ \psi$ for some $s_\mu \in k^*$. Set $s = (s_\mu)$. □

For a $\mathbb{T}$-equivariant embedding $\mathcal{T} \subset (G/P)^{sf}_a \subset V^{sf}$ we define

$$\mathcal{Z} = \{s \in \mathbb{V} | s\mathcal{T} \subset (G/P)_a\}.$$

Let $V^\times \subset V$ be the complement to the union of all weight hyperplanes, and let $\mathcal{T}^\times = \mathcal{T} \cap V^\times$ and $(G/P)^\times_a = (G/P)_a \cap V^\times$. Then it is easy to see that

$$\mathcal{Z} = \bigcap_{t \in \mathcal{T}^\times} t^{-1}(G/P)^\times_a.$$

Recall that $\pi_1 : V \to V'$ denotes the natural G'-equivariant projection.

Suppose that X is the blow-up of X' at a point M, and $x \in V'$ is any point in the fibre of $\mathcal{T}' \to X'$ over M. By Corollary 3.44 for any $y \in (G'/P')_a^\circ$ there is a $\mathbb{T}$-equivariant embedding $\varphi : \mathcal{T} \to (G/P)_a^{sf}$ such that $\pi_1 \circ \varphi$ is the composition $\mathcal{T} \to \mathrm{Bl}_{Tx}(\mathcal{T}') \to \mathcal{T}'$ followed by the action of yx^{-1}.

Lemma 3.49 *Assume that $\mathcal{T}$ is embedded into $(G/P)_a^{sf}$ by the map φ defined by some $y \in (G'/P')_a^\circ$. If $r = 4$, then $\mathcal{Z} = \mathbb{T}$. If $r = 5, 6,$ or 7, then $\pi_1(\mathcal{Z}) = y^{-1}(G'/P')_a^\circ$ and $\pi_1 : \mathcal{Z} \to y^{-1}(G'/P')_a^\circ$ is a torsor with the structure group D.*

Proof For $r = 4$ the statement follows easily from Corollary 3.40. So we assume $r > 4$.

First, we will show that $y^{-1}(G'/P')_a^\circ \subset \pi_1(\mathcal{Z})$. By Corollary 3.44 for any $y' \in (G'/P')_a^\circ$ there exists an embedding $\psi : \mathcal{T} \to V^{sf}$ such that $\pi_1 \circ \psi$ is the composition $\mathcal{T} \to \mathrm{Bl}_{Tx}(\mathcal{T}') \to \mathcal{T}'$ followed by the action of $y'x^{-1}$. Since $\psi = z\varphi$ for some $z \in \mathcal{Z}$, we have $\pi_1 \circ \psi = \pi_1(z)\pi_1 \circ \varphi$, hence we obtain $\pi_1(z) = y'y^{-1}$.

Next we will show that $\pi_1(\mathcal{Z}) \subset y^{-1}(G'/P')_a^\circ$. Recall that we have $\pi_1(\varphi(\mathcal{T}) \cap H_\omega) = Ty$. Indeed, the morphism π_1 blows down the exceptional divisor $\varphi(\mathcal{T}) \cap H_\omega$ to the orbit Ty, cf. Lemma 3.43 and formula (3.7). Thus for any $z \in \mathcal{Z}$ we have

$$\pi_1(z(\varphi(\mathcal{T})) \cap H_\omega) = \pi_1(z)Ty = Ty'$$

for some $y' \in \mathbb{V}'$. On the other hand, we have $\pi_1(H_\omega \cap (G/P)_a) = (G'/P')_a$ by formula (3.8), hence $y' \in (G'/P')_a$. Therefore, we have $\pi_1(z) = y^{-1}y'$ for some $y' \in (G'/P')_a$. Note that the image $\psi(\mathcal{T}) = z\varphi(\mathcal{T}) \subset (G/P)_a^{sf}$ does not lie in any coordinate hyperplane. This ensures that $y' \in (G'/P')_a^\circ$.

Finally, $\pi_1 : \mathcal{Z} \to y^{-1}(G'/P')_a^\circ$ is a D-torsor since $\mathcal{Z} \subset t^{-1}(G/P)_a^\times$ for some $t \in \mathcal{T}^\times$, and we already know that $\pi_1 : (G/P)_a^\times \to V'^\times \setminus (G'/P')_a$ is a D-torsor from the proof of Theorem 3.34 and the remark that follows it. □

Corollary 3.50 *We have $\mathcal{Z}/\mathbb{T} = y^{-1}(G'/P')_a^\circ/T \simeq (G'/P')_a^\circ/T$. The dimension of $\mathcal{Z}$ is $2 + \dim G'/P'$ which equals 8, 12, or 18 for $r = 5$, 6, or 7, respectively.*

Theorem 3.51 *[22] There exist $r - 3$ points $z_0, \ldots, z_{r-4} \in \mathcal{Z}$, $z_0 = 1$ such that*

$$\mathcal{T} = \bigcap_{z \in \mathcal{Z}} z^{-1}(G/P)_a^{sf} = \bigcap_{i=0}^{r-4} z_i^{-1}(G/P)_a^{sf}.$$

Proof We give here only the main idea and refer to [22] for details. Note that for $r = 4$ the statement is trivial. We identify $\mathcal{T}$ with $\varphi(\mathcal{T})$ with the choice of φ as above. It is clear that

$$\mathcal{T} \subset \bigcap_{z\in\mathcal{Z}} z^{-1}(G/P)_a^{sf} \subset \bigcap_{i=0}^{r-4} z_i^{-1}(G/P)_a^{sf}.$$

To prove equality it suffices to show that $\pi_1(\mathcal{T}) = yx^{-1}\mathcal{T}'$ is dense in $\pi_1(\mathcal{S})$, where $\mathcal{S} = \bigcap_{i=0}^{r-4} z_i^{-1}(G/P)_a^{sf}$.

Let $\mu \in \hat{\mathbb{T}}$. For a Zariski closed $L \subset V'$ with ideal I_L let $I_\mu(L) = I_L \cap k[V']_\mu$. It is not hard to see from the inductive construction of $\mathcal{T}'$ that for any weight μ of V_2 we have

$$\dim k[V']_\mu = r-2,\ \dim I_\mu((G'/P')_a) = 1,\ \dim I_\mu(\mathcal{T}') = r-4.$$

Hence $\dim I_\mu(yx^{-1}\mathcal{T}') = r-4$.

On the other hand, after a suitable normalization of coordinates we obtain

$$\tilde{p}^\mu := u_\omega u_\mu + p_2^\mu \in I_{(G/P)_a} \cap k[V]_{\mu+\omega}.$$

It is possible to show (see [22]) that for generic $z_0, \dots, z_{r-4} \in \mathcal{Z}$, $z_0 = 1$, the set $\{z_i^{-1}\tilde{p}^\mu\}_{i=0,\dots,r-4}$ is linearly independent. Therefore one can find $c_1, \dots, c_{r-4} \in k$, such that $\{z_i^{-1}\tilde{p}^\mu - c_i\tilde{p}^\mu\}_{i=1,\dots,r-4}$ is a linearly independent set in $k[V']_\mu$. Thus, $I_\mu(\pi_1(\mathcal{S})) \subset I_\mu(yx^{-1}\mathcal{T}')$ has dimension $r-4$, hence $I_\mu(\pi_1(\mathcal{S})) = I_\mu(yx^{-1}\mathcal{T}')$. The statement follows. □

Now we can write down the equations for $\mathcal{T}$ in V^{sf} as well as for the closure of $\mathcal{T}$ in V. Indeed, one can apply $r-3$ dilatations (given by $z_0, \dots, z_{r-4} \in \mathcal{Z}$) to the equations of Theorem 3.51. Note that for $r = 4, 5, 6$, or 7 there are $5, 10, 27$, or 133 equations defining $(G/P)_a \subset V$ respectively. Thus, there are $5, 20, 81$, or 532 equations for the Zariski closure of $\mathcal{T}$ in V. Note that for $r \neq 7$ these numbers coincide with the numbers obtained in [14], [26] and [27]. For $r = 7$ we have more equations. On the other hand, it is shown in [22] that it is sufficient to use 504 equations to define $\mathcal{T}$ in V^{sf}.

We can deal with the situation of Section 6.1 in a similar way.

Theorem 3.52 *Let $G = Spin(2n)$, and let $(G/P)_a^{sf}$ and $\mathcal{T}$ be as in Theorem 3.46. Then $\mathcal{T} = z((G/P)_a^{sf}) \cap (G/P)_a^{sf}$ for some $z \in \mathbb{V}$.*

The proof of the theorem is similar to the case $r = 5$ in Theorem 3.51 and we leave it to the reader.

Let us return to the situation when X is a del Pezzo surface. Denote by X^e the union of the exceptional curves in X and let $X^\times = X \backslash X^e$. Let $\mathrm{Div}_{X^e}(X)$ be the group of divisors supported in X^e. The kernel of the natural surjective map $\mathrm{Div}_{X^e}(X) \to \mathbf{Pic}\, X$ is the subgroup of principal divisors supported in X^e. Thus, there is an exact sequence

$$0 \to k[X^\times]^*/k^* \to \mathrm{Div}_{X^e}(X) \to \mathbf{Pic}\, X \to 0. \tag{3.11}$$

Let $\mathbb{S} = \mathbb{V}/\mathbb{T}$ and $\mathcal{T}^\times = \mathcal{T} \cap \mathbb{V}$. The embedding $\mathcal{T}^\times \to \mathbb{V}$ induces the embedding $X^\times \to \mathbb{S}$ and therefore a homomorphism $\varphi : \hat{\mathbb{S}} \to k[X^\times]^*/k^*$. Note that $\mathbf{Pic}\, X \simeq \hat{\mathbb{T}}$. Furthermore, $\mathrm{Div}_{X^e}(X)$ is a free abelian group generated by the exceptional curves, hence it is isomorphic to $\hat{\mathbb{V}}$. By comparison of the exact sequence (3.11) and the exact sequence

$$0 \to \hat{\mathbb{S}} \to \hat{\mathbb{V}} \to \hat{\mathbb{T}} \to 0$$

we obtain that φ is an isomorphism.

Let $\tilde{\mathcal{Z}}$ be the image of $\mathcal{Z}$ under the natural projection $\mathbb{V} \to \mathbb{S}$. It is clear that

$$\tilde{\mathcal{Z}} = \{s \in \mathbb{S} | s(X^\times) \subset Y\}.$$

For a subset $A \subset \mathbb{S}$ let $P^n(A)$ denote the set of products of n elements in A.

Proposition 3.53 *[22] There exists a unique $s \in \mathbb{S}$ such that $P^{r-4}(X^\times) \subset s\tilde{\mathcal{Z}}$.*

For the proof see [22], Section 3.

3.7.1 Non-split case

Let k be not algebraically closed field and Γ denote the Galois group $\mathrm{Gal}(\bar{k}/k)$. Let Z be an algebraic variety over k and $\bar{Z}$ be obtained from Z by extending the ground field to $\bar{k}$. For any homomorphism $\sigma : \Gamma \to \mathrm{Aut}(Z)$ one can define a twisted form Z_σ obtained by twisting the Galois group action on $\bar{Z}$ by σ.

Let G be a split simple algebraic group over k, T be a maximal split torus and $W = N(T)/T$ be the Weyl group. For any $\sigma \in \mathrm{Hom}(\Gamma, W)$ one can define the twisted torus T_σ.

Theorem 3.54 *[10] For any $\sigma \in \mathrm{Hom}(\Gamma, W)$ the twisted torus T_σ is isomorphic to a maximal torus of G.*

Let a maximal parabolic subgroup $P \subset G$, the representation V, the homogeneous spaces G/P and $(G/P)_a$ be as in the case of an algebraically closed field. Since $(G/P)^{sf}$ is W-invariant, for any homomorphism $\sigma : \Gamma \to W$ the quotient $(G/P)^{sf}/T$ has a well defined twisted form Y_σ. Let $\mathbb{T}_\sigma$ be the extension of T_σ by the group of scalar operators in V. Let $(G/P)^{sf}_{a,\sigma}$ be the set of points $x \in (G/P)_a$ such that $T_\sigma x$ is closed in V and the stabilizer of x in $\mathbb{T}_\sigma$ is trivial.

It follows from Theorem 3.54 (for details see Section 4 of [22]) that a cocycle $\sigma \in Z^1(k, W)$ can be lifted to a cocycle in $Z^1(k, N(T))$ that gives rise to a trivial cocycle in $Z^1(k, G)$. Using this result, one can prove the following

Lemma 3.55 *[22] The varieties Y_σ and $(G/P)^{sf}_{a,\sigma}/\mathbb{T}_\sigma$ are isomorphic, and the twisted variety Y_σ has a k-point.*

Let X be a del Pezzo surface of degree $5, 4, 3$ or 2 over k. Then Γ permutes the exceptional curves on $\overline{X} = X \times_k \bar{k}$. Since W is the automorphism group of the incidence graph of the exceptional curves, one can define a homomorphism $\sigma_X : \Gamma \to W$. Note that σ_X is defined up to conjugation in W.

Theorem 3.56 *[22] Let $r = 4, 5, 6,$ or 7, and let X be a del Pezzo surface of degree $9-r$ over k that has a k-point. Let $\sigma = \sigma_X \in \mathrm{Hom}(\Gamma, W)$. There exists an embedding $X \to Y_\sigma$ such that the restriction of the morphism $(G/P)^{sf}_{a,\sigma} \to Y_\sigma$ to X is a universal torsor whose type is the isomorphism $\hat{\mathbb{T}}_\sigma \simeq \mathbf{Pic}\,\overline{X}$.*

Proof We just give a sketch and refer the reader to [22] for a complete proof.

The first step is to show that there exists the following exact sequence of tori

$$1 \to \mathbb{T}_\sigma \to \mathbb{V}_\sigma \to \mathbb{S}_\sigma \to 1.$$

Now define X^e and $X^\times$ as for an algebraically closed k. Since $\hat{\mathbb{S}}_\sigma = \bar{k}[\mathbb{S}_\sigma]^*/\bar{k}^*$ and the action of Γ on $\hat{\mathbb{S}}_\sigma$ is compatible with the action of Γ on $\mathrm{Div}_{\overline{X}^e}(\overline{X})$, the isomorphism $\varphi : \hat{\mathbb{S}}_\sigma \tilde{\longrightarrow} \bar{k}[X^\times]^*/\bar{k}^*$ is an isomorphism of Γ-modules.

A k-point $x_0 \in X^\times$ provides a splitting of the exact sequence of Γ-modules

$$0 \to \bar{k}^* \to \bar{k}[X^\times]^* \to \bar{k}[X^\times]^*/\bar{k}^* \to 0.$$

In particular, there is a Γ-invariant injective map $i : \bar{k}[X^\times]^*/\bar{k}^* \to$

$\bar{k}[X^\times]^*$. Since $i \circ \varphi$ is Γ-invariant, there exists a unique embedding $\psi : X^\times \to \mathbb{S}_\sigma$ such that $\psi(x_0) = 1$ and $\chi \circ \psi = i \circ \varphi(\chi)$ for all $\chi \in \hat{\mathbb{S}}_\sigma$.

Define a k-subvariety $\mathcal{L} \subset \mathbb{S}_\sigma$ by

$$\mathcal{L} = \{s \in \mathbb{S}_\sigma | s\psi(X^\times) \subset Y_\sigma\}.$$

It suffices to show that $\mathcal{L}$ has a k-point. By Proposition 3.53 there exists a unique $s \in \mathbb{S}_\sigma(\bar{k})$ such that $P^{r-4}(\psi(X^\times))(\bar{k}) \subset s\mathcal{L}(\bar{k})$. Since $P^{r-4}(\psi(X^\times))$ and $\mathcal{L}$ are subvarieties of $\mathbb{S}_\sigma$ defined over k, the uniqueness of s implies that s is a k-point. Pick up a k-point m on $\psi(X^\times)$. Then $s^{-1}m^{r-4}$ is a k-point on $\mathcal{L}$. □

3.8 A generalization of the blow-up theorem and del Pezzo surface of degree 1

In this section we give a brief review of the results of [23], where Batyrev's conjecture is proved for the del Pezzo surfaces of degree 1 over an algebraically closed field of characteristic zero.

3.8.1 Generalized blow-up theorem

Let $P \subset G$ be a maximal parabolic subgroup of a simple connected group G. We use the notation of Sections 4 and 5 but drop the assumption $\mathfrak{g} = \mathfrak{g}_{-1} \oplus \mathfrak{g}_0 \oplus \mathfrak{g}_1$. We also assume that $\alpha = \alpha_i$ is a long root. We define a grading $V = \bigoplus_{i=0}^m V_i$, and the maps $\phi : \mathfrak{g}_{-1} \to V_1$ and $p_i : S^i(V_1) \to V_i$ as in Section 3.5. In particular, $V_2 = V_2^+ \oplus V_2^-$, where $V_2^+ = p_2(S^2(V_1))$ and $V_2^- = \mathfrak{g}_{-2}V_0$. By $\pi_2^+ : (G/P)_a \to V_2^+$ we denote the restriction of G'-equivariant projection.

Let $G_{\leq -2} \subset G$ be the unipotent subgroup with Lie algebra $\mathfrak{g}_{\leq -2} = \bigoplus_{i \leq -2} \mathfrak{g}_i$, and let $K = G_{\leq -2} \rtimes D$. Set

$$(G/P)_a^+ := (G/P)_a \setminus (\pi_2^{-1}(0) \cup \pi_1^{-1}(0)).$$

It is straightforward that $(G/P)_a^+$ is K-invariant. Define the map $\pi : (G/P)_a^+ \to V_1 \times \mathbb{P}(V_2^+)$ as $\pi := (\pi_1, \bar{\pi}_2^+)$, where $\bar{\pi}_2^+$ is the composition of $V_2^+ \backslash \{0\} \to \mathbb{P}(V_2^+)$ with π_2^+. It is not hard to see that π is a K-equivariant map. Let $\mathcal{Y}$ be the image of π.

Theorem 3.57 *[23] We have the following statements.*

(a) $\pi : (G/P)_a^+ \to \mathcal{Y}$ *is a K-torsor;*

(b) $\pi_1 : \mathcal{Y} \to V_1 \setminus \{0\}$ *is the inverse morphism to the blowing-up* $V_1 \setminus \{0\}$ *at* $(G'/P')_a \setminus \{0\}$.

3.8.2 Grading of length 5

Now assume that

$$\mathfrak{g} = \mathfrak{g}_{-2} \oplus \mathfrak{g}_{-1} \oplus \mathfrak{g}_0 \oplus \mathfrak{g}_1 \oplus \mathfrak{g}_2.$$

In this case we give a general construction of a D-torsor inside $(G/P)_a$.

Theorem 3.58 *Let* $Z \subset V_1 \setminus \{0\}$ *be a smooth closed subvariety such that* $Z_0 := Z \cap (G'/P')_a$ *is smooth in* Z. *Let* $q : S^2(\mathfrak{g}_{-1}) \to \mathfrak{g}_{-2}$ *be a linear map such that* $[a, x] = 4q(a, x)$ *for any* $x \in \phi^{-1}Z_0$ *and any* $a \in \phi^{-1}(T_{x,Z})$. *Let*

$$\tilde{Z} = D\{\exp(\phi^{-1}(x) + q(\phi^{-1}(x)))v | x \in Z\} \cap (G/P)_a^+,$$

and let $\hat{Z}$ *be the Zariski closure of* $\tilde{Z}$ *in* $\pi^{-1}(Z) \cap (G/P)_a^+$. *Then* $\hat{Z}$ *is a variety with a free action of* D, *and the map* $\bar{\pi}_1 : \hat{Z}/D \to Z$ *is the inverse to the blowing-up of* Z *at* Z_0.

Remark It might be interesting to generalize Theorem 3.58 to any pair (G, G'). This potentially leads to new examples, even if the grading becomes infinite and $\mathfrak{g}$ is a Kac–Moody Lie algebra, say, E_n for $n > 8$. The orbit of a highest weight vector has the structure of an ind-variety and is given by infinitely many quadratic equations. What does it mean for the Cox ring?

3.8.3 Batyrev's conjecture for del Pezzo surfaces of degree 1

For the pair (E_8, E_7) we have $V \simeq \mathfrak{g}$, and

$$\mathfrak{g} = \mathfrak{g}_{-2} \oplus \mathfrak{g}_{-1} \oplus \mathfrak{g}_0 \oplus \mathfrak{g}_1 \oplus \mathfrak{g}_2,$$

where $\mathfrak{g}'$ is the Lie algebra E_7, and $\mathfrak{g}_{-1} \simeq \mathfrak{g}_1$ is the 56-dimensional representation of $\mathfrak{g}'$. Then $V_i = \mathfrak{g}_{2-i}$. Furthermore, $\mathfrak{g}_2 = kv$ is the line of highest weight vectors, and the map $[,] : \Lambda^2(\mathfrak{g}_1) \to \mathfrak{g}_2$ induces a G'-invariant symplectic form ζ on $\mathfrak{g}_1$ such that $\zeta(x, y)v = [x, y]$. It follows that $V_2^+ = \mathfrak{g}'$, and V_2^- is the one-dimensional center of $\mathfrak{g}_0$.

In [23] we use Theorem 3.58 to prove Batyrev's conjecture for any del Pezzo surface X of degree 1, by deducing it from the corresponding result for the del Pezzo surface X' of degree 2, obtained from X

by the contraction of an exceptional curve. Recall that $\kappa : X' \to \mathbb{P}^2 = \mathbb{P}(H^0(X', -K_{X'})^*)$ is a double covering with a smooth quartic branch curve C. Then X can be obtained from X' by blowing-up a point M which does not lie on the exceptional divisors or C. The following observation is essential for our proof, see Lemma 4.1 in [23]:

$$\kappa^* : T^*_{\kappa(M),\mathbb{P}^2} \to T^*_{M,X'} \text{ is an isomorphism.} \tag{3.12}$$

Let $f' : \mathcal{T}' \to X'$ be the universal torsor. We fix an embedding

$$\mathcal{T}' \subset (G'/P')_a \subset V_1 = \mathfrak{g}_1.$$

Our goal is to construct a closed subset $Z \subset V_1$ isomorphic to $\mathcal{T}'$, and a map $q : S^2(\mathfrak{g}_{-1}) \to \mathfrak{g}_{-2}$ satisfying the conditions of Theorem 3.58. Then $\mathcal{T} = \hat{Z} \cap (G/P)_a^{sf}$ will be a universal torsor over X.

We start with constructing Z. Pick up $x \in \mathcal{T}'$ such that $f'(x) = M$. As in Lemma 3.43 we can find $y \in (G'/P')_a$ such that $Z := yx^{-1}\mathcal{T}'$ meets $(G'/P')_a$ in Ty, and $\pi_1^{-1}(Z)$ does not lie in any coordinate hyperplane section (cf. Lemma 4.2 in [23]). Then $h := f' \circ xy^{-1} : Z \to X'$ is a universal torsor of type $\beta : \hat{\mathbb{T}}' \to \mathbf{Pic}\, X'$. Let $\chi = \beta^{-1}(-K_{X'})$. It is easy to see that $\chi(T') = 1$. One can identify $H^0(X', -K_{X'})$ with $k[Z]_\chi = S^2_\chi(\mathfrak{g}_1)/I_\chi(Z)$ via $h^* \circ \kappa^* : H^0(X', -K_{X'}) \to k[Z]_\chi$.

Now we proceed to constructing q. One can identify $T^*_{\kappa(M),\mathbb{P}(H^0(X',-K_{X'})^*)}$ with $\{r \in S^2_\chi(\mathfrak{g}_1)/I_\chi(Z) | r(y) = 0\}$. It is not hard to check that

$$\langle h^* \circ \kappa^*(r), a\rangle = r(y, a)$$

for any $a \in T_{y,Z}$. Now let $\xi \in T^*_{y,Z}$ be defined by $\langle \xi, a\rangle = \zeta(a, y)$. If $\xi = 0$ set $q = 0$. Assume that $\xi \neq 0$. It is easy to see that $\langle \xi, a\rangle = 0$ for any $a \in T_{y,Ty}$. Hence $\xi = h^*(\eta)$ for some $\eta \in T^*_{M,X'}$. By (3.12) we have $\eta = \kappa^*(r)$. So we have $r(a, y) = \zeta(a, y)$ for any $a \in T_{y,Z}$. If we set $q(x, y) = r([x, w], [y, w])w$ for a suitably normalized $w \in \mathfrak{g}_{-2}$, then q satisfies the requirements of Theorem 3.58.

Note that our construction of $\mathcal{T}$ implies that the ideal $I_{\mathcal{T}}$ is quadratic. Indeed, it follows from the fact that the varieties G/P, $Z = \mathcal{T}'$ and $\tilde{Z}$ are all defined by quadratic equations. We hope that the following analogue of Theorem 3.51 holds.

Conjecture 3.59 *Let H be the centralizer of $\mathbb{T}$ in $GL(V)$, and let*

$$\mathcal{Z} = \{z \in H | z(\mathcal{T}) \subset (G/P)_a\}.$$

Then we have

$$\mathcal{T} = \bigcap_{z \in \mathcal{Z}} z^{-1}(G/P)_a^{sf}.$$

References

[1] V.V. Batyrev and O.N. Popov. The Cox ring of a del Pezzo surface. In: *Arithmetic of higher-dimensional algebraic varieties* (Palo Alto, 2002), *Progr. Math.* **226** Birkhäuser, 2004, 85–103.

[2] N. Bourbaki. *Groupes et algèbres de Lie.* Chapitres IV-VIII. Masson, Paris, 1975, 1981.

[3] A-M. Castravet and J. Tevelev. Hilbert's 14th problem and Cox rings. *Compos. Math.* **142** (2006) 1479–1498.

[4] J-L. Colliot-Thélène et J-J. Sansuc. La descente sur les variétés rationnelles, II. *Duke Math. J.* **54** (1987) 375–492.

[5] U. Derenthal. Universal torsors of Del Pezzo surfaces and homogeneous spaces. *Adv. Math.* **213** (2007) 849–864.

[6] U. Derenthal. On the Cox ring of Del Pezzo surfaces. arXiv:math.AG/0603111

[7] I.V. Dolgachev. *Lectures on invariant theory.* Cambridge University Press, 2003.

[8] R. Friedman and J.W. Morgan. Exceptional groups and del Pezzo surfaces. In: Symposium in Honor of C. H. Clemens (Salt Lake City, UT, 2000), 101–116, *Contemp. Math.* **312** Amer. Math. Soc., Providence, RI, 2002.

[9] W. Fulton and J. Harris. *Representation theory.* Springer-Verlag, 1991.

[10] Ph. Gille. Type des tores maximaux des groupes semi-simples. *J. Ramanujan Math. Soc.* **19** (2004) 213–230.

[11] G. Harder. Halbeinfache Gruppenschemata über vollständigen Kurven. *Invent. Math.* **6** (1968) 107–149.

[12] R. Hartshorne. *Algebraic Geometry.* Graduate Texts in Mathematics **52**, Springer-Verlag, 1977.

[13] J. Humphreys. *Linear algebraic groups.* Graduate Texts in Mathematics **21**, Springer-Verlag, 1975.

[14] A. Laface and M. Velasco. Picard-graded Betti numbers and the defining ideals of Cox rings. *J. Algebra* **322** (2009) 353–372.

[15] W. Lichtenstein. A system of quadratics describing the orbit of the highest vector. *Proc. AMS* **84** (1982) 605–608.

[16] Yu.I. Manin. *Cubic forms.* 2nd. ed. North-Holland, 1986.

[17] D. Mumford, J. Fogarty, and F. Kirwan. *Geometric invariant theory.* 3rd enlarged edition. Springer-Verlag, 1994.

[18] A.L. Onishchik and E.B. Vinberg. *Lie groups and algebraic groups.* Springer-Verlag, 1990.

[19] V.L. Popov. Picard groups of homogeneous spaces of linear algebraic groups and one-dimensional homogeneous vector fiberings. (Russian) *Izv. Akad. Nauk SSSR* Ser. Mat. **38** (1974) 294–322.

[20] O.N. Popov. *Del Pezzo surfaces and algebraic groups.* Diplomarbeit, Universität Tübingen, 2001.

[21] V.V. Serganova and A.N. Skorobogatov. Del Pezzo surfaces and representation theory. *J. Algebra Number Theory* **1** (2007) 393–419.

[22] V.V. Serganova and A.N. Skorobogatov. On the equations for universal torsors over del Pezzo surfaces. *J. Inst. Math. Jussieu* **9** (2010) 203–223.

[23] V.V. Serganova and A.N. Skorobogatov. Adjoint representation of E8 and del Pezzo surfaces of degree 1 *Ann. Inst. Fourier* **61**, no. 6, (2011) 2337–2360.

[24] A.N. Skorobogatov. Automorphisms and forms of toric quotients of homogeneous spaces. *Mat. Sbornik* **200** (2009) 107–122 (Russian). English translation: *Sbornik: Mathematics* **198** (2009) 793–808.

[25] A. Skorobogatov. *Torsors and rational points.* Cambridge University Press, 2001.

[26] M. Stillman, D. Testa and M. Velasco. Gröbner bases, monomial group actions, and the Cox ring of del Pezzo surfaces. *J. Algebra* **316** (2007) 777–801.

[27] B. Sturmfels and Z. Xu. Sagbi bases of Cox–Nagata rings. *J. Eur. Math. Soc.* **12** (2010) 429–459.

[28] B. Sturmfels and M. Velasco. Blow-ups of P^{n-3} at n points and spinor varieties. *J. Commutative Algebra* **2** (2010) 223–244.

[29] D. Testa, A. Várilly-Alvarado and M. Velasco. Cox rings of degree one del Pezzo surfaces. *J. Algebra Number Theory* **3** (2009) 729–761.

PART TWO

CONTRIBUTED PAPERS

4

Torsors over Luna strata

Ivan V. Arzhantsev

Department of Higher Algebra, Faculty of Mechanics and Mathematics, Moscow State University

Abstract Let G be a reductive group and X_H be a Luna stratum on the quotient space $V//G$ of a rational G-module V. We consider torsors over X_H with both non-commutative and commutative structure groups. It allows us to describe the divisor class group and the Cox ring of a Luna stratum under mild assumptions. This approach gives a simple cause why many Luna strata are singular along their boundary.

Introduction

Consider a reductive group G and a rational finite-dimensional G-module V over an algebraically closed field $\mathbb{K}$ of characteristic zero. The inclusion of the algebra of invariants $\mathbb{K}[V]^G$ in the polynomial algebra $\mathbb{K}[V]$ gives rise to the quotient morphism $\pi\colon V \to V//G := \operatorname{Spec} \mathbb{K}[V]^G$. In [8], D. Luna introduced a stratification of the quotient space $X := V//G$ by smooth locally closed subvarieties. Every stratum is defined by the conjugacy class of the isotropy group H of a closed G-orbit in the fibre of π. With any such subgroup H one associates the group $W = N_G(H)/H$. Then the stratum X_H comes with a W-torsor $\pi\colon V^{\langle H\rangle} \to X_H$, where $V^{\langle H\rangle}$ is an open subset in the subspace V^H of H-fixed vectors.

The aim of this work is to use this observation to describe the Cox rings of Luna strata and to apply them to the study of geometric properties of the strata and their closures. It is well known that a smooth variety

Y with a finitely generated divisor class group $\mathrm{Cl}(Y)$ admits a canonical presentation as a geometric quotient of a quasiaffine variety by an action of a quasitorus Q. This presentation also is known as a universal torsor over Y. The quasiaffine variety mentioned above is the relative spectrum of the so-called Cox sheaf on Y and Q is a direct product of a torus and a finite abelian group such that the group of characters $\mathbb{X}(Q)$ is identified with $\mathrm{Cl}(Y)$.

The W-torsor over a stratum X_H is not a good candidate for a universal torsor, because the group W in general is not commutative. A naive idea is to consider the commutant $S = [W, W]$, the factor group $Q = W/S$, and to decompose the W-torsor into two steps, namely

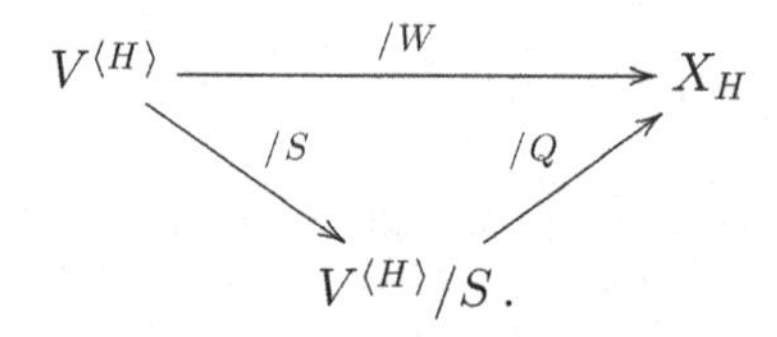

Note that the groups H, W, S, and Q are reductive. In Theorem 4.10 we show that this indeed gives a universal Q-torsor $V^{\langle H\rangle}/S \to X_H$ under some "codimension two" assumption on the stratum. In particular, the divisor class group of X_H is isomorphic to $\mathbb{X}(Q)$.

It turns out that in many cases a Luna stratum is singular along its boundary, i.e. the stratum coincides with the smooth locus of its closure in X, see [6, Theorem 1.2], [7, Theorem 7] and [15, Theorem 8.1]. In [6], J. Kuttler and Z. Reichstein use this property to prove that quite often the Luna stratification is intrinsic. The latter means that every automorphism of X as of an abstract affine variety preserves the stratification, possibly permuting the strata. In our terms the reason for X_H to be singular along its boundary may be formulated as follows. Since the canonical quotient presentation of the normalization $\mathrm{Norm}(\overline{X_H})$ of the closure of the stratum is given in Theorem 4.10 by the quotient morphism $p\colon V^H//S \to V^H//W$ with the acting group Q, it is natural to expect that $V^{\langle H\rangle}/S$ is the preimage of the principal Luna stratum in $V^H//W$ for the Q-variety $V^H//S$. Thus the fibre of p over a point $x \in V^H//W$ is a free Q-orbit if and only if x is in X_H, and this looks like a characterization of smooth points on $V^H//W$. We realize this approach in Theorem 4.15 under the assumption that the group W is commutative.

The text is organized as follows. In Sections 4.1 and 4.2 we recall basic facts about the Luna stratification and Cox rings respectively. The class of admissible Luna strata which we are going to deal with in the theorems is defined and discussed in Section 4.3. In particular, we prove in Proposition 4.8 that if H is the isotropy group of a point with a closed G-orbit in a module V, then the Luna stratum corresponding to H and to the G-module $V^{\oplus k}$ with $k \geq 2$ is admissible. The idea to obtain strata with good properties by replacing a module V by $V^{\oplus k}$ is taken from [6]. In Section 4.4 we formulate the main result (Theorem 4.10) and obtain some corollaries. The proof of Theorem 4.10 is given in Section 4.5. Finally, Section 4.6 is devoted to examples.

4.1 The Luna stratification

In this section we recall basic facts on the Luna stratification obtained in [8], see also [14, Section 6] and [16]. Let G be a reductive affine algebraic group over an algebraically closed field $\mathbb{K}$ of characteristic zero and V be a rational finite-dimensional G-module. Denote by $\mathbb{K}[V]$ the algebra of polynomial functions on V and by $\mathbb{K}[V]^G$ the subalgebra of G-invariants. Let $V/\!/G$ be the spectrum of the algebra $\mathbb{K}[V]^G$. The inclusion $\mathbb{K}[V]^G \subseteq \mathbb{K}[V]$ gives rise to a morphism $\pi\colon V \to V/\!/G$ called the *quotient morphism* for the G-module V. It is well known that the morphism π is a categorical quotient for the action of the group G on V in the category of algebraic varieties, see [14, 4.6]. In particular, π is surjective.

The affine variety $X := V/\!/G$ is irreducible and normal. It is smooth if and only if the point $\pi(0)$ is smooth on X. In the latter case X is an affine space. Every fibre of the morphism π contains a unique closed G-orbit. For any closed G-invariant subset $A \subseteq V$ its image $\pi(A)$ is closed in X. These and other properties of the quotient morphism may be found in [14, 4.6].

By Matsushima's criterion, if an orbit $G \cdot v$ is closed in V, then the isotropy group $\mathrm{Stab}(v)$ is reductive, see [11], [12], or [14, 4.7]. Moreover, there exists a finite collection $\{H_1, \ldots, H_r\}$ of reductive subgroups in G such that if an orbit $G \cdot v$ is closed in V, then $\mathrm{Stab}(v)$ is conjugate to one of these subgroups. This implies that every fibre of the morphism π contains a point whose isotropy group coincides with some H_i.

For every isotropy group H of a closed G-orbit in V the subset

$$V_H := \left\{ \begin{array}{c} w \in V\,;\ \text{there exists}\ v \in V\ \text{such that} \\ \overline{G \cdot w} \supset G \cdot v = \overline{G \cdot v}\ \text{and}\ \mathrm{Stab}(v) = H \end{array} \right\}$$

is G-invariant and locally closed in V. The image $X_H := \pi(V_H)$ turns out to be a smooth locally closed subset of X. In particular, X_H is a smooth quasiaffine variety.

By *stratification* of a variety X we mean a decomposition of X into disjoint union of smooth locally closed subsets.

Definition 4.1 The stratification

$$X = \bigsqcup_{i=1}^{r} X_{H_i}$$

is called the *Luna stratification* of the quotient space X.

There is a unique open dense stratum called the *principal stratum* of X. A stratum X_{H_i} is contained in the closure of a stratum X_{H_j} if and only if the subgroup H_i contains a subgroup conjugate to H_j. This induces a partial ordering of the set of strata compatible with the (reverse) ordering on the set of conjugacy classes.

Consider the subsets

$$V^H = \{v \in V\,;\, H \cdot v = v\} \quad \text{and}$$
$$V^{\langle H \rangle} = \{v \in V\,;\, \mathrm{Stab}(v) = H \,\text{and}\, G \cdot v = \overline{G \cdot v}\}.$$

Then $V^{\langle H \rangle}$ is an open subset of V^H. Moreover, the restriction of the morphism π to V^H maps $V^{\langle H \rangle}$ to X_H surjectively.

Let $N_G(H)$ be the normalizer of the subgroup H in G. Since H is reductive, the subgroup $N_G(H)$ is reductive and the connected component $N_G(H)^0$ is the product of H^0 and $C_G(H)^0$, where $C_G(H)^0$ is the connected component of the centralizer $C_G(H)$ of H in G, see [10, Lemma 1.1]. The group $N_G(H)$ preserves the subspace V^H. Moreover, the kernel of the $N_G(H)$-action on V^H is H, and we obtain an effective action of the (reductive) group $W := N_G(H)/H$ on V^H. By [9], for any point $v \in V^H$ the orbit $G \cdot v$ is closed in V if and only if the orbit $W \cdot v$ is closed in V^H. In particular, $V^{\langle H \rangle}$ is the union of closed free W-orbits in V^H.

The following definition, which first appeared in [13], plays an important role in Invariant Theory.

Definition 4.2 An action of a reductive group F on an affine variety Z is *stable* if there exists an open dense subset $U \subseteq Z$ such that the orbit $F \cdot z$ is closed in Z for any $z \in U$.

Equivalently, an F-action on Z is stable if the general fibre of the quotient morphism $\pi\colon Z \to Z//F := \operatorname{Spec} \mathbb{K}[Z]^F$ is an F-orbit, i.e. there is an open dense subset $U' \subseteq Z//F$ such that $\pi^{-1}(u)$ is an F-orbit for any $u \in U'$.

Since $V^{\langle H \rangle}$ is open in V^H, the action of W on V^H is stable. This action enjoys an additional nice property, namely, the isotropy group of the general point in V^H is trivial.

Summarizing, we observe that the restriction of the quotient morphism π to $V^{\langle H \rangle}$ defines a W-torsor $\pi\colon V^{\langle H \rangle} \to X_H$.

Restricting the algebra of invariants $\mathbb{K}[V]^G$ to the subspace V^H, we get a subalgebra $\operatorname{Res}_H(\mathbb{K}[V]^G)$ in $\mathbb{K}[V^H]^W$. The closure $\overline{X_H}$ may be identified with $\operatorname{Spec} \operatorname{Res}_H(\mathbb{K}[V]^G)$. By [9], the morphism

$$\pi_H : V^H//W \to \overline{X_H}$$

given by the inclusion $\operatorname{Res}_H(\mathbb{K}[V]^G) \subseteq \mathbb{K}[V^H]^W$ is the normalization morphism. For the principal stratum, the restriction Res_H is injective, the closure $\overline{X_H} = X$ is normal, and we obtain an isomorphism $\mathbb{K}[V]^G \to \mathbb{K}[V^H]^W$. Several results on normality of the closures of Luna strata and on strata of codimension one may be found in [16] and [17].

4.2 Cox rings and characteristic spaces

Let X be a normal algebraic variety with finitely generated divisor class group $\operatorname{Cl}(X)$. Assume that any regular invertible function $f \in \mathbb{K}[X]^\times$ is constant. Roughly speaking, the *Cox ring* of X may be defined as

$$R(X) := \bigoplus_{D \in \operatorname{Cl}(X)} \Gamma(X, \mathcal{O}_X(D)).$$

In order to obtain a multiplicative structure on $R(X)$ some technical work is needed, especially when the group $\operatorname{Cl}(X)$ has torsion. We refer for details to [1, Section 4] and [5]. In a similar way one defines the *Cox sheaf* $\mathcal{R}$, which is a sheaf of $\operatorname{Cl}(X)$-graded algebras on X. The Cox ring $R(X)$ is the ring of global sections of the sheaf $\mathcal{R}$. Assume that this sheaf is locally of finite type. Then the relative spectrum $\widehat{X} := \operatorname{Spec}_X(\mathcal{R})$ is a quasiaffine variety.

By definition, a *quasitorus*[1] is a commutative reductive algebraic group. It is not difficult to check that any quasitorus is a direct product of a torus and a finite abelian group. Since $\mathcal{R}$ is a sheaf of $\operatorname{Cl}(X)$-graded

[1] Also called a diagonalizable group.

algebras, the variety $\widehat{X}$ comes with an action of a quasitorus Q_X whose group of characters $\mathbb{X}(Q_X)$ is identified with $\mathrm{Cl}(X)$.

Now we need a notion from Geometric Invariant Theory. Let G be a reductive group and Z be a normal G-variety. An invariant morphism $\pi\colon Z \to X$ is a *good quotient* if π is affine and the pullback map $\pi^*\colon \mathcal{O}_X \to \mathcal{O}_Z^G$ is an isomorphism. An example of a good quotient is the quotient morphism $\pi\colon V \to V/\!/G$ discussed in Section 4.1. Moreover, for any open subset $U \subseteq V/\!/G$ the restriction $\pi\colon \pi^{-1}(U) \to U$ is a good quotient as well.

It turns out that the Q_X-action on $\widehat{X}$ admits a good quotient $p_X : \widehat{X} \to X$. This morphism is called the *characteristic space* over X. If the variety X is smooth, then the Q_X-action on $\widehat{X}$ is free, and the map $p_X : \widehat{X} \to X$ is a Q_X-torsor. Moreover, this is a *universal torsor* over X in the sense of [18].

The following definition appeared in [1, Definition 6.4.1], see also [5, Definition 1.37].

Definition 4.3 An action of a reductive group F on an affine variety Z is said to be *strongly stable* if there exists an open dense invariant subset $U \subseteq Z$ such that

1. the complement $Z \setminus U$ is of codimension at least two in Z;
2. the group F acts freely on U;
3. for every $z \in U$ the orbit $F \cdot z$ is closed in Z.

One may check that the Q_X-action on $\widehat{X}$ defined above is strongly stable with U being the preimage $p_X^{-1}(X_{\mathrm{reg}})$ of the smooth locus of X.

Definition 4.4 A normal variety Y equipped with an action of a quasitorus Q is called *Q-factorial* if any Q-invariant Weil divisor on Y is principal.

Again the characteristic space $\widehat{X}$ is a Q_X-factorial variety, see [1, Lemma 5.3.6] and [5, Theorem 1.14].

The following theorem given in [1, Theorem 6.4.3] and [5, Theorem 1.40] shows that strong stability and Q-factoriality define exactly the class of characteristic spaces. This result is our main tool in the proof of Theorem 4.10.

Theorem 4.5 *Let a quasitorus Q act on a normal quasiaffine variety $\mathcal{X}$ with a good quotient $p\colon \mathcal{X} \to X$. Assume that $\mathbb{K}[\mathcal{X}]^\times = \mathbb{K}^\times$ holds,*

$\mathcal{X}$ is Q-factorial and the Q-action is strongly stable. Then there is a commutative diagram

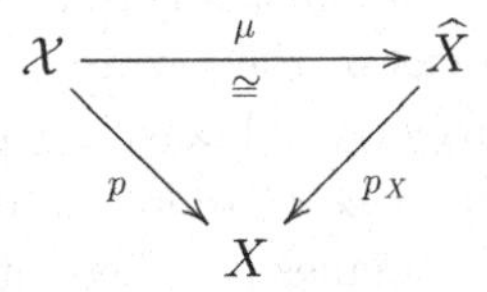

where the quotient space X is a normal variety with $\mathbb{K}[X]^\times = \mathbb{K}^\times$, we have $\mathrm{Cl}(X) = \mathbb{X}(Q)$, *the morphism $p_X : \widehat{X} \to X$ is a characteristic space over X, and the isomorphism $\mu\colon \mathcal{X} \to \widehat{X}$ is equivariant with respect to the actions of $Q = Q_X$.*

4.3 Admissible Luna strata

Let us fix settings and notation for the remaining part of the text. Let G be a reductive group, V be a rational finite-dimensional G-module, $\pi\colon V \to V/\!/G := \operatorname{Spec} \mathbb{K}[V]^G$ be the quotient morphism with the quotient space $X := V/\!/G$, and H be an isotropy group of a closed G-orbit in V. Define the group $W = N_G(H)/H$, its commutant $S := [W, W]$ and the factor group $Q := W/S$. Recall that the groups W and S are reductive. Moreover, S is semisimple provided W is connected. Being reductive and commutative, the group Q is a quasitorus.

We adapt Definition 4.3 to the situation we are dealing with.

Definition 4.6 A triple (G, V, H) is *admissible*, if the action of the group W on V^H is strongly stable. An *admissible* stratum is the Luna stratum X_H corresponding to an admissible triple (G, V, H).

Remark 4.7 A triple (G, V, H) is admissible if and only if the complement of $V^{\langle H\rangle}$ in V^H has codimension at least two. Indeed, we know from Section 4.1 that the action of W on $V^{\langle H\rangle}$ is free, all W-orbits on $V^{\langle H\rangle}$ are closed in V^H and the subset $V^{\langle H\rangle}$ is characterized by these properties. In other words, the subset $V^{\langle H\rangle}$ is the preimage of the principal Luna stratum in the W-module V^H. This shows that in order to check admissibility it suffices to work with principal strata.

Since $\overline{X_H} = \pi(V^H)$, the complement of an admissible stratum in its closure has codimension at least two. In this case the algebra of regular functions $\mathbb{K}[X_H]$ coincides with $\mathbb{K}[\mathrm{Norm}(\overline{X_H})]$, and the latter algebra is isomorphic to $\mathbb{K}[V^H]^W$, see Section 4.1. In particular, $\mathbb{K}[X_H]^\times = \mathbb{K}^\times$ holds for any admissible stratum.

Following [6] we say that a Luna stratum X_H is *singular along its boundary* if the singular locus of the closure $\overline{X_H}$ is precisely the complement $\overline{X_H} \setminus X_H$. This property is known for the principal stratum on the quotient variety of the space of representations of a quiver except for some low dimensional anomalies [7, Theorem 7]. Moreover, there are many modules where every Luna stratum is singular along its boundary, see [6, Theorem 1.2]. In particular, this is the case for the module $V^{\oplus r}$, where V is any rational G-module and $r \geq 2\dim(V)$, or where V is the adjoint module and $r \geq 3$. If the closure of a Luna stratum is normal, then "singular along its boundary" implies that the boundary has codimension at least two. This implication always holds for the principal stratum. If we know additionally that the morphism $\pi\colon V \to X$ does not contract invariant divisors[2], then the principal stratum singular along its boundary is admissible.

The following proposition provides us with a wide class of admissible strata.

Proposition 4.8 *Let V be a rational G-module and H be an isotropy group of a closed G-orbit in V. For any positive integer k consider the module $V^{\oplus k} = V \oplus \cdots \oplus V$ (k times) with the diagonal G-action. Then the subgroup H is an isotropy group of a closed G-orbit in $V^{\oplus k}$ and for $k \geq 2$ the triple $(G, V^{\oplus k}, H)$ is admissible.*

Proof Since H is an isotropy group of a closed G-orbit in V, there is a point $v \in V$ such that the orbit $G \cdot v$ is closed and the isotropy group $\mathrm{Stab}(v)$ coincides with H. The G-orbit of the point $(v, \ldots, v) \in V^{\oplus k}$ is contained in the diagonal $\Delta V \subseteq V^{\oplus k}$. Since $\Delta V \cong V$, this orbit is closed in ΔV and thus in $V^{\oplus k}$. This proves that H is an isotropy group of a closed G-orbit in $V^{\oplus k}$.

Consider the diagonal action of W on $(V^{\oplus k})^H = (V^H)^{\oplus k}$ with $k \geq 2$. We claim that if at least one component v_i of a point $v = (v_1, \ldots, v_k)$ is contained in $V^{\langle H \rangle}$, then v is contained in $(V^{\oplus k})^{\langle H \rangle}$. Indeed, assume that $v_1 \in V^{\langle H \rangle}$. Then the W-isotropy group of v_1, and thus of v, is trivial. The orbit $W \cdot v$ is contained in a closed subset $(W \cdot v_1) \times (V^H)^{\oplus(k-1)}$. If this orbit is not closed, its closure contains a W-orbit of smaller dimension. On the other hand, any point in $(W \cdot v_1) \times (V^H)^{\oplus(k-1)}$ has the trivial isotropy group.

[2] This is the case if G is semisimple.

We conclude that the complement of $(V^{\oplus k})^{\langle H\rangle}$ in $(V^{\oplus k})^H$ is contained in the set of points $(v_1, \dots, v_k)$, where every v_i is not in $V^{\langle H\rangle}$. Clearly, the latter set has codimension at least two in $(V^H)^{\oplus k}$. □

Finally let us consider the case when the group G is finite. By Remark 4.7 it suffices to know when the principal stratum is admissible. Recall that a linear operator is called a *pseudoreflection* if it has a finite order and its subspace of fixed vectors is a hyperplane.

Proposition 4.9 *Let G be a finite group and V be a G-module. Then the principal Luna stratum in X is admissible if and only if the image of G in* $\mathrm{GL}(V)$ *does not contain pseudoreflections.*

Proof Any G-orbit in V is closed, so the principal stratum is admissible if and only if the set of points in V with non-trivial isotropy groups has codimension at least two. This means that for every $g \in G$, $g \neq e$, its subspace of fixed vectors has codimension at least two, or, equivalently, g is not a pseudoreflection. □

4.4 The Cox ring of an admissible stratum

We preserve the settings of the previous section. The quasitorus Q acts on the quotient space $V^H/\!/S$ and this action defines an $\mathbb{X}(Q)$-grading on the algebra $\mathbb{K}[V^H]^S$.

Theorem 4.10 *Let G be a reductive group, V be a rational finite-dimensional G-module, and H be an isotropy group of a closed G-orbit in V. Assume that the stratum X_H is admissible. Then the Q-quotient morphism*

$$p\colon V^H/\!/S \longrightarrow V^H/\!/W \cong \mathrm{Norm}(\overline{X_H})$$

is a characteristic space over the normalization $\mathrm{Norm}(\overline{X_H})$. *Moreover, if $q\colon V^H \to V^H/\!/S$ is the S-quotient morphism and $Y := q(V^{\langle H\rangle})$, then the restriction of p to Y defines a Q-torsor*

$$p : Y \longrightarrow X_H,$$

which is a universal torsor over X_H. In particular, the divisor class group $\mathrm{Cl}(X_H)$ *may be identified with $\mathbb{X}(Q)$, and the Cox ring $R(X_H)$ is isomorphic to $\mathbb{K}[V^H]^S$ as an $\mathbb{X}(Q)$-graded ring.*

Corollary 4.11 *The Cox ring of an admissible Luna stratum is finitely generated.*

Proof Since the group S is reductive, the algebra of invariants $\mathbb{K}[V^H]^S$ is finitely generated. □

Corollary 4.12 *If a Luna stratum X_H is admissible and the group W is connected, then the group $\mathrm{Cl}(X_H)$ is free and the Cox ring $R(X_H)$ is factorial.*

Proof Since the group W is connected and reductive, the factor group Q is a torus, and the group $\mathbb{X}(Q) \cong \mathrm{Cl}(X_H)$ is a lattice. Moreover, the commutant S is semisimple and the algebra of invariants $\mathbb{K}[V^H]^S \cong R(X_H)$ is factorial, see [14, Theorem 3.17]. □

Remark 4.13 In more general settings, the Cox ring of a variety with a free divisor class group is a unique factorization domain, see. e.g. [5, Theorem 1.14].

Corollary 4.14 *Let (G, V, H) be an admissible triple. Then the divisor class group of the variety $\mathrm{Norm}(\overline{X_H})$ is isomorphic to $\mathbb{X}(Q)$, while the Picard group of $\mathrm{Norm}(\overline{X_H})$ is trivial.*

Proof The first statement follows directly from Theorem 4.10. For the second one we note that the origin in V^H is a W-fixed point which projects to a Q-fixed point on $V^H//S$. Since for the characteristic space $p\colon V^H//S \to \mathrm{Norm}(\overline{X_H})$ we have a Q-fixed point above, it follows from [5, Corollary 1.36] that the Picard group of $\mathrm{Norm}(\overline{X_H})$ is trivial. □

The following theorem shows that many admissible strata are singular along their boundary.

Theorem 4.15 *Let G be a reductive group, V be a rational finite-dimensional G-module, and H be an isotropy group of a closed G-orbit in V. Assume that the stratum X_H is admissible and the group W is commutative. Then the stratum X_H is singular along its boundary.*

Proof Under our assumptions the group S is trivial and so Q coincides with W. By Theorem 4.10, the quotient morphism $p\colon V^H \to V^H//W$ is the characteristic space over $V^H//W \cong \mathrm{Norm}(\overline{X_H})$. Since the variety V^H is smooth, a point $x \in \mathrm{Norm}(\overline{X_H})$ is smooth if and only if the fibre $p^{-1}(x)$ consists of a unique free W-orbit, see [4, Proposition 4.12]. This shows that the preimage of the smooth locus in $\mathrm{Norm}(\overline{X_H})$ coincides with $V^{\langle H\rangle}$ and so the smooth locus of $\mathrm{Norm}(\overline{X_H})$ is X_H. This implies that the smooth locus of $\overline{X_H}$ is X_H too. □

Finally let us specialize Theorem 4.10 to the case when the group G is a torus T. Any rational T-module V admits the weight decomposition

$$V = \bigoplus_{\lambda \in \mathbb{X}(T)} V(\lambda), \quad \text{where} \quad V(\lambda) = \{v \in V \,;\, t \cdot v = \lambda(t)v\}.$$

The *multiplicity* of a weight λ in V is defined as $\dim V(\lambda)$.

Proposition 4.16 *Let V be a rational T-module such that every weight in V has multiplicity at least two. Then every Luna stratum X_H is admissible, the divisor class group of X_H is a subgroup of the lattice $\mathbb{X}(T)$ and the Cox ring $R(X_H)$ is polynomial.*

Proof For any subgroup H in T which is an isotropy group of a closed G-orbit in V the group W is just T/H. Thus it is commutative, $S = \{e\}$, and $Q = W$. Since $V^H = \bigoplus V(\lambda)$ with $\lambda|_H = 1$, the multiplicity of any weight of the torus W in V^H is at least two. To get admissibility, it suffices to show that an effective stable linear W-action satisfying this condition is strongly stable. For any vector $v \in V^H$ define its weight system as the set of λ's such that the component of v in $V(\lambda)$ is non-zero. Then the isotropy group of v in W is trivial if and only if the weight system generates the lattice $\mathbb{X}(W)$. Moreover, the orbit $W \cdot v$ is closed in V^H if and only if the convex hull of the weight system contains zero in its relative interior. By assumption, these properties are fulfilled if all weight components of v are non-zero. As all weight multiplicities are at least two, the codimension of the complement of $V^{\langle H \rangle}$ in V^H is at least two as well, and the stratum X_H is admissible. Further,

$$\mathrm{Cl}(X_H) \cong \mathbb{X}(Q) = \mathbb{X}(W) = \mathbb{X}(T/H),$$

and $\mathbb{X}(T/H)$ is a subgroup of $\mathbb{X}(T)$. Finally,

$$R(X_H) \cong \mathbb{K}[V^H]^S = \mathbb{K}[V^H]$$

is a polynomial algebra. □

Remark 4.17 More generally, every Luna stratum as in Theorem 4.15 is a toric variety, so its Cox ring is polynomial [3]. The rays of the polyhedral cone defining the affine toric variety $V^H/\!/W$ are obtained via Gale duality from the weights of the quasitorus W on V^H, see [1, Chapter II, 2.1], while the stratum X_H may be recovered as the smooth locus of $V^H/\!/W$.

Corollary 4.18 *Under the assumptions of Proposition 4.16, every*

Luna stratum in $V//T$ is singular along its boundary. In particular, the Luna stratification in $V//T$ is intrinsic.

Proof The first statement follows from Theorem 4.15 and Proposition 4.16. For the second one, see [6, Lemma 3.1]. □

4.5 Proof of Theorem 4.10

We obtain the result by checking all the conditions of Theorem 4.5.

Proposition 4.19 *Let (G, V, H) be an admissible triple. Then the action of the quasitorus Q on $V^H//S$ is strongly stable.*

Proof We begin with the following general observation.

Lemma 4.20 *Let F be a reductive group, $S = [F, F]$ be its commutant and Q be the factor group F/S. If an action of the group F on an irreducible affine variety Z is strongly stable, then the induced action of Q on $Z//S$ is strongly stable as well.*

Proof Let U be an open subset of Z as in Definition 4.3. Then the group S acts on U freely and all S-orbits on U are closed in Z. Let $q: Z \to Z//S$ be the quotient morphism and $Y = q(U)$. Then Y is open in $Z//S$, $U = q^{-1}(Y)$ and since q is surjective, the complement of Y in $Z//S$ has codimension at least two. Moreover, for any $z \in U$ the image $q(F \cdot z)$ is a closed Q-orbit of the point $q(z)$ with trivial isotropy group. This proves that the group Q acts on Y freely and all Q-orbits on Y are closed in $Z//S$. □

Let us return to the proof of the proposition. By definition of an admissible triple the action of the group W on V^H is strongly stable, and Lemma 4.20 implies that the induced action of Q on $V^H//S$ is strongly stable as well. □

Proposition 4.21 *Let (G, V, H) be an admissible triple. Then the Q-variety $V^H//S$ is Q-factorial.*

Proof Take any Q-invariant Weil divisor D on $V^H//S$. Without loss of generality we may assume that D is effective. The preimage $q^*(D)$ of D under the quotient morphism $q: V^H \to V^H//S$ is a W-invariant divisor on V^H. Since V^H is an affine space, any such divisor is a principal divisor of some regular W-semiinvariant function. Such a function is S-invariant, thus it may be considered as a regular function on $V^H//S$ having D as its principal divisor. $\square$

Since $\mathbb{K}[\mathrm{Norm}(\overline{X_H})]^\times \cong \mathbb{K}[V^H//W]^\times = \mathbb{K}^\times$, we obtain the first statement of Theorem 4.10 from Theorem 4.5. The second one follows from the fact that the complement of X_H in $\mathrm{Norm}(\overline{X_H})$ has codimension at least two and [1, Lemma 5.1.2]. The proof of Theorem 4.10 is completed.

4.6 Examples

In this section we consider several examples. They illustrate Theorem 4.10 and show that the condition for a stratum to be admissible is essential.

Example 4.22 Let $G = \mathrm{SL}(2)$ and V be the tangent algebra of G considered as an adjoint G-module. This module has conjugacy classes of subgroups which are isotropy groups of closed G-orbits, namely the maximal torus T and the group $\mathrm{SL}(2)$ itself. Since V^T is an affine line and the complement of $V^{\langle T\rangle}$ in V^T is a point, the triple $(\mathrm{SL}(2), V, T)$ is not admissible. In this case the quotient space $X := V//G$ is an affine line $\mathbb{A}^1$, and the stratum X_T coincides with $\mathbb{A}^1 \setminus \{0\}$. The group $W = N_G(T)/T$ has order two, so $S = \{e\}$ and $Q = W$. The divisor class group $\mathrm{Cl}(\mathbb{A}^1 \setminus \{0\})$ is trivial, thus does not coincide with $\mathbb{X}(Q)$, while $\mathbb{K}[\mathbb{A}^1 \setminus \{0\}]^\times \neq \mathbb{K}^\times$ and the Cox ring of X_T is not defined.

Further, the principal stratum for $V \oplus V$ corresponds to $H = \{\pm E\}$. One easily checks that it is also not admissible. In this case $W = \mathrm{PSL}(2) = S$, Q is trivial, and

$$(V \oplus V)^H//S = (V \oplus V)//G \cong \mathbb{A}^3,$$

so all conclusions of Theorem 4.10 hold.

Example 4.23 Let us take an arbitrary connected semisimple group G and the adjoint G-module V. The principal stratum for V corresponds to $H = T$, where T is a maximal torus in G, and V^H is a Cartan subalgebra. In this case W is the (finite) Weyl group associated with the group G and Q is its maximal commutative factor. The principal stratum is not admissible, but starting from $k = 2$ the stratum for $V^{\oplus k}$ associated with $H = T$ is admissible. Its divisor class group does not depend on k and is isomorphic to $\mathbb{X}(Q) \cong Q$. In order to describe the Cox ring one should calculate the algebra of invariants of the diagonal action of the group $S = [W, W]$ on the direct sum of Cartan subalgebras. It is important to note that by [15, Theorem 8.1] the principal stratum for $V^{\oplus k}$ is singular along its boundary for any $k \geq 2$ provided G has no simple factors of rank 1.

Example 4.24 Take $G = \mathrm{SL}(n)$ and $V = \mathbb{K}^n \oplus (\mathbb{K}^n)^*$, where $\mathbb{K}^n$ is the tautological $\mathrm{SL}(n)$-module and $(\mathbb{K}^n)^*$ is its dual. The principal stratum for V corresponds to $H \cong \mathrm{SL}(n-1)$, and $W = Q$ is a one-dimensional torus. Since $V//G \cong \mathbb{A}^1$, the principal stratum is not admissible. But starting from $k = 2$ the subgroup H defines an admissible statum for $V^{\oplus k}$ whose divisor class group is isomorphic to $\mathbb{Z}$. By Theorem 4.15, this stratum is singular along its boundary.

Let us show that the Cox ring of an admissible Luna stratum is not always factorial, cf. Corollary 4.12. The example below is taken from [2, Section 3], where the Cox ring of the quotient space of an arbitrary finite linear group is described.

Example 4.25 Let $V = \mathbb{K}^2$ and G be the quaternion group

$$Q_8 = \left\{ \pm E, \pm \begin{pmatrix} i & 0 \\ 0 & -i \end{pmatrix}, \pm \begin{pmatrix} 0 & 1 \\ -1 & 0 \end{pmatrix}, \pm \begin{pmatrix} 0 & i \\ i & 0 \end{pmatrix} \right\},$$

where $i^2 = -1$. By Proposition 4.9, the principal Luna stratum in X corresponding to $H = \{e\}$ is admissible. As $S = [Q_8, Q_8] = \{\pm E\}$, the Cox ring of the principal stratum $\mathbb{K}[V]^S$ is the ring of functions on the two-dimensional quadratic cone. This ring is not factorial.

Note that Example 4.23 also leads to non-factorial Cox rings. This is already the case for $G = \mathrm{SL}(3)$ and $k = 2$, but the description of the Cox ring here is more complicated.

Acknowledgement

The preparation of this text was inspired by the workshop "Torsors: theory and applications", January 10–14, 2011, International Centre for Mathematical Sciences, Edinburgh, organized by Victor Batyrev and Alexei Skorobogatov. The final version was written during a stay of the author at the Institut Fourier, Grenoble. He wishes to thank these institutions for generous support and hospitality. Special thanks are due to the referee for the careful reading, useful remarks and suggestions.

References

[1] I.V. Arzhantsev, U. Derenthal, J. Hausen, and A. Laface. Cox rings. arXiv:1003.4229.

[2] I.V. Arzhantsev and S.A. Gaifullin. Cox rings, semigroups and automorphisms of affine algebraic varieties. Mat. Sbornik 201 (2010), no. 1, 3–24 (Russian); English transl.: Sbornik: Math. 201 (2010), no. 1, 1–21.

[3] D.A. Cox. The homogeneous coordinate ring of a toric variety. J. Alg. Geom. 4 (1995), no. 1, 17–50.

[4] J. Hausen. Cox rings and combinatorics II. Moscow Math. J. 8 (2008), no. 4, 711–757.

[5] J. Hausen. Three letures on Cox rings. This volume.

[6] J. Kuttler and Z. Reichstein. Is the Luna stratification intrinsic? Ann. Inst. Fourier (Grenoble) 58 (2008), no. 2, 689–721.

[7] L. Le Bruyn and C. Procesi. Semisimple representations of quivers. Trans. Amer. Math. Soc. 317 (1990). no. 2, 585–598.

[8] D. Luna. Slices étales. Bull. Soc. Math. France, Memoire 33 (1973), 81–105.

[9] D. Luna. Adhérences d'orbite et invariants. Invent. Math. 29 (1975), 231–238.

[10] D. Luna and R.W. Richardson. A generalization of the Chevalley restriction theorem. Duke Math. J. 46 (1979), no. 3, 487–496.

[11] Y. Matsushima. Espaces homogénes de Stein des groupes de Lie complexes. Nagoya Math. J. 16 (1960), 205–218.

[12] A.L. Onishchik. Complex hulls of compact homogeneous spaces. Dokl. Akad. Nauk SSSR 130 (1960), no. 4, 726–729 (Russian); English Transl.: Soviet Math. Dokl. 1 (1960), 88–91.

[13] V.L. Popov. Criteria for the stability of the action of a semisimple group on a factorial manifold. Izv. Akad. Nauk SSSR 34 (1970), 523–531 (Russian); English transl.: Math. USSR, Izv. 4 (1970), 527–535.

[14] V.L. Popov and E.B. Vinberg. Invariant Theory. Algebraic Geometry IV, Encyclopaedia Math. Sciences, vol. 55, Springer-Verlag Berlin, 1994, pp. 123–278.

[15] R.W. Richardson. Conjugacy classes of n-tuples in Lie algebras and algebraic groups. Duke Math. J. 57 (1988), no. 1, 1–35.

[16] G.W. Schwarz. Lifting smooth homotopies of orbit spaces. Inst. Hautes Études Sci. Publ. Math. 51 (1980), 37–135.

[17] D.A. Shmel'kin. On algebras of invariants and codimension 1 Luna strata for nonconnected groups. Geom. Dedicata 72 (1998), no. 2, 189–215.

[18] A.N. Skorobogatov. Torsors and rational points. Cambridge Tracts in Mathematics 144, Cambridge University Press, Cambridge, 2001.

5
Abélianisation des espaces homogènes et applications arithmétiques

Cyril Demarche
Université Pierre et Marie Curie (Paris 6)

Abstract Let X be a homogeneous space (not necessarily principal) of a connected algebraic group G (not necessarily linear) over a field K of characteristic zero. We construct a canonical abelian group $H^0_{\text{ab}}(K, X)$ defined as the hypercohomology group of a complex of semi-abelian varieties of length 3, and a natural canonical map called the abelianization map $\text{ab}^0_X : X(K) \to H^0_{\text{ab}}(K, X)$. This map generalizes Borovoi's abelianization maps for H^0 and H^1 of reductive groups. As an application, if K is a number field, we get a formula for the defects of weak and strong approximations on X in terms of the hypercohomology of an explicit complex of Galois modules.

Résumé Étant donné un espace homogène X d'un groupe algébrique connexe G (pas forcément linéaire) sur un corps K de caractéristique nulle, on construit un groupe abélien canoniquement associé à X et noté $H^0_{\text{ab}}(K, X)$, défini comme un groupe d'hypercohomologie d'un complexe de variétés semi-abéliennes de longueur 3. On dispose alors d'une application canonique, appelée application d'abélianisation, $\text{ab}^0_X : X(K) \to H^0_{\text{ab}}(K, X)$, qui est compatible aux applications d'abélianisation de Borovoi pour le H^0 et le H^1 d'un groupe réductif, et qui généralise ces applications. On obtient notamment, comme application de cette construction et dans le cas où K est un corps de nombres, une formule explicite décrivant les défauts d'approximation faible et forte sur X en termes de l'hypercohomologie d'un complexe de modules galoisiens.

5.1 Introduction

La cohomologie abélianisée des groupes réductifs sur un corps, introduite notamment par Borovoi (voir [Bor93] et [Bor98]) et Breen, est un outil important pour l'étude de l'arithmétique des groupes algébriques sur les corps locaux et les corps de nombres. Dans ce texte, on étend ces constructions à la cohomologie d'un morphisme de schémas en groupes sur une base quelconque. En particulier, on généralise certains résultats dus à Breen et Labesse dans [Lab99], à propos de la cohomologie d'un morphisme de groupes réductifs sur un corps.

Cette construction permet notamment de "dévisser" la cohomologie de certains morphismes de schémas en groupes à l'aide de la cohomologie des morphismes de schémas en groupes semi-simples simplement connexes, et de la cohomologie des complexes de schémas semi-abéliens de longueur 3.

L'une des motivations de cette construction générale est l'arithmétique des espaces homogènes de groupes algébriques sur les corps de nombres. Cela fournit en effet un cadre naturel pour déduire certaines propriétés arithmétiques des espaces homogènes à partir de propriétés similaires pour les variétés semi-abéliennes (les théorèmes de dualité pour les 1-motifs par exemple) et pour les groupes semi-simples simplement connexes (le théorème d'approximation forte par exemple).

Dans ce texte, on présente ainsi deux applications de cette cohomologie abélianisée des morphismes de groupes algébriques. On calcule le défaut d'approximation forte et le défaut d'approximation faible sur un espace homogène (d'un groupe connexe, à stabilisateurs connexes ou linéaires commutatifs) sur un corps de nombres, en termes de l'hypercohomologie galoisienne d'un complexe de modules galoisiens explicites.

Citons par exemple l'application suivante (voir corollaire 5.37). Pour cela, définissons quelques notations : si K est un corps de caractéristique nulle, et G/K est un groupe algébrique connexe, on note G^{lin} le sous-groupe linéaire maximal de G, G^{ab} la variété abélienne quotient G/G^{lin}, G^{u} le radical unipotent de G^{lin}, $G^{\mathrm{red}} := G^{\mathrm{lin}}/G^{\mathrm{u}}$, G^{ss} le sous-groupe dérivé de G^{red} et G^{sc} le revêtement simplement connexe de G^{ss}; enfin, G^{scu} désigne le produit fibré de G^{u} et G^{sc} au-dessus de G^{red}, et $\rho : G^{\mathrm{scu}} \to G$ le morphisme naturel. Pour finir, pour un groupe topologique A, on note $A^D := \mathrm{Hom}_{\mathrm{cont.}}(A, \mathbf{Q}/\mathbf{Z})$. Le résultat suivant donne une formule explicite pour le défaut d'approximation forte sur un espace homogène :

Théorème (Corollaire 5.37) *Soit k un corps de nombres, G un k-groupe connexe, S_0 un ensemble fini de places de k. Soit H un sous-k-groupe linéaire connexe de G, et soit $X := G/H$. On suppose que le* *groupe G^{sc} vérifie l'approximation forte hors de S_0 et que $Ш^1(G^{ab})$ est fini.*

On note $P^0(k,X) := \prod_{v\in\Omega_\infty} \pi_0(X(k_v)) \times \prod'_{v\in\Omega_f} X(k_v)$ où $\pi_0(X(k_v))$ désigne l'ensemble des composantes connexes de $X(k_v)$, Ω_∞ (resp. Ω_f) l'ensemble des places infinies (resp. finies) de k. On note $\widehat{C}_X$ le complexe $\widehat{C}_X := [\widehat{T}_{G^{\mathrm{lin}}} \to (G^{ab})^* \oplus \widehat{T}_H \oplus \widehat{T}_{G^{sc}} \to \widehat{T}_{H^{sc}}]$, où $\widehat{T}_F$ désigne le module des caractères d'un tore maximal T_F d'un k-groupe linéaire F.

Alors il existe une application naturelle

$$P^0(k,X) \xrightarrow{\theta} \left(\mathbf{H}^0(k,\widehat{C}_X)/\ Ш^0(\widehat{C}_X)\right)^D,$$

dont le noyau est exactement l'adhérence de $\left(\prod_{v\in S_0} \rho(G^{scu}(k_v))\right).X(k)$ dans $P^0(k,X)$. En outre, θ est surjective si le groupe G est linéaire.

Plus généralement, on obtient la description des défauts d'approximation faible et forte pour la cohomologie d'un morphisme de groupes algébriques connexes quelconque, ne correspondant pas nécessairement à un espace homogène (cas où le morphisme est injectif). On obtient également un résultat analogue dans le cas où le groupe H est supposé linéaire commutatif (et non plus connexe) : voir corollaire 5.44.

Le plan du texte est le suivant : la section 5.2 est consacrée à des rappels sur l'abélianisation de la cohomologie des groupes réductifs et à la généralisation aux groupes non linéaires. Dans les sections 5.3 et 5.4, on construit l'application d'abélianisation pour un morphisme de schémas en groupes sur une base quelconque, puis on s'intéresse au cas particulier d'un anneau d'entiers de corps de nombres à la section 5.5. La section 5.6 traite des théorèmes de dualité pour certains complexes de variétés semi-abéliennes. Les section 5.7 et 5.8 sont consacrées aux applications arithmétiques aux espaces homogènes.

Remerciements : je remercie très chaleureusement David Harari pour ses précieux commentaires. Je remercie également Philippe Gille pour certaines suggestions intéressantes, ainsi que le rapporteur pour ses remarques et commentaires.

5.2 Applications d'abélianisation pour les groupes connexes sur un corps

Dans tout le texte, K désigne un corps de caractéristique nulle.

L'objectif de cette section est de généraliser les constructions de Borovoi dans [Bor98] et [Bor93] pour la cohomologie galoisienne, de degré 0 et 1 des groupes réductifs sur K. Si G/K est un groupe algébrique connexe, on note G^{lin} le sous-groupe linéaire maximal de G, G^{ab} la variété abélienne quotient G/G^{lin}, G^{u} le radical unipotent de G^{lin}, $G^{\mathrm{red}} := G^{\mathrm{lin}}/G^{\mathrm{u}}$, G^{ss} le sous-groupe dérivé de G^{red} et G^{sc} le revêtement simplement connexe de G^{ss}.

Rappelons la construction pour un groupe réductif G. On dispose d'un morphisme naturel $\rho : G^{\mathrm{sc}} \to G$. On note Z le centre de G, Z^{sc} celui de G^{sc}, et T un K-tore maximal de G. On pose $T^{\mathrm{sc}} := \rho^{-1}(T)$, qui est un tore maximal de G^{sc}.

Deligne a montré (voir [Del79], 2.0.2) que le morphisme $[G^{\mathrm{sc}} \to G]$ définit un module croisé de Picard. On dispose des morphismes naturels suivants entre modules croisés :

$$\begin{array}{ccccc} Z^{\mathrm{sc}} & \longrightarrow & T^{\mathrm{sc}} & \longrightarrow & G^{\mathrm{sc}} \\ \downarrow{\scriptstyle\rho} & & \downarrow{\scriptstyle\rho} & & \downarrow{\scriptstyle\rho} \\ Z & \longrightarrow & T & \longrightarrow & G. \end{array}$$

Le lemme 3.8.1 de [Bor98] assure que ces morphismes de modules croisés sont des quasi-isomorphismes. Or on dispose d'un morphisme de modules croisés évident

$$G \to [G^{\mathrm{sc}} \to G]$$

duquel on déduit des applications en hypercohomologie :

$$\mathrm{ab}^i_G : H^i(K, G) \to \mathbf{H}^i(K, [T^{\mathrm{sc}} \to T]) =: H^i_{\mathrm{ab}}(K, G)$$

pour $i = 0, 1$. Étendons ces constructions à des K-groupes algébriques connexes (pas nécessairement linéaires).

Soit G un K-groupe algébrique connexe.

Dans [Dem11c], section 4.1.1, on a construit, sous l'hypothèse que G^{lin} soit réductif, une sous-variété semi-abélienne maximale SA_G de G, qui est une extension de G^{ab} par T_G. Cette construction utilise par exemple un théorème de Rosenlicht (voir [Ros56], corollaire 3 du théorème 12) qui assure l'existence de $D \subset G$ k-sous-groupe distingué connexe minimal tel que $G_{\mathrm{lin}} := G/D$ soit linéaire.

On définit le complexe de variétés semi-abéliennes

$$C_G := [T_G^{\mathrm{sc}} \to \mathrm{SA}_G].$$

Ce complexe généralise le complexe de tores $[T_G^{\mathrm{sc}} \to T_G]$ pour les groupes non linéaires. Par exemple, si G est réductif, on retrouve $C_G = [T_G^{\mathrm{sc}} \to T_G]$. Si G est une variété semi-abélienne, alors $C_G = [0 \to G]$ peut être considéré comme un 1-motif.

5.2.1 Application d'abélianisation en degré 0 et 1

Dans ce paragraphe, on construit les applications ab_G^0 et ab_G^1. On suppose d'abord G^{lin} réductif.

Le théorème 3.5.3 de [Bor98] et le lemme 4.3 de [Dem11c] assurent que les morphismes naturels

$$\mathbf{H}^i(K, [Z_G^{\mathrm{sc}} \to Z_G]) \to \mathbf{H}^i(K, [T_G^{\mathrm{sc}} \to \mathrm{SA}_G]) \to \mathbf{H}^i(K, [G^{\mathrm{sc}} \to G])$$

sont des isomorphismes de groupes pour $i = 0$ et 1. Ainsi le morphisme naturel de modules croisés $[1 \to G] \to [G^{\mathrm{sc}} \to G]$ induit-il des applications

$$\mathrm{ab}_G^0 : G(K) \to H_{\mathrm{ab}}^0(K, G) := \mathbf{H}^0(K, C_G)$$

$$\mathrm{ab}_G^1 : H^1(K, G) \to H_{\mathrm{ab}}^1(K, G) := \mathbf{H}^1(K, C_G)\,.$$

Exemples :

- Si G est réductif, on retrouve exactement les applications de Borovoi (cf [Bor98], 3.10).
- Si G est une variété semi-abélienne, alors $H_{\mathrm{ab}}^i(K, G) = H^i(K, G)$ et ab_G^i est l'identité.

Généralisons cette construction au cas où G^{lin} n'est pas réductif. Soit G^{u} le radical unipotent de G^{lin}, notons $G^{\mathrm{red}} := G^{\mathrm{lin}}/G^{\mathrm{u}}$ et $G' := G/G^{\mathrm{u}}$. On a un diagramme commutatif exact de groupes algébriques :

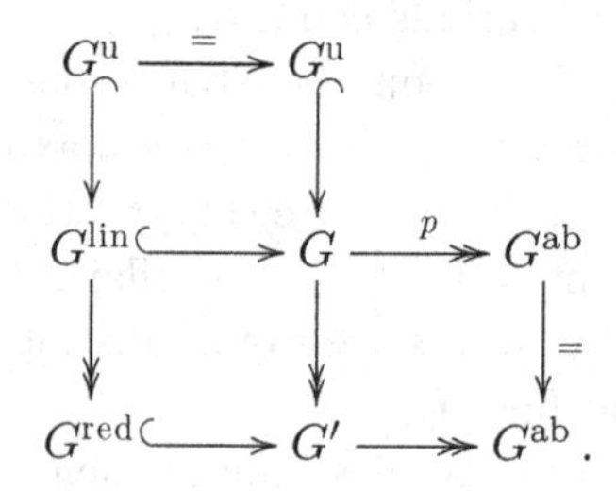

Alors $G'^{\mathrm{lin}} = G^{\mathrm{red}}$ est réductif, donc on peut définir $C_{G'}$ et $\mathrm{ab}_{G'}^i : H^i(K, G') \to H_{\mathrm{ab}}^i(K, G')$. On définit alors $\mathrm{ab}_G^i : H^i(K, G) \to H_{\mathrm{ab}}^i(K, G)$ comme la composée :

$$H^i(K, G) \to H^i(K, G') \xrightarrow{\mathrm{ab}_{G'}^i} H_{\mathrm{ab}}^i(K, G') =: H_{\mathrm{ab}}^i(K, G)\,.$$

5.2.2 Propriété des applications d'abélianisation

Les flèches ab_G^i définies à la section précédente vérifient les mêmes propriétés que dans le cas des groupes réductifs. En particulier, les applications ab_G^i sont fonctorielles en G et indépendantes du tore maximal choisi. Plus généralement, presque toutes les propriétés démontrées dans [Bor98] restent valable pour un groupe connexe G quelconque. Montrons par exemple le résultat suivant, où $\Omega_{\mathbb{R}}$ désigne l'ensemble des places réelles de k :

Proposition 5.1 *Soit k un corps de nombres, G un k-groupe algébrique connexe. Alors le diagramme commutatif suivant*

$$\begin{array}{ccc} H^1(k,G) & \xrightarrow{\mathrm{ab}_G^1} & H^1_{\mathrm{ab}}(k,G) \\ \downarrow{\scriptstyle \mathrm{loc}_\infty} & & \downarrow{\scriptstyle \mathrm{loc}'_\infty} \\ \prod_{v\in\Omega_{\mathbb{R}}} H^1(k_v,G) & \xrightarrow{\prod_{v\in\Omega_{\mathbb{R}}} \mathrm{ab}_{G_v}^1} & \prod_{v\in\Omega_{\mathbb{R}}} H^1_{\mathrm{ab}}(k_v,G) \end{array}$$

est cartésien.

Démonstration C'est un dévissage à partir du cas linéaire traité par Borovoi (théorème 5.11 de [Bor98]) : par définition du sous-groupe $D \subset G$, le sous-groupe $D \cap G^{\mathrm{ss}} \subset G^{\mathrm{ss}}$ est central, donc fini. Donc le morphisme naturel $G \to G/D$ induit un isomorphisme naturel entre G^{sc} et le revêtement semi-simple simplement connexe de G_{lin}. Par conséquent, le diagramme (18) de [Dem11c] assure que la suite exacte de k-groupes $1 \to D \to G \to G_{\mathrm{lin}} \to 1$ induit une suite exacte de complexes

$$0 \to D \to C_G \to C_{G_{\mathrm{lin}}} \to 0\,.$$

On considère le diagramme commutatif de la figure 5.1 ci-dessous, dont les colonnes sont exactes (on rappelle que D est central). À l'aide des théorèmes 5.11 de [Bor98] appliqué à G_{lin} (qui assure que la quatrième ligne est exacte), et en utilisant l'injectivité de la dernière flèche horizontale ainsi que la surjectivité du morphisme $H^0_{\mathrm{ab}}(k, G_{\mathrm{lin}}) \to \prod_{v\in\Omega_{\mathbb{R}}} \widehat{H}^0_{\mathrm{ab}}(k_v, G_{\mathrm{lin}})$ (conséquence de l'approximation faible aux places réelles : voir théorème 5.34 plus bas), on conclut par une chasse au diagramme. □

Remarque 5.2 En revanche, contrairement au cas des groupes linéaires,

$$\begin{array}{ccccc}
H^0(k, G_{\mathrm{lin}}) & \to & H^0_{\mathrm{ab}}(k, G_{\mathrm{lin}}) \times \prod_{v \in \Omega_{\mathbb{R}}} \widehat{H}^0(k_v, G_{\mathrm{lin}}) & \to & \prod_{v \in \Omega_{\mathbb{R}}} \widehat{H}^0_{\mathrm{ab}}(k_v, G_{\mathrm{lin}}) \\
\downarrow & & \downarrow \qquad\qquad \downarrow & & \downarrow \\
H^1(k, D) & \to & H^1(k, D) \times \prod_{v \in \Omega_{\mathbb{R}}} H^1(k_v, D) & \to & \prod_{v \in \Omega_{\mathbb{R}}} H^1(k_v, D) \\
\downarrow & & \downarrow \qquad\qquad \downarrow & & \downarrow \\
H^1(k, G) & \to & H^1_{\mathrm{ab}}(k, G) \times \prod_{v \in \Omega_{\mathbb{R}}} H^1(k_v, G) & \to & \prod_{v \in \Omega_{\mathbb{R}}} H^1_{\mathrm{ab}}(k_v, G) \\
\downarrow & & \downarrow \qquad\qquad \downarrow & & \downarrow \\
H^1(k, G_{\mathrm{lin}}) & \to & H^1_{\mathrm{ab}}(k, G_{\mathrm{lin}}) \times \prod_{v \in \Omega_{\mathbb{R}}} H^1(k_v, G_{\mathrm{lin}}) & \to & \prod_{v \in \Omega_{\mathbb{R}}} H^1_{\mathrm{ab}}(k_v, G_{\mathrm{lin}}) \\
\downarrow & & \downarrow \qquad\qquad \downarrow & & \\
H^2(k, D) & \to & H^2(k, D) \times \prod_{v \in \Omega_{\mathbb{R}}} H^2(k_v, D) & &
\end{array}$$

Figure 5.1

l'application

$$H^1(k, G) \to H^1_{\mathrm{ab}}(k, G) \times \prod_v H^1(k_v, G)$$

n'est pas injective en général. Voir par exemple la proposition 3.16 de [BCTS08].

5.3 Application d'abélianisation pour les morphismes de schémas en groupes connexes

L'objectif de cette section est de généraliser les flèches d'abélianisation (pour le degré 0 uniquement) au cas des morphismes de schémas en groupes connexes sur une base quelconque.

Notons d'abord que Labesse et Breen ont donné dans [Lab99], chapitre 1 et appendice B, une définition de la cohomologie galoisienne d'un morphisme de modules croisés et d'une application d'abélianisation pour un complexe de groupes réductifs sur un corps (voir [Lab99], section 1.8). Leur construction fait appel à la notion d'ensemble croisé (qui est plus générale que celle de morphisme de modules croisés), et leur définition est exprimée essentiellement en termes de cocycles.

Pour définir la cohomologie d'un morphisme de modules croisés, on ne fait pas ici appel à la notion d'ensemble croisé, mais seulement à la notion plus commune de module croisé, et au lien de cette notion avec la notion de gr-champ. Cela permet de définir les ensembles de cohomologie souhaités en termes de torseurs. Cette définition "géométrique" généralise naturellement la construction cocyclique de Labesse et Breen. On obtient ainsi par exemple une généralisation de l'application d'abélianisation de [Lab99], section 1.8, pour un morphisme de groupes réductifs.

Notons au passage que la construction et les propriétés des applications d'abélianisation de Borovoi pour les schémas en groupes réductifs ont récemment été étendues au cas d'un schéma de base quelconque par Gonzalez-Aviles (voir [GA11]), avec des techniques analogues à celles que l'on utilise dans la suite.

5.3.1 Torseurs sous un gr-champ

Pour plus de détails sur le sujet, on pourra consulter [Bre90], [Bre94] ou les sections 5.1 et 6.1 de [Ald08]. Soient S un schéma muni du site fppf ou du site étale (dans les applications, on se limitera essentiellement au site étale sur $S = \mathrm{Spec}(K)$). Soit $\mathscr{M}$ un gr-champ sur S (voir [Bre94], sections 1.1 et 1.2 pour la notion de champ et de gr-champ).
Exemple : un exemple fondamental de gr-champ, en vue de l'application à la cohomologie des morphismes de schémas en groupes, est le suivant : étant donné un module croisé $M := [\alpha : F \to G]$ sur S, on peut lui associer un gr-champ $\mathscr{M}$, défini par Breen (voir [Bre90], section 4.4), de manière fonctorielle en M.

On note $\otimes_{\mathscr{M}} : \mathscr{M} \times \mathscr{M} \to \mathscr{M}$ le foncteur définissant la loi de groupe sur $\mathscr{M}$, et $H^0(S, \mathscr{M})$ le groupe des sections globales de $\mathscr{M}$ modulo transformation.

On peut définir la notion de torseur sous $\mathscr{M}$ (voir [Bre90], section 6.1) : un champ en groupoïdes $\mathscr{P}$ sur S est un $\mathscr{M}$-torseur à droite s'il est muni d'un morphisme de champs

$$m : \mathscr{P} \times \mathscr{M} \to \mathscr{P}$$

compatible avec la structure de groupe sur $\mathscr{M}$, et de sorte que le morphisme

$$(\mathrm{pr}_{\mathscr{P}}, m) : \mathscr{P} \times \mathscr{M} \to \mathscr{P} \times \mathscr{P}$$

soit une équivalence. On demande également que cette action de $\mathscr{M}$ sur $\mathscr{P}$ vérifie des conditions de compatibilité naturelles (voir [Bre90], section 6.1), et qu'il existe un recouvrement $(U_i \to S)_{i\in I}$ de S tel que les catégories fibres $\mathscr{P}_{U_i}$ soient non vides. On dispose également de la notion naturelle de morphisme de $\mathscr{M}$-torseurs. On notera $\mathrm{TORS}(\mathscr{M})$ la catégorie des torseurs à droite sous $\mathscr{M}$, et $H^1(S,\mathscr{M})$ l'ensemble pointé des classes d'isomorphisme de $\mathscr{M}$-torseurs.

On dispose de la notion de produit contracté pour des torseurs sous des gr-champs (voir [Bre90], section 6.7) : étant donnés un champ en groupoïdes $\mathscr{P}$ muni d'une action à droite de $\mathscr{M}$ et un champ en groupoïdes $\mathscr{Q}$ muni d'une action à gauche de $\mathscr{M}$, on peut construire un champ en groupoïdes produit contracté, que l'on note $\mathscr{P}\wedge^{\mathscr{M}}\mathscr{Q}$. En particulier, si $\varphi:\mathscr{M}\to\mathscr{N}$ est un morphisme de gr-champs, le produit contracté induit un foncteur φ_*

$$\begin{array}{ccc} \mathrm{TORS}(\mathscr{M}) & \longrightarrow & \mathrm{TORS}(\mathscr{N}) \\ \mathscr{P} & \longmapsto & \varphi_*(\mathscr{P}) := \mathscr{P}\wedge^{\mathscr{M}}\mathscr{N} \end{array}$$

où $\mathscr{N}$ désigne le champ en groupoïdes $\mathscr{N}$ munit de l'action de $\mathscr{M}$ induite par φ.

Si $\mathscr{P}$ est un torseur à droite sous $\mathscr{M}$, on appelle trivialisation de $\mathscr{P}$ un morphisme de $\mathscr{M}$-torseurs $\sigma:\mathscr{P}\xrightarrow{\simeq}\mathscr{M}$, où $\mathscr{M}$ est considéré comme le torseur trivial sous $\mathscr{M}$.

Enfin, étant donnés $\mathscr{P}$ un torseur à droite sous $\mathscr{M}$ et U un recouvrement de S trivialisant $\mathscr{P}$, on peut leur associer un cocycle à valeurs dans $\mathscr{M}$ (voir par exemple [Ald08], section 5.1.2).

Désormais, on s'intéresse à des torseurs "sous un morphisme de gr-champs" : soit

$$f:\mathscr{M}\to\mathscr{N}$$

un morphisme de gr-champs, on définit suivant [Ald08], section 6.1, un $(\mathscr{M},\mathscr{N})$-torseur comme une paire $(\mathscr{P},\sigma)$ où $\mathscr{P}$ est un $\mathscr{M}$-torseur et $\sigma:\mathscr{P}\wedge^{\mathscr{M}}\mathscr{N}\xrightarrow{\simeq}\mathscr{N}$ est une trivialisation du $\mathscr{N}$-torseur $\mathscr{P}\wedge^{\mathscr{M}}\mathscr{N} = f_*(\mathscr{P})$, que l'on peut voir de manière équivalente comme un morphisme $\mathscr{M}$-équivariant $\mathscr{P}\to\mathscr{N}$. On notera $\mathrm{TORS}(\mathscr{M},\mathscr{N})$ la catégorie des $(\mathscr{M},\mathscr{N})$-torseurs.

À nouveau, à un tel torseur, on peut associer une classe d'équivalence de 1-cocycles, comme dans la section 6.1 de [Ald08].

On note $\mathbf{H}^0(K,[\mathscr{M}\xrightarrow{f}\mathscr{N}])$ l'ensemble pointé des classes d'isomorphisme de $(\mathscr{M},\mathscr{N})$-torseurs.

Lemme 5.3 *Soit $f:\mathscr{M}\to\mathscr{N}$ un morphisme de gr-champs. Alors on* *a une suite exacte naturelle d'ensembles pointés*

$$H^0(S,\mathscr{M}) \xrightarrow{f_*} H^0(S,\mathscr{N}) \to \mathbf{H}^0(S,[\mathscr{M} \xrightarrow{f} \mathscr{N}]) \to H^1(S,\mathscr{M}) \xrightarrow{f_*} H^1(S,\mathscr{N}).$$

Démonstration Les flèches sans nom sont définies de la façon suivante : l'application $H^0(S,\mathscr{N}) \to \mathbf{H}^0(S,[\mathscr{M} \xrightarrow{f} \mathscr{N}])$ envoie la classe d'une section globale $g \in H^0(S,\mathscr{N})$ sur la classe de la paire $(\mathscr{M},\sigma_g)$, où $\mathscr{M}$ est le $\mathscr{M}$-torseur trivial et σ_g est la trivialisation de $\mathscr{M} \wedge^{\mathscr{M}} \mathscr{N} = \mathscr{N}$ définie par la section g (i.e. σ_g est la translation par g sur $\mathscr{N}$). La flèche $\mathbf{H}^0(S,[\mathscr{M} \xrightarrow{f} \mathscr{N}]) \to H^1(S,\mathscr{M})$ est la flèche naturelle d'oubli de la section, qui envoie la classe d'un $(\mathscr{M},\mathscr{N})$-torseur $(\mathscr{P},\sigma)$ sur la classe du $\mathscr{M}$-torseur $\mathscr{P}$. Avec ces définitions, l'exactitude de la suite du lemme est immédiate. □

Remarque 5.4 En général, les trois derniers ensembles de cette suite ne sont pas des groupes. Cependant, on dispose toujours d'une action (à gauche) du groupe $H^0(S,\mathscr{N})$ sur l'ensemble $\mathbf{H}^0(S,[\mathscr{M} \xrightarrow{f} \mathscr{N}])$, définie de la façon suivante : si g est une section globale de $\mathscr{N}$, g définit un isomorphisme du $\mathscr{N}$-torseur trivial $\sigma_g : \mathscr{N} \to \mathscr{N}$ (par translation à gauche par g). Donc pour tout $(\mathscr{M},\mathscr{N})$-torseur $(\mathscr{P},\sigma)$, on définit un $(\mathscr{M},\mathscr{N})$-torseur $(\mathscr{Q},\tau) := g.(\mathscr{P},\sigma)$ où $\mathscr{Q} := \mathscr{P}$ et $\tau : \mathscr{P} \to \mathscr{N}$ est la composée $\mathscr{P} \xrightarrow{\sigma} \mathscr{N} \xrightarrow{\sigma_g} \mathscr{N}$. Cela permet de définir une action à gauche du groupe $H^0(S,\mathscr{N})$ sur l'ensemble $\mathbf{H}^0(S,[\mathscr{M} \xrightarrow{f} \mathscr{N}])$.

5.3.2 Torseurs sous un gr-champ tressé

Dans cette sous-section, on s'intéresse à des $(\mathscr{M},\mathscr{N})$-torseurs où le morphisme $f : \mathscr{M} \to \mathscr{N}$ est un morphisme de gr-champs tressés ("braided gr-stacks" en anglais), suivant la terminologie de [Bre94], section 1.8, ou [Bre92] (intuitivement, un gr-champ tressé est un gr-champ vérifiant une certaine condition de commutativité). Pour de tels gr-champs, le morphisme naturel :

$$\otimes_{\mathscr{M}} : \mathscr{M} \times \mathscr{M} \to \mathscr{M}$$

est un morphisme de gr-champs, c'est-à-dire que c'est un morphisme additif (voir [Bre94], 2.13, ou [AN09], 7.1.2). Par conséquent, on peut pousser en avant un torseur sous $\mathscr{M} \times \mathscr{M}$ pour obtenir un $\mathscr{M}$-torseur, via le foncteur

$$(\otimes_{\mathscr{M}})_* : \mathrm{TORS}(\mathscr{M} \times \mathscr{M}) \to \mathrm{TORS}(\mathscr{M}).$$

On peut donc ainsi définir une "gr-structure" sur TORS($\mathscr{M}$) par la formule suivante (voir [Bre94], 2.13) : si $\mathscr{P}$ et $\mathscr{Q}$ sont des $\mathscr{M}$-torseurs (à droite), on pose

$$\mathscr{P}.\mathscr{Q} := (\otimes_{\mathscr{M}})_* (\mathscr{P} \times \mathscr{Q})$$

où $\mathscr{P} \times \mathscr{Q}$ est muni de sa structure naturelle de $\mathscr{M} \times \mathscr{M}$-torseur. En particulier, cette opération définit une structure de groupe sur l'ensemble $H^1(S, \mathscr{M})$. De même, cette opération munit l'ensemble $\mathbf{H}^0(S, [\mathscr{M} \xrightarrow{f} \mathscr{N}])$ d'une structure de groupe. En effet, si $(\mathscr{P}, \sigma)$, $(\mathscr{Q}, \tau)$ sont deux $(\mathscr{M}, \mathscr{N})$-torseurs, on définit leur produit comme le $(\mathscr{M}, \mathscr{N})$-torseur $(\mathscr{P}.\mathscr{Q}, \sigma.\tau)$ où la trivialisation $\sigma.\tau$ est définie de la façon suivante : on dispose du morphisme $\sigma \times \tau : \mathscr{P} \times \mathscr{Q} \to \mathscr{N} \times \mathscr{N}$. On en déduit un morphisme

$$\sigma.\tau : (\otimes_{\mathscr{M}})_* (\mathscr{P} \times \mathscr{Q}) \to (\otimes_{\mathscr{N}})_* (\mathscr{N} \times \mathscr{N}) = \mathscr{N}$$

qui est bien une trivialisation de $\mathscr{P}.\mathscr{Q}$.

On vérifie que dans ce cas, la suite exacte du lemme 5.3 est une suite exacte de groupes.

Si de plus les gr-champs $\mathscr{M}$ et $\mathscr{N}$ sont de Picard (au sens de [Bre94], section 1.8), alors le groupe $\mathbf{H}^0(S, [\mathscr{M} \xrightarrow{f} \mathscr{N}])$ est un groupe abélien.

Exemple : Soit $S = \operatorname{Spec} K$. Soit $f : H \to G$ un morphisme de groupes algébriques connexes sur K. Le module croisé $[H^{\mathrm{sc}} \to H]$ (resp. $[G^{\mathrm{sc}} \to G]$) est alors de Picard (ou stable, selon la terminologie de Conduché dans [Con84], 3.11), et donc les gr-champs $\mathscr{M}_H$ et $\mathscr{M}_G$ associés à ces modules croisés (voir [Bre90], section 4.4) sont de Picard. Par conséquent, l'ensemble $\mathbf{H}^0(K, [\mathscr{M}_H \to \mathscr{M}_G])$ est canoniquement un groupe abélien.

Proposition 5.5 *Soit*

$$\begin{array}{ccc} M_1 & \xrightarrow{\phi_1} & M_1' \\ \downarrow f & & \downarrow f' \\ M_2 & \xrightarrow{\phi_2} & M_2' \end{array}$$

un diagramme commutatif de morphismes de S-modules croisés tressés. On note $\mathscr{M}_i$ (resp. $\mathscr{M}_i'$) le gr-champ associé au module croisé M_i. Si ϕ_1 et ϕ_2 sont des quasi-isomorphismes, le morphisme induit

$$\phi_* : \mathbf{H}^0(S, [\mathscr{M}_1 \to \mathscr{M}_2]) \to \mathbf{H}^0(S, [\mathscr{M}_1' \to \mathscr{M}_2'])$$

est un isomorphisme de groupes.

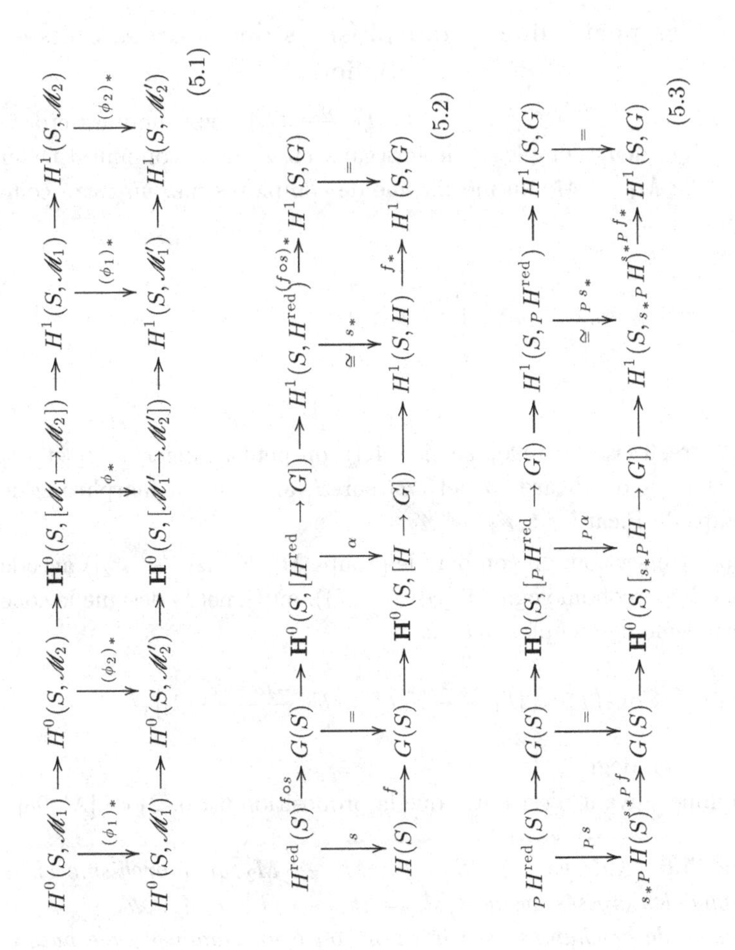

Figure 5.2 Diagrammes (5.1), (5.2) et (5.3).

Démonstration Considérons le diagramme (5.1) (voir figure 5.2) qui est un diagramme commutatif exact de groupes, dont les lignes proviennent du lemme 5.3.

On sait que les deux premiers et les deux derniers morphismes verticaux sont des isomorphismes (on peut par exemple identifier les groupes en question avec les groupes d'hypercohomologie des modules croisés correspondant, pour lesquels le résultat est connu : voir par exemple le théorème 3.5.3 de [Bor98] dans le cas des corps; on peut aussi appliquer le lemme 5.4.1 de [AN09]). Donc par le lemme des cinq, la flèche centrale est un isomorphisme, ce qui conclut la preuve. □

5.3.3 Cas particuliers : morphismes de modules croisés abéliens

Soient $M_1 = [F_1 \xrightarrow{\alpha_1} G_1]$ et $M_2 = [F_2 \xrightarrow{\alpha_2} G_2]$ deux modules croisés abéliens (i.e. deux complexes de faisceaux en groupes commutatifs sur S). Soit $f : M_1 \to M_2$ un morphisme de complexes, i.e. un carré commutatif

$$\begin{array}{ccc} F_1 & \xrightarrow{f_F} & F_2 \\ \downarrow{\scriptstyle \alpha_1} & & \downarrow{\scriptstyle \alpha_2} \\ G_1 & \xrightarrow{f_G} & G_2 . \end{array}$$

Suivant Breen (voir [Bre90], section 4.4), on peut associer à M_1 et M_2 des gr-champs de Picard $\mathscr{M}_1$ et $\mathscr{M}_2$ sur S, ainsi qu'un morphisme de gr-champs de Picard $f : \mathscr{M}_1 \to \mathscr{M}_2$.

Dans cette section, on compare le groupe $\mathbf{H}^0(S, [\mathscr{M}_1 \xrightarrow{f} \mathscr{M}_2])$ avec le groupe d'hypercohomologie $\mathbf{H}^0(S, \text{Cône}(f))$, où $\text{Cône}(f)$ désigne le cône du morphisme de complexes f, i.e.

$$\text{Cône}(f) := [F_1 \xrightarrow{\alpha_1 \oplus f_F} G_1 \oplus F_2 \xrightarrow{-f_G + \alpha_2} G_2] ,$$

avec F_1 est en degré -2.

Le lemme suivant n'est autre que la proposition 6.1.6.(2) de [Ald08] :

Lemme 5.6 (Aldrovandi) *Soit $f : M_1 \to M_2$ un morphisme entre des S-modules croisés abéliens $M_i := [F_i \xrightarrow{\alpha_i} G_i]$, et $f : \mathscr{M}_1 \to \mathscr{M}_2$ le morphisme de gr-champs associé. Alors on a un isomorphisme naturel de groupes abéliens*

$$\Lambda : \mathbf{H}^0(S, [\mathscr{M}_1 \xrightarrow{f} \mathscr{M}_2]) \cong \mathbf{H}^0(S, \text{Cône}(f)) .$$

Démonstration La bijection est construite dans [Ald08]. Il suffit donc de vérifier que c'est un morphisme de groupes. On peut le faire en utilisant les formules explicites de [AN10], section 7.4 dans le cas des modules croisés abéliens. □

Montrons un résultat de compatibilité :

Proposition 5.7 *Soit $f : M_1 \to M_2$ un morphisme de modules croisés abéliens, et $f : \mathscr{M}_1 \to \mathscr{M}_2$ le morphisme de gr-champs associé. Le dia-*

gramme suivant

$$\begin{array}{ccccc} H^0(S,\mathscr{M}_2) & \xrightarrow{\mathscr{T}} & \mathbf{H}^0(S,[\mathscr{M}_1 \to \mathscr{M}_2]) & \xrightarrow{\Delta} & H^1(S,\mathscr{M}_1) \\ \Lambda_2 \downarrow \cong & & \Lambda \downarrow \cong & & \Lambda_1 \downarrow \cong \\ \mathbf{H}^0(S,M_2) & \xrightarrow{T} & \mathbf{H}^0(S,\text{Cône}(f)) & \xrightarrow{\delta} & \mathbf{H}^1(S,M_1) \end{array}$$

est un diagramme commutatif de suites exactes de groupes abéliens, où la première ligne est la suite exacte du lemme 5.3, et la seconde ligne est la suite exacte d'hypercohomologie associée au triangle exact $M_1 \xrightarrow{f} M_2 \to \text{Cône}(f) \to M_1[1]$.

Démonstration La preuve est immédiate à partir des formules explicites pour les cocycles associés aux torseurs sous un gr-champ que l'on trouve par exemple dans [Ald08]. On utilise la suite exacte de complexes $0 \to M_2 \xrightarrow{T} \text{Cône}(f) \xrightarrow{\delta} M_1[1] \to 0$ provenant du diagramme de complexes suivant :

$$\begin{array}{ccccc} & & F_1 & \xrightarrow{\text{id}} & F_1 \\ & & \downarrow \alpha_1 \oplus f_F & & \downarrow \alpha_1 \\ F_2 & \xrightarrow{(0,\text{id})} & G_1 \oplus F_2 & \xrightarrow{\text{pr}_{G_1}} & G_1 \\ \downarrow \alpha_2 & & \downarrow -f_G + \alpha_2 & & \\ G_2 & \xrightarrow{\text{id}} & G_2 & & \end{array}$$

pour en déduire explicitement les morphismes $T : \mathbf{H}^0(S, M_2) \to \mathbf{H}^0(S, \text{Cône}(f))$ et $\delta : \mathbf{H}^0(S, \text{Cône}(f)) \to \mathbf{H}^1(S, M_1)$ en termes de cocycles. □

5.3.4 Construction de l'application d'abélianisation

Sur une base quelconque

Soit S un schéma muni du site fppf. Soit $f : H \to G$ un morphisme de S-schémas en groupes. On fait les hypothèses suivantes :

1. H et G sont lisses connexes quasi-séparés de présentation finie.
2. H est affine et il existe une suite exacte courte de schémas en groupes

$$1 \to H^{\mathrm{u}} \to H \xrightarrow{\pi} H^{\mathrm{red}} \to 1$$

avec H^{red} réductif et H^{u} unipotent (i.e. à fibres unipotentes). On suppose également cette extension scindée par un morphisme de groupes $s : H^{\mathrm{red}} \to H$ et que pour tout S-torseur P sous H, $H^1(S, {}_PH^{\mathrm{u}}) = 1$.

3. G est extension d'un schéma abélien G^{ab} par un schéma en groupes affine G^{lin}, lui-même extension d'un schéma en groupes réductif G^{red} par un sous-schéma en groupes unipotent G^{u} caractéristique.
4. Le morphisme $H^{\mathrm{u}} \to G/G^{\mathrm{u}}$ se factorise à travers $G^{\mathrm{sc}} \to G/G^{\mathrm{u}}$.

Définissons un complexe de S-schémas en groupes commutatif de longueur 3 associé à f et s, noté $\overline{C}_{f,s}$. Notons $Z_{H^{\mathrm{red}}}$ le centre de H^{red}, qui est un S-groupe de type multiplicatif, et Z'_G le centre de $G' := G/G^{\mathrm{u}}$ (représentable par un sous-schéma en groupes fermé de présentation finie de G'). On dispose alors d'un morphisme naturel $Z_{H^{\mathrm{red}}} \xrightarrow{s} H \xrightarrow{f} G \to G'$.

Or le rang unipotent des fibres de G' est nul, donc par [SGA3], exposé XVI, propositions 3.1 et 3.4, Z'_G est plat et le quotient G'/Z'_G est représentable par un S-schéma en groupes. D'où un morphisme de schémas en groupes induit par f et s : $Z_{H^{\mathrm{red}}} \xrightarrow{g} G'/Z'_G$. Alors le noyau de g est un sous-schéma en groupes de type multiplicatif de $Z_{H^{\mathrm{red}}}$, il est donc plat sur S, donc par [SGA3], exp. XVI, corollaire 2.3, l'image de g est représentable par un sous-schéma en groupes de type multiplicatif Z'_H de G'/Z'_G. Notons alors $Z_{H,G}$ le produit fibré de Z'_H et G' au-dessus de G'/Z'_G. Alors $Z_{H,G}$ est un sous-schéma en groupes commutatif de G', et le morphisme g induit un morphisme $g : Z_{H^{\mathrm{red}}} \to Z_{H,G}$.

Avec des notations évidentes, on dispose de morphismes induits par s et f : $H^{\mathrm{red}} \to G'^{\mathrm{lin}} = G^{\mathrm{red}}$ et $H^{\mathrm{sc}} \to G^{\mathrm{sc}}$, d'où finalement un carré commutatif de schémas en groupes commutatifs sur S

$$\begin{array}{ccc} Z_{H^{\mathrm{sc}}} & \xrightarrow{g^{\mathrm{sc}}} & Z^{\mathrm{sc}}_{H,G} \\ \downarrow{\scriptstyle \rho_H} & & \downarrow{\scriptstyle \rho_G} \\ Z_{H^{\mathrm{red}}} & \xrightarrow{g} & Z_{H,G} \, , \end{array}$$

où $Z^{\mathrm{sc}}_{H,G} := \rho_G^{-1}(Z_{H,G})$ et $\rho_G : G^{\mathrm{sc}} \to G'$ est le morphisme naturel. Notons $\mathscr{M}^{\mathrm{ab}}_H \xrightarrow{g} \mathscr{M}^{\mathrm{ab}}_G$ le morphisme de gr-champs sur S associé au morphisme de modules croisés abéliens $g : M^{\mathrm{ab}}_H \to M^{\mathrm{ab}}_G$ donné par le carré précédent.

Par définition, on note $\overline{C}_{f,s}$ (et parfois abusivement $\overline{C}_f$) le cône de ce morphisme de complexes verticaux, à savoir

$$\overline{C}_{f,s} := [Z_{H^{\mathrm{sc}}} \xrightarrow{\rho_H \oplus g^{\mathrm{sc}}} Z_{H^{\mathrm{red}}} \oplus Z^{\mathrm{sc}}_{H,G} \xrightarrow{-g+\rho_G} Z_{H,G}] \, ,$$

avec $Z_{H^{\mathrm{sc}}}$ en degré -2.

Construisons maintenant l'application d'abélianisation

$$\mathrm{ab}^0_{f,s} : \mathbf{H}^0(S,[H \xrightarrow{f} G]) \to \mathbf{H}^0(S,\overline{C}_{f,s}) =: H^0_{\mathrm{ab}}(S,[H \xrightarrow{f} G])$$

(on notera parfois abusivement $H^0_{\mathrm{ab}}(S,[H \xrightarrow{f} G])$ le groupe $\mathbf{H}^0(S,\overline{C}_{f,s})$).

On note $g : \mathscr{M}_H \to \mathscr{M}_G$ le morphisme de gr-champs sur S associé au morphisme de modules croisés $g : M_H \to M_G$ représenté par le carré

$$\begin{array}{ccc} H^{\mathrm{sc}} & \xrightarrow{g^{\mathrm{sc}}} & G^{\mathrm{sc}} \\ \downarrow{\scriptstyle \rho_H} & & \downarrow{\scriptstyle \rho_G} \\ H^{\mathrm{red}} & \xrightarrow{g} & G' . \end{array}$$

Les inclusions des centres dans les groupes respectifs assurent l'existence d'un diagramme commutatif naturel de modules croisés tressés :

$$\begin{array}{ccc} M_H^{\mathrm{ab}} & \xrightarrow{\phi_1} & M_H \\ \downarrow{\scriptstyle g} & & \downarrow{\scriptstyle g} \\ M_G^{\mathrm{ab}} & \xrightarrow{\phi_2} & M_G . \end{array}$$

On sait par le lemme 3.8.1 de [Bor98] (voir aussi le lemme 4.3 de [Dem11c]) que les morphismes ϕ_i sont des quasi-isomorphismes de modules croisés. Donc par la proposition 5.5, ce diagramme induit un isomorphisme de groupes abéliens

$$\phi_* : \mathbf{H}^0(S,[\mathscr{M}_H^{\mathrm{ab}} \to \mathscr{M}_G^{\mathrm{ab}}]) \cong \mathbf{H}^0(S,[\mathscr{M}_H \to \mathscr{M}_G]) . \tag{5.4}$$

Cas où H est réductif : si on suppose H réductif, on dispose du diagramme commutatif de modules croisés :

$$\begin{array}{ccc} [1 \to H] & \xrightarrow{(1,f)} & [1 \to G] \\ \downarrow{\scriptstyle (1,\mathrm{id})} & & \downarrow{\scriptstyle (1,\mathrm{id})} \\ [H^{\mathrm{sc}} \to H] & \xrightarrow{(f^{\mathrm{sc}},f)} & [G^{\mathrm{sc}} \to G'] \end{array}$$

qui induit par fonctorialité, une application naturelle

$$\mathbf{H}^0(S,[H \to G]) \to \mathbf{H}^0(S,[\mathscr{M}_H \to \mathscr{M}_G]) .$$

Finalement, sous cette hypothèse, l'isomorphisme (5.4) et le lemme 5.6 fournissent une application canonique

$$\mathrm{ab}^0_f : \mathbf{H}^0(S,[H \xrightarrow{f} G]) \to \mathbf{H}^0(S,\overline{C}_f) .$$

Proposition 5.8 *Pour tout $g \in G(S)$ et $x \in \mathbf{H}^0(S,[H \to G])$, on a*

$$\mathrm{ab}_f^0(g.x) = T(\mathrm{ab}_G^0(g)) + \mathrm{ab}_f^0(x)$$

où $T : H^0_{\mathrm{ab}}(S,G) \to \mathbf{H}^0(S,\overline{C}_f)$ est le morphisme induit par le triangle exact

$$C_H \xrightarrow{f} C_G \to \overline{C}_f \to C_H[1]\,.$$

Démonstration D'après la remarque 5.4, il est clair que les actions de $G(S)$ sur $\mathbf{H}^0(S,[H \to G])$ et de $H^0(S,\mathscr{M}_G)$ sur $\mathbf{H}^0(S,[\mathscr{M}_H \to \mathscr{M}_G])$ sont compatibles avec les morphismes naturels $G(S) \to H^0(S,\mathscr{M}_G)$ et $\mathbf{H}^0(S,[H \to G]) \to \mathbf{H}^0(S,[\mathscr{M}_H \to \mathscr{M}_G])$. En utilisant ensuite la proposition 5.7, on voit qu'il suffit de montrer que si l'on note $T' : H^0(S,\mathscr{M}_G) \to \mathbf{H}^0(S,[\mathscr{M}_H \to \mathscr{M}_G])$, alors on a pour tout $g \in H^0(S,\mathscr{M}_G)$ et tout $\alpha \in \mathbf{H}^0(S,[\mathscr{M}_H \to \mathscr{M}_G])$, $g.\alpha = T'(g)+\alpha$. Et ceci résulte du diagramme commutatif suivant, où $(\mathscr{P}, s : \mathscr{P} \to \mathscr{M}_G)$ désigne un représentant de α :

$$\begin{array}{ccccccc}
(\otimes_{\mathscr{M}_H})_*(\mathscr{M}_H \times \mathscr{P}) & \xrightarrow{f.s} & (\otimes_{\mathscr{M}_G})_*(\mathscr{M}_G \times \mathscr{M}_G) & \xrightarrow{\cong} & \mathscr{M}_G & \xrightarrow{g} & \mathscr{M}_G \\
\downarrow{\cong} & & & & \downarrow{=} & & \downarrow{=} \\
\mathscr{P} & & \xrightarrow{\quad\quad s \quad\quad} & & \mathscr{M}_G & \xrightarrow{g} & \mathscr{M}_G\,.
\end{array}$$

□

Cas général : on ne suppose plus H réductif. Alors le diagramme commutatif

$$\begin{array}{ccc}
H^{\mathrm{red}} & \xrightarrow{f \circ s} & G \\
\downarrow{s} & & \downarrow{=} \\
H & \xrightarrow{f} & G
\end{array}$$

induit un morphisme $\alpha : \mathbf{H}^0(S,[H^{\mathrm{red}} \to G]) \to \mathbf{H}^0(S,[H \to G])$.

Lemme 5.9 *Le morphisme $\alpha : \mathbf{H}^0(S,[H^{\mathrm{red}} \to G]) \to \mathbf{H}^0(S,[H \to G])$ est surjectif.*

Si deux éléments $x, y \in \mathbf{H}^0(S,[H^{\mathrm{red}} \to G])$ vérifient $\alpha(x) = \alpha(y)$, alors $\mathrm{ab}^0_{f\circ s}(x) = \mathrm{ab}^0_{f\circ s}(y)$.

Démonstration L'hypothèse (2) assure que l'application naturelle $H^1(S,H) \xrightarrow{\pi_*} H^1(S,H^{\mathrm{red}})$ est une bijection, de réciproque s_*. La surjectivité de α résulte alors d'une chasse au diagramme dans le diagramme exact (5.2) (voir figure 5.2).

Soient $x, y \in \mathbf{H}^0(S, [H^{\mathrm{red}} \to G])$ tels que $\alpha(x) = \alpha(y)$. Au vu de la proposition 5.14 qui suit (dans le cas où H est réductif), si (P, t) désigne un représentant de x, il suffit de montrer que $\mathrm{ab}^0_{f\circ s;x}(y') = 0$, où y' est l'image de y dans $\mathbf{H}^0(S, [{}_PH^{\mathrm{red}} \to G])$ obtenue après torsion de y par (P, t). Par hypothèse, ${}_P\alpha(y')$ est la classe neutre dans $\mathbf{H}^0(S, [{}_{s_*P}H \to G])$. Donc une chasse au diagramme dans (5.3) (voir figure 5.2), qui est obtenu en tordant le diagramme (5.2), assure qu'il existe $g \in G(S)$ (modulo ${}_PH^{\mathrm{red}}(S)$) tel que $g.y'$ est la classe neutre dans $\mathbf{H}^0(S, [{}_PH^{\mathrm{red}} \to G])$. Puisque ${}_P\alpha(y')$ est triviale, on a $g \in {}_{s_*P}H(S)$. Or les éléments g et ${}_Ps({}_{s_*P}\pi(g))$ dans ${}_{s_*P}H(S)$ diffèrent d'un élément $u \in {}_{s_*P}H^{\mathrm{u}}(S)$, de sorte que $g.y' = u.y'$ (puisque ${}_Ps({}_{s_*P}\pi(g))$ agit trivialement sur $\mathbf{H}^0(S, [{}_PH^{\mathrm{red}} \to G])$).

Par l'hypothèse (4), le morphisme ${}_{s_*P}H \to G/G^{\mathrm{u}}$ se factorise à travers $G^{\mathrm{sc}} \to G/G^{\mathrm{u}}$. Par conséquent, l'image dans $G(S)$ de l'élément $u \in {}_{s_*P}H^{\mathrm{u}}(S)$ provient d'un élément $g' \in G^{\mathrm{sc}}(S)$, donc $g'.y'$ est triviale dans $\mathbf{H}^0(S, [{}_PH^{\mathrm{red}} \to G])$. Donc $\mathrm{ab}^0_{f\circ s;x}(g'.y') = 0$. Or la proposition 5.8 assure que $\mathrm{ab}^0_{f\circ s;x}(g'.y') = T(\mathrm{ab}^0_G(g')) + \mathrm{ab}^0_{f\circ s;x}(y')$. Puisque $g' \in G^{\mathrm{sc}}(S)$, on a $\mathrm{ab}^0_G(g') = 0$, donc $\mathrm{ab}^0_{f\circ s;x}(y') = 0$, ce qui conclut la preuve. □

On peut désormais, à l'aide du lemme 5.9, définir l'application $\mathrm{ab}^0_{f,s}$ en toute généralité. Si $x \in \mathbf{H}^0(S, [H \xrightarrow{f} G])$, il existe $x' \in \mathbf{H}^0(S, [H^{\mathrm{red}} \xrightarrow{f\circ s} G])$ tel que $\alpha(x') = x$. Alors on pose $\mathrm{ab}^0_{f,s}(x) := \mathrm{ab}^0_{f\circ s}(x') \in \mathbf{H}^0(S, \overline{C}_{f,s})$. D'après le lemme 5.9, la classe $\mathrm{ab}^0_{f\circ s}(x') \in \mathbf{H}^0(S, \overline{C}_{f,s})$ ne dépend pas du relevé x' de x choisi, donc cela définit bien une application

$$\mathrm{ab}^0_{f,s} : \mathbf{H}^0(S, [H \xrightarrow{f} G]) \to \mathbf{H}^0(S, \overline{C}_{f,s}).$$

Remarque 5.10 En général, cette construction dépend du choix de la section s. En revanche, si l'on suppose que toutes les sections de la suite exacte

$$1 \to H^{\mathrm{u}} \to H \to H^{\mathrm{red}} \to 1$$

sont conjuguées par un élément de $H^{\mathrm{u}}(S)$, alors on montre que le complexe $\overline{C}_{f,s}$ et l'application $\mathrm{ab}^0_{f,s}$ ne dépendent pas de la section s choisie.

Théorème 5.11 *Le diagramme (5.5) de la figure 5.3 ci-dessous est commutatif à lignes exactes, où les morphismes de la première ligne sont les morphismes usuels, et la seconde ligne est extraite de la suite exacte longue d'hypercohomologie associée au triangle exact*

$$C_H \to C_G \to \overline{C}_{f,s} \to C_H[1].$$

En outre, étant donnés $g \in G(S)$ et $x \in \mathbf{H}^0(S, [H \to G])$, on a

$$\mathrm{ab}^0_{f,s}(g.x) = T(\mathrm{ab}^0_G(g)) + \mathrm{ab}^0_{f,s}(x) \in H^0_{\mathrm{ab}}(S, [H \to G]) .$$

Démonstration La commutativité du premier et du dernier carrés résulte de la fonctorialité en L des flèches ab^0_L et ab^1_L pour un schéma en groupes connexe L. Pour les deux carrés centraux, c'est une conséquence de la définition des applications ab^0_G, ab^0_f et ab^1_H, ainsi que de la proposition 5.7.

Concernant la dernière assertion du théorème, si H est réductif, c'est exactement la proposition 5.8. Pour le cas général, on se donne une section s de $H \to H^{\mathrm{red}}$. On a vu qu'alors le morphisme naturel $\mathbf{H}^0(S, [H^{\mathrm{red}} \xrightarrow{f \circ s} G]) \to \mathbf{H}^0(S, [H \to G])$ est surjectif. On relève x en $y \in \mathbf{H}^0(S, [H^{\mathrm{red}} \to G])$. On sait alors, grâce au cas déjà traité (H^{red} est réductif), que

$$\mathrm{ab}^0_{f \circ s}(g.y) = T_{f \circ s}(\mathrm{ab}^0_G(g)) + \mathrm{ab}^0_{f \circ s}(y) \in \mathbf{H}^0(S, [H^{\mathrm{red}} \to G]) .$$

Enfin, par définition, $\mathrm{ab}^0_{f,s}(x) = \mathrm{ab}^0_{f \circ s}(y)$, $\mathrm{ab}^0_{f,s}(g.x) = \mathrm{ab}^0_{f \circ s}(g.y)$ et $T_{f \circ s}(\mathrm{ab}^0_G(g)) = T(\mathrm{ab}^0_G(g))$. □

Étudions le comportement de l'application ab^0_f vis-à-vis de la torsion par un torseur (voir par exemple [Gir71], III.2.6 pour la définition de la torsion par un torseur. Dans ce texte, si P et X sont deux torseurs (à droite) sous H, on note ${}_PX$ le torseur noté $\Theta_P(X)$ dans la proposition III.2.6.1 de [Gir71]). On peut tordre l'ensemble $\mathbf{H}^0(S, [H \to G])$ de deux façons : par un H-torseur P ou par un (H, G)-torseur (P, t). Remarquons que le faisceau en groupes tordu ${}_PH$ est représentable puisque H/S est affine (voir [Ray70], théorème XI.3.1 ou [Mil80], théorème III.4.3).

Si P est un H-torseur, on dispose naturellement d'une application de torsion

$$\mathbf{H}^0(S, [H \xrightarrow{f} G]) \xrightarrow{\theta_P} \mathbf{H}^0(S, [{}_PH \xrightarrow{{}_Pf} {}_PG])$$

où l'image d'un (H, G)-torseur (Q, r) est le $({}_PH, {}_PG)$-torseur (Q', r') défini par $Q' := {}_PQ$ et $r' := {}_Pr : {}_PQ \to {}_PG$.

De même, si (P, t) est un (H, G) torseur, on dispose d'une application de torsion

$$\mathbf{H}^0(S, [H \xrightarrow{f} G]) \xrightarrow{\theta_{(P,t)}} \mathbf{H}^0(S, [{}_PH \xrightarrow{{}_{(P,t)}f} G])$$

où le morphisme ${}_{(P,t)}f$ est défini par la composée

$${}_PH \xrightarrow{{}_Pf} {}_{f_*P}G \xrightarrow{t_*} G .$$

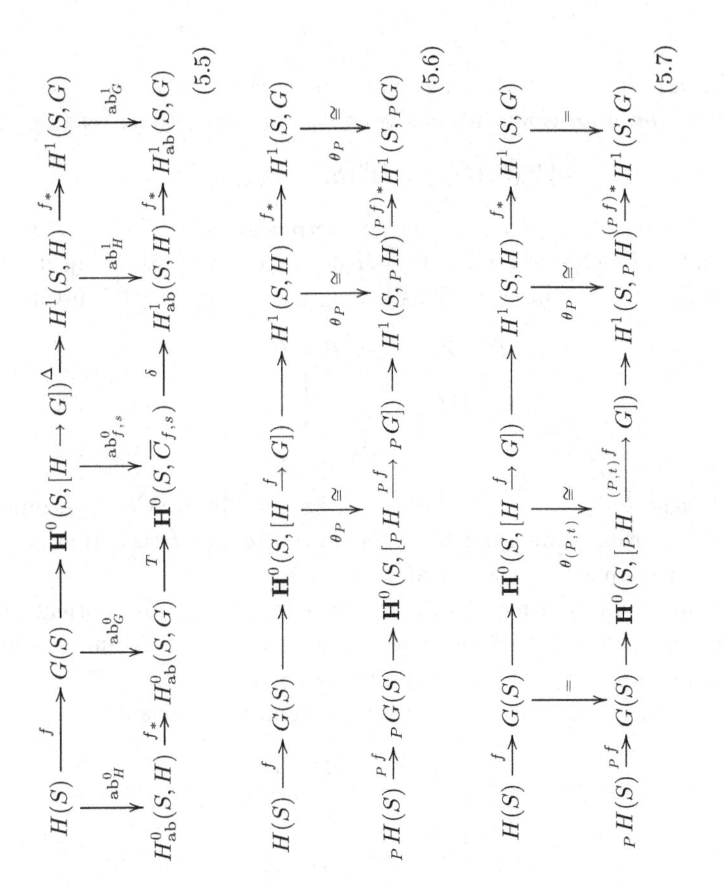

Figure 5.3 Diagrammes (5.5), (5.6) et (5.7).

Par définition, si (Q, r) est un (H, G)-torseur, on lui associe le torseur $(Q', r') := \theta_{(P,t)}(Q, r)$ tel que $Q' := {}_PQ$ et $r' : {}_PQ \xrightarrow{{}_P r} {}_{r_*P}G \xrightarrow{t_*} G$. Le lemme suivant est évident :

Lemme 5.12 1. *Soit P un H-torseur. Alors le diagramme (5.6) de la figure 5.3 est commutatif à lignes exactes, et ses colonnes sont des bijections.,*

2. *Soit (P, t) un (H, G)-torseur. Alors le diagramme (5.7) de la figure 5.3 est commutatif à lignes exactes, et ses colonnes sont des bijections.*

Lemme 5.13 *Soit P un H-torseur. Alors on a un isomorphisme canon-*

ique de complexes

$$\overline{C}_{f,s} \xrightarrow{\cong} \overline{C}_{{}_P f, {}_P s}\,,$$

où ${}_P f : {}_P H \to {}_P G$ *est obtenu par torsion à partir de* $f : H \to G$*. En particulier, on a un isomorphisme canonique de groupes abéliens*

$$\mathbf{H}^0(S, \overline{C}_{f,s}) \cong \mathbf{H}^0(S, \overline{C}_{{}_P f, {}_P s})\,.$$

Démonstration On remarque que le morphisme $P \times Z_H \to P \times H$ provenant de l'inclusion de Z_H dans H définit un morphisme injectif de groupes $Z_H \to {}_P H$ s'insérant dans le diagramme commutatif suivant

$$\begin{array}{ccc} P \times Z_H & \longrightarrow & P \times H \\ \downarrow{\scriptstyle \mathrm{pr}_2} & & \downarrow{\scriptstyle \pi} \\ Z_H & \longrightarrow & {}_P H\,, \end{array}$$

où π désigne le quotient par l'action diagonale de H. Ce diagramme permet d'identifier canoniquement Z_H au centre de ${}_P H$ (l'action de H sur Z_H par conjugaison est triviale).

De même, le morphisme $P \times Z_{H,G} \to P \times G$ induit après quotient par l'action diagonale de H un morphisme injectif $Z_{H,G} \to {}_P G$ qui identifie naturellement le groupe $Z_{H,G}$ au groupe $Z_{{}_P H, {}_P G}$.

En résumé, le quotient du diagramme commutatif suivant

$$\begin{array}{ccc} P \times Z_H & \longrightarrow & P \times H \\ \downarrow{\scriptstyle f} & & \downarrow{\scriptstyle f} \\ P \times Z_{H,G} & \longrightarrow & P \times G \end{array}$$

par l'action diagonale de H induit un diagramme commutatif de groupes

$$\begin{array}{ccc} Z_H & \longrightarrow & {}_P H \\ \downarrow{\scriptstyle f} & & \downarrow{\scriptstyle {}_P f} \\ Z_{H,G} & \longrightarrow & {}_P G\,. \end{array}$$

Ce dernier diagramme induit alors naturellement le carré commutatif suivant :

$$\begin{array}{ccc} Z_H & \xrightarrow{\cong} & Z_{{}_P H} \\ \downarrow{\scriptstyle f} & & \downarrow{\scriptstyle {}_P f} \\ Z_{H,G} & \xrightarrow{\cong} & Z_{{}_P H, {}_P G}\,. \end{array}$$

Le même raisonnement avec les groupes H^{sc} et G^{sc} assure finalement le résultat. □

Proposition 5.14 *Soit (P,t) un (H,G)-torseur. Alors on a un diagramme commutatif naturel*

$$\begin{array}{ccc}
\mathbf{H}^0(S,[H \xrightarrow{f} G]) & \xrightarrow{\theta_{(P,t)}} & \mathbf{H}^0(S,[{}_PH \xrightarrow{{}_{(P,t)}f} G]) \\
\downarrow \mathrm{ab}^0_{f,s} & & \downarrow \mathrm{ab}^0_{{}_Pf,{}_Ps} \\
\mathbf{H}^0(S,\overline{C}_{f,s}) & \xrightarrow{t_{(P,t)}} & \mathbf{H}^0(S,\overline{C}_{f,s}) \cong \mathbf{H}^0(S,\overline{C}_{{}_Pf,{}_Ps}),
\end{array}$$

où $t_{(P,t)}(\alpha) := \alpha - \mathrm{ab}^0_{f,s}(P,t)$ et l'isomorphisme en bas à droite provient du lemme 5.13.

Démonstration Soit (Q,r) un (H,G)-torseur. On note $(\mathscr{Q},\mathscr{R})$ et $(\mathscr{P},\mathscr{T})$ les $(\mathscr{M}_H,\mathscr{M}_G)$-torseurs associés à (Q,r) et (P,t). Alors l'image de (Q,r) dans $\mathbf{H}^0(S,[\mathscr{M}_H \to \mathscr{M}_G])$ est donnée par le $(\mathscr{M}_H,\mathscr{M}_G)$-torseur $\mathscr{Q} \xrightarrow{\mathscr{R}} \mathscr{M}_G$, et si l'on tord ce torseur par $(\mathscr{P},\mathscr{T})$, on obtient le $(\mathscr{M}_{{}_PH},\mathscr{M}_G)$-torseur ${}_{\mathscr{P}}\mathscr{Q} \to \mathscr{M}_G$, qui n'est autre que l'image du $({}_PH,G)$-torseur $\theta_{(P,t)}(Q,r) = ({}_PQ \to G)$ dans $\mathbf{H}^0(S,[\mathscr{M}_{{}_PH} \to \mathscr{M}_G])$. Par conséquent, la définition de l'application d'abélianisation, ainsi que le fait que pour des gr-champs abéliens, la torsion par un torseur induit sur le groupe de cohomologie la translation par l'opposé de la classe dudit torseur, assurent le résultat. □

Sur un corps de caractéristique nulle

Dans cette section, on montre que la construction générale précédente s'exprime en termes d'un complexe de variétés semi-abéliennes.

Soit K un corps de caractéristique nulle. On note $S := \mathrm{Spec}(K)$, et on considère $f : H \to G$ un morphisme de K-groupes algébriques connexes (on ne suppose pas H linéaire ici). Remarquons d'abord que si H est linéaire, alors les conditions (1), (2), (3) et (4) de la section 5.3.4 sont vérifiées.

Si H n'est pas linéaire, on considère l'extension suivante :

$$1 \to H^{\mathrm{u}} \to H^{\mathrm{lin}} \xrightarrow{p} H^{\mathrm{red}} \to 1$$

où H^{u} est le radical unipotent de H^{lin}. On sait que cette extension est scindée par un morphisme $s : H^{\mathrm{red}} \to H^{\mathrm{lin}}$ (voir [SGA3], XXVI.2.3). Définissons $H' \subset H$ comme le sous-groupe de H engendré par D_H et $s(H^{\mathrm{red}})$. Alors H' se surjecte dans H^{ab}, et le noyau de $H' \to H^{\mathrm{ab}}$ s'identifie au sous-groupe de H engendré par $s(H^{\mathrm{red}})$ et $H^{\mathrm{lin}} \cap D_H$. Or

$H^{\text{lin}} \cap D_H$ est un sous-groupe central de H^{lin}, donc il est contenu dans $s(H^{\text{red}})$, donc on a une suite exacte courte naturelle :

$$1 \to s(H^{\text{red}}) \to H' \to H^{\text{ab}} \to 1 .$$

Par conséquent, H'^{lin} est réductif. En outre, on vérifie que le morphisme $H \to H/H^{\text{u}}$ induit un isomorphisme $H' \xrightarrow{\cong} H/H^{\text{u}}$, c'est-à-dire que le diagramme commutatif suivant définit une section notée également s du morphisme $H \to H/H^{\text{u}}$:

$$\begin{array}{ccccccccc} 1 & \longrightarrow & H^{\text{u}} & \longrightarrow & H & \longrightarrow & H/H^{\text{u}} & \longrightarrow & 1 \\ & & & & \uparrow & \nearrow_{\cong} & & & \\ & & & & H' . & & & & \end{array} \tag{5.8}$$

La section s définit un carré de variétés semi-abéliennes naturel suivant, noté $\widetilde{C}_{f,s}$, vu comme un morphisme dans la catégorie des complexes de modules galoisiens, entre les complexes $[T_H^{\text{sc}} \to \text{SA}_H]$ et $[T_G^{\text{sc}} \to \text{SA}_G]$:

$$\begin{array}{ccc} T_H^{\text{sc}} & \xrightarrow{g^{\text{sc}}} & T_G^{\text{sc}} \\ \downarrow{\rho_H} & & \downarrow{\rho_G} \\ \text{SA}_H & \xrightarrow{g} & \text{SA}_G . \end{array}$$

On note $C_{f,s}$ le cône de ce morphisme, à savoir le complexe

$$C_{f,s} := [T_{H^{\text{sc}}} \xrightarrow{\rho_H \oplus g^{\text{sc}}} \text{SA}_H \oplus T_{G^{\text{sc}}} \xrightarrow{-g+\rho_G} \text{SA}_G]$$

avec $T_{H^{\text{sc}}}$ en degré -2.

On vérifie que les inclusions de $Z_{H/H^{\text{u}}} \subset \text{SA}_H$, $Z_{H^{\text{sc}}} \subset T_{H^{\text{sc}}}$, $Z_{H,G} \subset \text{SA}_G$ et $Z_{H,G}^{\text{sc}} \subset Z_{G^{\text{sc}}}$ induisent des quasi-isomorphismes de modules croisés abéliens $[Z_{H^{\text{sc}}} \to Z_{H/H^{\text{u}}}] \to [T_{H^{\text{sc}}} \to \text{SA}_H]$ et $[Z_{H,G}^{\text{sc}} \to Z_{H,G}] \to [G^{\text{sc}} \to \text{SA}_G]$. Ces morphismes induisent donc un quasi-isomorphisme canonique (dans le cas où le groupe H est linéaire)

$$\overline{C}_{f,s} \to C_{f,s} .$$

On peut donc voir l'application d'abélianisation construite plus haut comme une application

$$\text{ab}^0_{f,s} : \mathbf{H}^0(K, [H \to G]) \to \mathbf{H}^0(K, C_{f,s}) .$$

Remarquons en particulier que le quasi-isomorphisme $\overline{C}_{f,s} \to C_{f,s}$ canonique assure que le groupe $\mathbf{H}^0(K, C_{f,s})$ et l'application $\text{ab}^0_{f,s} : \mathbf{H}^0(K, [H \to G]) \to \mathbf{H}^0(K, C_{f,s})$ ne dépendent pas des tores maximaux choisis.

Remarque 5.15 Dans le cas $S = \mathrm{Spec}(K)$ considéré ici, toutes les sections de la suite exacte

$$1 \to H^{\mathrm{u}} \to H^{\mathrm{lin}} \to H^{\mathrm{red}} \to 1$$

sont conjuguées, donc $\mathrm{ab}^0_{f,s}$ et $\mathbf{H}^0(K, C_{f,s})$ ne dépendent pas de la section s choisie.

On peut alors comparer l'application ab^0_f avec les applications d'abélianisation de Borovoi (et celles de la section 5.2). En effet, on dispose d'un diagramme commutatif exact

$$\begin{array}{ccccc} G(K) & \longrightarrow & \mathbf{H}^0(K, [H \to G]) & \longrightarrow & H^1(K, H) \\ \downarrow{\scriptstyle \mathrm{ab}^0_G} & & \downarrow{\scriptstyle \mathrm{ab}^0_f} & & \downarrow{\scriptstyle \mathrm{ab}^1_H} \\ H^0_{\mathrm{ab}}(K, G) & \longrightarrow & \mathbf{H}^0(K, C_f) & \longrightarrow & H^1_{\mathrm{ab}}(K, H), \end{array}$$

où la ligne inférieure est un morceau de la suite exacte longue d'hypercohomologie provenant de la suite exacte de complexes

$$0 \to C_G \to C_f \to C_H[1] \to 1 .$$

Remarque 5.16 Si $K = \mathbb{R}$, on peut donner une version "modifiée à la Tate" de cette application, comme dans [Dem11a] 3.1, pour obtenir une application canonique

$$\widehat{\mathrm{ab}^0_f} : \widehat{H}^0(\mathbb{R}, [H \to G]) \to \widehat{\mathbf{H}}^0(\mathbb{R}, C_f)$$

où $\widehat{H}^0(\mathbb{R}, [H \to G])$ est défini dans [Dem11a], 3.1, et $\widehat{\mathbf{H}}^0(\mathbb{R}, C_f)$ est le groupe d'hypercohomologie modifiée à la Tate du complexe C_f. En outre, cette application $\widehat{\mathrm{ab}^0_f}$ est compatible avec l'application ab^0_f, via les flèches naturelles $\mathbf{H}^0(\mathbb{R}, [H \to G]) \to \widehat{\mathbf{H}}^0(\mathbb{R}, [H \to G])$ et $\mathbf{H}^0(\mathbb{R}, C_f) \to \widehat{\mathbf{H}}^0(\mathbb{R}, C_f)$.

5.3.5 Application à l'abélianisation des espaces homogènes

Revenons au cas général d'un schéma de base S, et de deux schémas en groupes H et G satisfaisant les hypothèses (1), (2), (3) et (4) de la section 5.3.4. On suppose que le morphisme $f : H \to G$ est une immersion fermée d'un sous-groupe H de G, de sorte que le quotient $X := G/H$ soit représentable par un S-schéma. On note $\pi : G \to X$ le morphisme quotient.

Dans ce contexte, la proposition III.3.1.1 de [Gir71] assure que l'on a une bijection naturelle

$$X(S) \cong \mathbf{H}^0(S, [H \xrightarrow{f} G]) .$$

Par conséquent, l'application $\mathrm{ab}^0_{f,s}$ induit naturellement une application notée $\mathrm{ab}^0_{X,s}$ (notée par fois abusivement ab^0_X) de la forme

$$\mathrm{ab}^0_{X,s} : X(S) \to H^0_{\mathrm{ab}}(S, X) := \mathbf{H}^0(S, \overline{C}_{X,s}) ,$$

où $\overline{C}_{X,s} = \overline{C}_X$ est le complexe $\overline{C}_{f,s}$ défini plus haut.

On peut alors traduire les résultats précédents dans ce contexte particulier des espaces homogènes. Par exemple, on dispose de la suite exacte fondamentale suivante (voir théorème 5.11):

Théorème 5.17 *On a un diagramme commutatif naturel à lignes exactes :*

$$\begin{array}{ccccccccc}
H(S) & \xrightarrow{f} & G(S) & \xrightarrow{\pi} & X(S) & \xrightarrow{\Delta} & H^1(S,H) & \xrightarrow{f_*} & H^1(S,G) \\
\downarrow{\scriptstyle \mathrm{ab}^0_H} & & \downarrow{\scriptstyle \mathrm{ab}^0_G} & & \downarrow{\scriptstyle \mathrm{ab}^0_X} & & \downarrow{\scriptstyle \mathrm{ab}^1_H} & & \downarrow{\scriptstyle \mathrm{ab}^1_G} \\
H^0_{\mathrm{ab}}(S,H) & \xrightarrow{f_*} & H^0_{\mathrm{ab}}(S,G) & \xrightarrow{T} & H^0_{\mathrm{ab}}(S,X) & \xrightarrow{\delta} & H^1_{\mathrm{ab}}(S,H) & \xrightarrow{f_*} & H^1_{\mathrm{ab}}(S,G)
\end{array}$$

où les morphismes de la première ligne sont les morphismes usuels, et la seconde ligne est extraite de la suite exacte longue d'hypercohomologie associée au triangle exact

$$C_H \to C_G \to \overline{C}_X \to C_H[1] .$$

En outre, étant donnés $g \in G(S)$ et $x \in X(S)$, on a

$$\mathrm{ab}^0_X(g.x) = T(\mathrm{ab}^0_G(g)) + \mathrm{ab}^0_X(x) \in H^0_{\mathrm{ab}}(S, X) .$$

On dispose aussi d'une formule relative au changement de point de base dans X. La construction de ab^0_X est associée à une identification de l'espace homogène X avec le quotient G/H. Si on se donne un point $x \in X(S)$, le morphisme $\pi_x : G \to X$ défini par $\pi_x(g) := g.x$ permet d'identifier X avec le quotient de G par le stabilisateur H_x de x dans G. De cette façon, on peut construire pour tout $x \in X(S)$, un complexe $\overline{C}_{X,x}$, un groupe $H^0_{\mathrm{ab}}(S, X; x) := \mathbf{H}^0(S, \overline{C}_{X,x})$ et une application

$$\mathrm{ab}^0_{X,x} : X(S) \to H^0_{\mathrm{ab}}(S, X; x) .$$

Proposition 5.18 *Soit X comme plus haut. Soient $x_1, x_2 \in X(S)$.*

Alors il existe un isomorphisme canonique de groupes abéliens $H^0_{\mathrm{ab}}(S, X; x_2) \xrightarrow{\cong} H^0_{\mathrm{ab}}(S, X; x_1)$ *tel que le diagramme naturel*

$$\begin{array}{ccc} X(S) & \xrightarrow{\mathrm{ab}^0_{X,x_2}} & H^0_{\mathrm{ab}}(S, X; x_2) \\ {\scriptstyle \mathrm{ab}^0_{X,x_1}} \downarrow & & \downarrow \cong \\ H^0_{\mathrm{ab}}(S, X; x_1) & \xrightarrow{t_{x_1,x_2}} & H^0_{\mathrm{ab}}(S, X; x_1) \end{array}$$

soit commutatif, où $t_{x_1,x_2} : H^0_{\mathrm{ab}}(S, X; x_1) \to H^0_{\mathrm{ab}}(S, X; x_1)$ *est la translation par* $-\mathrm{ab}^0_{X,x_1}(x_2)$ *(i.e.* $t_{x_1,x_2}(\alpha) := \alpha - \mathrm{ab}^0_{X,x_1}(x_2)$*).*

Démonstration On commence par rappeler le lemme suivant :

Lemme 5.19 *Notons* $\pi : G \to X$ *la projection.*

Alors pour tout $x \in X(S)$*, on a un isomorphisme canonique de S-schémas en groupes*

$$_{\pi^{-1}(x)}H \xrightarrow{\cong} \mathrm{Stab}_G(x)$$

entre le groupe H tordu par le torseur $\pi^{-1}(x)$ *et le stabilisateur* $\mathrm{Stab}_G(x)$ *de x dans G.*

Démonstration On note $i : H \to G$ l'immersion fermée de H dans G et $P := \pi^{-1}(x)$. On définit un morphisme (de S-schémas) $\psi : P \times_S H \to G$ par la formule suivante :

$$\psi(p, h) := p.i(h).p^{-1} .$$

Alors

- le morphisme ψ est à valeurs dans le sous-schéma en groupes $\mathrm{Stab}_G(x)$ de G.
- le morphisme ψ se factorise par l'action diagonale de H sur $P \times_S H$ définie par $h'.(p, h) := (p.h'^{-1}, h'.h.h'^{-1})$, donc ψ induit un morphisme $\psi' : {}_P H \to \mathrm{Stab}_G(x)$.

Enfin, on vérifie que le morphisme ψ' ainsi défini est un isomorphisme de schémas en groupes. □

Avec ce lemme, la proposition 5.18 est une conséquence de la proposition 5.14. □

5.3.6 Application d'abélianisation sur un corps local ou sur un corps de nombres

Proposition 5.20 *Soit K un corps local non archimédien. Alors l'application*

$$\mathrm{ab}^0_f : \mathbf{H}^0(K, [H \to G]) \to H^0_{\mathrm{ab}}(K, [H \to G])$$

est surjective.

Démonstration On considère le diagramme commutatif à lignes exactes :

$$\begin{array}{ccccccccc} H(K) & \xrightarrow{f} & G(K) & \longrightarrow & \mathbf{H}^0(K,[H \to G]) & \longrightarrow & H^1(K,H) & \xrightarrow{f} & H^1(K,G) \\ \downarrow{\scriptstyle \mathrm{ab}^0_H} & & \downarrow{\scriptstyle \mathrm{ab}^0_G} & & \downarrow{\scriptstyle \mathrm{ab}^0_f} & & \downarrow{\scriptstyle \mathrm{ab}^1_H} & & \downarrow{\scriptstyle \mathrm{ab}^1_G} \\ H^0_{\mathrm{ab}}(K,H) & \to & H^0_{\mathrm{ab}}(K,G) & \to & H^0_{\mathrm{ab}}(K,[H \to G]) & \to & H^1_{\mathrm{ab}}(K,H) & \to & H^1_{\mathrm{ab}}(K,G). \end{array}$$

Or on sait que le morphisme $\mathrm{ab}^1_H : H^1(K,H) \to H^1_{\mathrm{ab}}(K,H)$ est surjectif : voir [Bor98], théorème 5.4 dans le cas où H est réductif; pour le cas général, on remarque que le morphisme $H^1(K,H) \to H^1(K, H/H^{\mathrm{u}})$ est bijectif par trivialité de la cohomologie des groupes unipotents, puis on conclut par dévissage au cas des variétés abéliennes et des groupes réductifs.

De même, le morphisme $\mathrm{ab}^1_G : H^1(K,G) \to H^1_{\mathrm{ab}}(K,G)$ est injectif : voir [Bor98], corollaire 5.4.1 pour le cas où G est réductif; pour le cas général, on utilise à nouveau la bijectivité de $H^1(K,G) \to H^1(K, G/G^{\mathrm{u}})$ et un dévissage au cas des groupes réductifs et des variétés abéliennes. Enfin, le morphisme $\mathrm{ab}^0_G : G(K) \to H^0_{\mathrm{ab}}(K,G)$ est surjectif : voir [Bor98], proposition 5.1 quand G est réductif; il suffit de remarquer que $G(K)$ se surjecte dans $G/G^{\mathrm{u}}(K)$ et de faire un dévissage pour le cas général. On conclut alors la preuve par une chasse au diagramme évidente, en utilisant la proposition 5.8. □

Remarque 5.21 Cette proposition est en fait valable sur tout corps K de caractéristique nulle, de dimension cohomologique ≤ 2, tel que pour toute extension finie L/K, indice et exposant des L-algèbres simples centrales coïncident, et tel que la dimension cohomologique de l'extension abélienne maximale de K est ≤ 1. La preuve précédente s'étend en effet à ce cadre ([CT08], théorème 8.4.(ii)) grâce aux cas connus de la conjecture II de Serre (voir [CTGP04], théorème 1.2). Par exemple, cette proposition est valable si K est un corps de nombres totalement imaginaire, ou une extension finie du corps $\mathbf{C}((x,y))$.

Proposition 5.22 *Soit K un corps local non archimédien. Soient $x, y \in \mathbf{H}^0(K, [H \to G])$. Alors $\mathrm{ab}^0_f(x) = \mathrm{ab}^0_f(y)$ si et seulement si il existe $g \in G^{\mathrm{scu}}(K)$ tel que $y = g.x$.*

Démonstration Tout d'abord, la condition est bien suffisante, grâce à la proposition 5.8. Montrons sa nécessité. On note également x un (H, G)-torseur représentant x. Par la proposition 5.14, on sait que $\mathrm{ab}^0_{{}_x f}({}_x y) = 0$ dans $H^0_{\mathrm{ab}}(K, [{}_x H \to G])$. Donc l'image $h \in H^1(K, {}_x H)$ de ${}_x y$ s'envoie sur 0 par $\mathrm{ab}^1_{{}_x H}$. Or l'application $\mathrm{ab}^1_{{}_x H}$ est injective, donc h est la classe triviale dans $H^1(K, {}_x H)$.

Donc par la suite exacte du théorème 5.11, ${}_x y$ provient d'un élément $g_0 \in G(K)$. Par commutativité du diagramme du théorème 5.11, l'image de $\mathrm{ab}^0_G(g_0)$ dans $H^0_{\mathrm{ab}}(K, [{}_x H \to G])$ est nulle, donc $\mathrm{ab}^0_G(g_0)$ se relève en $h'_0 \in H^0_{\mathrm{ab}}(K, {}_x H)$. Or le morphisme

$$\mathrm{ab}^0_{{}_x H} : H^0(K, {}_x H) \to H^0_{\mathrm{ab}}(K, {}_x H)$$

est surjectif, donc h'_0 se relève en $h_0 \in {}_x H(K)$. Posons alors $g_1 := g_0.{h_0}^{-1} \in G(K)$. Il est clair que $\mathrm{ab}^0_G(g_1) = 0$, donc g_1 est l'image d'un élément g de $G^{\mathrm{scu}}(K)$. Enfin, par construction, on a

$$g.x = (g_0.{h_0}^{-1}).x = g_0.x = y$$

car h_0^{-1} est dans le stabilisateur ${}_x H$ de x, ce qui conclut la preuve. □

Le corollaire suivant généralise le corollaire 5.2 de [Bor98] (cas où $H = 1$) et le corollaire 5.4.1 de [Bor98] (cas où $G = 1$).

Corollaire 5.23 *Soit K un corps local non archimédien. L'application ab^0_f induit une bijection*

$$\mathrm{ab}^0_f : G^{\mathrm{scu}}(K) \backslash \mathbf{H}^0(K, [H \to G]) \xrightarrow{\cong} H^0_{\mathrm{ab}}(K, [H \to G])$$

Démonstration C'est la conjonction des propositions 5.20 et 5.22. □

Théorème 5.24 *Soit k un corps de nombres. On suppose les groupes H et G linéaires. On rappelle que $\Omega_{\mathbb{R}}$ désigne l'ensemble des places réelles de k. Alors le diagramme suivant*

$$\begin{array}{ccc}
\mathbf{H}^0(k, [H \to G]) & \xrightarrow{\mathrm{ab}^0_f} & H^0_{\mathrm{ab}}(k, [H \to G]) \\
\downarrow{\scriptstyle \mathrm{loc}_\infty} & & \downarrow{\scriptstyle \mathrm{loc}'_\infty} \\
\prod_{v \in \Omega_{\mathbb{R}}} \widehat{\mathbf{H}}^0(k_v, [H \to G]) & \xrightarrow{\prod_{v \in \Omega_{\mathbb{R}}} \mathrm{ab}^0_{f_v}} & \prod_{v \in \Omega_{\mathbb{R}}} \widehat{H}^0_{\mathrm{ab}}(k_v, [H \to G])
\end{array}$$

est cartésien, où les applications $\mathrm{ab}^0_{f_v}$ sont définies à la remarque 5.16.

Démonstration Notons $\mathrm{ab}^0_\infty := \prod_{v \in \Omega_\mathbb{R}} \mathrm{ab}^0_{f_v}$.

Soient $x' \in H^0_{\mathrm{ab}}(k, [H \to G])$ et $x_\infty \in \prod_{v \in \Omega_\mathbb{R}} \widehat{\mathbf{H}}^0(k_v, [H \to G])$ tels que $\mathrm{loc}'_\infty(x') = \mathrm{ab}^0_\infty(x_\infty)$. On cherche un élément $x \in \mathbf{H}^0(k, [H \to G])$ tel que $\mathrm{ab}^0_f(x) = x'$ et $\mathrm{loc}_\infty(x) = x_\infty$.

Pour cela, on considère le diagramme commutatif à lignes exactes (les lignes proviennent du théorème 5.11) :

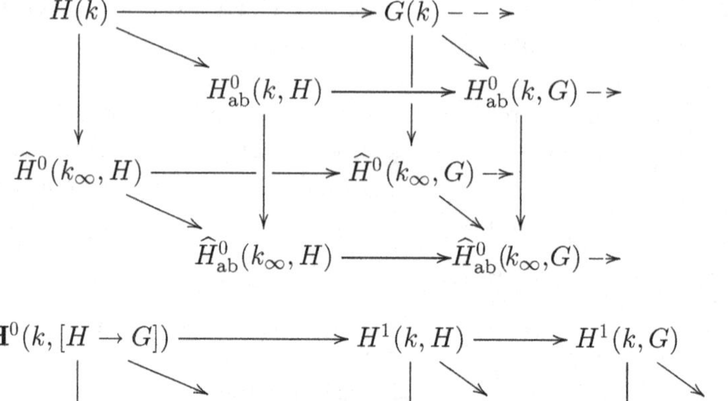

$$
\begin{array}{ccccc}
\mathbf{H}^0(k,[H\to G]) & \longrightarrow & H^1(k,H) & \longrightarrow & H^1(k,G) \\
& \searrow & & \searrow & & \searrow \\
& H^0_{\mathrm{ab}}(k,[H\to G]) & \longrightarrow & H^1_{\mathrm{ab}}(k,H) & \longrightarrow & H^1_{\mathrm{ab}}(k,G) \\
\downarrow & & \downarrow & & \downarrow \\
\widehat{H}^0(k_\infty,[H\to G]) & \longrightarrow & H^1(k_\infty,H) & \longrightarrow & H^1(k_\infty,G) \\
& \searrow \quad \downarrow & & \searrow \quad \downarrow & & \searrow \quad \downarrow \\
& \widehat{H}^0_{\mathrm{ab}}(k_\infty,[H\to G]) & \longrightarrow & H^1_{\mathrm{ab}}(k_\infty,H) & \longrightarrow & H^1_{\mathrm{ab}}(k_\infty,G).
\end{array}
\tag{5.9}
$$

Notons h' l'image de x' dans $H^1_{\mathrm{ab}}(k, H)$ et h_∞ celle de x_∞ dans $H^1(k_\infty, H)$. Alors par commutativité du diagramme, les éléments h' et h_∞ ont même image dans $H^1_{\mathrm{ab}}(k_\infty, H)$. Donc par le théorème 5.11 de [Bor98] (voir aussi la proposition 5.1), il existe $h \in H^1(k, H)$ tel que $\mathrm{loc}_\infty(h) = h_\infty$ et $\mathrm{ab}^1_H(h) = h'$. En reprenant le diagramme (5.9), on constate que l'image de h dans $H^1(k, G)$ s'envoie sur 0 dans $H^1_{\mathrm{ab}}(k, G)$ et sur la classe triviale dans $H^1(k_\infty, G)$. Donc en utilisant à nouveau le théorème 5.11 de [Bor98], appliqué à G, on en déduit que $g \in H^1(k, G)$ est la classe triviale. Donc par exactitude du diagramme (5.9), h se relève en $x^0 \in \mathbf{H}^0(k, [H \to G])$. Alors, en notant $x^0_\infty := \mathrm{loc}_\infty(x)$ et $x^{0\prime} := \mathrm{ab}^0_f(x^0)$, on constate par commutativité de (5.9) que $x^{0\prime}$ et x' ont même image h' dans $H^1_{\mathrm{ab}}(k, H)$, et que x^0_∞ et x_∞ ont même image h_∞ dans $H^1(k_\infty, H)$. Donc par exactitude, il existe $g' \in H^0_{\mathrm{ab}}(k, G)$ et $g_\infty \in \widehat{H}^0(k_\infty, G)$ tels que $x' = T(g') + x^{0\prime}$ dans $H^0_{\mathrm{ab}}(k, [H \to G])$ et $x_\infty = g_\infty . x^0_\infty$ dans

$\widehat{H}^0(k_\infty, [H \to G])$. Les éléments $\mathrm{loc}'_\infty(g')$ et $\mathrm{ab}_\infty(g_\infty)$ de $\widehat{H}^0_{\mathrm{ab}}(k_\infty, G)$ ont même image $x^0_\infty{}' - x'_\infty$ dans $\widehat{H}^0_{\mathrm{ab}}(k_\infty, [H \to G])$, donc il existe $\overline{h}'_\infty \in \widehat{H}^0_{\mathrm{ab}}(k_\infty, H)$ tel que

$$f(\overline{h}'_\infty) = \mathrm{loc}'_\infty(g') - \mathrm{ab}_\infty(g_\infty)\,.$$

On utilise alors le théorème 5.34 pour trouver $\overline{h}' \in H^0_{\mathrm{ab}}(k, H)$ tel que $\mathrm{loc}'_\infty(\overline{h}') = \overline{h}'_\infty$. On a alors un diagramme commutatif à lignes exactes :

$$\begin{array}{ccccc} G(k) & \xrightarrow{\mathrm{ab}^0_G} & H^0_{\mathrm{ab}}(k, G) & \xrightarrow{\partial} & H^1(k, G^{\mathrm{scu}}) \\ \downarrow{\scriptstyle \mathrm{loc}_\infty} & & \downarrow{\scriptstyle \mathrm{loc}'_\infty} & & \downarrow{\scriptstyle \mathrm{loc}^{\mathrm{scu}}_\infty} \\ \widehat{H}^0(k_\infty, G) & \xrightarrow{\mathrm{ab}^0_G} & \widehat{H}^0_{\mathrm{ab}}(k_\infty, G) & \longrightarrow & H^1(k_\infty, G^{\mathrm{scu}})\,. \end{array}$$

Or par construction, le point $g' - f(\overline{h}') \in H^0_{\mathrm{ab}}(k, G)$ s'envoie par l'application ∂ sur une classe $\Delta \in H^1(k, G^{\mathrm{scu}})$ telle que $\mathrm{loc}_\infty(\Delta)$ est la classe triviale dans $H^1(k_\infty, G^{\mathrm{scu}})$. Or $\mathrm{loc}^{\mathrm{scu}}_\infty$ a un noyau trivial (principe de Hasse pour les groupes semi-simple simplement connexes : voir [PR94], théorème 6.6), donc Δ est triviale, donc le diagramme précédent assure que $g' - f(\overline{h}') = \mathrm{ab}^0_G(\overline{g})$ pour un $\overline{g} \in G(k)$. Or on a par construction l'égalité $\mathrm{ab}_\infty(g_\infty) = \mathrm{loc}'_\infty(g' - f(\overline{h}'))$ dans $\widehat{H}^0_{\mathrm{ab}}(k_\infty, G)$, donc on en déduit que $\mathrm{ab}^0_G(\mathrm{loc}_\infty(\overline{g})) = \mathrm{ab}^0_G(g_\infty)$ dans $\widehat{H}^0_{\mathrm{ab}}(k_\infty, G)$. Or le morphisme

$$\mathrm{ab}^0_G : \widehat{H}^0(k_\infty, G) \to \widehat{H}^0_{\mathrm{ab}}(k_\infty, G)$$

est injectif, puisque $H^0(k_\infty, G^{\mathrm{scu}})$ est connexe, donc on en déduit que $\mathrm{loc}_\infty(\overline{g}) = g_\infty$. Finalement, on pose

$$x := \overline{g}.x^0 \in \mathbf{H}^0(k, [H \to G])\,.$$

Alors par construction, grâce à la commutativité du diagramme (5.9), le point x vérifie les deux conditions souhaitées, à savoir $\mathrm{ab}^0_f(x) = x'$ et $\mathrm{loc}_\infty(x) = x_\infty$, d'où le théorème. □

5.3.7 Abélianisation et résolutions flasques

Sur un corps K, on va donner une définition équivalente de l'application d'abélianisation, pour un morphisme de groupes linéaires connexes, en termes de résolutions flasques des groupes H et G. Pour les définitions et les propriétés des résolutions flasques, on renvoie aux travaux de Colliot-Thélène dans [CT08].

On se donne $f : H \to G$ un morphisme de K-groupes algébriques linéaires connexes. Soient

$$1 \to S_{H'} \to H' \to H \to 1$$

$$1 \to S_{G'} \to G' \to G \to 1$$

des résolutions flasques de H et G : $S_{H'}$ et $S_{G'}$ sont des K-tores flasques, H' et G' sont des groupes quasi-triviaux, et les extensions précédentes sont centrales. On note $\mathscr{F}_{H'}$ le gr-champ associé au module croisé $[S_{H'} \to H']$, et de même pour $\mathscr{F}_{G'}$. Par la preuve de la proposition 6.6 de [CT08], il existe deux morphismes $f_1 : S_{H'} \to S_{G'}$ et $f_2 : H' \to G'$ de sorte que le diagramme suivant commute :

$$\begin{array}{ccccccccc} 1 & \longrightarrow & S_{H'} & \longrightarrow & H' & \longrightarrow & H & \longrightarrow & 1 \\ & & \downarrow{\scriptstyle f_1} & & \downarrow{\scriptstyle f_2} & & \downarrow{\scriptstyle f} & & \\ 1 & \longrightarrow & S_{G'} & \longrightarrow & G' & \longrightarrow & G & \longrightarrow & 1\,. \end{array}$$

On vérifie que le morphisme naturel

$$\mathbf{H}^0(K, [\mathscr{F}_{H'} \xrightarrow{f_*} \mathscr{F}_{G'}]) \to \mathbf{H}^0(K, [H \to G])$$

est une bijection Considérons désormais le diagramme commutatif suivant :

$$\begin{array}{ccc} S_{H'} & \xrightarrow{f_1} & S_{G'} \\ \downarrow & & \downarrow \\ P_{H'} & \xrightarrow{f_2} & P_{G'} \end{array}$$

où $P_{H'}$ (resp. $P_{G'}$) est le K-tore quasi-trivial ${H'}^{\mathrm{tor}}$ (resp. ${G'}^{\mathrm{tor}}$). On note $\overline{C}_{H',G',f}$ le cône de ce carré de tores. Si $\mathscr{F}^{\mathrm{ab}}_{H'}$ désigne le gr-champ associé au module croisé abélien $[S_{H'} \to P_{H'}]$ (idem pour G'), on a des morphismes naturels de gr-champs $\mathscr{F}_{H'} \to \mathscr{F}^{\mathrm{ab}}_{H'}$ et $\mathscr{F}_{G'} \to \mathscr{F}^{\mathrm{ab}}_{G'}$, de sorte que l'on a un morphisme canonique

$$\mathbf{H}^0(K, [\mathscr{F}_{H'} \xrightarrow{f_*} \mathscr{F}_{G'}]) \to \mathbf{H}^0(K, [\mathscr{F}^{\mathrm{ab}}_{H'} \xrightarrow{f_*} \mathscr{F}^{\mathrm{ab}}_{G'}])\,.$$

Enfin, le groupe $\mathbf{H}^0(K, [\mathscr{F}^{\mathrm{ab}}_{H'} \xrightarrow{f_*} \mathscr{F}^{\mathrm{ab}}_{G'}])$ s'identifie naturellement au groupe d'hypercohomologie $\mathbf{H}^0(K, \overline{C}_{H',G',f})$ (voir lemme 5.6 ou [Ald08], proposition 6.1.6). On a construit une application

$$\overline{\mathrm{ab}}^0_{H',G',f} : \mathbf{H}^0(K, [H \to G]) \to \mathbf{H}^0(K, \overline{C}_{H',G',f})\,.$$

On vérifie que cette construction ne dépend pas des résolutions H' et G' choisies (voir proposition 3.2.(iv) de [CT08]), ni des morphismes f_1 et f_2 (voir [CT08], preuve de la proposition 6.6).

Enfin, la proposition A.1.(iv) de [CT08] assure que les complexes $\overline{C}_{H',G',f}$ et C_f sont naturellement quasi-isomorphes, et on a ainsi défini une application

$$\overline{\mathrm{ab}}_f^0 : \mathbf{H}^0(K, [H \to G]) \to \mathbf{H}^0(K, C_f) = H^0_{\mathrm{ab}}(K, [H \to G]) .$$

Un calcul assure que cette flèche coïncide avec la flèche ab_f^0 définie plus haut.

5.4 Application d'abélianisation pour les espaces homogènes à stabilisateurs abéliens

Dans cette section, on ne suppose plus H connexe, mais on lui impose d'être commutatif, extension d'un schéma abélien par un groupe linéaire commutatif, lui-même extension d'un schéma en groupes de type multiplicatif par un sous-schéma en groupes unipotent caractéristique. On note H^{u} le radical unipotent de H^{lin}. On dispose des morphismes quotients suivants : $H \to H^{\mathrm{tma}} := H/H^{\mathrm{u}}$ et $G \to G^{\mathrm{sab}}$. Le groupe H^{tma} est une extension d'une variété abélienne H^{ab} par un groupe de type multiplicatif $H^{\mathrm{tm}} = H^{\mathrm{lin}}/H^{\mathrm{u}}$; on a également un morphisme naturel $H^{\mathrm{tma}} \to G^{\mathrm{sab}}$ induit par l'inclusion de H dans G (le morphisme $H^{\mathrm{u}} \to G^{\mathrm{sab}}$ est trivial par [SGA3], exposé XVII, lemme 6.2.5). De la même manière que dans la section précédente, on a un morphisme d'abélianisation naturel

$$\mathrm{ab}^{0}{}'_f : \mathbf{H}^0(S, [H \to G]) \to \mathbf{H}^0(S, [H^{\mathrm{tma}} \to G^{\mathrm{sab}}]) .$$

Le diagramme suivant est alors commutatif (C_f^{ab} désigne le complexe $[H^{\mathrm{tma}} \to G^{\mathrm{sab}}]$):

$$\begin{array}{ccccc}
G(S) & \longrightarrow & \mathbf{H}^0(S, [H \to G]) & \longrightarrow & H^1(S, H) \\
\downarrow{\scriptstyle \mathrm{ab}^{0}{}'_G} & & \downarrow{\scriptstyle \mathrm{ab}^{0}{}'_f} & & \downarrow{\scriptstyle \mathrm{ab}^{1}{}'_H} \\
H^0(S, G^{\mathrm{sab}}) & \xrightarrow{T} & \mathbf{H}^0(S, C_f^{\mathrm{ab}}) & \longrightarrow & H^1(S, H^{\mathrm{tma}})
\end{array}$$

où la seconde ligne provient du triangle exact dans la catégorie dérivée :

$$H^{\mathrm{tma}} \to G^{\mathrm{sab}} \to C_f^{\mathrm{ab}} \to H^{\mathrm{tma}}[1] .$$

On a également une formule de compatibilité analogue au théorème 5.11, à savoir que l'on a, pour tout $g \in G(S)$, $x \in \mathbf{H}^0(S, [H \to G])$:

$$\mathrm{ab}^{0'}_f(g.x) = T(\mathrm{ab}^0_G(g)) + \mathrm{ab}^{0'}_f(x)\,.$$

Si le groupe H est connexe, l'application $\mathrm{ab}^{0'}_f$ est compatible avec l'application ab^0_f définie dans les sections précédentes, au sens où $\mathrm{ab}^{0'}_f$ est la composée

$$\mathbf{H}^0(S, [H \to G]) \xrightarrow{\mathrm{ab}^0_f} \mathbf{H}^0(K, \overline{C}_f) \xrightarrow{t_*} \mathbf{H}^0(K, C_f^{\mathrm{ab}})$$

où le morphisme t_* provient du morphisme de complexes naturel $\overline{C}_f \to C_f^{\mathrm{ab}}$.

Enfin, mentionnons que l'on dispose aussi dans ce contexte de l'analogue de la proposition 5.14.

5.5 Applications d'abélianisation sur un anneau d'entiers

Soit k un corps de nombres, $\mathscr{O}_k$ son anneau des entiers et U un ouvert de $\mathrm{Spec}(\mathscr{O}_k)$. L'objectif de cette section est de montrer que si U est suffisamment petit, les constructions de la section 5.3.4 (avec les tores maximaux) s'étendent sur l'ouvert U.

À partir de cette section, et jusqu'à la fin du texte, la cohomologie considérée est *la cohomologie étale*, sauf mention explicite du contraire. Remarquons que puisque tous les schémas en groupes considérés sont supposés lisses, la cohomologie étale coïncide avec la cohomologie fppf.

5.5.1 Cas d'un stabilisateur connexe

Soient $f : H \to G$ un morphisme de k-groupes connexes. On suppose H linéaire réductif. Quitte à réduire U, on peut supposer qu'il existe $\mathscr{H} \xrightarrow{f} \mathscr{G}$ un morphisme de schémas en groupes lisses connexes sur U, de sorte que H (resp. G, resp. $f : H \to G$) s'identifie à la fibre générique de $\mathscr{H}$ (resp. $\mathscr{G}$, resp. $f : \mathscr{H} \to \mathscr{G}$). Quitte à réduire encore U, on suppose que $\mathscr{H}$ est réductif sur U, et que $\mathscr{G}$ est extension d'un schéma abélien par un groupe affine, lui-même extension scindée d'un schéma en groupes réductif par un schéma en groupes unipotent. On vérifie alors que l'on satisfait les hypothèses (1), (2), (3) et (4) : pour les hypothèses (1) et (3), on peut par exemple utiliser [Gro66], théorème 8.10.5, [SGA3], exposé VI-B proposition 9.2 et [Ana73], théorème 4.C. Les hypothèses (2) et (4) sont évidentes puisque $\mathscr{H}$ est réductif sur U. On définit $T_{\mathscr{H}^{sc}}$ (resp. $T_{\mathscr{G}^{sc}}$, resp. $T_{\mathscr{H}}$, resp. $\mathrm{SA}_{\mathscr{G}}$) comme l'adhérence schématique de T_H^{sc} (resp. T_G^{sc}, resp. T_H, resp. SA_G) dans $\mathscr{H}^{sc}$ (resp. $\mathscr{G}^{sc}$, resp. $\mathscr{H}$, resp. $\mathscr{G}/\mathscr{G}^{u}$). Alors, quitte à réduire encore U, on peut supposer que les schémas en groupes ainsi obtenus sont des sous-schémas semi-abéliens maximaux des schémas en groupes correspondants. On dispose alors de morphismes d'abélianisation

$$H^i(U, \mathscr{H}) \to \mathbf{H}^i(U, [T_{\mathscr{H}^{sc}} \to T_{\mathscr{H}}])$$

$$H^i(U, \mathscr{G}) \to \mathbf{H}^i(U, [T_{\mathscr{G}^{sc}} \to \mathrm{SA}_{\mathscr{G}}])$$

pour $i = 0$ et 1.

On obtient aussi une application naturelle

$$\begin{aligned}\mathrm{ab}_f^0 : \mathbf{H}^0(U, [\mathscr{H} \to \mathscr{G}]) \to H^0_{\mathrm{ab}}(U, [\mathscr{H} \to \mathscr{G}]) \\ := \mathbf{H}^0(U, [T_{\mathscr{H}^{sc}} \to T_{\mathscr{H}} \oplus T_{\mathscr{G}^{sc}} \to \mathrm{SA}_{\mathscr{G}}])\end{aligned}$$

fonctorielle en U et qui coïncide avec l'application d'abélianisation de la section 5.3.4, au vu des quasi-isomorphismes (sur U suffisamment petit) $[Z_{\mathscr{H}}^{sc} \to Z_{\mathscr{H}}] \to [T_{\mathscr{H}}^{sc} \to T_{\mathscr{H}}]$ et $[Z_{\mathscr{H},\mathscr{G}}^{sc} \to Z_{\mathscr{H},\mathscr{G}}] \to [T_{\mathscr{G}}^{sc} \to \mathrm{SA}_{\mathscr{G}}]$.

De même si H n'est pas supposé réductif, on a une interprétation analogue de l'application d'abélianisation en termes de schémas semi-abéliens, sur un ouvert U assez petit. Il faut tout de même vérifier les hypothèses (1), (2), (3) et (4) de la section 5.3.4. Comme auparavant, les hypothèses (1) et (3) sont vérifiées dans ce contexte, si U est suffisamment petit. De même, l'hypothèse (4) est satisfaite puisque sur k, le morphisme $H^{u} \to G/G^{u}$ se factorise à travers $G^{sc} \to G/G^{u}$, donc c'est aussi le cas sur un ouvert assez petit. Vérifions l'hypothèse (2). Si U est assez petit, $\mathscr{H}$ est bien extension scindée d'un schéma en groupes réductif par un schéma en groupes unipotent. Montrons que pour tout U-torseur P sous $\mathscr{H}$, on a $H^1(U, {}_P\mathscr{H}^{u}) = 1$. Par le théorème 2.1 de [Nis84], le noyau de $H^1(U, {}_P\mathscr{H}^{u}) \to H^1(k, {}_PH^{u}) = 1$ s'identifie au groupe des classes de ${}_P\mathscr{H}^{u}$ sur U. Or ${}_PH^{u}$ est unipotent, donc il vérifie l'approximation forte sur U, donc le groupe des classes de ${}_P\mathscr{H}^{u}$ sur U est trivial, donc $H^1(U, {}_P\mathscr{H}^{u}) = 1$.

Donc l'hypothèse (2) est vérifiée. On peut donc définir l'application $\mathrm{ab}^0_{f,s}$.

Et on a donc bien dans ce contexte une description de l'application $\mathrm{ab}^0_{f,s}$ sur l'ouvert U en termes d'un complexe de schémas semi-abéliens sur U.

5.5.2 Cas d'un stabilisateur abélien

Soit $f : H \to G$ un morphisme de k-groupes, avec H linéaire commutatif et G connexe. Si U est un ouvert suffisamment petit, on étend la situation pour obtenir un morphisme $f : \mathscr{H} \to \mathscr{G}$ entre un U-schéma en groupes affine lisse commutatif et un U-schéma en groupes lisse connexe $\mathscr{G}$. Quitte à réduire U, on peut supposer que le groupe $\mathscr{H}$ est isomorphe au produit direct $\mathscr{H}^{\mathrm{u}} \times \mathscr{H}^{\mathrm{m}}$, où $\mathscr{H}^{\mathrm{u}}$ est le radical unipotent de $\mathscr{H}$ et $\mathscr{H}^{\mathrm{m}}$ est un groupe de type multiplicatif, quotient de $\mathscr{H}$.

Comme dans le cas des corps, on dispose d'une application naturelle, définie pour un ouvert U suffisamment petit :

$$\mathrm{ab}^{0\prime}{}_f : \mathbf{H}^0(U, [\mathscr{H} \to \mathscr{G}]) \to \mathbf{H}^0(U, [\mathscr{H}^{\mathrm{m}} \to \mathscr{G}^{\mathrm{sab}}]) .$$

Enfin, notons que puisque H^{u} est unipotent, il vérifie l'approximation forte, donc le groupe $H^1(U, \mathscr{H}^{\mathrm{u}})$ est trivial, et donc on a une bijection naturelle

$$H^1(U, \mathscr{H}) \cong H^1(U, \mathscr{H}^{\mathrm{m}}) .$$

De même, si v une place de k associée à un point fermé de U, alors les applications

$$H^1(\mathscr{O}_v, \mathscr{H}) \to H^1(\mathscr{O}_v, \mathscr{H}^{\mathrm{m}})$$

$$H^1(k_v, H) \to H^1(k_v, H^{\mathrm{m}})$$

sont des bijections.

5.6 Théorèmes de dualité pour certains complexes de 1-motifs

Dans cette section, on généralise à certains complexes de longueur 3 les théorèmes de dualité pour les complexes de tores de longueur 2 obtenus dans [Dem11a].

Définition 5.25 *Un complexe de longueur* 3 *sur* S *est un complexe de* S*-schémas en groupes lisses commutatifs, de longueur* 3*, i.e. un complexe*

$$\left[M_1 \xrightarrow{f_1} M_2 \xrightarrow{f_2} M_3\right]$$

où les M_i *sont des* S*-schémas en groupes lisses commutatifs,* M_1 *étant en degré* -2.

Remarque 5.26 On voit un tel objet dans la catégorie des complexes bornés de faisceaux fppf sur S, ou parfois dans la catégorie dérivée associée à cette catégorie abélienne.

En particulier, soit un diagramme commutatif de morphismes de S-groupes commutatifs

$$\begin{array}{ccc} G_1 & \xrightarrow{f_1} & G_1' \\ \downarrow{\rho} & & \downarrow{\rho'} \\ G_2 & \xrightarrow{f_2} & G_2' \end{array},$$

on peut le voir comme un complexe (de longueur 2) de complexes de groupes commutatifs (de longueur 2) $[G_1 \xrightarrow{\rho} G_2] \xrightarrow{f} [G_1' \xrightarrow{\rho'} G_2']$, et alors son cône

$$[G_1 \xrightarrow{f_1 \oplus \rho} G_1' \oplus G_2 \xrightarrow{\rho' - f_2} G_2']$$

est un exemple de complexe de longueur 3.

Hypothèse : on suppose dans toute la section 5.6 que Ker f_1 *est fini*, que les groupes M_1 *et* M_2 *sont des schémas en groupes de type multiplicatif* et que M_3 *est un schéma semi-abélien.*

Dans toute la suite de cette section, si C désigne un complexe de longueur 3, on note $\widehat{C}$ le complexe dual, à savoir le cône du complexe

$$\left[\widehat{M_3} \xrightarrow{\widehat{f_2}} \widehat{M_2} \xrightarrow{\widehat{f_1}} \widehat{M_1}\right]$$

où $\widehat{M_i}$ désigne le 1-motif dual du 1-motif (éventuellement avec torsion, au sens de [Jos09]) associé à M_i, et où $\widehat{M_3}$ est en degré -2. Explicitement, $\widehat{C}$ est représenté par le complexe

$$\widehat{C} := [\widehat{M_3^{\mathrm{lin}}} \to ({M_3}^{\mathrm{ab}})^* \oplus \widehat{M_2} \to \widehat{M_1}]\,,$$

avec $\widehat{M_3^{\mathrm{lin}}}$ en degré -2. On dispose d'un accouplement naturel $C \otimes^{\mathbf{L}} \widehat{C} \to \mathbb{G}_m[2]$ qui induit notamment des accouplements en cohomologie

$$\mathbf{H}^i(S, C) \times \mathbf{H}^{-i}(S, \widehat{C}) \to H^2(S, \mathbb{G}_m) = \mathrm{Br}(S)\,.$$

Comme dans [Dem11a], pour un tel complexe C sur un corps de nombres k, on note, pour toute place archimédienne v de k, $\mathbf{H}^0(k_v, C)$ le groupe d'hypercohomologie *modifié* à la Tate, et $\mathbf{P}^0(k, C)$ le produit restreint des $\mathbf{H}^0(k_v, C)$ pour toutes les places v de k.

5.6.1 Topologie

Dans ce paragraphe, K est un corps local, et C est un complexe de longueur 3 sur K. On cherche à munir $\mathbf{H}^0(K,C)$ d'une topologie naturelle.

Notons $C_1 := \left[M_1 \xrightarrow{f_1} M_2\right]$, complexe de longueur 2. On considère le triangle exact évident

$$C_1 \to M_3 \to C \to C_1[1] .$$

On en déduit une suite exacte en hypercohomologie :

$$\mathbf{H}^0(K,C_1) \xrightarrow{f_2'} H^0(K,M_3) \to \mathbf{H}^0(K,C) \to \mathbf{H}^1(K,C_1) . \tag{5.10}$$

Lemme 5.27 *Le morphisme $f_2' : \mathbf{H}^0(K,C_1) \to H^0(K,M_3)$ est d'image fermée dans $H^0(K,M_3)$ (muni de la topologie induite par celle de K).*

Démonstration On regarde le diagramme commutatif à lignes exactes suivant

$$\begin{array}{ccccc} M_1(K) & \xrightarrow{f_1} & M_2(K) & \xrightarrow{\varpi} & Q \\ & & \downarrow{\scriptstyle f_2} & & \downarrow{\scriptstyle f_2''} \\ & & M_3(K) & \xrightarrow{=} & M_3(K) , \end{array}$$

où Q est le conoyau de f_1, muni de la topologie quotient. Alors $\mathrm{Im}(f_2'') = \mathrm{Im}(f_2)$, et cette dernière est fermée (voir [Mar91], Corollaire I.2.1.3 et [PR94], corollaire 1 à la proposition 3.3), donc f_2'' est d'image fermée. On regarde alors le diagramme commutatif suivant, à lignes exactes :

$$\begin{array}{ccccc} Q & \xhookrightarrow{i} & \mathbf{H}^0(K,C_1) & \xrightarrow{\partial} & H^1(K,M_1) \\ \downarrow{\scriptstyle f_2''} & & \downarrow{\scriptstyle f_2'} & & \\ M_3(K) & \xrightarrow{=} & M_3(K). & & \end{array}$$

Or par définition de la topologie sur $\mathbf{H}^0(K,C_1)$, Q s'identifie à un sous-groupe fermé d'indice fini dans $\mathbf{H}^0(K,C_1)$ ($H^1(K,M_1)$ est fini car M_1 est de type multiplicatif). Et f_2'' est d'image fermée, donc $\mathrm{Im}(f_2')$ est un sous-groupe de $M_3(K)$ contenant comme sous-groupe d'indice fini le sous-groupe fermé $\mathrm{Im}(f_2'')$. Donc $\mathrm{Im}(f_2')$ est un sous-groupe fermé de $M_3(K)$. □

La suite exacte initiale (5.10) identifie le conoyau de f_2' avec un sous-groupe d'indice fini de $\mathbf{H}^0(K,C)$ ($\mathbf{H}^1(K,C_1)$ est fini grâce au triangle exact $\mathrm{Ker}(f_1)[1] \to C_1 \to \mathrm{Coker}(f_1) \to \mathrm{Ker}(f_1)[2]$). On munit ce sous-groupe de la topologie quotient, et cela définit bien une topologie sur le groupe $\mathbf{H}^0(K, C)$. Cette topologie est séparée grâce au fait que $\mathrm{Im}(f_2')$ est un sous-groupe fermé de $H^0(K, M_3)$ (voir lemme 5.27), et le morphisme $H^0(K, M_3) \to \mathbf{H}^0(K, C)$ est alors un morphisme ouvert.

De façon analogue, pour un morphisme de K-groupes $f : H \to G$, on munit l'ensemble $\mathbf{H}^0(K, [H \to G])$ de la topologie naturelle induite par l'action du groupe topologique $G(K)$: $U \subset \mathbf{H}^0(K, [H \to G])$ est ouvert si et seulement si l'intersection de U avec toute orbite de $G(K)$ est un ouvert de cette orbite (lorsqu'elle est munie de la topologie quotient). Autrement dit, on munit $\mathbf{H}^0(K, [H \to G])$ de la topologie somme disjointe des orbites sous $G(K)$, chaque orbite étant munie de la topologie quotient.

Alors $\mathbf{H}^0(K, [H \to G])$ est un espace topologique séparé, et les morphismes

$$G(K) \to \mathbf{H}^0(K, [H \to G]) \to H^1(K, H)$$

sont continus et ouverts ($H^1(K, H)$ est muni de la topologie discrète). On vérifie alors que si $f : \mathscr{H} \to \mathscr{G}$ est un morphisme de $\mathscr{O}_K$-schémas en groupes de type fini, l'image de $\mathbf{H}^0(\mathscr{O}_K, [\mathscr{H} \to \mathscr{G}])$ dans $\mathbf{H}^0(K, [H \to G])$ est un ouvert; on vérifie aussi que l'application d'abélianisation ab_f^0 : $\mathbf{H}^0(K, [H \to G]) \to H^0_{\mathrm{ab}}(K, [H \to G])$ est continue.

Enfin, on vérifie que si H est un K-sous-groupe de G, et si $X := G/H$, les topologies sur $X(K)$ et $\mathbf{H}^0(K, [H \to G])$ coïncident via l'identification $X(K) \cong \mathbf{H}^0(K, [H \to G])$.

5.6.2 Suites exactes de Poitou-Tate

Dans cette section, k est un corps de nombres. On reprend les notations de [Dem11b], à savoir que si A est un groupe topologique, on note $A^\wedge$ le complété de A pour la topologie des sous-groupes ouverts distingués d'indice fini; si en outre A est commutatif, on note ${}_nA$ le sous-groupe de n-torsion de A, et $A_\wedge := \varprojlim_n A/nA$. Enfin, A^D désigne le groupe des morphismes de groupes continus $A \to \mathbf{Q}/\mathbf{Z}$.

Proposition 5.28 *Soit C un complexe de longueur 3 sur k, avec $Ш^1({M_3}^{\mathrm{ab}})$ fini. On suppose le complexe C exact en degré -1 Alors on a une suite exacte de groupes abéliens :*

$$\mathbf{H}^0(k, C)^\wedge \to \mathbf{P}^0(k, C)^\wedge \to \left(\mathbf{H}^0(k, \widehat{C})\right)^D \to Ш^0(\widehat{C})^D \to 0\,.$$

Démonstration On considère le diagramme suivant, commutatif,

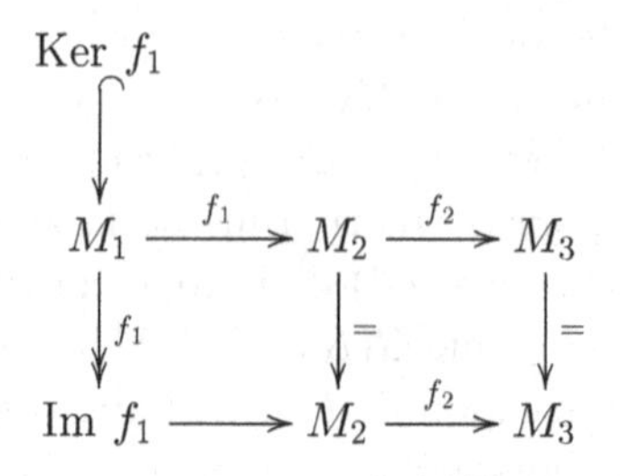

dont les colonnes sont exactes. La ligne inférieure de ce diagramme est naturellement quasi-isomorphe au complexe de longueur 2 [Coker $f_1 \to M_3$], lequel est quasi-isomorphe à la variété semi-abélienne quotient $M := M_3/\text{Coker } f_1$. En résumé, on a donc un triangle exact dans la catégorie dérivée de la forme :

$$\text{Ker } f_1[2] \to C \to M \to \text{Ker } f_1[3]. \tag{5.11}$$

On en déduit le diagramme commutatif (5.12) de la figure 5.4 (voir plus bas), où

$$P^{-1}(k, M) = \prod_{v \text{ réelle}} H^{-1}(k_v, M) \text{ et } P^3(k, \text{Ker } f_1) = \prod_{v \text{ réelle}} H^3(k_v, \text{Ker } f_1).$$

Le morphisme $P^{-1}(k, M) \to \mathbf{H}^2(k, \widehat{M})^D$ est défini via la dualité locale aux places réelles pour les 1-motifs (voir [HS05], Proposition 2.9).

La finitude de $H^3(k, \text{Ker } f_1)$ assure, via l'appendice de [HS05], que la suite

$$\mathbf{H}^0(k, C)^\wedge \to H^0(k, M)^\wedge \to H^3(k, \text{Ker } f_1)$$

est exacte. De même, considérons le morphisme $P^2(k, \text{Ker } f_1) \to \mathbf{P}^0(k, C)$ et notons Q son conoyau (muni de la topologie quotient). Alors la finitude de $P^3(k, \text{Ker } f_1)$ assure que le morphisme $Q^\wedge \to P^0(k, M)^\wedge$ reste injectif, et donc on conclut (en utilisant l'appendice de [HS05]) que

$$P^2(k, \text{Ker } f_1)^\wedge \to \mathbf{P}^0(k, C)^\wedge \to P^0(k, M)^\wedge$$

est exacte. Les groupes $\mathbf{H}^2(k, \widehat{M})$, $H^0(k, \widehat{\text{Ker } f_1})$ et $\mathbf{H}^0(k, \widehat{C})$ étant discrets, on a une suite exacte

$$\mathbf{H}^2(k, \widehat{M})^D \to H^0(k, \widehat{\text{Ker } f_1})^D \to \mathbf{H}^0(k, \widehat{C})^D.$$

Le morphisme $H^3(k, \text{Ker } f_1) \to P^3(k, \text{Ker } f_1)$ est un isomorphisme (voir [Mil06], théorème I.2.13). En outre, puisque M est une variété

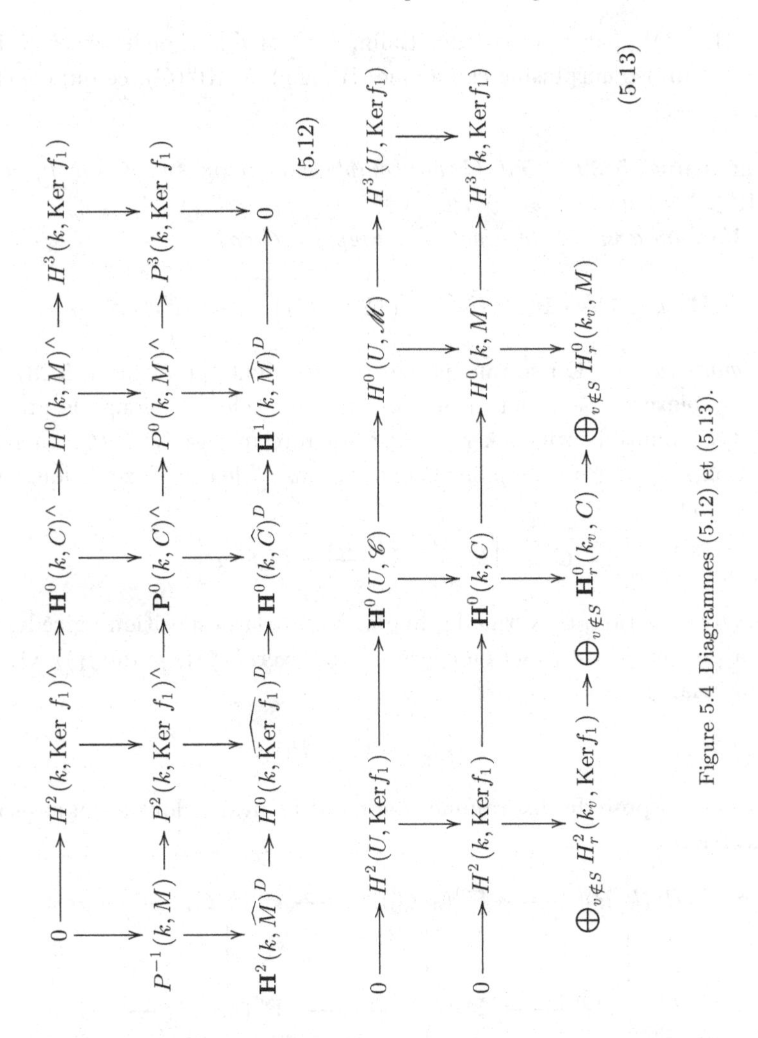

Figure 5.4 Diagrammes (5.12) et (5.13).

semi-abélienne, un dévissage au cas des tores et des variétés abéliennes, à l'aide du théorème 5.6 de [HS05], de la proposition 5.9 de [HS05] et du corollaire I.6.24 de [Mil06], assure que le morphisme $P^{-1}(k, M) \to \mathbf{H}^2(k, \widehat{M})^D$ est un isomorphisme.

Donc la proposition découle de l'exactitude des lignes du diagramme (5.12), et des suites de Poitou-Tate pour M (voir théorème 5.6 de [HS05]) et Ker f_1 (deuxième colonne du diagramme : on peut compléter la suite de Poitou-Tate usuelle car $H^0(k, \widehat{\mathrm{Ker}\, f_1})$ est fini). En ce qui concerne le conoyau du morphisme $\mathbf{P}^0(k, C)^\wedge \to \mathbf{H}^0(k, \widehat{C})^D$, il s'identifie par le lemme du serpent au conoyau de $P^0(k, M)^\wedge \to \mathbf{H}^1(k, \widehat{M})^D$, qui est égal à $Ш^1(\widehat{M})^D$ par Poitou-Tate. Enfin, le dual du triangle exact (5.11) fournit un isomorphisme canonique $Ш^1(\widehat{M}) \cong Ш^0(\widehat{C})$, ce qui conclut la preuve. □

Théorème 5.29 *Soit C un complexe de longueur 3 sur k, avec $Ш^1(M_3{}^{\mathrm{ab}})$ fini.*

Alors on a une suite exacte de groupes abéliens :

$$\mathbf{H}^0(k, C)^\wedge \to \mathbf{P}^0(k, C)^\wedge \to \left(\mathbf{H}^0(k, \widehat{C})\right)^D \to Ш^0(\widehat{C})^D \to 0\,.$$

Démonstration On se ramène au cas précédent (proposition 5.28), où le complexe C est exact en degré -1. Pour cela, on plonge le groupe de type multiplicatif Coker f_1 dans un tore quasi-trivial P : on note i : Coker $f_1 \to P$ un tel plongement. Notons C' le complexe de longueur 3 suivant :

$$C' := \left[M_1 \xrightarrow{f_1} M_2 \xrightarrow{f_2 \oplus i} M_3 \oplus P\right]\,.$$

Par construction, C' vérifie les hypothèses de la proposition précédente, à savoir que C' est exact en degré -1 (i.e. $\mathrm{Ker}(f_2 \oplus i) = \mathrm{Im}(f_1)$). Or on a un triangle exact naturel

$$P \to C' \to C \to P[1]\,,$$

donc on dispose du diagramme commutatif suivant, dont les lignes sont exactes :

$$\begin{array}{ccccccc}
H^0(k, P)^\wedge & \longrightarrow & \mathbf{H}^0(k, C')^\wedge & \longrightarrow & \mathbf{H}^0(k, C)^\wedge & \longrightarrow & 0 \\
\downarrow & & \downarrow & & \downarrow & & \\
P^0(k, P)^\wedge & \longrightarrow & \mathbf{P}^0(k, C')^\wedge & \longrightarrow & \mathbf{P}^0(k, C)^\wedge & \longrightarrow & 0 \\
\downarrow & & \downarrow & & \downarrow & & \\
H^2(k, \widehat{P})^D & \longrightarrow & \mathbf{H}^0(k, \widehat{C'})^D & \longrightarrow & \mathbf{H}^0(k, \widehat{C})^D & \longrightarrow & 0\,.
\end{array}$$

Par la proposition 5.28, la deuxième colonne est exacte, et P étant quasi-trivial, par Poitou-Tate, la flèche $P^0(k, P)^\wedge \to H^2(k, \widehat{P})^D$ est surjective, donc une chasse au diagramme assure l'exactitude de la troisième colonne. En outre, la surjectivité de $P^0(k, P)^\wedge \to H^2(k, \widehat{P})^D$ assure que les conoyaux des morphismes $\mathbf{P}^0(k, C')^\wedge \to \mathbf{H}^0(k, \widehat{C'})^D$ et $\mathbf{P}^0(k, C)^\wedge \to \mathbf{H}^0(k, \widehat{C})^D$ sont canoniquement isomorphes; or le premier est naturellement isomorphe (via la proposition 5.28) au groupe $Ш^0(\widehat{C'})^D$, et il ne reste donc qu'à identifier $Ш^0(\widehat{C})$ et $Ш^0(\widehat{C'})$. On utilise le triangle exact

$$\widehat{C} \to \widehat{C'} \to \widehat{P}[2] \to \widehat{C}[1],$$

qui assure que les lignes du diagramme suivant sont exactes :

$$\begin{array}{ccccccc} H^1(k,\widehat{P}) = 0 & \longrightarrow & \mathbf{H}^0(k,\widehat{C}) & \longrightarrow & \mathbf{H}^0(k,\widehat{C'}) & \longrightarrow & H^2(k,\widehat{P}) \\ \downarrow & & \downarrow & & \downarrow & & \downarrow \\ P^1(k,\widehat{P}) = 0 & \longrightarrow & \mathbf{P}^0(k,\widehat{C}) & \longrightarrow & \mathbf{P}^0(k,\widehat{C'}) & \longrightarrow & \mathbf{P}^2(k,\widehat{P}). \end{array}$$

Or P est quasi-trivial, donc $Ш^2(k,\widehat{P}) = 0$, donc le lemme du serpent assure que le morphisme naturel $Ш^0(\widehat{C}) \to Ш^0(\widehat{C'})$ est un isomorphisme, ce qui conclut la preuve du théorème. □

Désormais, on donne une version de ces résultats sur un ouvert $U = \mathrm{Spec}(\mathcal{O}_{k,S})$ de $\mathrm{Spec}(\mathcal{O}_k)$. Soit $\mathscr{C} = [\mathscr{M}_1 \xrightarrow{f_1} \mathscr{M}_2 \xrightarrow{f_2} \mathscr{M}_3]$ un complexe de longueur 3 sur U : les U-schémas en groupes $\mathscr{M}_i$ sont lisses commutatifs, $\mathscr{M}_1$ et $\mathscr{M}_2$ sont de type multiplicatif, $\mathrm{Ker}(f_1)$ est fini et $\mathscr{M}_3$ est un schéma semi-abélien. On suppose toujours $Ш^1({M_3}^{\mathrm{ab}})$ fini. On définit les groupes $\mathscr{P}^i_S(\mathscr{C})$:

$$\mathscr{P}^i_S(\mathscr{C}) := \prod_{v\in S} \mathbf{H}^i(k_v, C) \times \prod_{v\notin S} \mathbf{H}^i(\mathcal{O}_v, \mathscr{C}), \tag{5.14}$$

et comme dans [Har08], section 2, on note pour $v \notin S$, $\mathbf{H}^i_r(k_v, C) := \mathbf{H}^i(k_v, C)/\mathbf{H}^i(\mathcal{O}_v, \mathscr{C})$.

On s'intéresse au diagramme commutatif suivant :

$$\begin{array}{ccccc} \mathbf{H}^0(U,\mathscr{C})^\wedge & \longrightarrow & \mathscr{P}^0_S(\mathscr{C})^\wedge & \longrightarrow & \mathbf{H}^0(k,\widehat{C})^D \\ \downarrow & & \downarrow i & & \downarrow = \\ \mathbf{H}^0(k,C)^\wedge & \longrightarrow & \mathbf{P}^0(k,C)^\wedge & \longrightarrow & \mathbf{H}^0(k,\widehat{C})^D \\ \downarrow & & \downarrow & & \\ (\bigoplus_{v\notin S} \mathbf{H}^0_r(k_v,C))^\wedge & \xrightarrow{=} & (\bigoplus_{v\notin S} \mathbf{H}^0_r(k_v,C))^\wedge. & & \end{array} \tag{5.15}$$

Montrons les propriétés suivantes, quand l'ouvert U est assez petit :

- i est injective.
- La première colonne est exacte.

Proposition 5.30 *Pour U suffisamment petit, la suite suivante*

$$\mathbf{H}^0(U,\mathscr{C}) \to \mathbf{H}^0(k,C) \to \bigoplus_{v\notin S} \mathbf{H}^0_r(k_v,C)$$

est exacte.

Démonstration • On suppose que le complexe $\mathscr{C}$ est exact en degré -1. Le triangle exact

$$\text{Ker } f_1[2] \to \mathscr{C} \to \mathscr{M} \to \text{Ker } f_1[3],$$

où $\mathscr{M}$ est un U-schéma semi-abélien, induit le diagramme commutatif (5.13) dans la Figure 5.4. Les deux premières lignes sont exactes, le morphisme $H^3(U,\text{Ker}f_1) \to H^3(k,\text{Ker}f_1)$ est injectif ($\text{Ker}f_1$ est fini, donc la proposition II.2.9 et le théorème I.4.10.(c) de [Mil06] identifient canoniquement ces deux groupes à $\bigoplus_{v \text{ réelle}} H^3(k_v,\text{Ker}f_1)$); en outre, le morphisme $\bigoplus_{v\notin S} H^2_r(k_v,\text{Ker}f_1) \to \bigoplus_{v\notin S} \mathbf{H}^0_r(k_v,C)$ est injectif (par une chasse au diagramme facile utilisant l'injectivité de $H^0(\mathscr{O}_v,\mathscr{M}) \to H^0(k_v,M)$ et de $H^2(k_v,\text{Ker}f_1) \to \mathbf{H}^0(k_v,C)$ pour $v \notin S$). Alors l'exactitude de la deuxième colonne est une conséquence immédiate de ces propriétés et du fait que la troisième colonne est exacte; ce dernier fait est démontré à la fin de la preuve de la proposition 1 de [Har08]. D'où la proposition dans ce cas particulier.

- Montrons désormais le cas général. Pour cela, comme dans la preuve du théorème 5.29, on sait qu'il existe un triangle exact $\mathscr{P} \to \mathscr{C}' \to \mathscr{C} \to \mathscr{P}[1]$, où $\mathscr{P}$ est un U-tore quasi-trivial et $\mathscr{C}'$ vérifie les hypothèses du premier point (voir par exemple [CTS87], Proposition 1.3 pour l'existence de $\mathscr{P}$). On a alors le diagramme commutatif suivant

$$\begin{array}{ccccccc}
H^0(U,\mathscr{P}) & \longrightarrow & \mathbf{H}^0(U,\mathscr{C}') & \longrightarrow & \mathbf{H}^0(U,\mathscr{C}) & \longrightarrow & 0 \\
\downarrow & & \downarrow & & \downarrow & & \\
H^0(k,P) & \longrightarrow & \mathbf{H}^0(k,C') & \longrightarrow & \mathbf{H}^0(k,C) & \longrightarrow & 0 \\
\downarrow & & \downarrow & & \downarrow & & \\
\bigoplus_{v\notin S} H^0_r(k_v,P) & \longrightarrow & \bigoplus_{v\notin S} \mathbf{H}^0_r(k_v,C') & \longrightarrow & \bigoplus_{v\notin S} \mathbf{H}^0_r(k_v,C) & &
\end{array}$$

dont les trois lignes sont exactes (la troisième par une chasse au diagramme facile). Alors le cas précédent assure l'exactitude de la deuxième colonne, et pour U assez petit la flèche $H^0(k,P) \to$

$\bigoplus_{v \notin S} H^0_r(k_v, P)$ est surjective (le nombre de classes de P est fini), donc une chasse au diagramme assure la proposition dans le cas général.

□

Proposition 5.31 *Pour U suffisamment petit, la suite suivante est exacte :*

$$\mathbf{H}^0(U, \mathscr{C})^\wedge \to \mathbf{H}^0(k, C)^\wedge \to (\bigoplus_{v \notin S} \mathbf{H}^0_r(k_v, C))^\wedge .$$

Démonstration Il s'agit essentiellement de compléter la suite exacte de la proposition précédente. Pour cela, il suffit de montrer que le morphisme
$\mathbf{H}^0(k, C) \to \bigoplus_{v \notin S} \mathbf{H}^0_r(k_v, C)$ a un conoyau fini.

- On suppose d'abord que le complexe $\mathscr{C}$ est exact en degré -1. On a un triangle exact

$$\text{Ker } f_1[2] \to \mathscr{C} \to \mathscr{M} \to \text{Ker } f_1[3] ,$$

d'où le diagramme commutatif (5.16) (voir figure 5.5), à lignes exactes (l'exactitude de la seconde ligne résulte d'une chasse au diagramme utilisant la trivialité des groupes $H^3(k_v, \text{Ker } f_1)$ et $H^3(\mathscr{O}_v, \text{Ker } f_1)$).

Alors la finitude de $H^3(k, \text{Ker } f_1)$, la finitude des conoyaux des première et troisième flèches verticales (via Poitou-Tate pour les groupes finis pour la première, et via le lemme 3 de [Har08] pour la troisième) et le lemme du serpent assurent la finitude du conoyau de la deuxième flèche verticale, et donc la proposition dans ce cas.
- Cas général : on sait qu'il existe un triangle exact $\mathscr{P} \to \mathscr{C}' \to \mathscr{C} \to \mathscr{P}[1]$ où $\mathscr{P}$ est un U-tore quasi-trivial et $\mathscr{C}'$ vérifie les hypothèses du premier point. On considère alors le diagramme commutatif suivant à lignes exactes :

$$\begin{array}{ccccccc} H^0(k, P) & \longrightarrow & \mathbf{H}^0(k, C') & \longrightarrow & \mathbf{H}^0(k, C) & \longrightarrow & 0 \\ \downarrow & & \downarrow & & \downarrow & & \\ \bigoplus_{v \notin S} H^0_r(k_v, P) & \longrightarrow & \bigoplus_{v \notin S} \mathbf{H}^0_r(k_v, C') & \longrightarrow & \bigoplus_{v \notin S} \mathbf{H}^0_r(k_v, C) & \longrightarrow & 0 . \end{array}$$

Par le cas précédent, la deuxième flèche verticale a un conoyau fini, donc par le lemme du serpent, la troisième également, ce qui conclut la preuve.

□

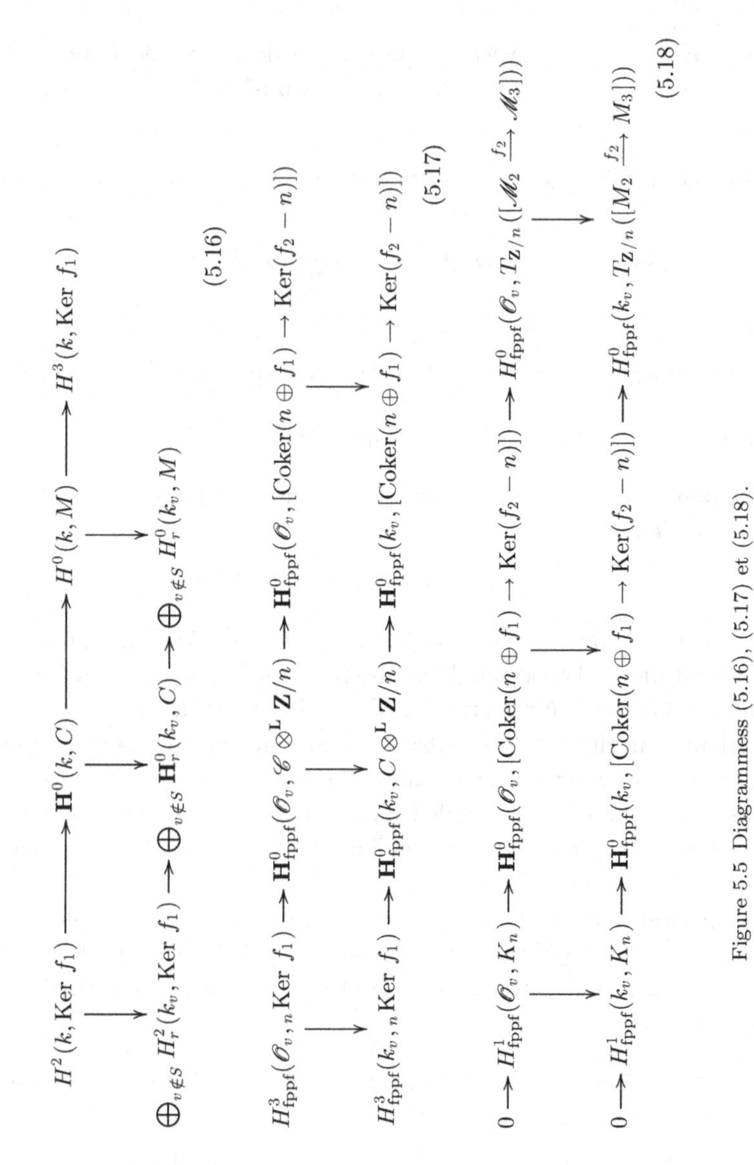

Figure 5.5 Diagrammess (5.16), (5.17) et (5.18).

Proposition 5.32 *Pour S assez grand, le morphisme $i : \mathscr{P}^0_S(\mathscr{C})^\wedge \to \mathbf{P}^0(k, C)^\wedge$ est injectif.*

Démonstration Considérons le triangle exact suivant

$$\mathscr{C} \xrightarrow{n} \mathscr{C} \to \mathscr{C} \otimes^{\mathbf{L}} \mathbf{Z}/n \to \mathscr{C}[1]\,.$$

Suivant la preuve du lemme 5.3 de [HS05], pour montrer que $\mathscr{P}^0_S(\mathscr{C})_\wedge \to \mathbf{P}^0(k,C)_\wedge$ est injectif, il suffit de montrer que pour toute place v hors de S et pour tout entier n non nul, le morphisme

$$\mathbf{H}^0_{\mathrm{fppf}}(\mathscr{O}_v, \mathscr{C} \otimes^{\mathbf{L}} \mathbf{Z}/n) \to \mathbf{H}^0_{\mathrm{fppf}}(k_v, C \otimes^{\mathbf{L}} \mathbf{Z}/n)$$

est injectif (voir la preuve de la proposition 2.5 de [Dem11a] et du lemme 5.21 de [Dem11b]), puis d'utiliser le fait que $\mathscr{C}$ est lisse sur U pour identifier $\mathbf{H}^0_{\mathrm{fppf}}(\mathscr{O}_v, \mathscr{C})$ et $\mathbf{H}^0_{\text{ét}}(\mathscr{O}_v, \mathscr{C})$.

Soit $n \in \mathbf{N}^*$. Puisque $\mathscr{M}_3$ est un U-schéma semi-abélien, le morphisme $f_2 - n : \mathscr{M}_2 \oplus \mathscr{M}_3 \to \mathscr{M}_3$ est surjectif. Par conséquent, on dispose du triangle exact suivant :

$${}_n\mathrm{Ker}\ f_1[3] \to \mathscr{C} \otimes^{\mathbf{L}} \mathbf{Z}/n \to [\mathrm{Coker}(n \oplus f_1) \to \mathrm{Ker}(f_2 - n)] \to {}_n\mathrm{Ker}\ f_1[4]\,.$$

On en déduit donc le diagramme commutatif à lignes exactes (5.17) (voir figure 5.5), pour toute place $v \notin S$. Or par le lemme III.1.1 de [Mil06], le groupe $H^3_{\mathrm{fppf}}(\mathscr{O}_v, {}_n\mathrm{Ker}\ f_1)$ est trivial. Donc l'injectivité de la deuxième flèche verticale est une conséquence de celle de la troisième. Pour montrer celle-ci, on utilise le triangle exact suivant :

$$K_n[1] \to [\mathrm{Coker}(n \oplus f_1) \to \mathrm{Ker}(f_2 - n)] \to T_{\mathbf{Z}/n}([\mathscr{M}_2 \xrightarrow{f_2} \mathscr{M}_3]) \to K_n[2]$$

où K_n est un U-schéma en groupes de type multiplicatif fini, quotient de ${}_n\mathrm{Ker}\ f_2$ et, par définition, $T_{\mathbf{Z}/n}(A) := H^0(A[-1] \otimes^{\mathbf{L}} \mathbf{Z}/n)$ pour tout complexe A de faisceaux fppf (voir [Dem11b], définition 2.2). On considère alors le diagramme commutatif à lignes exactes (5.18) de la figure 5.5.

La troisième flèche verticale est injective, et la première aussi puisque $H^1_v(\mathscr{O}_v, K_n) = 0$ (le groupe K_n est fini plat sur $\mathscr{O}_v$: voir lemme III.1.1 de [Mil06]). Donc on en déduit que le morphisme

$$\begin{aligned}\mathbf{H}^0_{\mathrm{fppf}}(\mathscr{O}_v, [\mathrm{Coker}(n \oplus f_1) \to \mathrm{Ker}(f_2 - n)]) \\ \to \mathbf{H}^0_{\mathrm{fppf}}(k_v, [\mathrm{Coker}(n \oplus f_1) \to \mathrm{Ker}(f_2 - n)])\end{aligned}$$

est injectif, ce qui implique l'injectivité du morphisme

$$\mathbf{H}^0_{\mathrm{fppf}}(\mathscr{O}_v, \mathscr{C} \otimes \mathbf{Z}/n) \to \mathbf{H}^0_{\mathrm{fppf}}(k_v, C \otimes \mathbf{Z}/n)\,.$$

On en déduit alors que l'on a une injection $\mathscr{P}^0_S(\mathscr{C})_\wedge \to \mathbf{P}^0(k,C)_\wedge$. Or on vérifie facilement que $\mathscr{P}^0_S(\mathscr{C})_\wedge \cong \mathscr{P}^0_S(\mathscr{C})^\wedge$ et que le morphisme $\mathbf{P}^0(k,C)_\wedge \to \mathbf{P}^0(k,C)^\wedge$ est injectif (voir preuve de la proposition 5.4 de [HS05]), donc cela conclut la preuve. □

Théorème 5.33 *Soit $\mathscr{C}$ un complexe de longueur 3 sur U, avec $Ш^1({M_3}^{\mathrm{ab}})$ fini. Alors, quitte à réduire U, on a une suite exacte de groupes abéliens :*

$$\mathbf{H}^0(U,\mathscr{C})^\wedge \to \mathscr{P}_S^0(\mathscr{C})^\wedge \to \mathbf{H}^0(k,\widehat{C})^D .$$

Démonstration Il s'agit juste d'une chasse au diagramme dans le diagramme (5.15) considéré plus haut : l'exactitude de la deuxième ligne (théorème 5.29), de la première colonne (proposition 5.31) et l'injectivité de i (proposition 5.32) assurent l'exactitude de la première ligne, d'où le théorème. □

Terminons cette section par un résultat d'approximation faible aux places infinies :

Théorème 5.34 *Soit C un complexe de longueur 3 sur k, tel que M_3 est linéaire (donc un k-tore). Alors le morphisme*

$$\mathbf{H}^0(k,C) \to \prod_{v\in S_\infty} \mathbf{H}^0(k_v,C)$$

est surjectif.

Démonstration • On suppose d'abord que C est exact en degré -1. On a un triangle exact

$$\mathrm{Ker}\ f_1[2] \to C \to M \to \mathrm{Ker}\ f_1[3]$$

où M est un quotient de M_3, donc un k-tore. D'où le diagramme commutatif exact suivant :

$$\begin{array}{ccccccc}
H^2(k,\mathrm{Ker}\ f_1) & \longrightarrow & \mathbf{H}^0(k,C) & \longrightarrow & H^0(k,M) & \longrightarrow & H^3(k,\mathrm{Ker}\ f_1) \\
\downarrow & & \downarrow & & \downarrow & & \downarrow \cong \\
P^2_\infty(k,\mathrm{Ker}\ f_1) & \longrightarrow & \mathbf{P}^0_\infty(k,C) & \longrightarrow & P^0_\infty(k,M) & \longrightarrow & P^3_\infty(k,\mathrm{Ker}\ f_1) .
\end{array}$$

Or dans ce diagramme aux lignes exactes, la première et la troisième colonnes sont surjectives (pour la première, c'est une conséquence de la suite de Poitou-Tate pour Ker f_1 : voir [NSW08], proposition 9.2.1. Pour la troisième, c'est une conséquence de l'approximation faible aux places infinies pour un tore : voir [San81], corollaire 3.5.(iii)), et la dernière est un isomorphisme, donc une chasse au diagramme assure la surjectivité de la deuxième flèche verticale.

• Cas général : il existe un triangle exact $P \to C' \to C \to P[1]$ où P est un tore quasi-trivial et où C' vérifie les hypothèses du point précédent.

Le résultat est alors évident par le cas précédent et la surjectivité de $\mathbf{P}^0_\infty(k, C') \to \mathbf{P}^0_\infty(k, C)$.

□

5.7 Le défaut d'approximation forte dans les espaces homogènes

L'objectif ici est de combiner les résultats des sections précédentes pour calculer le défaut d'approximation forte pour un espace homogène. On pourra consulter [BD11] pour un calcul indépendant de ce défaut en termes d'obstruction de Brauer-Manin.

Rappelons quelques notations usuelles : si G est un groupe connexe sur un corps de nombres k, pour un ensemble fini de places S, on note G_S le groupe $\prod_{v \in S} G(k_v)$ et $\rho : G^{\mathrm{scu}} \to G$ le morphisme naturel. Si X est une variété algébrique sur k, on note $P^0(k, X)$ l'ensemble des points adéliques modifiés de X, à savoir le produit

$$P^0(k, X) := \prod_{v \in \Omega_\infty} \pi_0(X(k_v)) \times \prod_{v \in \Omega_f}{}' X(k_v)$$

où $\pi_0(X(k_v))$ désigne l'ensemble des composantes connexes de $X(k_v)$, Ω_∞ (resp. Ω_f) l'ensemble des places infinies (resp. finies) de k, et $\prod'_{v \in \Omega_f} X(k_v)$ désigne le produit restreint des $X(k_v)$ par rapport aux $\mathscr{X}(\mathscr{O}_v)$ (pour un modèle $\mathscr{X}$ de X sur un ouvert de $\mathrm{Spec}(\mathscr{O}_k)$). On a donc une surjection naturelle $X(\mathbf{A}_k) \to P^0(k, X)$.

Rappelons enfin que pour un morphisme $f : H \to G$ de k-groupes algébriques, on dispose de l'espace topologique adélique $\mathbf{P}^0(k, [H \to G])$ muni de sa topologie de produit restreint (voir section 5.6.1 pour la définition de la topologie sur un corps local).

5.7.1 Espaces homogènes à stabilisateurs connexes

Théorème 5.35 *Soit k un corps de nombres, G un k-groupe connexe, H un k-groupe linéaire connexe, S_0 un ensemble fini de places de k. Soit $f : H \to G$ un morphisme de k-groupes. On suppose que le groupe G^{sc} vérifie l'approximation forte hors de S_0 et que $Ш^1(G^{\mathrm{ab}})$ est fini.*

Alors il existe une application naturelle (surjective si G est linéaire)

$$\mathbf{P}^0(k, [H \to G]) \xrightarrow{\theta} \left(\mathbf{H}^0(k, \widehat{C}_f)/\ Ш^0(\widehat{C}_f)\right)^D,$$

dont le noyau est exactement l'adhérence forte $\overline{\rho(G^{\mathrm{scu}}_{S_0}).\mathbf{H}^0(k,[H \to G])}$ *de* $\rho(G^{\mathrm{scu}}_{S_0}).\mathbf{H}^0(k,[H \to G])$ *dans* $\mathbf{P}^0(k,[H \to G])$.

Remarque 5.36 Le cas particulier $H = 1$ fournit une généralisation du corollaire 3.20 de [Dem11a] au cas des groupes non linéaires. Le cas particulier $G = 1$ n'est autre que la suite exacte de Kottwitz-Borovoi (voir [Bor98], théorème 5.16 et [Dem11a], théorème 5.1, deuxième ligne de la première suite exacte).

Corollaire 5.37 *Soit k un corps de nombres, G un k-groupe connexe, S_0 un ensemble fini de places de k. Soit H un sous-k-groupe linéaire connexe de G, et soit $X := G/H$. On suppose que le groupe G^{sc} vérifie l'approximation forte hors de S_0 et que $\text{Ш}^1(G^{\mathrm{ab}})$ est fini. On note $C_X := C_f$, où $f : H \to G$ est l'inclusion naturelle.*

Alors il existe une application naturelle (surjective si G est linéaire) $P^0(k,X) \xrightarrow{\theta} \left(\mathbf{H}^0(k,\widehat{C}_X)/\ \text{Ш}^0(\widehat{C}_X)\right)^D$, *dont le noyau est exactement l'adhérence forte* $\overline{\rho(G^{\mathrm{scu}}_{S_0}).X(k)}$ *de* $\rho(G^{\mathrm{scu}}_{S_0}).X(k)$ *dans* $P^0(k,X)$.

Démonstration du corollaire 5.37 Il suffit d'appliquer le théorème 5.35 à l'inclusion $f : H \to G$ de H dans G, et d'identifier $\mathbf{H}^0(k,[H \to G])$ avec $X(k)$. □

Remarque 5.38

1. L'hypothèse de linéarité de H dans le corollaire n'est pas une restriction importante : en effet, tout espace homogène d'un groupe connexe G à stabilisateur connexe peut être muni d'une structure d'espace homogène d'un groupe connexe G' à stabilisateur *linéaire* connexe.
2. On peut formuler ce corollaire sous la forme d'une suite exacte naturelle d'ensembles pointés
$$1 \to \overline{\rho(G^{\mathrm{scu}}_{S_0}).X(k)} \to P^0(k,X) \xrightarrow{\theta} \left(\mathbf{H}^0(k,\widehat{C}_X)/\ \text{Ш}^0(\widehat{C}_X)\right)^D \to 1\,.$$
3. Pour toute place archimédienne v de k, le groupe $G^{\mathrm{scu}}(k_v)$ est connexe, donc par définition de l'ensemble $P^0(k,X)$, on peut remplacer dans l'énoncé du corollaire l'adhérence de $\rho(G^{\mathrm{scu}}_{S_0}).X(k)$ par l'adhérence de $\rho(G^{\mathrm{scu}}_{S_0^f}).X(k)$, où S_0^f est l'ensemble des places finies contenues dans S_0. Par exemple, si $S_0 \subset \Omega_\infty$, alors le corollaire 5.37 décrit l'adhérence de $X(k)$ dans $P^0(k,X)$.
4. Au vu du résultat principal de [Dem11c], le groupe $\mathbf{H}^0(k,\widehat{C}_X)/\ \text{Ш}^0(\widehat{C}_X)$ s'identifie au sous-groupe $\mathrm{Br}_a(X,G)$ du groupe de Brauer $\mathrm{Br}(X)/\mathrm{Br}(k)$ introduit dans [BD11]. On peut donc voir

le corollaire 5.37 comme une version cohomologique explicite du résultat principal de [BD11].

Démonstration du théorème 5.35 Soit $U = \mathrm{Spec}(\mathscr{O}_{k,S})$ un ouvert de $\mathrm{Spec}(\mathscr{O}_k)$ de bonne réduction pour H, G et f, avec S contenant S_0 et les places archimédiennes de k. On suppose U suffisamment petit pour pouvoir appliquer les résultats des sections 5.5.1 et 5.6. On note $\mathscr{H}$, $\mathscr{G}$ et f des modèles respectifs de H, G et f sur U. On note C_H, C_G, C_f les complexes de k-groupes introduits à la section 5.5, et $\mathscr{C}_H$, $\mathscr{C}_G$ et $\mathscr{C}_f$ leurs analogues sur U.

On considère alors le diagramme commutatif suivant (les flèches obliques sont les flèches d'abélianisation pour H, G et f, les flèches étiquetées θ sont définies via la dualité locale ou l'obstruction de Brauer-Manin), construit notamment à partir du théorème 5.17 :

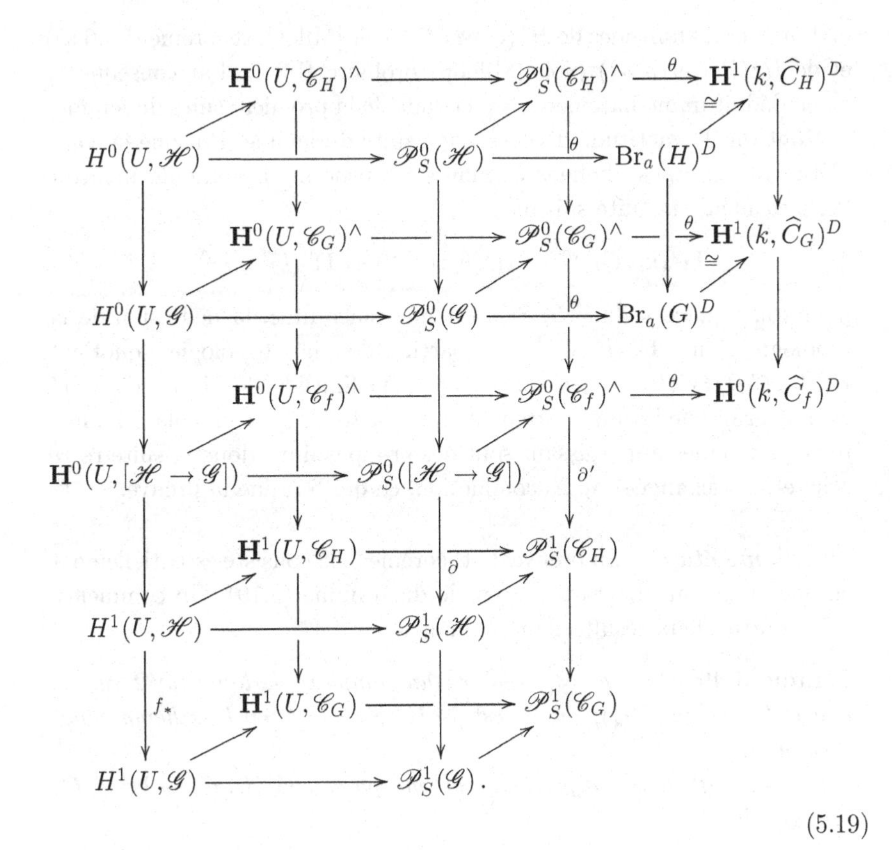

(5.19)

On rappelle que les groupes $\mathscr{P}^i_S(\mathscr{C}_H)$, $\mathscr{P}^i_S(\mathscr{C}_G)$ et $\mathscr{P}^i_S(\mathscr{C}_f)$ ont été définis en (5.14).

Première étape : Montrons l'exactitude de deux suites extraites du diagramme (5.19) :

Lemme 5.39 *Les suites suivantes*

$$\mathbf{H}^0(U,\mathscr{C}_G)^\wedge \to \mathbf{H}^0(U,\mathscr{C}_f)^\wedge \to \mathbf{H}^1(U,\mathscr{C}_H)$$

$$\mathscr{P}^0_S(\mathscr{C}_H)^\wedge \to \mathscr{P}^0_S(\mathscr{C}_G)^\wedge \to \mathscr{P}^0_S(\mathscr{C}_f)^\wedge$$

sont exactes.

Démonstration On montre d'abord la finitude de $\mathbf{H}^1(U,\mathscr{C}_H)$. Il suffit de considérer le triangle exact suivant

$$\mathrm{Ker}(\rho_{\mathscr{H}})[1] \to \mathscr{C}_H \to \mathscr{H}^{\mathrm{tor}} \to \mathrm{Ker}(\rho_{\mathscr{H}})[2]$$

et d'utiliser les finitudes de $H^1(U,\mathscr{H}^{\mathrm{tor}})$ (voir [Mil06], théorème II.4.6.a)) et de $H^2(U,\mathrm{Ker}(\rho_{\mathscr{H}}))$ (voir [Mil06], corollaire II.3.3). Par conséquent, on en déduit immédiatement l'exactitude de la première suite du lemme.

Montrons l'exactitude de la seconde suite du lemme. Puisque la complétion de groupes abéliens commute au produit, il suffit de montrer l'exactitude de la suite suivante :

$$\mathbf{H}^0(k_v,C_H)^\wedge \to \mathbf{H}^0(k_v,C_G)^\wedge \to \mathbf{H}^0(k_v,C_f)^\wedge$$

pour toute place v dans S. Si v est une place finie, la définition de la topologie sur $\mathbf{H}^0(k_v,C_f)$ à partir de la topologie quotient $\mathbf{H}^0(k_v,C_G)/\mathbf{H}^0(k_v,C_H)$ assure, grâce à la finitude de $\mathbf{H}^1(k_v,C_H)$ (H est linéaire), que la suite précédente est exacte. Si v est une place infinie, tous les groupes apparaissant sont des groupes finis, donc la suite reste exacte (et inchangée) après complétion, ce qui termine la preuve. □

Deuxième étape : La preuve du théorème 5.35 consiste essentiellement en une chasse au diagramme dans le diagramme (5.19). On commence par montrer deux résultats préliminaires :

Lemme 5.40 *Soit* $\mathscr{C} = [\mathscr{T} \to \mathscr{S}]$ *un complexe de longueur 2 sur un ouvert* U *de* $\mathrm{Spec}(\mathscr{O}_k)$*, où* $\mathscr{T}$ *est un* U*-tore et* $\mathscr{S}$ *un* U*-schéma semi-abélien.*

Si U *est suffisamment petit, alors le morphisme* $\mathbf{H}^1(U,\mathscr{C}) \to \mathbf{H}^1(k,C)$ *est injectif.*

Démonstration Montrons d'abord que le groupe $\mathrm{Ker}(\mathbf{H}^1(U,\mathscr{C}) \to \mathbf{H}^1(k,C))$ est annulé par un entier N indépendant de U : par un argument de restriction-corestriction, on peut supposer que les tores $\mathscr{T}$ et $\mathscr{S}^{\mathrm{lin}}$ sont déployés sur U. Or le morphisme $\mathrm{Br}(U) \to \mathrm{Br}(k)$ est injectif, donc par Hilbert 90, pour montrer que le noyau $\mathrm{Ker}(\mathbf{H}^1(U,\mathscr{C}) \to \mathbf{H}^1(k,C))$ est trivial, il suffit de montrer l'injectivité de $H^1(U,\mathscr{S}) \to H^1(k,S)$. On voit alors $\mathscr{S}$ comme une extension d'un schéma abélien $\mathscr{A}$ par le U-tore déployé $\mathscr{S}^{\mathrm{lin}}$. On dispose du diagramme commutatif suivant

$$\begin{array}{ccccc} 0 = H^1(U,\mathscr{S}^{\mathrm{lin}}) & \longrightarrow & H^1(U,\mathscr{S}) & \longrightarrow & H^1(U,\mathscr{A}) \\ & & \downarrow & & \downarrow \\ 0 = H^1(k,S^{\mathrm{lin}}) & \longrightarrow & H^1(k,S) & \longrightarrow & H^1(k,A)\,, \end{array}$$

où l'égalité $H^1(U,\mathscr{S}^{\mathrm{lin}}) = 0$ est valable si U est assez petit, car le groupe $\mathrm{Pic}(U)$ est trivial pour U assez petit. La preuve du lemme II.5.5 de [Mil06] assure que la dernière flèche verticale est injective, donc le morphisme $H^1(U,\mathscr{S}) \to H^1(k,S)$ est injectif. Cela assure donc qu'il existe un entier N tel que pour tout ouvert U assez petit, $\mathrm{Ker}(\mathbf{H}^1(U,\mathscr{C}) \to \mathbf{H}^1(k,C))$ est annulé par N. Soit alors U assez petit pour avoir la propriété précédente, et de sorte que N soit inversible sur U. Alors la proposition 5.3 de [Dem11b] assure que le morphisme $\mathbf{H}^1(U,\mathscr{C}) \to \mathbf{H}^1(k,C)$ est injectif. □

Proposition 5.41 *Soit $\mathscr{H}$ un schéma en groupes affine lisse connexe sur un ouvert de* $\mathrm{Spec}(\mathscr{O}_k)$. *Alors pour tout ouvert U de* $\mathrm{Spec}(\mathscr{O}_k)$ *suffisamment petit, le diagramme suivant*

$$\begin{array}{ccc} H^1(U,\mathscr{H}) & \longrightarrow & \mathscr{P}^1_S(\mathscr{H}) \\ \downarrow{\scriptstyle \mathrm{ab}^1_{\mathscr{H}}} & & \downarrow{\scriptstyle \mathrm{ab}^1_{\mathscr{H}}} \\ H^1_{\mathrm{ab}}(U,\mathscr{H}) & \longrightarrow & \mathscr{P}^1_S(\mathscr{C}_H) \end{array}$$

est cartésien.

Démonstration Le théorème 5.11.(i) de [Bor98] assure que le diagramme suivant est cartésien :

$$\begin{array}{ccc} H^1(k,H) & \longrightarrow & P^1(k,H) \\ \downarrow{\scriptstyle \mathrm{ab}^1_H} & & \downarrow{\scriptstyle \mathrm{ab}^1_H} \\ H^1_{\mathrm{ab}}(k,H) & \longrightarrow & \mathbf{P}^1(k,C_H)\,. \end{array}$$

La proposition est alors une conséquence immédiate du corollaire A.8 de [GP08] (voir également le classique lemme 4.1.3 de [Har67]), de l'injectivité de $\mathscr{P}^1_S(\mathscr{H}) \to P^1(k, H)$ et du lemme 5.40. □

Poursuivons la deuxième étape de la preuve du théorème. Soit $(x_v) \in \mathbf{P}(k, [H \to G])$ d'image nulle dans $\mathbf{H}^0(k, \widehat{C}_f)^D$. L'objectif de cette étape et de la suivante est de montrer que (x_v) est dans l'adhérence de $\rho(G^{\text{scu}}_{S_0}).\mathbf{H}^0(k, [H \to G])$ dans $\mathbf{P}^0(k, [H \to G])$.

Tout d'abord, quitte à réduire U, on peut supposer que $(x_v) \in \mathscr{P}^0_S([\mathscr{H} \to \mathscr{G}])$.

Grâce au théorème 5.33, l'élément $(x'_v) := \text{ab}^0_f(x_v) \in \mathscr{P}^0_S(\mathscr{C}_f)^\wedge$ se relève en un élément $x' \in \mathbf{H}^0(U, \mathscr{C}_f)^\wedge$. On note h' l'image de x' dans $\mathbf{H}^1(U, \mathscr{C}_H)$. Par commutativité du diagramme (5.19), $(h_v) := \partial((x_v)) \in \mathscr{P}^1_S(\mathscr{H})$ et h' ont même image (h'_v) dans $\mathscr{P}^1_S(\mathscr{C}_H)$. La proposition 5.41 assure alors qu'il existe $h \in H^1(U, \mathscr{H})$ relevant $(h_v) \in \mathscr{P}^1_S(\mathscr{H})$ de sorte que $\text{ab}^1_{\mathscr{H}}(h) = h'$. Montrons que l'image $g := f_*(h) \in H^1(U, \mathscr{G})$ de h est triviale, quitte à réduire encore U.

Pour cela, on considère le diagramme commutatif suivant :

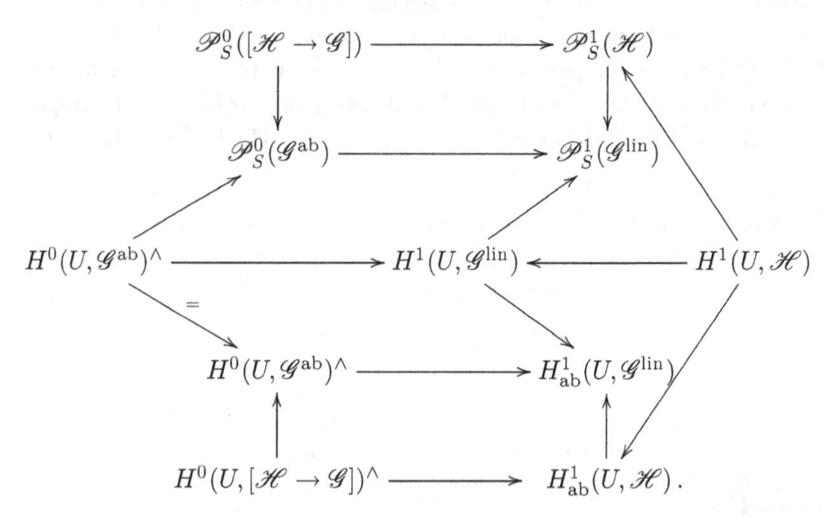

Notons que la classe g provient d'une classe $g' \in H^1(U, \mathscr{G}^{\text{lin}})$.

Notons g_0 l'image de x' par le morphisme $H^0(U, [\mathscr{H} \to \mathscr{G}])^\wedge \to H^0(U, \mathscr{G}^{\text{ab}})^\wedge$, et g'_0 l'image de g_0 dans $H^1(U, \mathscr{G}^{\text{lin}})$. Alors par construction, $(g_{0,v}) \in \mathscr{P}^0_S(\mathscr{G}^{\text{ab}})$ est l'image de (x_v) par $\mathscr{P}^0_S([\mathscr{H} \to \mathscr{G}]) \to \mathscr{P}^0_S(\mathscr{G}^{\text{ab}})$. Alors, par commutativité du diagramme, les classes g'_0 et g' dans $H^1(U, \mathscr{G}^{\text{lin}})$ ont même image dans $\mathscr{P}^1_S(\mathscr{G}^{\text{lin}})$ et dans $H^1_{\text{ab}}(U, \mathscr{G}^{\text{lin}})$. Par conséquent, en utilisant le théorème 5.11 de [Bor98] , on en déduit que $g' = g'_0$ dans $H^1(k, G^{\mathrm{lin}})$. Donc cela assure finalement que $g' \in H^1(k, G^{\mathrm{lin}})$ provient d'un élément de $H^0(k, G^{\mathrm{ab}})^\wedge$, donc aussi d'un élément de $H^0(k, G^{\mathrm{ab}})$. Donc l'image de g dans $H^1(k, G)$ est triviale. Donc quitte à réduire U, on peut supposer que g est trivial dans $H^1(U, \mathscr{G})$.

Alors, par exactitude de la première colonne du diagramme (5.19), h se relève en un point $\overline{x} \in \mathbf{H}^0(U, [\mathscr{H} \to \mathscr{G}])$. Alors les points (x_v) et $(\overline{x}_v)$ dans $\mathscr{P}^0_S([\mathscr{H} \to \mathscr{G}])$ ont même image (h_v) dans $\mathscr{P}^1_S(\mathscr{H})$, donc par exactitude de la troisième colonne du diagramme, il existe un point $(\overline{g}_v) \in \mathscr{P}^0_S(\mathscr{G})$ tel que $(x_v) = (\overline{g}_v).(\overline{x}_v)$ dans $\mathscr{P}^0_S([\mathscr{H} \to \mathscr{G}])$.

Troisième étape : On a construit à l'étape précédente un point $(\overline{g}_v) \in \mathscr{P}^0_S(\mathscr{G})$ tel que $(x_v) = (\overline{g}_v).(\overline{x}_v)$. Dans cette étape, on approxime le point $(\overline{g}_v)$ ainsi obtenu par un point rationnel de G, grâce aux résultats sur le défaut d'approximation forte dans les groupes connexes.

On considère le morphisme tordu par (un représentant de) $\overline{x}$, noté ${}_{\overline{x}}f : {}_{\overline{x}}\mathscr{H} \to \mathscr{G}$. On peut construire l'analogue du diagramme (5.19) pour ${}_{\overline{x}}f$, avec l'application $\mathrm{ab}^0_{\overline{x}f}$ décrite dans la proposition 5.18. On utilise cette proposition dans la suite, sans nécessairement la mentionner.

On note $(\overline{g}'_v) := (\mathrm{ab}^0_{\mathscr{G}}(\overline{g}_v)) \in \mathscr{P}^0(\mathscr{C}_G)$. Par commutativité du diagramme, les points x' et $\overline{x}' := \mathrm{ab}^0_{\overline{x}f}(\overline{x})$ dans $\mathbf{H}^0(U, \mathscr{C}_f; \overline{x})$ ont même image $\widetilde{h}'$ dans $\mathbf{H}^1(U, \mathscr{C}_{\overline{x}H})$, donc par exactitude, ces deux points diffèrent d'un élément $\widetilde{g}' \in \mathbf{H}^0(U, \mathscr{C}_G)^\wedge$. Par commutativité de l'analogue du diagramme (5.19), avec ${}_{\overline{x}}\mathscr{H}$ à la place de $\mathscr{H}$, les éléments $(\overline{g}'_v)$ et $(\widetilde{g}'_v)$ (qui sont dans $\mathscr{P}^0_S(\mathscr{C}_G)^\wedge$) ont même image dans $\mathscr{P}^0_S(\mathscr{C}_f; \overline{x})$ (à savoir (x'_v)). Par conséquent, il existe $(\overline{h}'_v) \in \mathscr{P}^0_S(\mathscr{C}_{\overline{x}H})^\wedge$ tel que $(\overline{h}'_v)$ s'envoie dans $\mathscr{P}^0_S(\mathscr{C}_G)^\wedge$ sur la différence $(\widetilde{g}'_v) - (\overline{g}'_v)$.

Or la preuve du théorème 3.19 et le lemme 2.7 de [Dem11a] assurent que $P^0(k, {}_{\overline{x}}H)$ et $\mathbf{P}^0(k, C_{\overline{x}H})^\wedge$ ont même image par θ dans $\mathbf{H}^1(k, \widehat{C}_{\overline{x}H})^D$. Donc il existe $(\overline{h}_v) \in P^0(k, {}_{\overline{x}}H)$ tel que $\theta((\overline{h}_v)) = \theta((\overline{h}'_v))$ dans $\mathbf{H}^1(k, \widehat{C}_{\overline{x}H})^D$. Considérons alors le point $(\overline{p}_v) := (\overline{g}_v).(\overline{h}_v)^{-1} \in P^0(k, G)$. Par commutativité du cube supérieur droit de (5.19) et par construction des divers points, il est clair que $\theta((\overline{p}_v)) = 0$ dans $\mathrm{Br}_a(G)^D$. Donc grâce au résultat pour les groupes connexes (voir corollaire 3.20 de [Dem11a] dans le cas linéaire; le cas général peut se démontrer de la même façon en remplaçant la proposition 2.5 de [Dem11a] par le théorème 5.33), on en déduit que l'élément $(\overline{p}_v) \in P^0(k, G)$ est dans l'adhérence de $\rho(G^{\mathrm{scu}}_{S_0}).G(k)$. Donc par commutativité du diagramme, le point $(x_v) = (\overline{p}_v).(\overline{x}_v) = \pi_{\overline{x}}((\overline{p}_v))$ est produit d'un élément de $\overline{\rho(G^{\mathrm{scu}}_{S_0}).G(k)}$ par un élément de $\mathbf{H}^0(U,[\mathscr{H}\to\mathscr{G}])$. En particulier, le point (x_v) est dans l'adhérence de $\rho(G^{\mathrm{scu}}_{S_0}).\mathbf{H}^0(k,[H\to G])$ dans $\mathscr{P}^0_S([\mathscr{H}\to\mathscr{G}])$.

D'où finalement une suite exacte, en passant à la limite inductive sur U :

$$\overline{\rho(G^{\mathrm{scu}}_{S_0}).\mathbf{H}^0(k,[H\to G])}\to\mathbf{P}^0(k,[H\to G])\xrightarrow{\theta}\mathbf{H}^0(k,\widehat{C}_f)^D\,.$$

Quatrième étape : on suppose G linéaire. L'objectif de cette étape de la preuve est de calculer le conoyau de l'application $\theta:\mathbf{P}^0(k,[H\to G])\to\mathbf{H}^0(k,\widehat{C}_f)^D$. Pour cela, on montre que $\mathbf{P}^0(k,C_f)$ et $\mathbf{P}^0(k,[H\to G])$ ont même image par θ. Considérons le diagramme suivant :

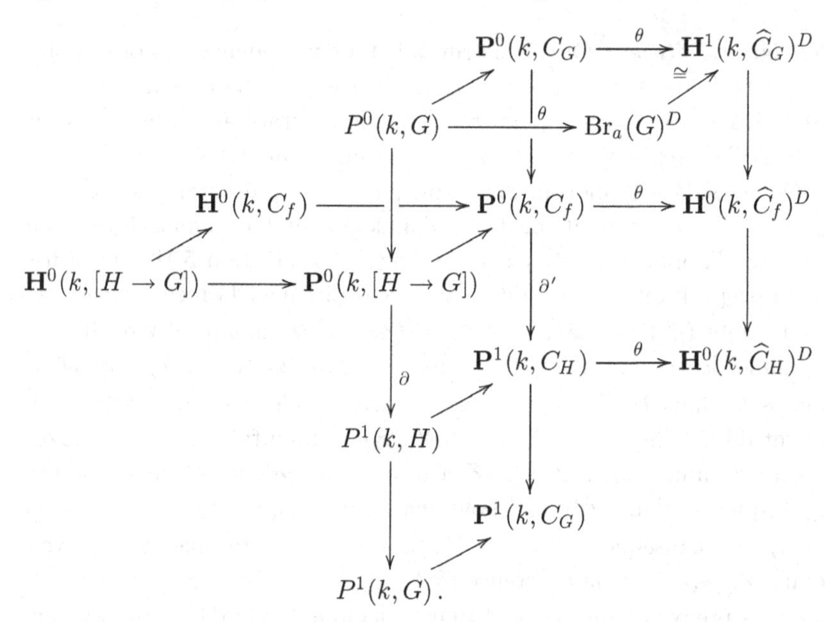

Soit alors $\gamma\in\mathbf{H}^0(k,\widehat{C}_f)^D$ et $\alpha\in P^0(k,C_f)$ tel que $\theta(\alpha)=\gamma$. Par le théorème 5.34, on sait que le morphisme $\mathbf{H}^0(k,C_f)\to\mathbf{P}^0_\infty(k,C_f)$ est surjectif. Donc il existe $\alpha'\in\mathbf{H}^0(k,C_f)$ tel que $\alpha'_\infty=\alpha_\infty$. Notons alors $\alpha^0:=\alpha-\alpha'\in\mathbf{P}^0(k,C_f)$. On a toujours $\theta(\alpha^0)=\gamma$. On note β l'image de α^0 dans $P^1(k,C_H)$ et γ' celle de γ dans $\mathbf{H}^0(k,\widehat{C}_H)^D$. Par le théorème 5.11 de [Bor98], il existe $\tau\in P^1(k,H)$, trivial aux places infinies, tel que $\mathrm{ab}^0_H(\tau)=\beta$. Notons g l'image de τ dans $P^1(k,G)$.

Par construction, g_∞ est trivial, et g s'envoie sur 0 dans $P^1(k,C_G)$. Donc par le même théorème 5.11 de [Bor98], g est trivial. Par conséquent, τ se relève en un élément $x'\in\mathbf{P}^0(k,[H\to G])$. Par commutativité du diagramme, on voit que $(x'_v)\in\mathbf{P}^0(k,C_f)$ s'envoie sur β dans $\mathbf{P}^1(k,C_H)$, donc par exactitude de la troisième colonne, $\alpha^0 - (x'_v)$ se relève en un élément g' dans $\mathbf{P}^0(k, C_G)$. La preuve du théorème 3.19 de [Dem11a] assure que $\mathbf{P}^0(k, C_G)$ et $P^0(k, G)$ ont même image dans $\mathbf{H}^1(k, \widehat{C}_G)^D$ par le morphisme θ. Donc il existe $g^0 \in P^0(k, G)$ tel que $\theta(g^0) = \theta(g')$ dans $\mathbf{H}^1(k, \widehat{C}_G)^D$. Finalement, l'élément $g^0.x' \in \mathbf{P}^0(k, [H \to G])$ vérifie bien $\theta(g^0.x') = \gamma$. Cela assure donc que $\mathbf{P}^0(k, [H \to G])$ et $\mathbf{P}^0(k, C_f)$ ont même image dans $\mathbf{H}^0(k, \widehat{C}_f)^D$. Or pour des raisons topologiques, $\mathbf{P}^0(k, C_f)$ et $\mathbf{P}^0(k, C_f)^\wedge$ ont même image également par θ; enfin, le conoyau de $\theta : \mathbf{P}^0(k, C_f)^\wedge \to \mathbf{H}^0(k, \widehat{C}_f)^D$ est $Ш^0(\widehat{C}_f)^D$ (voir théorème 5.29). Cela termine la preuve du théorème 5.35. □

5.7.2 Espaces homogènes à stabilisateurs abéliens

Dans cette section, on s'intéresse au cas où le groupe H est abélien, non nécessairement connexe.

Théorème 5.42 *Soit k un corps de nombres, G un k-groupe connexe tel que G^{ss} est simplement connexe, H un k-groupe linéaire commutatif, S_0 un ensemble fini de places de k. Soit $f : H \to G$ un morphisme de k-groupes. On suppose que le groupe G^{sc} vérifie l'approximation forte hors de S_0 et que $Ш^1(G^{ab})$ est fini.*

Alors la dualité locale définit une application (surjective si G est linéaire) $\mathbf{P}^0(k, [H \to G]) \xrightarrow{\theta} \left(\mathbf{H}^1(k, \widehat{C}_f^{ab}) / Ш^1(\widehat{C}_f^{ab})\right)^D$ dont le noyau est l'adhérence forte $\overline{\rho(G_{S_0}^{scu}).\mathbf{H}^0(k, [H \to G])}$ de $\rho(G_{S_0}^{scu}).\mathbf{H}^0(k, [H \to G])$ dans $\mathbf{P}^0(k, [H \to G])$.

Remarque 5.43 Le cas particulier $G = 1$ n'est autre que la suite exacte de Poitou-Tate en degré 1 pour les groupes linéaires commutatifs (voir la deuxième ligne de la première suite de [Dem11b], théorème 6.3).

Corollaire 5.44 *Soit k un corps de nombres, G un k-groupe connexe tel que G^{ss} est simplement connexe, S_0 un ensemble fini de places de k. Soit H un sous-k-groupe linéaire commutatif de G, et soit $X := G/H$. On suppose que le groupe G^{sc} vérifie l'approximation forte hors de S_0 et que $Ш^1(G^{ab})$ est fini. On note $C_X^{ab} := C_f^{ab}$, où $f : H \to G$ est l'inclusion naturelle.*

Alors la dualité locale définit une application (surjective si G est linéaire) $P^0(k, X) \xrightarrow{\theta} \left(\mathbf{H}^1(k, \widehat{C}_X^{ab}) / Ш^1(\widehat{C}_X^{ab})\right)^D$ dont le noyau est l'adhérence forte $\overline{\rho(G_{S_0}^{scu}).X(k)}$ de $\rho(G_{S_0}^{scu}).X(k)$ dans $P^0(k, X)$.

Démonstration du corollaire 5.44 Il suffit d'appliquer le théorème 5.42 à l'inclusion $f : H \to G$ de H dans G, et d'identifier $\mathbf{H}^0(k, [H \to G])$ avec $X(k)$. □

Remarque 5.45 En utilisant l'isomorphisme $\mathrm{Br}_1(X)/\mathrm{Br}(k) \cong \mathbf{H}^1(k, \widehat{C}_X^{\mathrm{ab}})$ (voir [BvH11], théorème 7.2), ainsi que la comparaison entre l'accouplement de Brauer-Manin et l'accouplement $P^0(k, C_X^{\mathrm{ab}}) \times \mathbf{H}^1(k, \widehat{C}_X^{\mathrm{ab}}) \to \mathbf{Q}/\mathbf{Z}$ induit par le cup-produit local (voir [Dem09], lemme 4.5.1, ou [HS08], fin de la section 6 pour le cas des tores), ce corollaire généralise le théorème 4.5 de [CTX09] et le théorème 4.3 de [HS11], où le groupe G est supposé semi-simple simplement connexe. En résumé, le corollaire 5.44 affirme que l'obstruction de Brauer-Manin algébrique à l'approximation forte sur X est la seule.

Démonstration du théorème 5.42 Soit $U = \mathrm{Spec}(\mathscr{O}_{k,S})$ un ouvert de $\mathrm{Spec}(\mathscr{O}_k)$ de bonne réduction pour H, G et f, avec S contenant S_0 et les places archimédiennes de k. On suppose U suffisamment petit pour pouvoir appliquer les résultats des sections 5.6 et 5.5.2. On note $\mathscr{H}$, $\mathscr{G}$ et f des modèles respectifs de H, G et f sur U. Les hypothèses sur G assurent que C_G est quasi-isomorphe à G^{sab}.

Dans toute la preuve du théorème, on considère le diagramme commutatif de la figure 5.6.

Première étape **:** On se donne $(x_v) \in \mathbf{P}^0(k, [H \to G])$ d'image nulle dans $\mathbf{H}^1(k, \widehat{C}_f^{\mathrm{ab}})^D$. Quitte à réduire l'ouvert U, on peut supposer que $(x_v) \in \mathscr{P}_S^0([\mathscr{H} \to \mathscr{G}])$. Grâce au théorème 5.33, l'élément $(x'_v) := \mathrm{ab}_f^{0'}(x_v) \in \mathscr{P}_S^0(\mathscr{C}_f^{\mathrm{ab}})^\wedge$ se relève en un élément $x' \in \mathbf{H}^0(U, \mathscr{C}_f^{\mathrm{ab}})^\wedge$. Le groupe $H^1(U, \mathscr{H}^{\mathrm{m}})$ étant fini, on peut considérer l'image h' de x' dans $H^1(U, \mathscr{H}^{\mathrm{m}})$. Or les morphismes $H^1(U, \mathscr{H}) \to H^1(U, \mathscr{H}^{\mathrm{m}})$ et $\mathscr{P}_S^1(\mathscr{H}) \to \mathscr{P}_S^1(\mathscr{H}^{\mathrm{m}})$ sont des bijections grâce à la section 5.5.2. Donc on peut voir h' dans $H^1(U, \mathscr{H})$ et considérer alors $g := f_*(h') \in H^1(U, \mathscr{G})$.

Comme dans la preuve du théorème 5.35, on montre que quitte à réduire U, on peut supposer que g est trivial dans $H^1(U, \mathscr{G})$. Par exactitude de la première colonne du diagramme, h' se relève alors en un point $\overline{x} \in \mathbf{H}^0(U, [\mathscr{H} \to \mathscr{G}])$.

Les points (x_v) et $(\overline{x}_v)$ dans $\mathscr{P}_S^0([\mathscr{H} \to \mathscr{G}])$ ont alors même image (h_v) dans $\mathscr{P}_S^1({}_{\overline{x}}\mathscr{H})$, donc par exactitude de la troisième colonne du diagramme, il existe un point $(\overline{g}_v) \in \mathscr{P}_S^0(\mathscr{G})$ tel que $(x_v) = (g_v).(\overline{x}_v)$ dans $\mathscr{P}_S^0([\mathscr{H} \to \mathscr{G}])$.

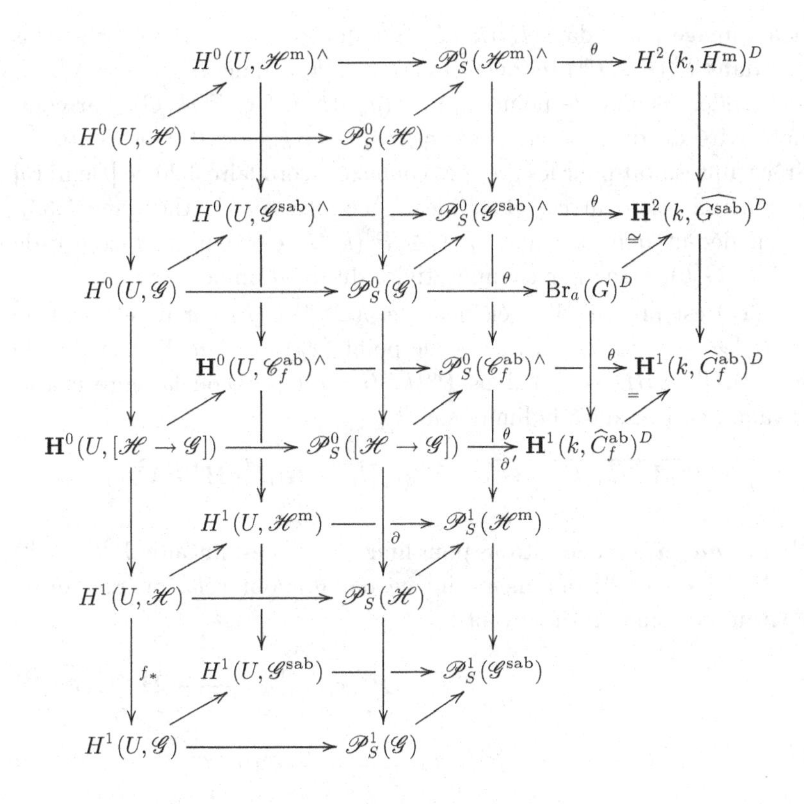

Figure 5.6

Deuxième étape : À partir de maintenant, on considère le morphisme ${}_{\overline{x}}f : {}_{\overline{x}}\mathscr{H} \to \mathscr{G}$ tordu par $\overline{x}$. Comme dans la preuve du théorème 5.35, on utilise dans la suite (parfois sans la citer) l'analogue évident de la proposition 5.18 dans le contexte du stabilisateur abélien.

On note $(\overline{g}'_v) := (\mathrm{ab}^0_G(\overline{g}_v)) \in \mathscr{P}^0_S(\mathscr{G}^{\mathrm{sab}})$. Par commutativité du diagramme, les points x' et $\overline{x}' := \mathrm{ab}^0_{f,\overline{x}}(\overline{x})$ dans $\mathbf{H}^0(U, \mathscr{C}^{\mathrm{ab}}_{\overline{x}f})$ ont même image $\widetilde{h}'$ dans $H^1(U, {}_{\overline{x}}\mathscr{H}^{\mathrm{m}})$, donc par exactitude, ces deux points diffèrent d'un élément $\widetilde{g}' \in H^0(U, \mathscr{G}^{\mathrm{sab}})^{\wedge}$.

Les éléments $(\overline{g}'_v)$ et $(\widetilde{g}'_v)$ (qui sont dans $\mathscr{P}^0_S(\mathscr{G}^{\mathrm{sab}})^{\wedge}$) ont même image dans $\mathscr{P}^0_S(\mathscr{C}^{\mathrm{ab}}_{\overline{x}f})$ (à savoir (x'_v)). Par conséquent, il existe $(\overline{h}'_v) \in \mathscr{P}^0_S({}_{\overline{x}}\mathscr{H}^{\mathrm{m}})^{\wedge}$ tel que $(\overline{h}'_v)$ s'envoie dans $\mathscr{P}^0_S(\mathscr{G}^{\mathrm{sab}})^{\wedge}$ sur la différence $(\widetilde{g}'_v) - (\overline{g}'_v)$.

Or le morphisme $\mathscr{P}^0_S({}_{\overline{x}}\mathscr{H}) \to \mathscr{P}^0_S({}_{\overline{x}}\mathscr{H}^{\mathrm{m}})$ est surjectif (car $H^1(\mathscr{O}_v, {}_{\overline{x}}\mathscr{H}^{\mathrm{u}})$ est trivial), et les groupes $P^0(k, {}_{\overline{x}}H^{\mathrm{m}})$ et $P^0(k, {}_{\overline{x}}H^{\mathrm{m}})^{\wedge}$ ont même image par θ dans $H^2(k, \widehat{_{\overline{x}}H^{\mathrm{m}}})^D$, donc on peut trouver un point $(\overline{h}_v)$ dans $P^0(k, {}_{\overline{x}}H^{\mathrm{m}})$ tel que $\theta((\overline{h}_v)) = \theta((\overline{h}'_v))$ dans $H^2(k, \widehat{_{\overline{x}}H^{\mathrm{m}}})^D$.

Considérons alors le point $(\overline{p}_v) := (\overline{g}_v).(\overline{h}_v)^{-1} \in P^0(k, G)$. Par commutativité du diagramme, il est clair que $\theta((\overline{p}_v)) = 0$ dans $\mathrm{Br}_a(G)^D$. Grâce au résultat pour les groupes connexes (corollaire 3.20 de [Dem11a] et généralisation au cas non linéaire en utilisant le théorème 5.33), on en déduit que l'élément $(\overline{p}_v) \in P^0(k, G)$ est dans l'adhérence de $\rho(G^{\mathrm{scu}}_{S_0}).G(k)$. Donc par commutativité du diagramme, le point $(x_v) = (\overline{p}_v).(\overline{x}_v)$ est produit d'un élément de $\overline{\rho(G^{\mathrm{scu}}_{S_0}).G(k)}$ par un élément de $\mathbf{H}^0(U, [\mathscr{H} \to \mathscr{G}])$. En particulier, le point (x_v) est dans l'adhérence de $\rho(G^{\mathrm{scu}}_{S_0}).\mathbf{H}^0(k, [H \to G])$ dans $\mathbf{P}^0(k, [H \to G])$. D'où la suite exacte suivante, en passant à la limite sur U :

$$\overline{\rho(G^{\mathrm{scu}}_{S_0}).\mathbf{H}^0(k, [H \to G])} \to \mathbf{P}^0(k, [H \to G]) \xrightarrow{\theta} \mathbf{H}^0(k, \widehat{C}^{\mathrm{ab}}_f)^D .$$

Troisième étape : Montrons pour finir que si G est linéaire, $\mathbf{P}^0(k, C^{\mathrm{ab}}_f)$ et $\mathbf{P}^0(k, [H \to G])$ ont même image par θ. Pour cela, on regarde le diagramme commutatif suivant :

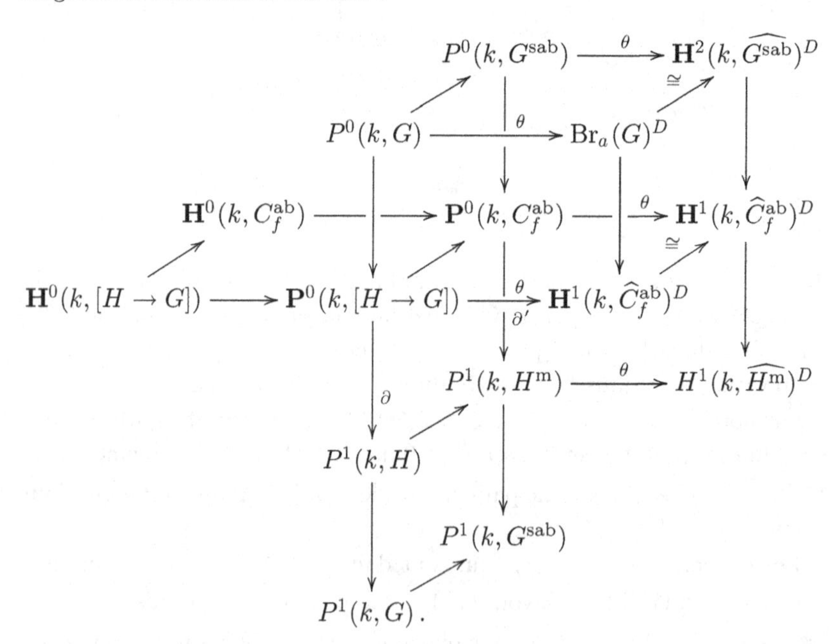

Soit $\gamma \in \mathbf{H}^1(k, \widehat{C}^{\mathrm{ab}}_f)^D$ et $\alpha \in P^0(k, C^{\mathrm{ab}}_f)$ tel que $\theta(\alpha) = \gamma$. On sait que le morphisme $\mathbf{H}^0(k, C^{\mathrm{ab}}_f) \to \mathbf{P}^0_{\infty}(k, C^{\mathrm{ab}}_f)$ est surjectif (voir théorème 5.34). Donc il existe $\alpha' \in \mathbf{H}^0(k, C^{\mathrm{ab}}_f)$ tel que $\alpha'_{\infty} = \alpha_{\infty}$. Notons alors

$\alpha^0 := \alpha - \alpha' \in \mathbf{P}^0(k, C_f^{\mathrm{ab}})$. On a toujours $\theta(\alpha^0) = \gamma$. On note β l'image de α^0 dans $P^1(k, H^m)$ et γ' celle de γ dans $H^1(k, \widehat{H^{\mathrm{m}}})^D$.

Par trivialité de la cohomologie des groupes unipotents, il existe $\tau \in P^1(k, H)$, trivial aux places infinies, tel que $\mathrm{ab}_H^{0\,\prime}(\tau) = \beta$. Notons alors g l'image de τ dans $P^1(k, G)$. Par construction, g_∞ est trivial, et g s'envoie sur 0 dans $P^1(k, G^{\mathrm{sab}})$. Donc par le théorème 5.11 de [Bor98], g est trivial. Par conséquent, τ se relève en un élément $x' \in \mathbf{P}^0(k, [H \to G])$. L'élément $(x'_v) \in \mathbf{P}^0(k, C_f^{\mathrm{ab}})$ s'envoie sur β dans $P^1(k, H^{\mathrm{m}})$, donc par exactitude de la troisième colonne, $\alpha^0 - (x'_v)$ se relève en un élément g' dans $P^0(k, G^{\mathrm{sab}})$. Or par le cas des groupes (voir la preuve du théorème 3.19 de [Dem11a]; on rappelle que G est linéaire ici), on sait que $P^0(k, G^{\mathrm{sab}})$ et $P^0(k, G)$ ont même image dans $H^1(k, \widehat{G^{\mathrm{sab}}})^D$ par le morphisme θ. Donc il existe $g^0 \in P^0(k, G)$ tel que $\theta(g^0) = \theta(g')$ dans $H^1(k, \widehat{G^{\mathrm{sab}}})^D$. Finalement, l'élément $g^0.x' \in \mathbf{P}^0(k, [H \to G])$ vérifie bien $\theta(g^0.x') = \gamma$. Cela assure donc que $\mathbf{P}^0(k, [H \to G])$ et $\mathbf{P}^0(k, C_f^{\mathrm{ab}})$ ont même image dans $\mathbf{H}^1(k, \widehat{C}_f^{\mathrm{ab}})^D$. Or pour des raisons topologiques, $\mathbf{P}^0(k, C_f^{\mathrm{ab}})$ et $\mathbf{P}^0(k, C_f^{\mathrm{ab}})^\wedge$ ont même image également par θ; et on a calculé le conoyau de $\theta : \mathbf{P}^0(k, C_f^{\mathrm{ab}})^\wedge \to \mathbf{H}^1(k, \widehat{C}_f^{\mathrm{ab}})^D$ au théorème 5.29 et c'est exactement $Ш^1(\widehat{C}_f^{\mathrm{ab}})^D$. Cela termine la preuve du théorème. □

5.8 Le défaut d'approximation faible sur les espaces homogènes

L'objectif de cette section est d'appliquer les méthodes d'abélianisation pour calculer le défaut d'approximation faible sur l'espace homogène $X = G/H$. On obtient des résultats qui généralisent ceux de Borovoi dans [Bor99]. On adopte la convention suivante : si S est un ensemble de places d'un corps de nombres k, $H^i(k_S, .)$ désigne $\prod_{v \in S} H^i(k_v, .)$, avec la convention que si v est une place archimédienne, $H^i(k_v, .)$ désigne le groupe de cohomologie modifié à la Tate. De même, si X est une k-variété algébrique, on note $X(k_S) := \prod_{v \in S} X(k_v)$, en remplaçant, pour les places archimédiennes v, l'ensemble $X(k_v)$ par l'ensemble des composantes connexes de $X(k_v)$; en outre, Ω désigne l'ensemble des places de k. Enfin, pour un complexe de modules galoisiens C, on définit $Ш^i_S(C) := \mathrm{Ker}(\mathbf{H}^i(k, C) \to \prod_{v \notin S} \mathbf{H}^i(k_v, C))$, et $Ш^i_\omega(C) := \{\alpha \in \mathbf{H}^i(k, C) : \alpha_v = 0$ pour presque toute place $v\}$.

5.8.1 Cas du stabilisateur connexe

Théorème 5.46 *Soit k un corps de nombres, G un k-groupe connexe, H un k-groupe linéaire connexe et soit $f : H \to G$ un morphisme. On suppose* $Ш^1(G^{\mathrm{ab}})$ *fini.*

Alors il existe une application naturelle (surjective si G est linéaire) $\mathbf{H}^0(k_\Omega, [H \to G]) \xrightarrow{\theta} \left(Ш^0_\omega(k, \widehat{C}_f)/ Ш^0(\widehat{C}_f)\right)^D$, *dont le noyau est l'adhérence faible* $\overline{\mathbf{H}^0(k, [H \to G])}^f$ *de* $\mathbf{H}^0(k, [H \to G])$ *dans* $\mathbf{H}^0(k_\Omega, [H \to G])$.

En outre, pour tout ensemble fini de places S de k contenant les places archimédiennes, l'application θ induit une application (surjective si G est linéaire) $\theta_S : \mathbf{H}^0(k_S, [H \to G]) \to \left(Ш^0_S(k, \widehat{C}_f)/ Ш^0(\widehat{C}_f)\right)^D$ *dont le noyau est l'adhérence* $\overline{\mathbf{H}^0(k, [H \to G])}^S$ *de* $\mathbf{H}^0(k, [H \to G])$ *dans* $\mathbf{H}^0(k_S, [H \to G])$.

Corollaire 5.47 *Soit k un corps de nombres, G un k-groupe connexe, H un sous-k-groupe linéaire connexe de G, et soit $X := G/H$. On suppose* $Ш^1(G^{\mathrm{ab}})$ *fini.*

Alors il existe une application naturelle (surjective si G est linéaire) $X(k_\Omega) \xrightarrow{\theta} \left(Ш^0_\omega(k, \widehat{C}_X)/ Ш^0(\widehat{C}_X)\right)^D$ *dont le noyau est l'adhérence faible* $\overline{X(k)}^f$ *de $X(k)$ dans $X(k_\Omega)$.*

En outre, pour tout ensemble fini de places S de k contenant les places archimédiennes, l'application θ induit une application (surjective si G est linéaire) $\theta_S : X(k_S) \to \left(Ш^0_S(k, \widehat{C}_X)/ Ш^0(\widehat{C}_X)\right)^D$ *dont le noyau est l'adhérence* $\overline{X(k)}^S$ *de $X(k)$ dans $X(k_S)$.*

Remarque 5.48 Ce corollaire est à rapprocher des résultats analogues obtenus par Borovoi aux théorèmes 1.3, 1.11 et au corollaire 1.12 de [Bor99].

Démonstration du théorème 5.46 On commence par montrer une variante du théorème 5.29

Proposition 5.49 *Soit $C = [M_1 \to M_2 \to M_3]$ un complexe de longueur 3 sur k, et S un ensemble fini de places de k. Si* $Ш^1({M_3}^{\mathrm{ab}})$ *est fini, alors on a une suite exacte de groupes abéliens :*

$$\mathbf{H}^0(k, C)^\wedge \to \prod_{v \in S} \mathbf{H}^0(k_v, C)^\wedge \to \left(Ш^0_S(k, \widehat{C})\right)^D \to Ш^0(k, \widehat{C})^D \to 0 .$$

Démonstration Tout d'abord, on remarque que l'exactitude du morceau

$$\prod_{v \in S} \mathbf{H}^0(k_v, C)^\wedge \to \left(Ш^0_S(k, \widehat{C}) \right)^D \to Ш^0(k, \widehat{C})^D \to 0$$

est une conséquence de l'exactitude de $\mathbf{P}^0(k, C)^\wedge \to \mathbf{H}^0(k, \widehat{C})^D \to Ш^0(k, \widehat{C})^D \to 0$ que l'on a montrée au théorème 5.29, via le diagramme commutatif suivant

$$\begin{array}{ccc} \mathbf{P}^0(k, C)^\wedge & \longrightarrow & \mathbf{H}^0(k, \widehat{C})^D \\ \downarrow & & \downarrow \\ \prod_{v \in S} \mathbf{H}^0(k_v, C)^\wedge & \longrightarrow & Ш^0_S(k, \widehat{C})^D . \end{array}$$

Par conséquent, il reste à montrer l'exactitude en $\prod_{v \in S} \mathbf{H}^0(k_v, C)^\wedge$.

- On suppose le complexe C exact en degré -1. On a alors un triangle exact

$$\text{Ker } f_1[2] \to C \to M \to \text{Ker } f_1[3] ,$$

où $M := M_3/\text{Coker } f_1$. On en déduit le diagramme commutatif (5.20) de la figure 5.7. La finitude de $H^3(k, \text{Ker } f_1)$ assure, via l'appendice de [HS05], que la suite

$$\mathbf{H}^0(k, C)^\wedge \to H^0(k, M)^\wedge \to H^3(k, \text{Ker } f_1)$$

est exacte. De même, considérons le morphisme $\prod_{v \in S} H^2(k_v, \text{Ker } f_1) \to \prod_{v \in S} \mathbf{H}^0(k_v, C)$ et notons Q son conoyau (muni de la topologie quotient). Alors la finitude de $\prod_{v \in S} H^3(k_v, \text{Ker } f_1)$ assure que le morphisme $Q^\wedge \to \prod_{v \in S} H^0(k_v, M)^\wedge$ reste injectif, et donc on conclut (en utilisant l'appendice de [HS05]) que la suite suivante est exacte :

$$\prod_{v \in S} H^2(k_v, \text{Ker } f_1)^\wedge \to \prod_{v \in S} \mathbf{H}^0(k_v, C)^\wedge \to \prod_{v \in S} H^0(k, M)^\wedge .$$

Or le morphisme $H^3(k, \text{Ker } f_1) \to \prod_{v \in S} H^3(k_v, \text{Ker } f_1)$ est un isomorphisme (voir [Mil06], théorème I.2.13), ainsi que le morphisme $\prod_{v \in S} H^{-1}(k_v, M) \to \mathbf{H}^2(k, \widehat{M})^D$ (par dévissage à partir du théorème I.4.10 de [Mil06] et de la proposition 5.9 de [HS05]), donc la proposition découle de l'exactitude des lignes du diagramme, et des suites de Cassels-Tate pour la variété semi-abélienne M (proposition 5.3 de [HS08]) et Ker f_1 (première colonne du diagramme : on peut compléter la suite de Cassels-Tate usuelle). Cela conclut la preuve dans ce cas particulier.

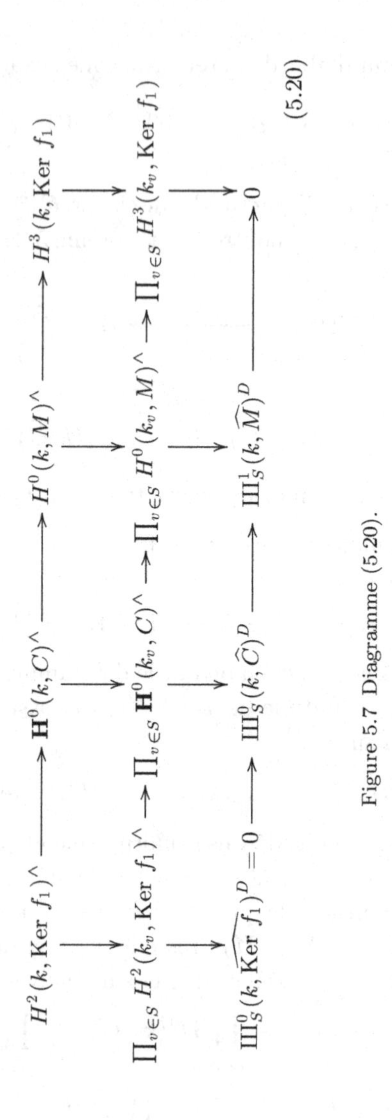

Figure 5.7 Diagramme (5.20).

- Cas général. On peut plonger Coker f_1, qui est un groupe de type multiplicatif, dans un tore quasi-trivial P : on note i : Coker $f_1 \to P$ un tel plongement. Notons alors C' le complexe

$$C' := \left[M_1 \xrightarrow{f_1} M_2 \xrightarrow{f_2 \oplus i} M_3 \oplus P \right] .$$

Alors C' vérifie les hypothèses du cas précédent, et on a un triangle exact naturel $P \to C' \to C \to P[1]$. Donc on dispose du diagramme

$$\begin{array}{ccccccc}
H^0(k,P)^\wedge & \longrightarrow & \mathbf{H}^0(k,C')^\wedge & \longrightarrow & \mathbf{H}^0(k,C)^\wedge & \longrightarrow & 0 \\
\downarrow & & \downarrow & & \downarrow & & \\
\prod_{v \in S} H^0(k_v,P)^\wedge & \longrightarrow & \prod_{v \in S} \mathbf{H}^0(k_v,C')^\wedge & \longrightarrow & \prod_{v \in S} \mathbf{H}^0(k_v,C)^\wedge & \longrightarrow & 0 \\
\downarrow & & \downarrow & & \downarrow & & \\
\text{Ш}^2_S(k,\widehat{P})^D & \longrightarrow & \text{Ш}^0_S(k,\widehat{C'})^D & \longrightarrow & \text{Ш}^0_S(k,\widehat{C})^D & \longrightarrow & 0
\end{array}$$

Figure 5.8

commutatif de la figure 5.8, dont les lignes sont exactes (l'exactitude de la dernière ligne résulte d'une chasse au diagramme facile). Par le cas précédent, la deuxième colonne est exacte, et P étant quasi-trivial, par Poitou-Tate, la flèche $\prod_{v \in S} H^0(k_v,P)^\wedge \to \text{Ш}^2_S(k,\widehat{P})^D$ est surjective, donc une chasse au diagramme assure l'exactitude de la troisième colonne, ce qui conclut la preuve de la proposition.

□

Poursuivons la preuve du théorème 5.46 : on s'intéresse d'abord au cas particulier où $H = 1$. Montrons la seconde partie du théorème dans ce cas : on considère le diagramme

$$\begin{array}{ccccc}
H^0(k,G^{\mathrm{sc}})^\wedge & \longrightarrow & \prod_{v \in S} H^0(k_v,G^{\mathrm{sc}})^\wedge & & \\
\downarrow & & \downarrow & & \\
H^0(k,G)^\wedge & \longrightarrow & \prod_{v \in S} H^0(k_v,G)^\wedge & \xrightarrow{\theta} & \text{Ш}^1_S(k,\widehat{C}_G)^D \\
\downarrow & & \downarrow & & \downarrow = \\
\mathbf{H}^0(k,C_G)^\wedge & \longrightarrow & \prod_{v \in S} \mathbf{H}^0(k_v,C_G)^\wedge & \xrightarrow{\theta} & \text{Ш}^1_S(k,\widehat{C}_G)^D \\
\downarrow & & \downarrow & & \\
H^1(k,G^{\mathrm{sc}}) & \longrightarrow & \prod_{v \in S} H^1(k_v,G^{\mathrm{sc}}) . & &
\end{array} \tag{5.21}$$

Soit $(g_v) \in \prod_{v \in S} H^0(k_v,G)$ tel que $\theta((g_v)) = 0$. Alors, par la proposition 5.49, l'image de $(\mathrm{ab}^0_G(g_v))$ dans $\prod_{v \in S} \mathbf{H}^0(k_v,C_G)^\wedge$ se relève en un élément g' dans $\mathbf{H}^0(k,C_G)^\wedge$. Alors le diagramme précédent assure que l'image de g' dans $H^1(k,G^{\mathrm{sc}})$ est dans $\text{Ш}^1_{\Omega \setminus S}(k,G^{\mathrm{sc}}) = \text{Ш}^1(k,G^{\mathrm{sc}}) = 1$, donc g' se relève en $\tilde{g} \in H^0(k,G)^\wedge$. Par exactitude de la deuxième colonne, l'élément $(\tilde{g}_v)^{-1}.(\mathrm{ab}^0_G(g_v)) \in \prod_{v\in S} H^0(k_v,G)^\wedge$ se relève en un élément $(h_v) \in \prod_{v\in S} H^0(k_v,G^{\mathrm{sc}})^\wedge$. On utilise le lemme :

Lemme 5.50 *La flèche naturelle*

$$\left(\overline{H^0(k,G^{\mathrm{sc}})}^S\right)^\wedge \to \prod_{v\in S} H^0(k_v,G^{\mathrm{sc}})^\wedge$$

est surjective.

Démonstration C'est une conséquence immédiate de l'approximation faible sur G^{sc}. □

Avec ce lemme, l'élément (h_v) provient d'un élément $h \in \left(\overline{H^0(k,G^{\mathrm{sc}})}^S\right)^\wedge$. On considère alors l'image $\rho(h) \in \left(\overline{H^0(k,G)}^S\right)^\wedge$: par fonctorialité, l'élément $\rho(h).\tilde{g} \in \left(\overline{H^0(k,\mathscr{G})}^S\right)^\wedge$ s'envoie alors sur (g_v) dans $\prod_{v\in S} H^0(k_v,G)^\wedge$. On a donc montré que la suite suivante

$$\left(\overline{H^0(k,G)}^S\right)^\wedge \to \prod_{v\in S} H^0(k_v,G)^\wedge \to (\text{Ш}^1_S(k,\widehat{C}_G))^D$$

est exacte. Comme dans le corollaire 3.18 de [Dem11a], on en déduit l'exactitude de la suite

$$\overline{H^0(k,G)}^S \to \prod_{v\in S} H^0(k_v,G) \to (\text{Ш}^1_S(k,\widehat{C}_G))^D ,$$

d'où la seconde partie du théorème 5.46 dans le cas où $H = 1$.

Montrons ce même résultat dans le cas général : on considère le diagramme commutatif de la figure 5.9, où S est un ensemble fini de places de k contenant S_∞. On suit alors exactement la structure de la preuve du théorème 5.35, pour montrer la seconde partie du théorème 5.46 via une chasse au diagramme dans le grand diagramme précédent, en utilisant la proposition 5.49, le théorème 5.11 de [Bor98] et le cas $H = 1$ montré plus haut. Pour conclure la preuve, on a besoin du lemme suivant :

Lemme 5.51 *Soit H un k-groupe linéaire connexe, S un ensemble fini de places de k contenant S_∞. Alors on a l'égalité suivante :*

$$\mathrm{Im}\left(H^0(k_S,H)\xrightarrow{\theta}\mathbf{H}^1(k,\widehat{C}_H)^D\right)=\mathrm{Im}\left(\mathbf{H}^0(k_S,C_H)^\wedge\xrightarrow{\theta}\mathbf{H}^1(k,\widehat{C}_H)^D\right).$$

Démonstration Tout d'abord, il est clair que les images de

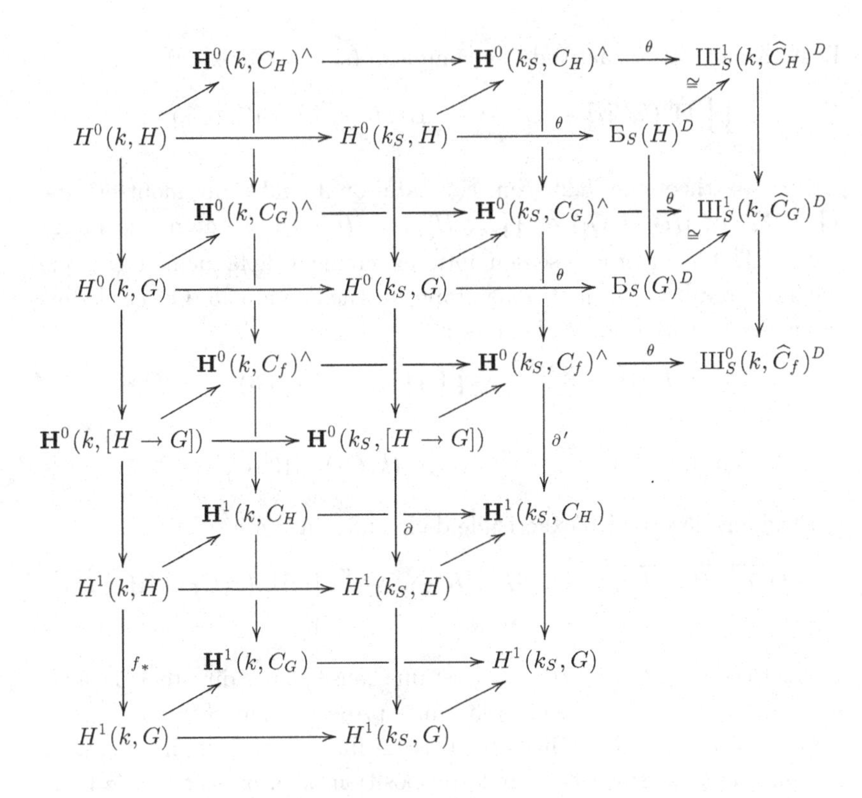

Figure 5.9

$\prod_{v \in S} \mathbf{H}^0(k_v, C_H)^\wedge$ et $\prod_{v \in S} \mathbf{H}^0(k_v, C_H)$ par θ coïncident. Il suffit donc de montrer que

$$\mathrm{Im}\Big(H^0(k_S, H) \xrightarrow{\theta} \mathbf{H}^1(k, \widehat{C}_H)^D\Big) = \mathrm{Im}\Big(\mathbf{H}^0(k_S, C_H) \xrightarrow{\theta} \mathbf{H}^1(k, \widehat{C}_H)^D\Big).$$

En s'inspirant de la preuve du théorème 3.19 de [Dem11a], on constate qu'il suffit pour cela de montrer que l'application ab^1_H induit une injection

$$\mathrm{ab}^1_H : Ш^1_{\Omega \setminus S}(k, H) \hookrightarrow Ш^1_{\Omega \setminus S}(k, C_H).$$

C'est une conséquence du théorème 5.12 de [Bor98]. □

Ce lemme permet donc de conclure la preuve de l'exactitude de la suite suivante

$$1 \to \overline{\mathbf{H}^0(k, [H \to G])}^S \to \prod_{v \in S} \mathbf{H}^0(k_v, [H \to G]) \xrightarrow{\theta_S} Ш^0_S(\widehat{C}_f)^D .$$

Reste à montrer l'exactitude du complexe (si G est linéaire)

$$\prod_{v\in S} \mathbf{H}^0(k_v,[H\to G]) \xrightarrow{\theta_S} Ш^0_S(\widehat{C}_f)^D \to Ш^0(\widehat{C}_f)^D .$$

Grâce au théorème 5.29, on constate qu'il suffit de montrer que $\prod_{v\in S} \mathbf{H}^0(k_v,[H\to G])$ et $\prod_{v\in S} H^0_{\mathrm{ab}}(k_v,[H\to G])$ ont même image dans $Ш^0_S(\widehat{C}_f)^D$. Et ceci se démontre exactement de la même façon que dans la preuve de la quatrième étape du théorème 5.35. On en déduit finalement l'exactitude de la suite :

$$1 \to \overline{\mathbf{H}^0(k,[H\to G])}^S \to \prod_{v\in S} \mathbf{H}^0(k_v,[H\to G]) \xrightarrow{\theta_S} \Big(Ш^0_S(k,\widehat{C}_f)/\, Ш^0(\widehat{C}_f)\Big)^D \to 1 .$$

Étudions désormais l'exactitude de la suite suivante

$$1 \to \overline{\mathbf{H}^0(k,[H\to G])}^f \to \mathbf{H}^0(k_\Omega,[H\to G]) \xrightarrow{\theta} \Big(Ш^1_\omega(k,\widehat{C}_f)/\, Ш^1(\widehat{C}_f)\Big)^D \to 1 .$$

L'exactitude en $\mathbf{H}^0(k_\Omega,[H\to G])$ est une conséquence directe de la suite exacte précédente en passant à la limite projective sur S.

Dans le cas où G est linéaire, il reste finalement à montrer que le morphisme θ est surjectif. Par la proposition 5.49, on sait que le morphisme $\theta' : \prod_{v\in\Omega} \mathbf{H}^0(k_v,C_f) \to \Big(Ш^0_\omega(k,\widehat{C}_f)/\, Ш^0(\widehat{C}_f)\Big)^D$ est surjectif. Par conséquent il suffit de montrer que les ensembles $\mathbf{H}^0(k_\Omega,[H\to G])$ et $\prod_{v\in\Omega} \mathbf{H}^0(k_v,C_f)$ ont même image dans $\Big(Ш^0_\omega(k,\widehat{C}_f)\Big)^D$. Pour cela, on considère le diagramme de la figure 5.10. Soit $\gamma \in Ш^0_\omega(k,\widehat{C}_f)^D$ et $\alpha \in \prod_{v\in\Omega} \mathbf{H}^0(k_v,C_f)$ tel que $\theta(\alpha)=\gamma$. Par le théorème 5.34, on sait que le morphisme $\mathbf{H}^0(k,C_f) \to \prod_{v\in S_\infty} \mathbf{H}^0(k_v,C_f)$ est surjectif. Donc il existe $\alpha' \in \mathbf{H}^0(k,C_f)$ tel que $\alpha'_\infty = \alpha_\infty$. Notons alors $\alpha^0 := \alpha - \alpha' \in \prod_{v\in\Omega} \mathbf{H}^0(k_v,C_f)$. On a toujours $\theta(\alpha^0)=\gamma$. On note β l'image de α^0 dans $\prod_{v\in\Omega} \mathbf{H}^1(k,C_H)$ et γ' celle de γ dans $Ш^0_\omega(k,\widehat{C}_H)^D$. Par le théorème 5.11 de [Bor98], il existe $\tau \in \prod_{v\in\Omega} H^1(k_v,H)$, trivial aux places infinies, tel que $\mathrm{ab}^0_H(\tau)=\beta$. Notons alors g l'image de τ dans $\prod_{v\in\Omega} H^1(k,G)$.

Par construction, g_∞ est trivial, et g s'envoie sur 0 dans $\prod_{v\in\Omega} \mathbf{H}^1(k,C_G)$. Donc par le même théorème 5.11 de [Bor98], g est trivial. Par conséquent, τ se relève en un élément $x' \in \prod_{v\in\Omega} \mathbf{H}^0(k_v,[H\to G])$. Par commutativité du diagramme, on voit que $(x'_v) \in$

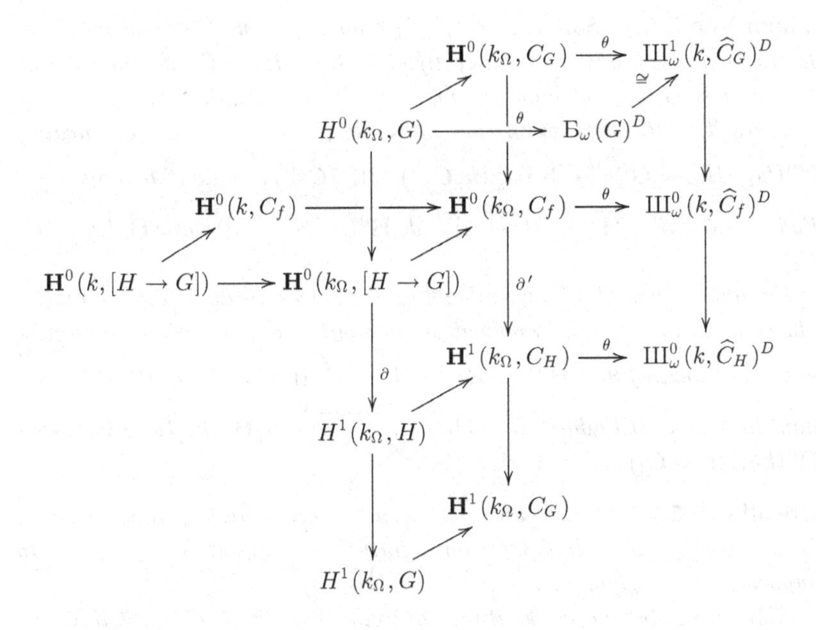

Figure 5.10

$\prod_{v\in\Omega} \mathbf{H}^0(k_v, C_f)$ s'envoie sur β dans $\prod_{v\in\Omega} \mathbf{H}^1(k_v, C_H)$, donc par exactitude de la troisième colonne, $\alpha^0 - (x'_v)$ se relève en un élément g' dans $\prod_{v\in\Omega} \mathbf{H}^0(k_v, C_G)$. Or $\prod_{v\in\Omega} \mathbf{H}^0(k_v, C_G)$ et $\prod_{v\in\Omega} H^0(k_v, G)$ ont même image dans $Ш^1_\omega(k, \widehat{C}_G)^D$ par le morphisme θ (la preuve est exactement la même que celle du théorème 3.19 de [Dem11a]). Donc il existe $g^0 \in \prod_{v\in\Omega} H^0(k_v, G)$ tel que $\theta(g^0) = \theta(g')$ dans $Ш^1_\omega(k, \widehat{C}_G)^D$. Finalement, l'élément $g^0.x' \in \prod_{v\in\Omega} \mathbf{H}^0(k_v, [H \to G])$ vérifie bien $\theta(g^0.x') = \gamma$. Cela assure donc que $\prod_{v\in\Omega} \mathbf{H}^0(k_v, [H \to G])$ et $\prod_{v\in\Omega} \mathbf{H}^0(k, C_f)$ ont même image dans $Ш^1_\omega(k, \widehat{C}_f)^D$. Or pour des raisons topologiques, $\prod_{v\in\Omega} \mathbf{H}^0(k_v, C_f)$ et $\prod_{v\in\Omega} \mathbf{H}^0(k_v, C_f)^\wedge$ ont même image également par θ; on a calculé le conoyau de $\theta : \prod_{v\in\Omega} \mathbf{H}^0(k_v, C_f)^\wedge \to Ш^0_\omega(k, \widehat{C}_f)^D$ à la proposition 5.49, et c'est exactement $Ш^0(\widehat{C}_f)^D$. Cela termine la preuve du théorème 5.46. □

5.8.2 Cas du stabilisateur abélien

On peut également démontrer de manière analogue les versions à stabilisateur abélien des énoncés précédents, à savoir :

Théorème 5.52 *Soit k un corps de nombres, G un k-groupe connexe, H un k-groupe linéaire commutatif et soit $f : H \to G$ un morphisme. On suppose G^{ss} simplement connexe et $Ш^1(G^{\mathrm{ab}})$ fini.*

Alors il existe une application naturelle (surjective si G est linéaire) $\mathbf{H}^0(k_\Omega, [H \to G]) \xrightarrow{\theta} \left(Ш^1_\omega(k, \widehat{C}_f^{\mathrm{ab}}) / Ш^1(\widehat{C}_f^{\mathrm{ab}}) \right)^D$, dont le noyau est l'adhérence faible $\overline{\mathbf{H}^0(k, [H \to G])}^f$ de $\mathbf{H}^0(k, [H \to G])$ dans $\mathbf{H}^0(k_\Omega, [H \to G])$.

En outre, pour tout ensemble fini de places S de k contenant les places archimédiennes, l'application θ induit une application (surjective si G est linéaire) $\theta_S : \mathbf{H}^0(k_S, [H \to G]) \to \left(Ш^1_S(k, \widehat{C}_f^{\mathrm{ab}}) / Ш^1(\widehat{C}_f^{\mathrm{ab}}) \right)^D$ dont le noyau est l'adhérence $\overline{\mathbf{H}^0(k, [H \to G])}^S$ de $\mathbf{H}^0(k, [H \to G])$ dans $\mathbf{H}^0(k_S, [H \to G])$.

Corollaire 5.53 *Soit k un corps de nombres, G un k-groupe connexe, H un sous-k-groupe linéaire commutatif de G, et soit $X := G/H$. On suppose G^{ss} simplement connexe et $Ш^1(G^{\mathrm{ab}})$ fini.*

Alors il existe une application naturelle (surjective si G est linéaire) $X(k_\Omega) \xrightarrow{\theta} \left(Ш^1_\omega(k, \widehat{C}_X^{\mathrm{ab}}) / Ш^1(\widehat{C}_X^{\mathrm{ab}}) \right)^D$ dont le noyau est exactement l'adhérence faible $\overline{X(k)}^f$ de $X(k)$ dans $X(k_\Omega)$.

En outre, pour tout ensemble fini de places S de k contenant les places archimédiennes, l'application θ induit une application (surjective si G est linéaire) $\theta_S : X(k_S) \to \left(Ш^1_S(k, \widehat{C}_X^{\mathrm{ab}}) / Ш^1(\widehat{C}_X^{\mathrm{ab}}) \right)^D$ dont le noyau est exactement l'adhérence $\overline{X(k)}^S$ de $X(k)$ dans $X(k_S)$.

Références

[Ald08] E. Aldrovandi – «2-gerbes bound by complexes of *gr*-stacks, and cohomology», *J. Pure Appl. Algebra* **212** (2008), no. 5, p. 994–1038.

[AN09] E. Aldrovandi et B. Noohi – «Butterflies. I. Morphisms of 2-group stacks», *Adv. Math.* **221** (2009), no. 3, p. 687–773.

[AN10] —, «Butterflies II: torsors for 2-group stacks», *Adv. Math.* **225** (2010), no. 2, p. 922–976.

[Ana73] S. Anantharaman – «Schémas en groupes, espaces homogènes et espaces algébriques sur une base de dimension 1», Sur les groupes algébriques, Soc. Math. France, Paris, 1973, p. 5–79. Bull. Soc. Math. France, Mém. 33.

[BCTS08] M. Borovoi, J.-L. Colliot-Thélène et A. N. Skorobogatov – «The elementary obstruction and homogeneous spaces», *Duke Math. J.* **141** (2008), no. 2, p. 321–364.

[BD11] M. Borovoi et C. Demarche – «Manin obstruction to strong approximation for homogeneous spaces», *Comment. Math. Helv.* (2011), à paraître.

[Bor93] M. Borovoi – «Abelianization of the second nonabelian Galois cohomology», *Duke Math. J.* **72** (1993), no. 1, p. 217–239.

[Bor98] — , «Abelian Galois cohomology of reductive groups», *Mem. Amer. Math. Soc.* **132** (1998), no. 626, p. viii+50.

[Bor99] — , «The defect of weak approximation for homogeneous spaces», *Ann. Fac. Sci. Toulouse Math. (6)* **8** (1999), no. 2, p. 219–233.

[Bre90] L. Breen – «Bitorseurs et cohomologie non abélienne», The Grothendieck Festschrift, Vol. I, Progr. Math., vol. 86, Birkhäuser Boston, Boston, MA, 1990, p. 401–476.

[Bre92] — , «Théorie de Schreier supérieure», *Ann. Sci. École Norm. Sup. (4)* **25** (1992), no. 5, p. 465–514.

[Bre94] — , «On the classification of 2-gerbes and 2-stacks», *Astérisque* (1994), no. 225, p. 160.

[BvH11] M. Borovoi et J. van Hamel – «Extended equivariant Picard complexes and homogeneous spaces», *Transform. Groups* **17** (2012), p. 51–86.

[Con84] D. Conduché – «Modules croisés généralisés de longueur 2», *J. Pure Appl. Algebra* **34** (1984), no. 2-3, p. 155–178.

[CT08] J.-L. Colliot-Thélène – «Résolutions flasques des groupes linéaires connexes», *J. Reine Angew. Math.* **618** (2008), p. 77–133.

[CTGP04] J.-L. Colliot-Thélène, P. Gille et R. Parimala – «Arithmetic of linear algebraic groups over 2-dimensional geometric fields», *Duke Math. J.* **121** (2004), no. 2, p. 285–341.

[CTS87] J.-L. Colliot-Thélène et J.-J. Sansuc – «Principal homogeneous spaces under flasque tori: applications», *J. Algebra* **106** (1987), no. 1, p. 148–205.

[CTX09] J.-L. Colliot-Thélène et F. Xu – «Brauer-Manin obstruction for integral points of homogeneous spaces and representation by integral quadratic forms», *Compos. Math.* **145** (2009), no. 2, p. 309–363, With an appendix by Dasheng Wei and Fei Xu.

[Del79] P. Deligne – «Variétés de Shimura: interprétation modulaire, et techniques de construction de modèles canoniques», Automorphic forms, representations and L-functions (Proc. Sympos. Pure Math., Oregon State Univ., Corvallis, Ore., 1977), Part 2, Proc. Sympos. Pure Math., XXXIII, Amer. Math. Soc., Providence, R.I., 1979, p. 247–289.

[Dem09] C. Demarche – «Méthodes cohomologiques pour l'étude des points rationnels sur les espaces homogènes», Thèse de l'université Paris-Sud, 2009.

[Dem11a] — , «Le défaut d'approximation forte dans les groupes linéaires connexes», *Proc. Lond. Math. Soc. (3)* **102** (2011), no. 3, p. 563–597.

[Dem11b] — , «Suites de Poitou-Tate pour les complexes de tores à deux termes», *Int. Math. Res. Not. IMRN* (2011), no. 1, p. 135–174.

[Dem11c] — , «Une formule pour le groupe de Brauer algébrique d'un torseur», *J. Algebra* **347** (2011), p. 96–132.

[GA11] C. GONZALEZ-AVILES – «Quasi-abelian crossed modules and non-abelian cohomology», prépublication, 2011.

[Gir71] J. GIRAUD – *Cohomologie non abélienne*, Springer-Verlag, Berlin, 1971, Die Grundlehren der mathematischen Wissenschaften, Band 179.

[GP08] P. GILLE et A. PIANZOLA – «Isotriviality and étale cohomology of Laurent polynomial rings», *J. Pure Appl. Algebra* **212** (2008), no. 4, p. 780–800.

[Gro66] A. GROTHENDIECK – «Éléments de géométrie algébrique. IV. Étude locale des schémas et des morphismes de schémas. III», *Inst. Hautes Études Sci. Publ. Math.* (1966), no. 28, p. 255.

[Har67] G. HARDER – «Halbeinfache Gruppenschemata über Dedekindringen», *Inventiones Mathematicae* **4** (1967), p. 165–191.

[Har08] D. HARARI – «Le défaut d'approximation forte pour les groupes algébriques commutatifs», *Algebra and Number Theory* **2** (2008), no. 5, p. 595–611.

[HS05] D. HARARI et T. SZAMUELY – «Arithmetic duality theorems for 1-motives», *J. reine angew. Math.* **578** (2005), p. 93–128, et *Corrigenda for "Aritmetic duality theorems for 1-motives"*, disponible sur http://www.math.u-psud/~harari/errata/corrigcrelle.pdf.

[HS08] —, «Local-global principles for 1-motives», *Duke Math. J.* **143** (2008), no. 3, p. 531–557.

[HS11] D. HARARI et A. SKOROBOGATOV – «Descent theory for open varieties», *this volume.*

[Jos09] P. JOSSEN – «The arithmetic of 1-motives», Thèse, 2009.

[Lab99] J.-P. LABESSE – «Cohomologie, stabilisation et changement de base», *Astérisque* (1999), no. 257, p. vi+161, Appendix A by Laurent Clozel and Labesse, and Appendix B by Lawrence Breen.

[Mar91] G. A. MARGULIS – *Discrete subgroups of semisimple Lie groups*, Ergebnisse der Mathematik und ihrer Grenzgebiete (3) [Results in Mathematics and Related Areas (3)], vol. 17, Springer-Verlag, Berlin, 1991.

[Mil80] J. S. MILNE – *Étale cohomology*, Princeton Mathematical Series, vol. 33, Princeton University Press, Princeton, N.J., 1980.

[Mil06] — , *Arithmetic duality theorems*, second éd., BookSurge, LLC, Charleston, SC, 2006.

[Nis84] Y. A. NISNEVICH – «Espaces homogènes principaux rationnellement triviaux et arithmétique des schémas en groupes réductifs sur les anneaux de Dedekind», *C. R. Acad. Sci. Paris Sér. I Math.* **299** (1984), no. 1, p. 5–8.

[NSW08] J. NEUKIRCH, A. SCHMIDT et K. WINGBERG – *Cohomology of number fields*, second éd., Grundlehren der Mathematischen Wissenschaften [Fundamental Principles of Mathematical Sciences], vol. 323, Springer-Verlag, Berlin, 2008.

[PR94] V. Platonov et A. Rapinchuk – *Algebraic groups and number theory*, Pure and Applied Mathematics, vol. 139, Academic Press Inc., Boston, MA, 1994, Translated from the 1991 Russian original by Rachel Rowen.

[Ray70] M. Raynaud – *Faisceaux amples sur les schémas en groupes et les espaces homogènes*, Lecture Notes in Mathematics, Vol. 119, Springer-Verlag, Berlin, 1970.

[Ros56] M. Rosenlicht – «Some basic theorems on algebraic groups», *Amer. J. Math.* **78** (1956), p. 401–443.

[San81] J.-J. Sansuc – «Groupe de Brauer et arithmétique des groupes algébriques linéaires sur un corps de nombres», *J. reine angew. Math.* **327** (1981), p. 12–80.

[SGA3] M. Demazure et A. Grothendieck – *Schémas en groupes*, Séminaire de Géométrie Algébrique du Bois Marie 1962/64 (SGA 3). Dirigé par M. Demazure et A. Grothendieck. Lecture Notes in Mathematics, Vol. 151-152-153, Springer-Verlag, Berlin, 1970.

6

Gaussian rational points on a singular cubic surface

Ulrich Derenthal

Mathematisches Institut, Ludwig-Maximilians-Universität München

Felix Janda

Departement Mathematik, ETH Zürich

Abstract Manin's conjecture predicts the asymptotic behavior of the number of rational points of bounded height on algebraic varieties. For toric varieties, it was proved by Batyrev and Tschinkel via height zeta functions and an application of the Poisson formula. An alternative approach to Manin's conjecture via universal torsors was used, mainly over the field $\mathbb{Q}$ of rational numbers so far. In this note, we give a proof of Manin's conjecture over the Gaussian rational numbers $\mathbb{Q}(i)$ and over other imaginary quadratic number fields with class number 1 for the singular toric cubic surface defined by $x_0^3 = x_1x_2x_3$.

Let $S \subset \mathbb{P}^3$ be the cubic surface defined over $\mathbb{Q}$ by the equation

$$x_0^3 = x_1x_2x_3.$$

It is rational, toric and contains precisely three singularities and three lines. Over any number field K, its set of K-rational points is clearly infinite. Let H be the Weil height on S, defined as

$$H(\mathbf{x}) = \prod_{\nu \in M_K} \max_{j \in \{0,\dots,3\}} \|x_j\|_\nu$$

where $\mathbf{x} = (x_0 : \ldots : x_3) \in S(K)$ with $x_0, \ldots, x_3 \in K$, the set of places of K is denoted as M_K and $\|\cdot\|_\nu$ is the (suitably normalized; see Section 6.2) norm at the place ν. The total number of K-rational points of bounded height on S is dominated by the number of easily countable points on the three lines. Therefore, we restrict our attention to K-rational points in the complement U of the lines on S.

A much more general conjecture of Manin [FMT89] predicts in the case of S that the number

$$N_{U,K,H}(B) = \#\{\mathbf{x} \in U(K) \mid H(\mathbf{x}) \leq B\}$$

of K-rational points of bounded height outside the lines behaves asymptotically as

$$N_{U,K,H}(B) \sim c_{S,K,H} B(\log B)^6,$$

as $B \to \infty$. A conjectural interpretation of the leading constant $c_{S,K,H} > 0$ was given by Peyre [Pey95] and refined by Batyrev and Tschinkel [BT98b].

Making use of the torus action on toric varieties to study the height zeta functions and to apply the Poisson formula, Manin's conjecture was proved for toric varieties over any number field by Batyrev and Tschinkel [BT98a]; see [BT98b, §5.3] for the application of this result to our cubic surface S.

For varieties without such an action of an algebraic group, an alternative approach using *universal torsors* was suggested by Salberger [Sal98]. He gave a second proof of Manin's conjecture over $\mathbb{Q}$ in the case of split toric varieties; see [Sal98, Example 11.50] for its application to S.

For the singular cubic surface S as above, Manin's conjecture over $\mathbb{Q}$ was also proved directly by Fouvry [Fou98], Heath-Brown and Moroz [HBM99], de la Bretèche [Bre98], de la Bretèche and Swinnerton-Dyer [BSD07] and Bhowmik, Essouabri, Lichtin [BEL07], using elementary or classical analytic number theoretic techniques and a parameterization of rational points closely related to universal torsors in some cases.

The basic example of a universal torsor applied to point counting is the following: To estimate the number

$$N_{\mathbb{P}^n,\mathbb{Q},H}(B) = \#\{\mathbf{x} \in \mathbb{P}^n(\mathbb{Q}) \mid H(\mathbf{x}) \leq B\}$$

of rational points of bounded height in n-dimensional projective space $\mathbb{P}^n$, the natural first step is the observation that any such $\mathbf{x}$ is represented uniquely up to sign by an $(n+1)$-tuple of coprime integers $(x_0, \ldots, x_n)$ subject to the condition $\max\{|x_0|, \ldots, |x_n|\} \leq B$. Geometrically, this corresponds to the fact that the open subset $\mathbb{A}^{n+1} \setminus \{0\}$ of $(n+1)$-dimensional affine space is a universal torsor over $\mathbb{P}^n$.

Based on this, Schanuel [Sch64], [Sch79] proved Manin's conjecture for projective spaces over arbitrary number fields. Over number fields other than $\mathbb{Q}$, no other proof of Manin's conjecture via universal torsors is known to us.

The purpose of this note is to begin the generalization of universal torsor techniques from $\mathbb{Q}$ to more general number fields. A first candidate is Manin's conjecture for the toric cubic surface S over the field $\mathbb{Q}(i)$ of Gaussian rational numbers because its class number is 1 and its ring of integers contains only finitely many units. It turns out that it is not too hard to generalize from $\mathbb{Q}(i)$ to the following setting:

Theorem 6.1 *Let K be an imaginary quadratic number field whose class number is 1, let w_K be the number of units in its ring of integers $\mathcal{O}_K$, and let d_K be the square root of the absolute value of its discriminant (cf. Table 6.1). Let $S \subset \mathbb{P}^3$ be the cubic surface defined by $x_0^3 = x_1x_2x_3$. Let U be the complement of the three lines on S. Then*

$$N_{U,K,H}(B) \sim c_{S,K,H}B(\log B)^6 + O(B(\log B)^5)$$

as $B \to \infty$, with

$$c_{S,K,H} = \frac{2^7\pi^9}{6!w_K^7d_K^9}\prod_p\left(1-\frac{1}{\|p\|_\infty}\right)^7\left(1+\frac{7}{\|p\|_\infty}+\frac{1}{\|p\|_\infty^2}\right)$$

where the product runs over all primes in $\mathcal{O}_K$ up to units.

n	-1	-2	-3	-7	-11	-19	-43	-67	-163
w_K	4	2	6	2	2	2	2	2	2
d_K	2	$2\sqrt{2}$	$\sqrt{3}$	$\sqrt{7}$	$\sqrt{11}$	$\sqrt{19}$	$\sqrt{43}$	$\sqrt{67}$	$\sqrt{163}$

Table 6.1 $K = \mathbb{Q}(\sqrt{n})$ *with class number* $h_K = 1$.

We will see in Section 6.2 that this result agrees with the conjectures of Manin, Peyre, Batyrev and Tschinkel.

Recently, Frei [Fre12] generalized our work to arbitrary number fields, removing our restriction to class number 1 and finite groups of units in the ring of integers.

Acknowledgements The authors are grateful to Tim Browning and the referee for helpful remarks. The first named author was supported by grant 200021_124737/1 of the Schweizer Nationalfonds and by grant DE 1646/2-1 of the Deutsche Forschungsgemeinschaft.

6.1 Geometry

In this section, we collect some facts on the geometry of our singular cubic surface S. The construction of its minimal desingularization as a blow-up of the projective plane in six points will be used in Section 6.3 to construct a parameterization of the K-rational points by integral points on a universal torsor.

Let $\mathcal{S}$ be the model of S over $\mathbb{Z}$ defined by the equation $x_0^3 = x_1x_2x_3$ in $\mathbb{P}^3_{\mathbb{Z}}$. We will consider the minimal desingularization $\widetilde{\mathcal{S}}$ of $\mathcal{S}$, which is obtained from $\mathbb{P}^2_{\mathbb{Z}}$ by a sequence of six blow-ups of points. All statements below will be true not only over $\mathbb{Z}$ but, suitably rephrased, over any field. In what follows, all statements involving variables j, k, l are meant to hold for all

$$(j,k,l) \in \{(1,2,3),(2,3,1),(3,1,2)\}.$$

The surface $\mathcal{S}$ is singular, with precisely three singularities

$$p_1 = (0:1:0:0), \quad p_2 = (0:0:1:0), \quad p_3 = (0:0:0:1).$$

They are rational double points of type $\mathbf{A}_2$ in the $\mathbf{ADE}$-classification. It contains precisely three lines $\ell_j = \{x_0 = x_j = 0\}$ through p_k and p_l.

The surface $\mathcal{S}$ is toric. Indeed, an action of a two-dimensional torus on $\mathcal{S}$ is given by

$$\mathbb{G}_{\mathrm{m},\mathbb{Z}}^2 \times \mathcal{S} \to \mathcal{S}, \quad (\mathbf{t},\mathbf{x}) \mapsto \mathbf{t}\cdot\mathbf{x} = (x_0 : t_1x_1 : t_2x_2 : (t_1t_2)^{-1}x_3),$$

giving an isomorphism from $\mathbb{G}_{\mathrm{m},\mathbb{Z}}^2$ to the open dense orbit

$$\mathcal{U} = \mathcal{S} \setminus (\ell_1 \cup \ell_2 \cup \ell_3) = \{\mathbf{x} \in \mathcal{S} \mid x_0x_1x_2x_3 \neq 0\}$$

of $(1:1:1:1)$, say. The corresponding fan can be found in Figure 6.1.

Resolving the singularity p_j gives two exceptional divisors $E_{k,l}$ (meeting the strict transform E_l of ℓ_l) and $E_{l,k}$ (meeting the strict transform E_k of ℓ_k). We obtain the minimal desingularization $\pi : \widetilde{\mathcal{S}} \to \mathcal{S}$, where the Picard group of $\widetilde{\mathcal{S}}$ is free of rank 7. The six curves $E_{j,k}, E_{k,j}$ are (-2)-curves (rational curves with self-intersection number -2), while

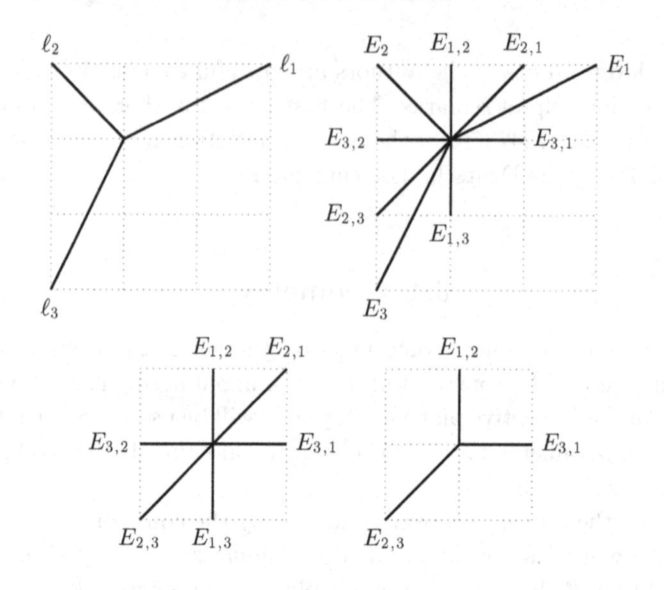

Figure 6.1 Fans of $\mathcal{S}, \widetilde{\mathcal{S}}, \widetilde{\mathcal{S}}_1, \mathbb{P}^2_{\mathbb{Z}}$, respectively.

the three transforms E_j of the lines are (-1)-curves (rational curves with self-intersection number -1). There are no other negative curves (rational curves with negative self-intersection number) of $\widetilde{\mathcal{S}}$. The negative curves correspond precisely to the rays in the fan of $\widetilde{\mathcal{S}}$ in Figure 6.1.

The surface $\mathcal{S}$ is rational, via the birational map

$$\begin{aligned}\phi : \mathcal{S} \dashrightarrow \mathbb{P}^2_{\mathbb{Z}}, \ \mathbf{x} \mapsto (x_0^2 : x_0x_1 : x_1x_2) &= (x_2x_3 : x_0^2 : x_0x_2) \\ &= (x_0x_3 : x_1x_3 : x_0^2),\end{aligned}$$

(where the three expressions coincide where they are defined), with inverse

$$\psi : \mathbb{P}^2_{\mathbb{Z}} \dashrightarrow \mathcal{S}, \quad \mathbf{z} \mapsto (z_0z_1z_2 : z_1^2z_2 : z_2^2z_0 : z_0^2z_1).$$

Indeed, ϕ and ψ restrict to isomorphisms between the open subsets $\mathcal{U} \subset \mathcal{S}$ and $\mathcal{V} = \{\mathbf{z} \in \mathbb{P}^2_{\mathbb{Z}} \mid z_0z_1z_2 \neq 0\} \subset \mathbb{P}^2_{\mathbb{Z}}$.

Then the following diagram commutes, where $\pi_0 : \widetilde{\mathcal{S}} \to \mathbb{P}^2_{\mathbb{Z}}$ is the blow-up of $\mathbb{P}^2_{\mathbb{Z}}$ in six points in *almost general position* [DP80]:

$$\begin{array}{ccc} \widetilde{\mathcal{S}} & & \\ \downarrow \pi & \searrow^{\pi_0} & \\ \mathcal{S} & \overset{\phi}{\dashrightarrow} & \mathbb{P}^2_{\mathbb{Z}}\,. \end{array}$$

More precisely, π_0 maps

- $E_1, E_{2,1}$ to $(1:0:0)$,
- $E_2, E_{3,2}$ to $(0:1:0)$,
- $E_3, E_{1,3}$ to $(0:0:1)$,
- $E_{2,3}, E_{3,1}, E_{1,2}$ to $\{z_0 = 0\}, \{z_1 = 0\}, \{z_2 = 0\}$, respectively.

Conversely, using the same symbol for divisors on $\widetilde{\mathcal{S}}$ and their projections and strict transforms on $\mathbb{P}^2_{\mathbb{Z}}$ and the intermediate $\widetilde{\mathcal{S}}_1$ (for example, on $\mathbb{P}^2_{\mathbb{Z}}$, we have $E_{2,3} = \{z_0 = 0\}$, $E_{3,1} = \{z_1 = 0\}$ and $E_{1,2} = \{z_2 = 0\}$) we obtain

$$\pi_0 : \widetilde{\mathcal{S}} \xrightarrow{\pi_2} \widetilde{\mathcal{S}}_1 \xrightarrow{\pi_1} \mathbb{P}^2_{\mathbb{Z}},$$

where $\widetilde{\mathcal{S}}_1$ is a smooth sextic del Pezzo surface, by

- blowing up the three points $E_{j,k} \cap E_{k,l} \in \mathbb{P}^2_{\mathbb{Z}}$ with exceptional divisors $E_{l,k}$, respectively, to obtain $\pi_1 : \widetilde{\mathcal{S}}_1 \to \mathbb{P}^2_{\mathbb{Z}}$;
- blowing up the three points $E_{k,j} \cap E_{l,j} \in \widetilde{\mathcal{S}}_1$ with exceptional divisors E_j, respectively, to obtain $\pi_2 : \widetilde{\mathcal{S}} \to \widetilde{\mathcal{S}}_1$.

This gives $\widetilde{\mathcal{S}}$ with three (-1)-curves E_j and six (-2)-curves $E_{j,k}, E_{k,j}$. Contracting the (-2)-curves via the anticanonical map gives $\pi : \widetilde{\mathcal{S}} \to \mathcal{S} \subset \mathbb{P}^3_{\mathbb{Z}}$.

6.2 The leading constant

For a smooth Fano variety defined over a number field K, Peyre [Pey95, Conjecture 2.3.1] gave a conjectural interpretation of the leading constant in Manin's conjecture. This was generalized to Fano varieties with at worst canonical singularities by Batyrev and Tschinkel [BT98b, §3.4 Step 4]. We will see that our theorem agrees with this prediction.

We start by collecting all number theoretic notation we need for this section. Let (r_K, s_K) be the number of real resp. pairs of complex embeddings of K, and let $q_K = r_K + s_K - 1$. Let $\mathcal{O}_K$ be its ring of integers. Let $\mathcal{O}_K^{\neq 0} = \mathcal{O}_K \setminus \{0\}$. Let w_K be the number of roots of unity in $\mathcal{O}_K$, and let R_K be the regulator of K. Let d_K denote the square root of the absolute value of the discriminant of K.

The set M_K of places of K consists of the archimedian places $M_{K,\infty}$ and the non-archimedian places $M_{K,f}$. For $\nu \in M_K$, let K_ν be the completion of K at ν and, for $\nu \in M_{K,f}$, let $\mathbb{F}_\nu$ be the residue field. For any $\nu \in M_{K,f}$, we define a norm by $\|x\|_\nu = |N_{K_\nu/\mathbb{Q}_p}(x)|_p$ for all $x \in K_\nu$, where p is the characteristic of $\mathbb{F}_\nu$ and $|\cdot|_p$ is the usual norm on $\mathbb{Q}_p$. For any $\nu \in M_{K,\infty}$ corresponding to a real embedding $\sigma : K \to \mathbb{R}$, we define $\|x\|_\nu = |\sigma(x)|$ for all $x \in K_\nu$, where $|\cdot|$ is the usual absolute value on $\mathbb{R}$. For any $\nu \in M_{K,\infty}$ corresponding to a pair of complex embeddings σ, σ', we define $\|x\|_\nu = |\sigma(x)|^2$, where $|\cdot|$ is the usual absolute value on $\mathbb{C}$.

We compute the expected asymptotic behavior of $N_{U,K,H}(B)$ with respect to the very ample anticanonical metrized sheaf $-\mathcal{K}_S = (-K_S, \|\cdot\|_\nu)$ [BT98b, Definition 3.1.3], where the family of ν-adic metrics corresponds to our anticanonical height function H.

We use the minimal desingularization $\pi : \widetilde{S} \to S$ and its integral model $\pi : \widetilde{\mathcal{S}} \to \mathcal{S}$ constructed in Section 6.1. As the $\mathbf{A}_2$-singularities on S are rational double points, we have $\pi^*(K_S) = K_{\widetilde{S}}$. Therefore, the $-\mathcal{K}_S$-index [BT98b, Definition 2.2.4] of S is 1, and the $-\mathcal{K}_S$-rank [BT98b, Definition 2.3.11] of S is $\operatorname{rk}\operatorname{Pic}(\widetilde{S}) = 7$. Hence the expected asymptotic formula according to Manin's conjecture is (in the notation of [BT98b, §3.4 Step 4])

$$N_{U,K,H}(B) = \frac{\gamma_{-\mathcal{K}_S}(U)}{6!}\delta_{-\mathcal{K}_S}(U)\tau_{-\mathcal{K}_S}(U)B(\log B)^6(1+o(1)).$$

The cohomological factor of the expected leading constant is

$$\delta_{-\mathcal{K}_S}(U) = \#H^1(\operatorname{Gal}(\overline{\mathbb{Q}}/K), \operatorname{Pic}(\widetilde{S}_{\overline{\mathbb{Q}}})) = 1,$$

as $\operatorname{Gal}(\overline{\mathbb{Q}}/K)$ acts trivially on $\operatorname{Pic}(\widetilde{S}_{\overline{\mathbb{Q}}})$ since $\widetilde{S}$ is split over K.

The factor $\gamma_{-\mathcal{K}_S}(U)/6!$ [BT98b, Definition 2.3.16] is simply $\alpha(\widetilde{S})$ as in [Pey95, Définition 2.4]. By [DJT08, Theorem 1.3], we have

$$\frac{\gamma_{-\mathcal{K}_S}(U)}{6!} = \alpha(\widetilde{S}) = \frac{\alpha(S_0)}{\#W(3\mathbf{A}_2)} = \frac{1}{120\cdot(3!)^3} = \frac{1}{36\cdot 6!},$$

where S_0 is a smooth cubic surface with $\alpha(S_0) = 1/120$ by [Der07, Theorem 4] and $W(3\mathbf{A}_2)$ is the Weyl group of the root system $3\mathbf{A}_2$ associated to the singularities of S.

Next, we compute the Tamagawa number

$$\tau_{-\mathcal{K}_S}(U) = \lim_{s\to 1}(s-1)^7 L(s, \operatorname{Pic}(\widetilde{S}_{\overline{\mathbb{Q}}})) \int_{\widetilde{S}(\mathbb{A}_K)} \omega_{-\mathcal{K}_S}$$

[BT98b, Definition 3.3.10], where the set $\widetilde{S}(\mathbb{A}_K)$ of adelic points coincides with the closure of $\widetilde{S}(K)$ in it since $\widetilde{S}$ satisfies weak approximation. Here, we have used that every non-archimedean valuation ν is a good valuation in the sense of [BT98b, Definition 3.3.5] since the reduction of the model $\widetilde{\mathcal{S}}$ of $\widetilde{S}$ at any finite place of K is a smooth projective variety.

Since $\widetilde{S}$ is split, the Frobenius morphism associated to every non-archimedean place ν corresponding to a prime ideal $\mathfrak{p}$ acts trivially on $\operatorname{Pic}(\widetilde{S}_{\overline{\mathbb{F}}_\nu})$ of rank 7. Therefore, $L_\nu(s, \operatorname{Pic}(\widetilde{S}_{\overline{\mathbb{Q}}})) = (1-\mathfrak{N}\mathfrak{p}^{-s})^{-7}$ (cf. [Pey95, §2.2.3]), and $L(s, \operatorname{Pic}(\widetilde{S}_{\overline{\mathbb{Q}}})) = \prod_{\nu \in M_{K,f}} L_\nu(s, \operatorname{Pic}(\widetilde{S}_{\overline{\mathbb{Q}}})) = \zeta_K(s)^7$. So

$$\lim_{s\to 1}(s-1)^7 L(s, \operatorname{Pic}(\widetilde{S}_{\overline{\mathbb{Q}}})) = \lim_{s\to 1}(s-1)^7 \zeta_K(s)^7 = \left(\frac{2^{r_K}(2\pi)^{s_K} h_K R_K}{w_K d_K}\right)^7$$

by the analytic class number formula.

Furthermore, by [BT98b, Definition 3.3.9],

$$\int_{\widetilde{S}(\mathbb{A}_K)} \omega_{-\mathcal{K}_S} = d_K^{-\dim(\widetilde{S})} \prod_{\nu \in M_K} \lambda_\nu^{-1} d_\nu(U)$$

where $\lambda_\nu = L_\nu(1, \operatorname{Pic}(\widetilde{S}_{\overline{\mathbb{Q}}}))$ for all (good) non-archimedean places and $\lambda_\nu = 1$ for the archimedean places. It remains to compute the local densities $d_\nu(U)$ defined in [BT98b, Remark 3.3.2].

We compute the archimedean densities on the open subset $U = \{x_0 \neq 0\}$ of S, defined by the cubic equation $f(x_0, \ldots, x_3) = x_0^3 - x_1x_2x_3$, as

$$d_\nu(U) = \int_{S(K_\nu)} \omega_{-\mathcal{K}_S}(g) = \begin{cases} 36, & \nu \text{ real,} \\ 36\pi^2, & \nu \text{ complex.} \end{cases}$$

Indeed, we apply [Pey95, Lemme 5.4.4] and see via the birational morphism $\rho : U \to \mathbb{A}^2$ defined by $(x_0 : x_1 : x_2 : x_3) \mapsto (x_1/x_0, x_2/x_0)$ that

$$d_\nu(U) = \int_{(K_\nu^\times)^2} \frac{1}{\max\{1, \|y_1\|_\nu, \|y_2\|_\nu, \|(y_1y_2)^{-1}\|_\nu\} \cdot \|-y_1y_2\|_\nu} \mathrm{d}y_{1,\nu} \mathrm{d}y_{2,\nu}.$$

A straightforward computation gives the values above. For complex ν, the Haar measure $\mathrm{d}y_{i,\nu}$ on K_ν is normalized as *twice* the usual Lebesgue measure obtained from regarding $K_\nu \cong \mathbb{C}$ as $\mathbb{R}^2$, as in [Pey95, §1.1]. For real ν, the value $d_\nu(S) = 36$ can also be found in [BT98b, §5.3].

As every non-archimedean place ν corresponding to a prime ideal $\mathfrak{p}$ is good, we can apply [BT98b, Theorem 3.3.7] to compute

$$d_\nu(U) = \frac{\#\widetilde{S}(\mathbb{F}_\mathfrak{p})}{\mathfrak{N}\mathfrak{p}^2} = 1 + \frac{7}{\mathfrak{N}\mathfrak{p}} + \frac{1}{\mathfrak{N}\mathfrak{p}^2},$$

since the norm $\mathfrak{N}\mathfrak{p}$ is the cardinality of the residue field $\mathbb{F}_\nu$ of K_ν. Indeed, for any finite field $\mathbb{F}_q$, the surface $\widetilde{S}_{\mathbb{F}_q}$ is the blow-up of $\mathbb{P}^2_{\mathbb{F}_q}$ in six $\mathbb{F}_q$-rational points, and any such blow-up replaces one $\mathbb{F}_q$-rational point by a rational curve containing $q+1$ points over $\mathbb{F}_q$. Since $\#\mathbb{P}^2_{\mathbb{Z}}(\mathbb{F}_q) = q^2+q+1$, we obtain the result. See also [Lou10, Lemma 2.3].

In total, the expected leading constant is

$$\frac{9^{q_K}}{4\cdot 6!}\left(\frac{2^{r_K}(2\pi)^{s_K}}{d_K}\right)^9\left(\frac{h_K R_K}{w_K}\right)^7\prod_{\mathfrak{p}}\left(1-\frac{1}{\mathfrak{N}\mathfrak{p}}\right)^7\left(1+\frac{7}{\mathfrak{N}\mathfrak{p}}+\frac{1}{\mathfrak{N}\mathfrak{p}^2}\right).$$

For imaginary quadratic number fields K with class number $h_K = 1$, we have $(r_K, s_K) = (0, 1)$, so $q_K = 0$. Since the number w_K of units in $\mathcal{O}_K$ is finite, its regulator R_K is 1. We denote the archimedean place as $\nu = \infty$. Let $\mathbb{N}_K$ be a fundamental domain for $\mathcal{O}_K^{\neq 0}$ modulo the action of the units. We identify each prime ideal $\mathfrak{p}$ with its unique generator $p \in \mathbb{N}_K$, with $\mathfrak{N}\mathfrak{p} = \|p\|_\infty$. We see that the expected leading constant of [BT98b] coincides with $c_{S,K,H}$ in our main theorem.

6.3 Passage to a universal torsor

We follow the strategy of [DT07]. This leads to a parameterization of rational points on S by integral points in $\mathbb{A}^9$ that is similar to the one used in [HBM99], but with a different set of coprimality conditions. We could construct coprimality conditions as in [HBM99], but we believe our conditions are more closely connected to the geometry of $\widetilde{S}$ and easier to work with. We note that our coprimality conditions are analogous to the ones obtained by Salberger [Sal98, 11.5] for toric varieties over $\mathbb{Q}$.

In the following, any statement involving j, k, l is meant to hold for all

$$(j, k, l) \in \{(1, 2, 3), (2, 3, 1), (3, 1, 2)\}.$$

A parameterization of K-rational points on $U \subset S$ is obtained via the map ψ defined in Section 6.1. The isomorphism $\psi_{|V} : V \to U$ induces a map

$$\Psi_0 : (\mathcal{O}_K^{\neq 0})^3 \to S(K), \quad \mathbf{y} \mapsto \Psi_0(\mathbf{y}) = (\Psi_0(\mathbf{y})_0 : \ldots : \Psi_0(\mathbf{y})_3)$$

where $\mathbf{y} = (y_{2,3}, y_{3,1}, y_{1,2})$ and

$$\Psi_0(\mathbf{y})_0 = y_{1,2}y_{3,1}y_{2,3}, \ \Psi_0(\mathbf{y})_j = y_{j,k}y_{l,j}^2.$$

This induces a w_K-to-1 map from

$$\{(y_{2,3}, y_{3,1}, y_{1,2}) \in (\mathcal{O}_K^{\neq 0})^3 \mid H(\Psi_0(\mathbf{y})) \leq B, \ \gcd(y_{1,2}, y_{2,3}, y_{3,1}) = 1\}$$

to

$$N_0(B) = \{\mathbf{x} \in U(K) \mid H(\mathbf{x}) \leq B\}.$$

However, this parameterization is not good enough to start counting integral elements in a region in $\mathcal{O}_K^3$ because the height condition is not as easy as one might hope since $\gcd(y_{1,2}y_{2,3}y_{3,1}, y_{1,2}y_{3,1}^2, y_{2,3}y_{1,2}^2, y_{3,1}y_{2,3}^2)$ (taken here and always in $\mathcal{O}_K$) may be non-trivial even if $\gcd(y_{1,2}, y_{2,3}, y_{3,1}) = 1$.

Motivated by the construction of $\widetilde{S}$ as the blow-up of $\mathbb{P}^2_{\mathbb{Z}}$ in intersection points of certain divisors, we modify this as follows.

In the first step, let $y_{l,k} = \gcd(y_{j,k}, y_{k,l})$. Write $y_{j,k} = y'_{j,k}y_{k,j}y_{l,k}$. Then $\gcd(y'_{j,k}, y'_{k,l}) = \gcd(y_{l,k}, y_{k,j}) = \gcd(y_{k,j}, y'_{k,l}) = 1$. Now we drop the $'$ again for notational simplicity. We obtain a map

$$\Psi_1 : (\mathcal{O}_K^{\neq 0})^6 \to S(K), \quad \mathbf{y} \mapsto \Psi_1(\mathbf{y}) = (\Psi_1(\mathbf{y})_0 : \ldots : \Psi_1(\mathbf{y})_3),$$

where $\mathbf{y} = (y_{1,2}, y_{2,1}, y_{1,3}, y_{3,1}, y_{2,3}, y_{3,2})$ and

$$\Psi_1(\mathbf{y})_0 = y_{1,2}y_{2,1}y_{1,3}y_{3,1}y_{2,3}y_{3,2}, \quad \Psi_1(\mathbf{y})_j = y_{j,k}y_{j,l}y_{k,j}^2y_{l,j}^2.$$

We note that the coprimality conditions can be expressed as follows: For $(u, v) \in \{(1,2), (2,1), (1,3), (3,1), (2,3), (3,2)\}$, we have $\gcd(y_u, y_v) = 1$ if and only if the divisors E_u and E_v do not intersect on $\widetilde{S}_1$, which holds if and only if the corresponding rays in the fan of $\widetilde{S}_1$ (Figure 6.1) are not neighbors.

Since the $y_{k,j}$ are unique up to units in $\mathcal{O}_K$, the map Ψ_1 induces a w_K^4-to-1 map from

$$\left\{ \mathbf{y} \in (\mathcal{O}_K^{\neq 0})^6 \;\middle|\; \begin{array}{c} H(\Psi_1(\mathbf{y})) \leq B, \text{coprimality as in} \\ \text{the fan of } \widetilde{S}_1 \text{ in Figure 6.1} \end{array} \right\}$$

to $N_0(B)$.

In the second step, let $y_j = \gcd(y_{k,j}, y_{l,j})$. As before, we obtain a map

$$\Psi_2 : (\mathcal{O}_K^{\neq 0})^9 \to S(K), \quad \mathbf{y} \mapsto \Psi_2(\mathbf{y}) = (\Psi_2(\mathbf{y})_0 : \ldots : \Psi_2(\mathbf{y})_3),$$

where $\mathbf{y} = (y_1, y_2, y_3, y_{1,2}, y_{2,1}, y_{1,3}, y_{3,1}, y_{2,3}, y_{3,2})$ and

$$\Psi_2(\mathbf{y})_0 = y_1y_2y_3y_{1,2}y_{2,1}y_{1,3}y_{3,1}y_{2,3}y_{3,2}, \quad \Psi_2(\mathbf{y})_j = y_j^3y_{j,k}y_{j,l}y_{k,j}^2y_{l,j}^2.$$

This induces a w_K^7-to-1 map from

$$\left\{ \mathbf{y} \in (\mathcal{O}_K^{\neq 0})^9 \;\middle|\; \begin{array}{c} H(\Psi_2(\mathbf{y})) \leq B, \text{coprimality as in} \\ \text{the fan of } \widetilde{S} \text{ in Figure 6.1} \end{array} \right\}$$

to $N_0(B)$.

Now we note that

$$H(\Psi_2(\mathbf{y})) = \max\{\|\Psi_2(\mathbf{y})_1\|_\infty, \|\Psi_2(\mathbf{y})_2\|_\infty, \|\Psi_2(\mathbf{y})_3)\|_\infty\}$$

because the coprimality conditions imply that $\Psi_2(\mathbf{y})_0, \dots, \Psi_2(\mathbf{y})_3$ are coprime for any $\mathbf{y}$ satisfying the coprimality conditions, and the archimedian norm of $\Psi_2(\mathbf{y})_0$ cannot be larger than all other three. Indeed, the second observation follows from $\Psi_2(\mathbf{y})_0^3 = \Psi_2(\mathbf{y})_1\Psi_2(\mathbf{y})_2\Psi_2(\mathbf{y})_3$. For the first observation, we note that any prime may divide at most two variables whose corresponding rays in Figure 6.1 are neighbors, and one checks that for each such pair of variables, there is one monomial in which these variables do not occur.

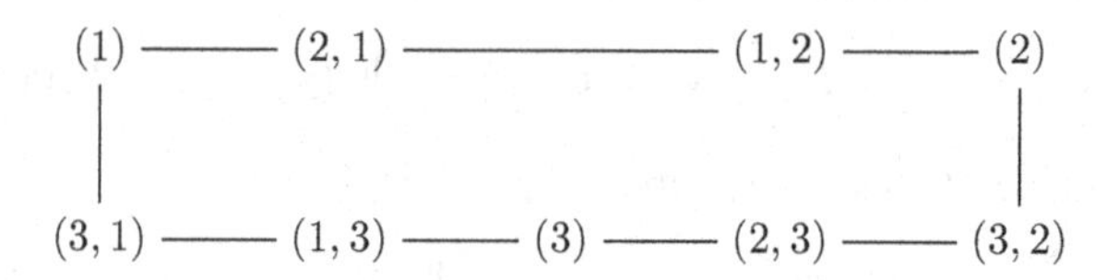

Figure 6.2 Graph $G = (V, E)$ encoding coprimality conditions.

We reformulate the coprimality conditions as follows, using the graph $G = (V, E)$ with nine vertices $V = \{(1), (2, 1), \dots\}$ and nine edges $E = \{\{(1), (2, 1)\}, \{(2, 1), (1, 2)\}, \dots\}$ in Figure 6.2.

Lemma 6.2 *Let $G = (V, E)$ be the graph in Figure 6.2. Let E' be the set all pairs $\{u, v\}$ of vertices $u, v \in V$ which are not adjacent in the graph.*

We have

$$N_{U,K,H}(B) = \frac{1}{w_K^7} \sum_{\substack{\mathbf{y} \in (\mathcal{O}_K^{\neq 0})^V \cap M(B) \\ \gcd(y_u, y_v) = 1 \text{ for all } \{u,v\} \in E'}} 1,$$

where $M(B)$ is the set of all $\mathbf{y} \in \mathbb{C}^V$ with

$$\|y_j^3 y_{j,k} y_{j,l} y_{k,j}^2 y_{l,j}^2\|_\infty \le B$$

for all $(j, k, l) \in \{(1, 2, 3), (2, 3, 1), (3, 1, 2)\}$.

6.4 Möbius inversions

Having found a suitable parameterization of K-rational points by points over $\mathcal{O}_K$ in an open subset of $\mathbb{A}^9$ in Lemma 6.2, the main problem is essentially to estimate the number of lattice points in the region described by the height conditions. This is done in Lemma 6.3; its proof is deferred to Section 6.5. Here, we remove the coprimality conditions by a Möbius inversion and recover the non-archimedian densities.

Applying Möbius inversion over all elements of E' to the expression in Lemma 6.2 gives

$$N_{U,K,H}(B) = \frac{1}{w_K^7} \sum_{\mathbf{d}\in\mathbb{N}_K^{E'}} \prod_{\alpha\in E'} \mu(d_\alpha) \sum_{\substack{\mathbf{y}\in(\mathcal{O}_K^{\neq 0})^V\cap M(B) \\ d_{\{u,v\}}\mid y_u, y_v\ \forall\{u,v\}\in E'}} 1.$$

We collect all terms dividing some y_v to obtain

$$N_{U,K,H}(B) = \frac{1}{w_K^7} \sum_{\mathbf{d}\in\mathbb{N}_K^{E'}} \prod_{\alpha\in E'} \mu(d_\alpha) \sum_{\substack{\mathbf{y}\in(\mathcal{O}_K^{\neq 0})^V\cap M(B) \\ r_v\mid y_v\ \forall v\in V}} 1,$$

where r_v is defined as the lowest common multiple of the d_α with $\alpha \in E'$ and $v \in \alpha$. This sum can be estimated as follows; see Section 6.5 for the proof.

Lemma 6.3 *For* $\mathbf{r} \in \mathbb{N}_K^V$, *let*

$$R_1 = \prod_{v\in V} \|r_v\|_\infty, \quad R_2 = \prod_{\substack{j,k\in\{1,2,3\} \\ j\neq k}} \|r_{j,k}\|_\infty^{2/3} \prod_{j\in\{1,2,3\}} \|r_j\|_\infty (\max_j \|r_j\|_\infty)^{-1/2}.$$

Then

$$\sum_{\substack{\mathbf{y}\in(\mathcal{O}_K^{\neq 0})^V\cap M(B) \\ r_v\mid y_v\ \forall v\in V}} 1 = \frac{2^7\pi^9}{6!d_K^9}\frac{B}{R_1}(\log(B))^6 + O\left(\frac{B}{R_2}(\log(B))^5\right).$$

Combining this with Lemma 6.2 gives

$$N_{U,K,H}(B) = \frac{2^7\pi^9}{6!w_K^7 d_K^9}\omega B(\log(B))^6 + O(\rho B(\log B)^5),$$

where

$$\omega = \sum_{\mathbf{d}\in\mathbb{N}_K^{E'}} \prod_{\alpha\in E'} \mu(d_\alpha)\frac{1}{R_1}, \quad \rho = \sum_{\mathbf{d}\in\mathbb{N}_K^{E'}} \prod_{\alpha\in E'} |\mu(d_\alpha)|\frac{1}{R_2}$$

with R_1 and R_2 depending on $\mathbf{d}$.

To show that ω and ρ are well-defined it will be enough to show the convergence of the defining sum of ρ since $|\omega| \leq \rho$. The Euler factor of ρ corresponding to some prime $p \in \mathbb{N}_K$ is $1 + O(\|p\|_\infty^{-7/6})$. Indeed, the factor will have only finitely many non-vanishing summands since $\mu(p^e) = 0$ for all $e \geq 2$. For $\mathbf{d} = (1, \ldots, 1)$, we have $R_2 = 1$. For $\mathbf{d}$ with

$d_\alpha = p$ for at least one $\alpha = \{u, v\} \in E'$, we have $r_u = r_v = p$ and therefore $R_2 \geq \|r_u\|_\infty^a \|r_v\|_\infty^b \geq \|p\|_\infty^{7/6}$ where $a, b \in \{\frac{1}{2}, \frac{2}{3}, 1\}$, with at most one of them equal to $\frac{1}{2}$.

Let us now calculate the Euler factors A_p of ω for some prime $p \in \mathbb{N}_K$. Let $A \in \mathbb{Z}[x]$ be the polynomial

$$A(x) = \sum_{\widetilde{\mathbf{d}} \in \{0,1\}^{E'}} \prod_{\alpha \in E'} \widetilde{\mu}(\widetilde{d}_\alpha) x^{\sum_{v \in V} \widetilde{r}_v}.$$

Here, $\widetilde{\mathbf{r}} \in \{0,1\}^V$ is defined depending on $\widetilde{\mathbf{d}}$ as follows: For any $v \in V$, the number $\widetilde{r}_v$ is the maximum of all $\widetilde{d}_\alpha$ with $v \in \alpha$. Furthermore, $\widetilde{\mu}$ is defined by

$$\widetilde{\mu}(n) = \begin{cases} 1, & n = 0, \\ -1, & n = 1. \end{cases}$$

Then $A_p = A(\|p\|_\infty^{-1})$.

By further Möbius inversions, we have

$$\begin{aligned} A(x) &= \sum_{\widetilde{\mathbf{k}} \in \{0,1\}^V} x^{\sum_{v \in V} \widetilde{k}_v} \sum_{\substack{\widetilde{\mathbf{d}} \in \{0,1\}^{E'} \\ \widetilde{\mathbf{r}} = \widetilde{\mathbf{k}}}} \prod_{\alpha \in E'} \widetilde{\mu}(\widetilde{d}_\alpha) \\ &= \sum_{\widetilde{\mathbf{n}} \in \{0,1\}^V} \prod_{v \in V} e(\widetilde{n}_v) \sum_{\substack{\widetilde{\mathbf{d}} \in \{0,1\}^{E'} \\ \widetilde{d}_\alpha \leq \widetilde{n}_v \text{ if } v \in \alpha}} \prod_{\alpha \in E'} \widetilde{\mu}(\widetilde{d}_\alpha), \end{aligned}$$

where the function $e : \{0,1\} \to \mathbb{Q}[x]$ defined by

$$e(n) = \begin{cases} 1 - x, & n = 0, \\ x, & n = 1 \end{cases}$$

is chosen such that

$$\sum_{k \in \{0,1\}} x^k F(k) = \sum_{n \in \{0,1\}} e(n) \sum_{0 \leq s \leq n} F(s)$$

for any function $F : \{0,1\} \to \mathbb{Z}$; this is applied above $\#V$ times. We have also used that $\widetilde{r}_v \leq \widetilde{n}_v$ if $\widetilde{d}_\alpha \leq \widetilde{n}_v$ for all $\alpha \in E'$ containing $v \in V$.

Note that since $\widetilde{\mu}(0) + \widetilde{\mu}(1) = 0$ for a fixed $\widetilde{\mathbf{n}} \in \{0,1\}^V$, the sum over $\widetilde{\mathbf{d}}$ will vanish if two vertices of $\{v \in V \mid \widetilde{n}_v = 1\}$ can be joined by a line in E'. So it will not vanish only if either all $\widetilde{n}_v$ are 0, exactly one of the nine $\widetilde{n}_v$ is equal to 1 or exactly two $\widetilde{n}_v$, $\widetilde{n}_w$ are 1 where $\{v, w\}$ is one of the nine edges E. So we have

$$A(x) = (1-x)^9 + 9(1-x)^8 x + 9(1-x)^7 x^2 = (1-x)^7 \cdot (1 + 7x + x^2)$$

and finally

$$\omega = \prod_p (1 - \|p\|_\infty^{-1})^7 (1 + 7\|p\|_\infty^{-1} + \|p\|_\infty^{-2}).$$

Up to the proof of Lemma 6.3, this completes the proof of our main theorem.

6.5 Estimations of lattice points

In this section, we prove Lemma 6.3. We proceed as in [HBM99]. First, we rewrite the sum as

$$\sum_{\substack{\mathbf{y} \in (\mathcal{O}_K^{\neq 0})^V \cap M(B) \\ r_v | y_v \ \forall v \in V}} 1 = \sum_{\substack{\mathbf{z} \in (\mathcal{O}_K^{\neq 0})^V \\ \|z_j\|_\infty \leq C_j}} 1,$$

where we define

$$\zeta_j = z_{j,k} z_{j,l} z_{k,j}^2 z_{l,j}^2, \quad C_j = B^{1/3} \|\zeta_j r_j^3 r_{j,k} r_{j,l} r_{k,j}^2 r_{l,j}^2\|_\infty^{-1/3}$$

for any $\{j,k,l\} = \{1,2,3\}$.

From here, unless stated otherwise, we use the convention that whenever j,k appears in a statement, we mean all $j,k \in \{1,2,3\}$ with $j \neq k$, and whenever j shows up, we mean all $j \in \{1,2,3\}$.

To sum over z_j for $j = 1,2,3$, one can use the following estimate on the number of integers in $\mathcal{O}_K^{\neq 0}$ in the circle $\mathrm{Ci}(C) = \{z \in \mathbb{C} \mid \|x\|_\infty \leq C\}$ of radius $\sqrt{C}$ in the complex plane. It seems interesting to note that [HBM99, §4] finds it convenient to use the similar estimate $C + O(\sqrt{C})$ instead of $C + O(1)$ (which is not available in our case) for the number of natural numbers smaller than C.

Lemma 6.4 *For any positive $C \in \mathbb{R}$, we have*

$$\#(\{x \in \mathcal{O}_K^{\neq 0}\} \cap \mathrm{Ci}(C)) = \frac{2\pi}{d_K} C + O(\sqrt{C}).$$

Proof The theorem follows in the case $C \geq C_0$ for some $C_0 > 0$ by the theorem on lattice points in homogenously expanding sets since the area of the circle is πC and the area of a fundamental domain is $d_K/2$. The missing point in the origin can be accounted for in the error term in this case. If $C < 1$, then $\mathrm{Ci}(C) \cap \mathcal{O}_K^{\neq 0} = \emptyset$ and we have $(2\pi/d_K)C = O(\sqrt{C})$, so the lemma is also true in this case. Finally, if $1 \leq C \leq C_0$, then

$\mathrm{Ci}(C) \subseteq \mathrm{Ci}(C_0)$ and $\sqrt{C} \geq 1$. Therefore, we can choose the implied constant in the error term to be greater than $\#(\mathrm{Ci}(C_0) \cap \mathcal{O}_K^{\neq 0}) + (2\pi/d_K)C_0$, which establishes the lemma in the remaining case. □

Therefore, with $B_j = B\|r_{j,k}r_{j,l}r_{k,j}^2r_{l,j}^2\|_\infty^{-1}$ for all $\{j,k,l\} = \{1,2,3\}$,

$$\sum_{\substack{\mathbf{z} \in (\mathcal{O}_K^{\neq 0})^V \\ \|z_j\|_\infty \leq C_j}} 1 = \sum_{\substack{z_{j,k} \in \mathcal{O}_K^{\neq 0} \\ \|\zeta_j\|_\infty \leq B_j}} \prod_{j=1}^{3} \left(\frac{2\pi}{d_K} C_j + O(\sqrt{C_j}) \right)$$

$$= \sum_{\substack{z_{j,k} \in \mathcal{O}_K^{\neq 0} \\ \|\zeta_j\|_\infty \leq B_j}} \left\{ \frac{2^3\pi^3}{d_K^3} C_1C_2C_3 + O\left(C_1C_2C_3 \max_j \left(C_j^{-1/2} \right) \right) \right\}$$

$$= \frac{2^3\pi^3 B}{d_K^3 R_1} \mathcal{M}(B, \mathbf{r}) + O\left(\frac{B^{5/6}}{R_2} \mathcal{R}(B, \mathbf{r}) \right),$$

where

$$\mathcal{M}(B, \mathbf{r}) = \sum_{\substack{z_{j,k} \in \mathcal{O}_K^{\neq 0} \\ \|\zeta_j\|_\infty \leq B_j}} \prod_{j,k} \|z_{j,k}\|_\infty^{-1}$$

and

$$\mathcal{R}(B, \mathbf{r}) = \sum_{\substack{z_{j,k} \in \mathcal{O}_K^{\neq 0} \\ \|\zeta_j\|_\infty \leq B_j}} \prod_{j,k} \|z_{j,k}\|_\infty^{-1} \max_j \|\zeta_j\|_\infty^{\frac{1}{6}}.$$

Let us begin with the estimation of the error term $\mathcal{R}$. Because of symmetry it will be no loss to assume that $\|\zeta_1\|_\infty \geq \|\zeta_2\|_\infty, \|\zeta_3\|_\infty$. Then

$$\mathcal{R}(B, \mathbf{r}) = \sum_{\substack{z_{j,k} \in \mathcal{O}_K^{\neq 0} \\ \|\zeta_j\|_\infty \leq B_j}} \|z_{2,1}z_{3,1}\|_\infty^{-2/3} \|z_{2,3}z_{3,2}\|_\infty^{-1} \|z_{1,2}z_{1,3}\|_\infty^{-5/6}$$

$$\ll \sum_{\substack{z_{j,k} \in \mathcal{O}_K^{\neq 0}\ \forall j \neq 1 \\ \|z_{j,k}\|_\infty \leq B\ \forall j \neq 1}} \|z_{2,1}z_{3,1}\|_\infty^{-2/3} \|z_{2,3}z_{3,2}\|_\infty^{-1} \sum_{\substack{u \in \mathcal{O}_K^{\neq 0} \\ \|u\|_\infty \leq U}} \frac{d(u)}{\|u\|_\infty^{5/6}}$$

where U is defined as $U = B\|z_{2,1}z_{3,1}\|_\infty^{-2}$ and d is the divisor function in $\mathcal{O}_K^{\neq 0}$.

Here, we need the following auxiliary result, which we will use again later.

Lemma 6.5 *For all sufficiently large B, we have*

$$\sum_{\substack{x\in\mathcal{O}_K^{\neq 0}\\ \|x\|_\infty\le B}} \|x\|_\infty^\alpha = \begin{cases} O(B^{\alpha+1}), & -1<\alpha\le 0,\\ O(\log B), & \alpha=-1.\end{cases}$$

Proof For any $n\in\mathbb{N}$, let

$$a_n = \#\{x\in\mathcal{O}_K^{\neq 0} \mid \|x\|_\infty = n\}.$$

By the Abel summation formula, we have

$$\sum_{\substack{x\in\mathcal{O}_K^{\neq 0}\\ \|x\|_\infty\le B}} \|x\|_\infty^\alpha = \sum_{1\le n\le B} a_n n^\alpha = B^\alpha \sum_{1\le n\le B} a_n - \alpha\int_1^B x^{\alpha-1}\sum_{1\le n\le x} a_n\,\mathrm{d}x.$$

We apply Lemma 6.4 to the sums over n. The first term is $O(B^{\alpha+1})$. For $-1<\alpha\le 0$, the second term is $O(B^{\alpha+1})$ as well; for $\alpha=-1$, it is $O(\log B)$. □

Using Lemma 6.5 twice, the inner sum can be estimated elementarily as

$$\sum_{\|u\|_\infty\le U}\frac{d(u)}{\|u\|_\infty^{5/6}} = \sum_{\|u\|_\infty\le U}\sum_{v\mid u}\|u\|_\infty^{-5/6} = \sum_{\|v\|_\infty\le U}\|v\|_\infty^{-5/6}\sum_{w\le U\|v\|_\infty^{-1}}\|w\|_\infty^{-5/6}$$
$$= O\left(U^{1/6}\sum_{\|v\|_\infty\le B}\|v\|_\infty^{-1}\right) = O(U^{1/6}\log B).$$

Inserting this into the original expression for $\mathcal{R}(B,\mathbf{r})$ and applying Lemma 6.5 again gives

$$\mathcal{R}(B,\mathbf{r}) \ll B^{1/6}\log B \sum_{\substack{z_{j,k}\in\mathcal{O}_K^{\neq 0}\ \forall j\neq 1\\ \|z_{j,k}\|_\infty\le B\ \forall j\neq 1}} \|z_{2,1}z_{3,1}z_{2,3}z_{3,2}\|_\infty^{-1}$$
$$\ll B^{1/6}\log B\left(\sum_{\substack{z\in\mathcal{O}_K^{\neq 0}\\ \|z\|_\infty\le B}}\|z\|_\infty^{-1}\right)^4 \ll B^{1/6}(\log B)^5.$$

Now it will be enough to show that

$$\mathcal{M}(B,\mathbf{r}) = \frac{2^4\pi^6}{6!d_K^6}(\log B)^6 + O(R_3(\log B)^5)$$

where we define $R_3 = \prod_{j,k}\|r_{j,k}\|_\infty^{1/3}$. Assume this is done for the case of

$r_{j,k} = 1$ for all j, k and all $B \geq B_0$ for some $B_0 > 1$. Then on the one hand

$$\mathcal{M}(B, \mathbf{r}) \leq \mathcal{M}(B, (1, \ldots, 1)) = \frac{2^4 \pi^6}{6! d_K^6} (\log B)^6 + O((\log B)^5)$$

and on the other hand

$$\mathcal{M}(B, \mathbf{r}) \geq \mathcal{M}(B/R_3^6, (1, \ldots, 1)) = \frac{2^4 \pi^6}{6! d_K^6} (\log(B/R_3^6))^6 + O((\log(B/R_3^6))^5)$$

for all $\mathbf{r}$ with $R_3 \leq (B/B_0)^{1/6}$. This gives the required estimate in this case since there is a constant C such that $\log R_3 \leq C R_3^{1/6}$ for any $R_3 \geq 1$. Otherwise we notice that the error term dominates the main term.

It therefore remains to estimate $\mathcal{M}(B) = \mathcal{M}(B, (1, \ldots, 1))$. In this case, $B_j = B$.

Lemma 6.6 *Let*

$$N(B) = \{\mathbf{z} \in \mathbb{C}^6 \mid \|z_{j,k}\|_\infty \geq 1, \|\zeta_j\|_\infty \leq B\},$$

where the ζ_j are defined as before. Define the integral

$$\mathcal{I}(B) = \left(\frac{2}{d_K}\right)^6 \int_{N(B)} \prod_{j,k} \frac{dz_{j,k}}{\|z_{j,k}\|_\infty}.$$

Then $\mathcal{M}(B) = \mathcal{I}(B) + O((\log B)^5)$ for all sufficiently large B.

Proof We fix a fundamental domain F of the lattice corresponding to $\mathcal{O}_K$ in $\mathbb{C}$; its area is $d_K/2$. Our goal is to compare the terms $\prod_{j,k} \|z_{j,k}\|_\infty^{-1}$ of the sum defining $\mathcal{M}(B)$ with integrals over translations of F. For an upper bound for $\mathcal{M}(B)$ compared to $\mathcal{I}(B)$, we must choose a translation $F(z)$ of F whose elements are closer to 0 than $z \in \mathcal{O}_K$. For an upper bound for $\mathcal{I}(B)$ compared to $\mathcal{M}(B)$, we must choose a translation $F'(z)$ of F whose elements are further away from 0 than $z \in \mathcal{O}_K$. Furthermore, we must be careful to stay away from the ball $\|z\|_\infty \leq 1$ and from the real and imaginary axes.

Let R be the smallest rectangle whose sides (of real length l_r resp. imaginary length l_i) are parallel to the real and imaginary axes and that contains F. For any $z \in \mathbb{C}$ with real part $|\Re(z)| \geq 1 + l_r$ (resp. $|\Re(z)| \geq 1$) and imaginary part $|\Im(z)| \geq 1 + l_i$ (resp. $|\Im(z)| \geq 1$), let $R(z)$ (resp. $R'(z)$) be the unique translation of the rectangle R with the following property: The point $z \in \mathbb{C}$ is the corner with the largest (resp. smallest) distance to $0 \in \mathbb{C}$ of $R(z)$ (resp. $R'(z)$). Let $F(z)$ (resp. $F'(z)$)

be the unique translation of F contained in $R(z)$ (resp. $R'(z)$). For any $x \in F(z)$ (resp. $x \in F'(z)$), we have $\|z\|_\infty \geq \|x\|_\infty$ (resp. $\|z\|_\infty \leq \|x\|_\infty$).

Let

$$E(B) = \{\mathbf{z} \in N(B) \mid |\Re(z_{j,k})| \geq 1 + l_r, |\Im(z_{j,k})| \geq 1 + l_i\}$$

and $G(B) = N(B) \setminus E(B)$. Let

$$G'(B) = \{\mathbf{z} \in \mathbb{C}^6 \mid 1 \leq \|z_{j,k}\|_\infty \leq B, |\Re(z_{1,2})| \leq 1 + l_r\}.$$

We note that $G(B)$ is contained in the union of $G'(B)$ with eleven other sets of a similar shape (with the analogous condition on $\Re(z_{j,k})$ or $\Im(z_{j,k})$) that we will be able to deal with in the same way as $G'(B)$.

First, we give an upper bound for $\mathcal{M}(B)$ in terms of $\mathcal{I}(B)$. We split $\mathcal{M}(B)$ into a sum over $E(B) \cap (\mathcal{O}_K^{\neq 0})^6$ giving the main term and a sum over $G(B) \cap (\mathcal{O}_K^{\neq 0})^6$ giving the error term. For the main term, we note that the sets $\prod_{j,k} F(z_{j,k})$ for all $\mathbf{z} \in E(B) \cap (\mathcal{O}_K^{\neq 0})^6$ are subsets of $N(B)$ whose pairwise intersections are null sets. As $\|z_{j,k}\|_\infty \geq \|x\|_\infty$ for any $x \in F(z_{j,k})$ with $\mathbf{z} \in E(B) \cap (\mathcal{O}_K^{\neq 0})^6$, we have $\|z_{j,k}\|_\infty^{-1} \leq \frac{2}{d_K} \int_{F(z_{j,k})} \|x\|_\infty^{-1} \mathrm{d}x$. Therefore,

$$\sum_{\mathbf{z} \in E(B) \cap (\mathcal{O}_K^{\neq 0})^6} \prod_{j,k} \|z_{j,k}\|_\infty^{-1} \leq \sum_{\mathbf{z} \in E(B) \cap (\mathcal{O}_K^{\neq 0})^6} \prod_{j,k} \frac{2}{d_K} \int_{F(z_{j,k})} \frac{\mathrm{d}x_{j,k}}{\|x_{j,k}\|} \leq \mathcal{I}(B).$$

For the error term, we deal with $G'(B)$ instead of $G(B)$, as mentioned before; here,

$$\sum_{\mathbf{z} \in G'(B) \cap (\mathcal{O}_K^{\neq 0})^6} \prod_{j,k} \|z_{j,k}\|_\infty^{-1} \ll (\log B)^5 \sum_{\substack{z_{1,2} \in \mathcal{O}_K^{\neq 0} \\ |\Re(z)| \leq 1 + l_r}} \|z_{1,2}\|_\infty^{-1} \ll (\log B)^5.$$

Indeed, in the first step, we use Lemma 6.5. In the second step, let N be the maximum number of lattice points in $\mathcal{O}_K$ in a box of real length $1+l_r$ and imaginary length 1. We note that the sum over $z_{1,2}$ is bounded because all $z_{1,2} \in \mathcal{O}_K^{\neq 0}$ with $|\Re(z_{1,2})| \leq 1+l_r$ and $|\Im(z_{1,2})| \leq 1$ contribute $\leq 4N$, and all $z_{1,2} \in \mathcal{O}_K$ with $k \leq |\Im(z_{1,2})| \leq k+1$ contribute $\leq 4Nk^{-2}$ (because $\|z_{1,2}\|_\infty \geq k^2$), which converges when summed over $k \in \mathbb{N}$. In total,

$$\mathcal{M}(B) \leq \mathcal{I}(B) + O((\log B)^5).$$

For the other direction, we note that, for any x with $|\Re(x)| \geq 1+l_r$ and $|\Im(x)| \geq 1 + l_i$, there is a $z \in \mathcal{O}_K$ such that $x \in F'(z)$, with $|\Re(z)| \geq 1$ and $|\Im(z)| \geq 1$ and $\|z\|_\infty \leq \|x\|_\infty$, by our construction of $F'(z)$. Therefore, for any $\mathbf{x} \in E(B)$, there is a $\mathbf{z} \in N(B) \cap (\mathcal{O}_K^{\neq 0})^6$ such that

$\mathbf{x} \in \prod_{j,k} F'(z_{j,k})$. Thus $E(B)$ is covered by $\bigcup_{\mathbf{z}\in N(B)\cap(\mathcal{O}_K^{\neq 0})^6} \prod_{j,k} F'(z_{j,k})$, and

$$\left(\frac{2}{d_K}\right)^6 \int_{E(B)} \prod_{j,k} \frac{\mathrm{d}x_{j,k}}{\|x_{j,k}\|_\infty}$$
$$\leq \sum_{\mathbf{z}\in N(B)\cap(\mathcal{O}_K^{\neq 0})^6} \prod_{j,k} \frac{2}{d_K} \int_{F(z_{j,k})} \frac{\mathrm{d}x_{j,k}}{\|x_{j,k}\|_\infty} \leq \mathcal{M}(B).$$

It remains to consider the integral over $G(B)$. Again, we just consider $G'(B)$. Here, we have

$$\int_{G'(B)} \prod_{j,k} \frac{\mathrm{d}x_{j,k}}{\|x_{j,k}\|_\infty} \ll (\log B)^5 \int_{1\leq\|x_{1,2}\|_\infty\leq B,\ |\Re(x_{1,2})|\leq 1+l_r} \frac{\mathrm{d}x_{1,2}}{\|x_{1,2}\|_\infty}$$
$$\ll (\log B)^5 \left(1 + 2(1+l_r)\int_1^\infty \frac{\mathrm{d}x}{x^2}\right) \ll (\log B)^5$$

because of $\int_{1\leq\|x_{j,k}\|_\infty\leq B} \|x_{j,k}\|_\infty^{-1} \mathrm{d}x_{j,k} \ll \log B$ and $\|x_{1,2}\|_\infty \geq |\Im(x_{1,2})|^2$ and the boundedness of the integral over all $x_{1,2}$ as above with $|\Im(x_{1,2})| \leq 1$. Therefore,

$$\mathcal{I}(B) \leq \mathcal{M}(B) + O((\log B)^5),$$

completing the proof. □

It remains to evaluate $\mathcal{I}(B)$. Using the rotation symmetries of its integrands, one can write

$$\mathcal{I}(B) = \left(\frac{2}{d_K}\right)^6 \pi^6 \int \frac{\mathrm{d}z_{1,2}}{z_{1,2}} \cdots \frac{\mathrm{d}z_{3,2}}{z_{3,2}},$$

where the integral runs now over all real $z_{j,k} \geq 1$ satisfying the three inequalities $z_{j,k}z_{j,l}z_{k,j}^2z_{l,j}^2 \leq B$ for $\{j,k,l\} = \{1,2,3\}$. Substituting $z_{j,k} = B^{t_{j,k}}$ shows that

$$\mathcal{I}(B) = V\frac{2^6\pi^6}{d_K^6}(\log B)^6,$$

where V denotes the integral

$$V = \int \mathrm{d}t_{1,2}\cdots\mathrm{d}t_{3,2}$$

over the six-dimensional convex polytope defined by the six inequalities $t_{j,k} \geq 0$ and the three inequalities

$$t_{j,k} + t_{j,l} + 2t_{k,j} + 2t_{l,j} \leq 1$$

for all $\{j,k,l\} = \{1,2,3\}$. The volume of this polytope is $V = (4 \cdot 6!)^{-1}$ [HBM99]. This completes the proof of Lemma 6.3.

References

[BEL07] G. Bhowmik, D. Essouabri, and B. Lichtin. Meromorphic continuation of multivariable Euler products. *Forum Math.*, 19(6):1111–1139, 2007.

[Bre98] R. de la Bretèche. Sur le nombre de points de hauteur bornée d'une certaine surface cubique singulière. Nombre et répartition de points de hauteur bornée (Paris, 1996), *Astérisque*, (251):51–77, 1998.

[BSD07] R. de la Bretèche and P. Swinnerton-Dyer. Fonction zêta des hauteurs associée à une certaine surface cubique. *Bull. Soc. Math. France*, 135(1):65–92, 2007.

[BT98a] V. V. Batyrev and Yu. Tschinkel. Manin's conjecture for toric varieties. *J. Algebraic Geom.*, 7(1):15–53, 1998.

[BT98b] V. V. Batyrev and Yu. Tschinkel. Tamagawa numbers of polarized algebraic varieties. Nombre et répartition de points de hauteur bornée (Paris, 1996), *Astérisque*, (251):299–340, 1998. 1996).

[Der07] U. Derenthal. On a constant arising in Manin's conjecture for del Pezzo surfaces. *Math. Res. Lett.*, 14(3):481–489, 2007.

[DJT08] U. Derenthal, M. Joyce, and Z. Teitler. The nef cone volume of generalized del Pezzo surfaces. *Algebra Number Theory*, 2(2):157–182, 2008.

[DP80] M. Demazure and H. C. Pinkham, editors. *Séminaire sur les Singularités des Surfaces*, volume 777 of *Lecture Notes in Mathematics*. Springer, Berlin, 1980. Held at the Centre de Mathématiques de l'École Polytechnique, Palaiseau, 1976–1977.

[DT07] U. Derenthal and Yu. Tschinkel. Universal torsors over del Pezzo surfaces and rational points. In *Equidistribution in number theory, an introduction*, volume 237 of *NATO Sci. Ser. II Math. Phys. Chem.*, pages 169–196. Springer, Dordrecht, 2007.

[FMT89] J. Franke, Yu. I. Manin, and Yu. Tschinkel. Rational points of bounded height on Fano varieties. *Invent. Math.*, 95(2):421–435, 1989.

[Fou98] É. Fouvry. Sur la hauteur des points d'une certaine surface cubique singulière. Nombre et répartition de points de hauteur bornée (Paris, 1996), *Astérisque*, (251):31–49, 1998.

[Fre12] C. Frei. Counting rational points over number fields on a singular cubic surface, to appear in *Algebra Number Theory*. arXiv:1204.0383

[HBM99] D. R. Heath-Brown and B. Z. Moroz. The density of rational points on the cubic surface $X_0^3 = X_1X_2X_3$. *Math. Proc. Cambridge Philos. Soc.*, 125(3):385–395, 1999.

[Lou10] D. Loughran. Manin's conjecture for a singular sextic del Pezzo surface. *J. Théor. Nombres Bordeaux*, 22(3):675–701, 2010.

[Pey95] E. Peyre. Hauteurs et mesures de Tamagawa sur les variétés de Fano. *Duke Math. J.*, 79(1):101–218, 1995.

[Sal98] P. Salberger. Tamagawa measures on universal torsors and points of bounded height on Fano varieties. Nombre et répartition de points de hauteur bornée (Paris, 1996), *Astérisque*, (251):91–258, 1998.

[Sch64] S. Schanuel. On heights in number fields. *Bull. Amer. Math. Soc.*, 70:262–263, 1964.

[Sch79] S. Schanuel. Heights in number fields. *Bull. Soc. Math. France*, 107(4):433–449, 1979.

7

Actions algébriques de groupes arithmétiques

Philippe Gille
C.N.R.S. et École normale supérieure (DMA)

Laurent Moret-Bailly
IRMAR, Université de Rennes 1

Abstract We obtain a general finiteness result for the H^1 of certain linear group schemes over the (S-)integers of global fields.

Résumé Nous obtenons un résultat général de finitude pour le H^1 de certains schémas en groupes linéaires sur les anneaux de (S-)entiers des corps globaux.

7.1 Introduction

Soit F un corps de nombres, d'anneau d'entiers A. Soient S un ensemble fini de places finies de F et A_S l'anneau des S-entiers de F, i.e. $A_S = \{x \in F \mid \forall v \notin S,\ v(x) \geq 0\}$. Pour toute place finie v de F, on note F_v le complété de F en v, $\overline{F_v}$ une clôture algébrique de F_v et $\overline{A_v}$ l'anneau des entiers de $\overline{F_v}$. Le but de cet article est de montrer le résultat suivant qui généralise le cas des classes de conjugaison dans les groupes arithmétiques étudié par Platonov [P], [PR, §8.1].

Théorème 7.1 *Soit G un A_S-schéma en groupes affine de type fini. Soit X un A_S-schéma plat, de type fini, muni d'une action*

$$G \times_{A_S} X \to X, \quad (g, x) \mapsto \rho(g).x.$$

Soit Z_0 un sous-A_S-schéma fermé de X, plat sur A_S. Soit $\mathrm{loc}\,(Z_0)$ *l'ensemble des sous-schémas fermés Z de X* (automatiquement plats sur A_S) *tels que, pour toute place finie* $v \notin S$, *il existe* $g_v \in G(\overline{A_v})$ *induisant un isomorphisme*

$$Z_0 \times_{A_S} \overline{A_v} \xrightarrow[\sim]{\rho(g_v)} Z \times_{A_S} \overline{A_v}.$$

Alors les orbites de $G(A_S)$ *sur* $\mathrm{loc}\,(Z_0)$ *sont en nombre fini, i.e* $G(A_S)\backslash \mathrm{loc}\,(Z_0)$ *est fini.*

Ceci répond à une question de E. Ullmo et A. Yafaev ; de façon précise, c'est le corollaire 7.17 qui intervient dans leur article sur la conjecture d'André-Oort sur les points spéciaux des variétés de Shimura [UY].

La démonstration s'appuie sur la théorie des espaces homogènes et la théorie de la cohomologie plate non abélienne.

7.2 Torseurs

Rappelons tout d'abord quelques définitions dans le cas d'une base affine noethérienne $\mathrm{Spec}(R)$ et d'un R-schéma en groupes affine G. Un G-espace formellement homogène principal (ou encore pseudo-torseur) est un R-schéma E muni d'une action à droite de G telle que le morphisme $(x,g) \mapsto (x, x.g) : E \times_R G \to E \times_R E$, soit un isomorphisme ([SGA3], IV.5.1). Un G-torseur E est un G-espace formellement homogène principal localement trivial pour la topologie fppf (fidèlement plate de présentation finie). Cela signifie qu'il existe une R-algèbre S fidèlement plate et de type fini telle que $E \times_R S \cong G \times_R S$.

Pour tout tel recouvrement S, on pose

$$Z^1(S/R, G) = \left\{ g \in G(S \otimes_R S) \;\middle|\; \begin{array}{c} p_{1,2}^*(g) p_{2,3}^*(g) = p_{1,3}^*(g) \\ \in G(S \otimes_R S \otimes_R S) \end{array} \right\}$$

$$\text{et} \quad H^1(S/R, G) := Z^1(S/R, G)/G(S),$$

où $G(S)$ agit sur $Z^1(S/R, G)$ par $g.z = p_1^*(g)\, z\, p_2^*(g)^{-1}$ ([Kn], chapitre III).

L'ensemble pointé $H^1(S/R, G)$ classifie les G-torseurs (ou, ce qui revient au même, les G-espaces formellement homogènes principaux) E sur $\mathrm{Spec}(R)$ qui sont trivialisés par le changement de base S/R ([M1], §III.4, page 120), i.e. vérifient $E \times_R S \xrightarrow{\sim} G \times_R S$.

Remarque 7.1 On suppose G fini étale. Puisque R est un anneau noethérien, il en est de même de S et de $S \otimes_R S$. En particulier, $G(S \otimes_R S)$

est un groupe fini. Par suite, les ensembles $Z^1(S/R,G)$ et $H^1(S/R,G)$ sont finis.

Toujours sous l'hypothèse G affine sur R noethérien, on définit alors l'ensemble $H^1_{\mathrm{fppf}}(R,G) := \varinjlim_S H^1(S/R,G)$ comme la limite sur les recouvrements fidèlement plats de type fini de R. L'ensemble pointé $H^1_{\mathrm{fppf}}(R,G)$ classifie les G-torseurs.

Remarque 7.2 La définition des torseurs donnée ici ne convient que pour les groupes affines. En général, la « bonne » notion est celle de faisceau (sur le site fppf de la base) muni d'une action de G, qui est localement isomorphe à G muni de son action par translations. Le H^1 défini plus haut classifie les torseurs en ce sens plus général ; il se trouve simplement que si G est affine, tout G-torseur est (représentable par) un schéma, par descente fidèlement plate de schémas affines.

Remarque 7.3 (torsion) Étant donné un G-torseur P, notons ${}^PG$ le schéma en groupes tordu associé (appelé « groupe adjoint » dans [Gi]) : c'est une forme intérieure de G, que l'on peut voir comme le schéma des G-automorphismes de P. On a alors une bijection canonique (la « torsion »)

$$H^1_{\mathrm{fppf}}(R,G) \longrightarrow H^1_{\mathrm{fppf}}(R,{}^PG),$$

associant à un G-torseur Q le schéma $\underline{\mathrm{Isom}}_G(P,Q)$ des G-isomorphismes de P sur Q ; elle envoie P sur le ${}^PG$-torseur trivial. Cette opération est fonctorielle : si $f : G \to G'$ est un morphisme de R-schémas en groupes affines, et si P' est le G'-torseur déduit de P par le changement de groupe f, on a un morphisme ${}^Pf : {}^PG \to {}^{P'}G'$ et un diagramme commutatif d'ensembles

$$\begin{array}{ccc} H^1_{\mathrm{fppf}}(R,G) & \xrightarrow{u} & H^1_{\mathrm{fppf}}(R,G') \\ {\scriptstyle v}\downarrow & & \downarrow{\scriptstyle v'} \\ H^1_{\mathrm{fppf}}(R,{}^PG) & \xrightarrow{{}^Pu} & H^1_{\mathrm{fppf}}(R,{}^{P'}G'). \end{array}$$

où les flèches verticales sont les bijections données par P et P' et les horizontales les morphismes (pointés) de changement de groupe $H^1(f)$ et $H^1({}^Pf)$. La remarque fondamentale, utilisée plusieurs fois dans la suite, est que l'application v induit une bijection de la fibre $u^{-1}(u([P])$ sur le noyau de ${}^Pu$. Ainsi, l'étude des fibres d'une application du type $H^1(f)$ se ramène à celle d'un noyau, pourvu que les hypothèses faites sur f soient préservées par torsion, ce qui est en pratique facile à vérifier. Tout ceci est fonctoriel par changement de base $\mathrm{Spec}(R') \to \mathrm{Spec}(R)$.

Dans la littérature, on considère généralement le cas où le groupe structural G est plat sur la base. Sur une base régulière de dimension un, le cas général se ramène au cas plat, comme nous allons le voir au paragraphe suivant.

7.3 Schémas sur une base de Dedekind

Soit B un schéma de Dedekind, c'est-à-dire un schéma noethérien régulier de dimension 1. On suppose B connexe, et l'on note η son point générique.

Si X est un B-schéma, nous noterons $\widetilde{X}$ l'adhérence schématique dans X de la fibre générique X_η. C'est un sous-schéma fermé de X ; en outre il est plat sur B, et c'est même l'unique sous-schéma fermé de X plat sur B et ayant la même fibre générique que X ([EGA4], (2.8.1)). On vérifie alors aisément les propriétés suivantes :

- la formation de $\widetilde{X}$ commute au changement de base plat, au sens suivant : si B' est un B-schéma plat, alors $\widetilde{X} \times_B B'$ est l'adhérence schématique de $X_\eta \times_B B'$ dans $X \times_B B'$;
- pour tout B-schéma plat B', on a $\widetilde{X}(B') = X(B')$;
- si X_1 et X_2 sont deux B-schémas, alors $\widetilde{X_1 \times_B X_2} = \widetilde{X_1} \times_B \widetilde{X_2}$;
- en particulier, si G est un B-schéma en groupes, alors $\widetilde{G}$ est un sous-schéma en groupes fermé de G ; en outre, si X est un G-torseur, alors $\widetilde{X}$ est stable sous $\widetilde{G}$ et est un $\widetilde{G}$-torseur pour cette action.

On voit ainsi que $X \mapsto \widetilde{X}$ est un foncteur, adjoint à droite de l'inclusion de la catégorie des B-schémas plats dans celle des B-schémas. Une conséquence des propriétés ci-dessus (notamment de la dernière) est la proposition suivante :

Proposition 7.4 *Soit B un schéma de Dedekind connexe, et soit G un B-schéma en groupes affine. Si $\widetilde{G} \subset G$ désigne comme ci-dessus l'adhérence schématique de la fibre générique de G, le foncteur naturel de changement de groupe est une équivalence de la catégorie des $\widetilde{G}$-torseurs vers celle des G-torseurs. Plus précisément, il admet pour quasi-inverse le foncteur $X \mapsto \widetilde{X}$.*

En particulier, l'application naturelle $H^1_{\mathrm{fppf}}(B, \widetilde{G}) \longrightarrow H^1_{\mathrm{fppf}}(B, G)$ est bijective.

Remarque 7.5 On peut aussi montrer la dernière assertion en utilisant directement la définition « à la Čech » du H^1 donnée au début : si

B' est un B-schéma plat et de type fini, on a en effet $Z^1(B'/B, G) = Z^1(B'/B, \widetilde{G})$ et $G(B') = \widetilde{G}(B')$, d'où $H^1(B'/B, G) = H^1(B'/B, \widetilde{G})$.

Remarque 7.6 Ici encore nous ne nous sommes limités aux groupes affines que pour éluder les questions de représentabilité des torseurs.

Proposition 7.7 *Soient B un schéma de Dedekind, G un B-schéma en groupes localement de type fini, H un sous-schéma en groupes de G, plat sur B. Alors l'immersion canonique de H dans G est un morphisme affine.*

En particulier, si G est affine sur B, il en est de même de H.

Démonstration. On peut supposer B affine connexe, de point générique η. Soit $\overline{H}$ l'adhérence schématique de H dans G. Alors $\overline{H}$ est plat sur B (comme adhérence schématique de sa fibre générique). De plus c'est un sous-schéma en groupes de G ; en effet, comme l'opération d'adhérence schématique commute au changement de base plat, $\overline{H} \times_B \overline{H}$ est l'adhérence schématique de $H \times_B H$ dans $G \times_B G$, d'où l'on déduit que $\overline{H}$ est stable par la loi de groupe.

On peut donc supposer, en remplaçant G par $\overline{H}$, que H est ouvert dans G et que $H_\eta = G_\eta$. Notons alors $f : G \to B$ le morphisme structural, Z le fermé complémentaire de H, z un point de Z, $b = f(z)$, $U = \operatorname{Spec}(A) \subset G$ un voisinage ouvert affine de z (de sorte que U est de type fini sur B). On a $H_\eta = G_\eta$, donc la fibre générique de Z est vide et l'image de $Z \cap U$ dans B est un ensemble fini de points fermés, que l'on peut supposer réduit à $\{b\}$. D'autre part, H_b est un sous-groupe ouvert (donc fermé) de G_b, de sorte que Z_b est une réunion de composantes connexes de G_b. Quitte à restreindre U, on peut donc supposer que $Z \cap U = G_b \cap U$, donc $H \cap U = \operatorname{Spec}(A[1/\pi])$ où π désigne une uniformisante en b.

On a donc montré que tout point de G admet un voisinage affine U tel que $H \cap U$ soit affine, cqfd. □

Remarque 7.8 Les auteurs ignorent si l'on peut se passer de la platitude de H. La proposition XXV.4.1 de [SGA3] impliquerait 7.7 sans cette hypothèse (du moins lorsque G est de type fini), mais elle n'est démontrée dans loc. cit. que pour H lisse ! Pour une autre généralisation de 7.7, voir [A], proposition 2.3.2.

Si H n'est pas plat, la preuve ci-dessus est en défaut car $\overline{H}$ n'est pas toujours un sous-schéma en groupes de G : voici un exemple communiqué par le rapporteur. Prenant pour B un trait, choisissons un B-groupe fini constant non trivial Γ et prenons $G = \Gamma \times_B \Gamma$. Il existe un sous-schéma en groupes H de G ayant pour fibre générique celle de $\Gamma \times_B \{e\}$ et pour fibre spéciale celle de $\{e\} \times_B \Gamma$; l'adhérence de H n'est alors pas un sous-groupe de G. (En revanche, H est bien affine).

Lemme 7.9 *Soit B un schéma de Dedekind connexe de point générique $\eta = \mathrm{Spec}(K)$. Soit G un B-schéma en groupes affine, fini et plat sur B. Alors la restriction $H^1_{\mathrm{fppf}}(B, G) \to H^1_{\mathrm{fppf}}(K, G)$ est injective.*

Démonstration. Par l'argument habituel de torsion (voir remarque 7.3), il suffit d'établir que la restriction a un noyau trivial. On se donne un G-torseur P qui est trivial au point générique. La théorie de la descente fidèlement plate (e.g. [EGA4, prop. 2.7.1.vii]) montre que P est propre sur B. Par hypothèse, $P(K) \neq \emptyset$, d'où $P(B) \neq \emptyset$ en appliquant le critère valuatif de propreté. On conclut que X est le G-torseur trivial. □

7.4 Finitude dans le cas local

Soit R un anneau d'entiers p-adiques (i.e. une extension finie de $\mathbf{Z}_p$) de corps des fractions K et de corps résiduel κ. On rappelle que si X est une K-variété algébrique, l'ensemble $X(K)$ est muni d'une topologie naturelle (par exemple [KS] p. 256). Si X est un R-schéma séparé de type fini, alors l'ensemble $X(R)$ est un ouvert compact de $X(K)$.

Lemme 7.10 *Soit G un R-schéma en groupes affine de type fini. Alors l'ensemble de cohomologie plate $H^1_{\mathrm{fppf}}(R, G)$ est fini.*

Démonstration. La proposition 7.4 ramène immédiatement au cas où G est plat sur R. Le groupe G admet une R-représentation fidèle $G \to \mathrm{GL}_n$ ([BT], §1.4.5), i.e. fait de G un sous-schéma fermé de GL_n. De plus, le quotient GL_n/G est représentable par un R-schéma X séparé ([A], théorème 4.C page 53). Par descente fidèlement plate, X est de présentation finie [SGA3, VI_B.9.2.xiii]. On a la suite exacte d'ensembles pointés

$$1 \to G(R) \to \mathrm{GL}_n(R) \to X(R) \xrightarrow{\varphi} H^1_{\mathrm{fppf}}(R, G) \to H^1_{\mathrm{fppf}}(R, \mathrm{GL}_n) = 1,$$

la dernière égalité étant le théorème 90 de Hilbert-Grothendieck. Par compacité de $X(R)$, il suffit alors de démontrer que les fibres de l'application caractéristique φ sont ouvertes, c'est-à-dire que les orbites $\mathrm{GL}_n(R).x$ sont ouvertes dans $X(R)$. Or, étant donné $x \in X(R)$, le morphisme $\mathrm{GL}_{n,K} \to X_K$, $g \mapsto g.x$, est lisse. Par suite, l'application continue $\mathrm{GL}_n(K) \to X(K)$ est ouverte. Vu que $\mathrm{GL}_n(R)$ est un ouvert de $\mathrm{GL}_n(K)$, il résulte que $\mathrm{GL}_n(R).x$ est un ouvert de $X(K)$ et a fortiori un ouvert de $X(R)$. □

Remarque 7.11 Sous les hypothèses de 7.10, l'ensemble $H^1_{\mathrm{fppf}}(K, G)$ est également fini, d'après Borel et Serre [BS, théorème 6.1]. Ce résultat ne sera pas utilisé ici, mais on en verra une version « géométrique » plus loin (proposition 7.19).

Lemme 7.12 *Sous les hypothèses du lemme 7.10, on suppose en outre que G est lisse sur R. Alors l'application naturelle*

$$H^1_{\mathrm{fppf}}(R, G) \to H^1_{\mathrm{fppf}}(\kappa, G_\kappa)$$

est injective.

En outre, si G_κ est connexe, alors $H^1_{\mathrm{fppf}}(\kappa, G_\kappa) = 1$ et par suite, d'après ce qui précède, on a $H^1_{\mathrm{fppf}}(R, G) = 1$.

Démonstration. Pour montrer la première assertion il suffit, par torsion, de montrer que l'application a un noyau trivial. Or un G-torseur X est en particulier un R-schéma lisse, donc l'application naturelle $X(R) \to X(\kappa)$ est surjective puisque R est hensélien. Donc si X_κ est trivial il en est de même de X. (La finitude de κ ne sert pas ici).

La seconde assertion n'est autre que le théorème de Lang (cf. [S], §II.2.3). □

7.5 Finitude dans le cas global

Soit F un corps de nombres. Soient S un ensemble fini de places finies de F et A_S l'anneau des S-entiers de F. Pour toute place finie v de F, on note F_v le complété de F en v, A_v son anneau d'entiers et κ_v son corps résiduel.

On note $\mathbb{A}_{f,S}$ l'anneau des S-adèles finis de F, c'est-à-dire le sous-anneau de $\prod_{v \notin S} F_v$ formé des éléments (x_v) tels que $x_v \in A_v$ pour presque tout v. Si G est un A_S-schéma en groupes affine de type fini, on pose alors

$$c(A_S, G) = \Big(\prod_{v \notin S} G(A_v) \Big) \backslash G(\mathbb{A}_{f,S}) \, / \, G(F).$$

C'est un ensemble fini ([B], Theorem 5.1) qui peut être décrit en termes de cohomologie plate puisque l'on a une bijection

$$c(A_S, G) \cong \mathrm{Ker}\Big[H^1_{\mathrm{fppf}}(A_S, G) \to H^1_{\mathrm{fppf}}(F, G) \times \prod_{v \notin S} H^1_{\mathrm{fppf}}(A_v, G)\Big].$$

Celle-ci induit, pour tout ensemble fini de places $S' \supset S$, une bijection entre sous-ensembles pointés

$$\Big(\prod_{v \in S' \setminus S} (G(A_v) \backslash G(F_v))\Big) \,/\, G(A_{S'}) \cong$$

$$\mathrm{Ker}\Big[H^1_{\mathrm{fppf}}(A_S, G) \to H^1_{\mathrm{fppf}}(A_{S'}, G) \times \prod_{v \in S' \setminus S} H^1_{\mathrm{fppf}}(A_v, G)\Big].$$

(Nisnevich, [N] Th. 2.1, voir [G], appendice).

Proposition 7.13 *Soit G un A_S-schéma en groupes affine de type fini. Alors l'ensemble de cohomologie plate $H^1_{\mathrm{fppf}}(A_S, G)$ est fini.*

Vu que $H^1_{\text{ét}}(A_S, G)$ s'injecte dans $H^1_{\mathrm{fppf}}(A_S, G)$, la proposition vaut aussi pour l'ensemble de cohomologie étale $H^1_{\text{ét}}(A_S, G)$.

Dans le cas où G/A_S est *lisse*, on note alors G^0 le sous-schéma en groupes de G des composantes neutres de G ([BT], §1.2.12). C'est un sous-schéma en groupes ouvert distingué de G, à fibres connexes[1], (nous dirons simplement « connexe »).

Démonstration de la proposition 7.13 :

Cas zéro : G est fini étale sur A_S. Un G-torseur est en particulier un A_S-schéma fini étale de rang n égal à celui de G. C'est donc le spectre d'une A_S-algèbre finie étale de rang n. Le théorème d'Hermite implique qu'il n'existe qu'un nombre fini de telles algèbres, à isomorphisme près, d'où un nombre fini de possibilités pour le schéma sous-jacent à un G-torseur. Enfin, pour chaque schéma X de ce type, l'ensemble des actions de G sur X est fini, d'où la conclusion.

Premier cas : G est lisse et connexe sur A_S. On considère les applications naturelles

$$H^1_{\mathrm{fppf}}(A_S, G) \xrightarrow{\alpha} H^1_{\mathrm{fppf}}(F, G) \xrightarrow{\beta} \prod_{v \notin S} H^1_{\mathrm{fppf}}(F_v, G).$$

L'application composée $\beta \circ \alpha$ se factorise par le produit $\prod_{v \notin S} H^1_{\mathrm{fppf}}(A_v, G)$ et est donc triviale en vertu du lemme 7.12.

D'autre part, d'après le théorème de Borel-Serre ([BS], théorème 7.1),

[1] L'hypothèse de lissité est nécessaire ici comme le montre l'exemple de μ_p sur $\mathbf{Z}$.

les fibres de β (en particulier son noyau) sont finies[2]. Il suffit donc de montrer que les fibres de α sont finies. Par torsion, il suffit même de démontrer que son noyau est fini. Or ce noyau est l'ensemble $c(A_S, G)$, qui est fini comme on l'a rappelé plus haut, d'où la finitude de $H^1_{\mathrm{fppf}}(A_S, G)$.

Second cas : G est extension d'un groupe fini étale par un groupe lisse connexe. Le A_S-groupe G/G^0 est alors fini étale et par suite, d'après le cas zéro, l'ensemble $H^1_{\mathrm{fppf}}(A_S, G/G^0)$ est fini ; il suffit donc de voir que les fibres de l'application naturelle

$$H^1_{\mathrm{fppf}}(A_S, G) \to H^1_{\mathrm{fppf}}(A_S, G/G^0)$$

sont finies, ou même, par torsion, que son noyau est fini. Or ce noyau est image de $H^1_{\mathrm{fppf}}(A_S, G^0)$ par l'application évidente, d'où la conclusion par le cas précédent.

Cas général. Il existe un ensemble fini $S' \supset S$ de « mauvaises » places tel que le schéma en groupes $G \times_{A_S} A_{S'}$ soit extension d'un $A_{S'}$-groupe constant tordu par un schéma en groupes lisse et connexe. L'argument de localisation du premier cas fonctionne de nouveau ici. On considère l'application de restriction

$$\Phi : H^1_{\mathrm{fppf}}(A_S, G) \to H^1_{\mathrm{fppf}}(A_{S'}, G) \times \prod_{v \in S' \setminus S} H^1_{\mathrm{fppf}}(A_v, G).$$

L'ensemble $H^1_{\mathrm{fppf}}(A_{S'}, G)$ est fini d'après le second cas et les $H^1_{\mathrm{fppf}}(A_v, G)$ sont des ensembles finis par le lemme 7.10. Il suffit donc de montrer que les fibres de Φ sont finies. Par torsion, on est ramené à voir que le noyau de Φ est fini. Or ce noyau est l'ensemble fini

$$\Big(\prod_{v \in S' \setminus S} G(A_v) \backslash G(F_v) \Big) / G(A_{S'}),$$

ce qui montre la finitude de $H^1_{\mathrm{fppf}}(A_S, G)$. □

7.6 Applications

On garde les notations précédant le théorème 7.1. On note $\mathcal{S}$ le « petit site fppf » de A_S, c'est-à-dire la catégorie des A_S-schémas plats de type fini, munie de la topologie fppf ; on appellera « A_S-faisceaux » les faisceaux sur $\mathcal{S}$.

[2] Sans hypothèse de connexité sur G/F, il est aussi vrai que l'image de l'application α est finie selon la proposition 4.4 de [HS].

Proposition 7.14 *Soit G un A_S-faisceau en groupes opérant sur un A_S-faisceau E, et soit $x \in E(A_S)$. On note Gx le sous-faisceau de E orbite de x (c'est-à-dire l'image du morphisme $G \to E$ envoyant g sur gx) et H le sous-faisceau en groupes de G stabilisateur de x.*

On suppose que H est représentable par un A_S-schéma en groupes affine de type fini. Alors l'ensemble quotient $G(A_S)\backslash(Gx)(A_S)$ est fini.

Démonstration. Comme Gx s'identifie au quotient $H\backslash G$, on a une injection naturelle

$$G(A_S)\backslash(Gx)(A_S) \hookrightarrow H^1_{\mathrm{fppf}}(A_S, H)$$

([Gi], III, corollaire 3.2.3). La proposition est donc une conséquence de 7.13. □

Démonstration du théorème 7.1. On applique la proposition 7.14 aux données suivantes :

- G est le groupe donné dans 7.1 ;
- pour tout A_S-schéma Y on note $\mathrm{Fer}_X(Y)$ l'ensemble des sous-schémas fermés de $X \times_{A_S} Y$ qui sont de présentation finie sur Y (condition vide si Y est noethérien) ; alors Fer_X est un faisceau pour la topologie fppf, et l'on prend pour E sa restriction au petit site $\mathcal{S}$;
- on prend pour x l'élément z_0 de $E(A_S)$ correspondant à Z_0.

Bien entendu G opère sur E via son action sur X. Le théorème est alors une conséquence immédiate de 7.14, une fois établis les deux lemmes suivants (du premier, on n'utilise d'ailleurs que l'inclusion $\mathrm{loc}(Z_0) \subset (Gz_0)(A_S)$) :

Lemme 7.15 *On a* $\mathrm{loc}(Z_0) = (Gz_0)(A_S)$.

Lemme 7.16 *Le stabilisateur de z_0 est représentable par un sous-schéma en groupes de G, plat et affine sur A_S.*

Démonstration du lemme 7.15. Il s'agit de voir que, pour un sous-schéma Z de X, les conditions suivantes sont équivalentes :

1. $Z \in \mathrm{loc}(Z_0)$;
2. il existe une A_S-algèbre B, fidèlement plate de type fini, et un élément g de $G(B)$, tels que $Z_B = gZ_{0,B}$.

Notons C la A_S-algèbre produit des $\overline{A_v}$, pour $v \notin S$. Supposons 1 : il existe alors $g \in G(C)$ tel que $gZ_{0,C} = Z_C$ comme sous-schémas de $X_C = X \times_{A_S} C$. Comme G est un A_S-schéma de type fini, il existe une sous-algèbre de type fini C_0 de C telle que $g \in G(C_0)$. Comme Z_{C_0} et Z_{0,C_0} (donc aussi gZ_{0,C_0}) sont des sous-schémas de présentation finie de X_{C_0}, il existe $C_1 \subset C$, de type fini sur A_S et contenant C_0, telle que $gZ_{0,C_1} = Z_{C_1}$. La condition 2 est donc satisfaite, avec $B = C_1$ (qui est bien fidèlement plate sur A_S comme sous-algèbre de C ; le fait que A_S soit un anneau de Dedekind est essentiel ici).

Réciproquement, si 2 est vérifiée, on en déduit 1 en observant que toute algèbre B comme en 2 admet un A_S-morphisme vers $\overline{A_v}$, pour tout $v \notin S$ (on pourra par exemple utiliser [EGA4], (14.5.4)). □

Démonstration du lemme 7.16. Désignons par un indice F la restriction des foncteurs considérés à la catégorie des F-schémas : ainsi, le schéma en groupes G_F opère sur le F-schéma X_F et sur le faisceau $\mathrm{Fer}_{X,F}$. D'après [DG], II, §1, théorème 3.6 b), le stabilisateur de $Z_{0,F}$ est représentable par un sous-schéma en groupes fermé H_F de G_F. Notons H l'adhérence schématique de H_F dans G, et montrons que H (vu comme A_S-faisceau) représente le stabilisateur de Z_0. Si Y est un A_S-schéma plat et $\gamma \in G(Y)$, on a en effet les équivalences :

$$\begin{aligned}
\gamma \in H(Y) \quad &\Leftrightarrow \quad \gamma_F \in H_F(Y_F) && \text{(définition de } H \text{ et platitude de } Y)\\
&\Leftrightarrow \quad \gamma_F Z_{0,F} = Z_{0,F} && \text{(définition de } H_F)\\
&\Leftrightarrow \quad \gamma Z_0 = Z_0 && (Z_0 \text{ est fermé et plat}),
\end{aligned}$$

d'où la représentabilité. La proposition 7.7 indique que H est affine sur A_S. □

Corollaire 7.17 *Soit G un A_S-schéma en groupes affine, plat et de type fini. Soit H_0 un sous-A_S-schéma en groupes fermé de G, plat sur A_S. Soit* loc (H_0) *l'ensemble des sous-groupes fermés H de G tels que, pour tout $v \notin S$, il existe $g_v \in G(\overline{A}_v)$ induisant un isomorphisme*

$$H_0 \times_{A_S} \overline{A_v} \xrightarrow[\sim]{\mathrm{ad}(g_v)} H \times_{A_S} \overline{A_v}.$$

Alors les orbites de $G(A_S)$ sur loc (H_0) *sont en nombre fini, c'est-à-dire qu'il n'existe qu'un nombre fini de $G(A_S)$-classes de conjugaison de tels sous-schémas en groupes.*

7.7 Cas géométrique

Modulo des hypothèses supplémentaires sur les groupes, les résultats précédents s'étendent au cas d'un anneau de Dedekind A de type fini sur $\mathbb{F}_p$ (p premier), et à ses complétés.

Dans ce qui suit, on notera F un corps global de caractéristique $p > 0$ (c'est-à-dire une extension de type fini et de degré de transcendance 1 de $\mathbb{F}_p$) on s'intéresse aux complétés de F en ses diverses places (cas local) et aux anneaux des courbes affines lisses connexes sur $\mathbb{F}_p$, de corps des fonctions F (cas global). Si v est une place de F, on notera F_v le complété de F en v et A_v son anneau d'entiers (noter cependant qu'il n'y a pas ici d'analogue de l'anneau A du cas arithmétique).

Le cas local requiert déjà une hypothèse additionnelle :

Lemme 7.18 *Soient v une place de F, et G un A_v-schéma en groupes affine de type fini de fibre générique lisse. Alors l'ensemble de cohomologie $H^1_{\text{fppf}}(A_v, G)$ est fini.*

La démonstration est similaire à celle du lemme 7.10 ; avec les mêmes notations, la lissité de G_K sert à assurer celle du morphisme d'orbite $\mathrm{GL}_{n,K} \to X_K$. Cette hypothèse est bien nécessaire : par exemple, les groupes

$$H^1(\mathbf{F}_p((t)), \mu_p) = \mathbf{F}_p((t))^\times / (\mathbf{F}_p((t)))^{\times p} \quad \text{et}$$

$$H^1(\mathbf{F}_p[[t]], \mu_p) = \mathbf{F}_p[[t]]^\times / (\mathbf{F}_p[[t)]])^{\times p}$$

ne sont pas finis.

Bien que cela ne soit pas strictement utilisé dans la suite, mentionnons le cas géométrique du théorème de Borel-Serre local (remarque 7.11).

Proposition 7.19 *Soient v une place de F, et G un F_v-schéma en groupes affine de type fini. On suppose que G est lisse, que G/G^0 est d'ordre premier à p et que le radical unipotent $R_u(G_{\overline{F}_v})$ de $G_{\overline{F}_v}$ est défini et déployé sur F_v comme sous-groupe unipotent de G. Alors $H^1(F_v, G)$ est fini.*

L'hypothèse sur G/G^0 est fondamentale ; ainsi, $H^1(\mathbf{F}_p((t)), \mathbb{Z}/p\mathbb{Z})$ est infini. Rappelons le lemme suivant [Sa, 1.13].

Lemme 7.20 *Soient k un corps quelconque et H un k-groupe affine lisse de type fini. Si U est un k-sous-groupe unipotent déployé (autre terminologie : k-résoluble) distingué de H, alors l'application canonique*

$$H^1(k, H) \xrightarrow{\pi} H^1(k, H/U)$$

est une bijection. C'est en particulier le cas pour k parfait et U le radical unipotent de G.

La démonstration originale étant canulée, nous y remédions ici.

Démonstration. Le groupe U admet une suite de composition centrale caractéristique dont les quotients sont des $\mathbb{G}_a^n$ [DG, IV.4.3.14]. Il suffit de traiter, par récurrence, le cas où $U = \mathbb{G}_a^n$. Toute k–forme galoisienne U' de $\mathbb{G}_a^n$ est isomorphe à $\mathbb{G}_a^n$ [O, §V.7, prop. b)] et vérifie donc $H^i(k, U') = 0$ pour tout $i > 0$. Si l'on tord U par $a \in Z^1(k_s/k, U(k_s))$ agissant par automorphismes intérieurs de H, on trouve donc $H^i(k, {}_aU) = 0$ et on en déduit l'injectivité de π (cf. [S], prop. 39, cor. 2). Comme U est un sous-groupe invariant abélien de G, on peut le tordre par $c \in Z^1(k_s/k, (H/U)(k_s))$ et, comme $H^2(k, {}_cU) = 0$, on en déduit la surjectivité de π (cf. [S], prop. 41, cor.). □

Démonstration de la proposition 7.19 :

Cas où G est fini (étale) d'ordre premier à p : soit F_v'/F_v une extension galoisienne telle que l'action de $\mathrm{Gal}(F_v')$ soit triviale sur $G(F_{v,s})$. Vu que les fibres de la restriction $H^1(F_v, G) \to H^1(F_v', G)$ sont finies (remarque 7.1), on peut supposer que $F_v' = F_v$, c'est-à-dire que G_{F_v} est constant. Alors $H^1(F_v, G) \cong \mathrm{Hom}_{ct}(\mathrm{Gal}(F_v), G)/\sim$. On a en fait $H^1(F_v, G) \cong \mathrm{Hom}_{ct}(\mathrm{Gal}(F_{v,mr}/F_v), G)/\sim$ où $F_{v,mr}$ désigne la clôture modérément ramifiée de F_v puisque le groupe d'inertie sauvage $\mathrm{Gal}(F_{v,s}/F_{v,mr})$ est un pro-p-groupe (voir [GMS, §II.7]). Or le groupe $\mathrm{Gal}(F_{v,mr}/F_v)$ est topologiquement engendré par deux éléments (*ibid*), d'où la finitude de $\mathrm{Hom}_{ct}(\mathrm{Gal}(F_{v,mr}/F_v), G)$ et de $H^1(F_v, G)$.

Cas où G est connexe : le lemme 7.20 permet de supposer que $R_u(G_{\overline{F}}) = 1$. On supposera donc G réductif. On sait [Ko, §1] que le groupe G admet une résolution $1 \to E \to \widetilde{G} \to G \to 1$ où E est un F_v-tore quasi-trivial et $\widetilde{G}$ une extension d'un F_v-tore T par le F_v-groupe semi-simple simplement connexe $D(\widetilde{G})$. Vu que $H^1(F_v, D(\widetilde{G})) = 1$ ([BT2], th. 4.7), le bord $H^1(F_v, G) \to H^2(F_v, E)$ a un noyau trivial. Tordre la suite exacte $1 \to D(\widetilde{G}) \to \widetilde{G} \to T \to 1$ par un 1-cocycle galoisien à valeurs dans $\widetilde{G}(F_{v,s})$ ne change pas les hypothèses faites sur cette suite ; en raisonnant fibre par fibre [S, §5.5], on obtient l'injectivité de $H^1(F_v, \widetilde{G}) \to H^1(F_v, T)$, d'où la finitude de $H^1(F_v, \widetilde{G})$. Ainsi le noyau du bord $H^1(F_v, G) \to H^2(F_v, E)$ est fini. Raisonnant une nouvelle fois par torsion, ce bord $H^1(F_v, G) \to H^2(F_v, E)$ a des fibres finies. Or l'extension centrale $1 \to E \to \widetilde{G} \to G \to 1$ est d'ordre fini d, donc l'image du bord $H^1(F_v, G) \to H^2(F_v, E)$ est inclus dans le groupe ${}_dH^2(F_v, E)$,

qui est fini, puisqu'il s'injecte dans un produit fini de groupes de Brauer d'extensions de F_v ([M2], th. III.6.9). On conclut que $H^1(F_v, \widetilde{G})$ est fini.
Cas général: le dévissage est alors le même que dans la preuve du théorème 6.2 de [BS]. □

Passons au cas global. Étant donné un ensemble fini S de places de F, on note $\mathbb{A}_{f,S}$ l'anneau des S-adèles de F, c'est-à-dire le sous-anneau de $\prod_{v \notin S} F_v$ formé des éléments (x_v) tels que $x_v \in A_v$ pour presque tout v. Si S est *non vide*, on notera A_S l'anneau des S-entiers de F, c'est-à-dire l'intersection des anneaux des $v \notin S$. Si C est la courbe projective lisse sur $\mathbb{F}_p$ de corps de fonctions F, on peut aussi définir A_S comme l'anneau de la courbe affine $C \setminus S$; les anneaux A_S sont exactement les $\mathbb{F}_p$-algèbres de type fini, normales, de corps des fractions F.

Si G désigne un A_S-schéma en groupes affine de type fini, on cherche à montrer la finitude de $H^1_{\text{fppf}}(A_S, G)$. La méthode utilisée dans le cas arithmétique faisait appel au théorème de Hermite, à la finitude des ensembles de classes $c(A_S, G)$ et au théorème de finitude de Borel-Serre. Le premier résultat n'est plus valable dans le cas géométrique et sera remplacé par le théorème de Chebotarev. Rappelons les analogues des deux autres dans la situation présente.

Comme dans le cas arithmétique, on pose

$$c(A_S, G) = \Big(\prod_{v \notin S} G(A_v)\Big) \setminus G(\mathbb{A}_{f,S}) \,/\, G(F).$$

Théorème 7.21 (Conrad [C, §5.1]) *Soient S un ensemble fini non vide de places de F et G un A_S-schéma en groupes affine de type fini. On suppose que G_F est lisse et que le radical unipotent $R_u(G_{\overline{F}})$ de $G_{\overline{F}}$ est défini et déployé sur F comme sous-groupe unipotent de G_F. Alors $c(A_S, G)$ est fini.*

Remarque 7.22 Il s'agit de l'énoncé de 2005. Utilisant la structure des groupes pseudo-réductifs (Conrad-Gabber-Prasad), Conrad a depuis étendu ce résultat à une classe plus vaste de groupes ainsi que tout ce qui suit d'ailleurs, cf. [C, §7.1].

Théorème 7.23 (Borel-Prasad) *Sous les hypothèses du théorème précédent, les fibres de l'application*

$$\omega_S : H^1(F, G) \to \prod_{v \notin S} H^1(F_v, G)$$

sont finies. De plus, si G_F est fini, ceci vaut plus généralement pour un ensemble S de densité nulle.

Dans le cas semi-simple, il s'agit du théorème B1 de [BP]. La preuve est fondée sur le principe de Hasse dû à Harder [H].

Démonstration du cas général du théorème 7.23.

Cas où G_F est fini: On peut supposer ici que S est fini ou infini de densité nulle. Soit F'/F une extension galoisienne telle que l'action de $\mathrm{Gal}(F')$ soit triviale sur $G(F_s)$ et satisfaisant $\gamma_{F'} = 1 \in H^1(F', G)$. Vu que les fibres de la restriction $H^1(F, G) \to H^1(F', G)$ sont finies (remarque 7.1), on peut supposer que $F' = F$, c'est-à-dire que G_F est constant. Le théorème de Chebotarev [J] montre alors que le noyau de $H^1(F, G) \to \prod_{v \notin S} H^1(F_v, G)$ est trivial ([BS], 7.3).

Cas où G_F est connexe : tout d'abord, il est loisible de se débarrasser du radical unipotent d'après le lemme 7.20. On supposera donc G_F réductif ; par torsion, il suffit en outre de voir que le noyau de ω_S est un ensemble fini. Ensuite, vu que les $H^1(F_v, G)$ sont finis (7.2), on peut prendre $S = \emptyset$. Il faut montrer que le noyau $\text{Ш}^1(F, G)$ de $\omega_\emptyset$ est fini. On sait que le groupe G_F admet une résolution $1 \to E \to \widetilde{G} \to G \to 1$ où E est un F-tore quasi-trivial et $\widetilde{G}$ une extension d'un F-tore T par le F-groupe semi-simple simplement connexe $D(\widetilde{G})$ [Ko, §1]. Vu que $H^1(F, E) = 1$ et $H^1(F_v, E) = 1$ pour toute place $v \in S$ et que $H^2(F, E)$ s'injecte dans $\oplus_v H^2(F_v, E)$ le diagramme commutatif exact d'ensembles pointés

$$\begin{array}{ccccccc}
1 & \longrightarrow & H^1(F, \widetilde{G}) & \longrightarrow & H^1(F, G) & \longrightarrow & H^2(F, E) \\
 & & \downarrow & & \downarrow & & \downarrow \\
1 & \longrightarrow & \prod_v H^1(F_v, \widetilde{G}) & \longrightarrow & \prod_v H^1(F, G) & \longrightarrow & \prod_v H^2(F_v, E)
\end{array}$$

indique que l'application $\text{Ш}^1(F, \widetilde{G}) \to \text{Ш}^1(F, G)$ est surjective. On est ramené au cas $\widetilde{G}$ où l'on va utiliser le principe de Hasse, i.e. $H^1(F, D(\widetilde{G})) = 1$. Ceci valant pour toute forme tordue de $D(\widetilde{G})$, il suit que l'application $\text{Ш}^1(F, \widetilde{G}) \to \text{Ш}^1(F, T)$ est injective. Or $\text{Ш}^1(F, T)$ est fini d'après le théorème de Poitou-Tate (cf. [M2], th. I.4.10). Il en est de même de $\text{Ш}^1(F, \widetilde{G})$.

Cas général: la preuve est alors la même que [BS], §7.10. □

La proposition 7.13 n'est plus vraie telle quelle (prendre $A_S = \mathbb{F}_p[t]$ et $G = \mathbb{Z}/p\mathbb{Z}$) ; on a cependant les énoncés suivants :

Proposition 7.24 *Soient S un ensemble fini non vide de places de F et G un A_S-schéma en groupes affine de type fini. On suppose que G_F est lisse et que le radical unipotent $R_u(G_{\overline{F}})$ de $G_{\overline{F}}$ est défini et déployé* *sur F comme sous-groupe unipotent de G_F. Alors, pour tout ensemble S' de places de F de densité nulle, les fibres de l'application naturelle*

$$H^1_{\mathrm{fppf}}(A_S, G) \to \prod_{v \notin S'} H^1_{\mathrm{fppf}}(F_v, G_F/G_F^0)$$

sont finies.

Corollaire 7.25 *Sous les hypothèses de la proposition 7.24, on suppose de plus que G_F/G_F^0 est d'ordre premier à p. Alors $H^1_{\mathrm{fppf}}(A_S, G)$ est fini.*

Démonstration du corollaire : soit $S' \supset S$ un ensemble fini de places tel que G soit lisse sur $A_{S'}$ et G/G^0 fini étale sur $A_{S'}$. Comme la flèche de 7.24 se factorise par $H^1_{\mathrm{fppf}}(A_{S'}, G/G^0)$, il suffit de voir que cet ensemble est fini. Par changement de base (et la remarque 7.1) on peut supposer que G/G^0 est constant, et c'est alors une conséquence immédiate du fait que le groupe fondamental « premier à p » de $A_{S'}$ est topologiquement de type fini [SGA1, corollaire XIII.2.12]. □

Démonstration de la proposition 7.24. Il existe un ensemble fini $S_0 \supset S$ de « mauvaises » places tel que le schéma en groupes $G \times_{A_S} A_{S_0}$ est lisse. Sans perte de généralité, on peut supposer que $S_0 \subset S'$.
Cas zéro : G est fini étale sur A_S. L'application considérée se factorise par l'application naturelle $H^1_{\mathrm{fppf}}(A_S, G) \to H^1_{\mathrm{fppf}}(F, G)$. Comme celle-ci est injective (lemme 7.9) l'assertion résulte du théorème 7.23.

Premier cas : G est lisse et connexe sur A_S. Il s'agit de voir que $H^1_{\mathrm{fppf}}(A_S, G)$ est fini ; l'argument est analogue à celui de la proposition 7.13 (premier cas) en remplaçant le théorème de Borel-Serre par le théorème 7.23 et en utilisant le théorème 7.21.

Second cas : G est extension d'un groupe fini étale par un groupe lisse connexe. L'application se factorise par $H^1_{\mathrm{fppf}}(A_S, G/G^0)$. Compte tenu du cas zéro, il suffit de montrer la finitude des fibres (ou seulement du noyau) de $H^1_{\mathrm{fppf}}(A_S, G) \to H^1_{\mathrm{fppf}}(A_S, G/G^0)$: comme dans la preuve de 7.13 cela résulte de la finitude de $H^1_{\mathrm{fppf}}(A_S, G^0)$, vue au premier cas.

Cas général. Suivant le second cas, l'énoncé vaut si l'on remplace S par S_0. Il suffit alors de voir que l'application de restriction $H^1_{\mathrm{fppf}}(A_S, G) \to H^1_{\mathrm{fppf}}(A_{S_0}, G)$ est à fibres finies. Comme dans la preuve de 7.13, on considère l'application

$$\Phi : H^1_{\mathrm{fppf}}(A_S, G) \to H^1_{\mathrm{fppf}}(A_{S_0}, G) \times \prod_{v \in S \setminus S_0} H^1_{\mathrm{fppf}}(A_v, G).$$

Celle-ci est à fibres finies, par le même argument que dans 7.13 (compte tenu du théorème 7.21), et il en est de même de la projection du produit de droite vers $H^1_{\mathrm{fppf}}(A_{S_0}, G)$ puisque les facteurs locaux $H^1_{\mathrm{fppf}}(A_v, G)$ sont finis (lemme 7.18), d'où la conclusion. □

La proposition 7.24 donne lieu à la variante suivante du théorème 7.1 :

Théorème 7.26 *Soit G un A_S-schéma en groupes affine de type fini. Soit X un A_S-schéma plat, de type fini, muni d'une action*

$$G \times_{A_S} X \to X, \quad (g, x) \mapsto \rho(g).x.$$

Soit Z_0 un sous-A_S-schéma fermé de X, plat sur A_S. Soit $\mathrm{loc}\,(Z_0)$ *l'ensemble des sous-schémas fermés Z de X* (automatiquement plats sur A_S) *tels que, pour toute place finie $v \notin S$, il existe $g_v \in G(A_v)$ induisant un isomorphisme*

$$Z_0 \times_{A_S} A_v \xrightarrow[\sim]{\rho(g_v)} Z \times_{A_S} A_v.$$

On suppose que le stabilisateur H de $Z_{0,F}$ est lisse et que le radical unipotent $R_u(H_{\overline{F}})$ de $H_{\overline{F}}$ est défini et déployé sur F comme sous-groupe unipotent de G_F. Alors les orbites de $G(A_S)$ sur $\mathrm{loc}\,(Z_0)$ *sont en nombre fini, i.e. $G(A_S)\backslash \mathrm{loc}\,(Z_0)$ est fini.*

Notons que la définition de l'ensemble $\mathrm{loc}\,(Z_0)$ est plus restrictive dans le cas géométrique. Ceci étant, l'énoncé plus fort vaut aussi dans le cas géométrique si l'on suppose de plus que H/H^0 est d'ordre premier à p en utilisant alors le corollaire 7.25.

Remerciements : Nous remercions Brian Conrad et Gopal Prasad pour leurs commentaires bienvenus.

Références

[A] S. Anantharaman, *Schémas en groupes, espaces homogènes et espaces algébriques sur une base de dimension 1*, Bull. Soc. Math. France, Mem. **33** (1973).

[B] A. Borel, *Some finiteness properties of adele groups over number fields*, Inst. Hautes Etudes Sci. Publ. Math. **16**, (1963), 5–30.

[BP] A. Borel et G. Prasad, *Finiteness theorems for discrete subgroups of bounded covolume in semi-simple groups*, Inst. Hautes Etudes Sci. Publ. Math. **69**, (1989), 119–171.

[BS] A. Borel et J.-P. Serre, *Théorèmes de finitude en cohomologie galoisienne*, Comment. Math. Helv. **39** (1964), 111–164.

[BT] F. Bruhat et J. Tits, *Groupes réductifs sur un corps local II*, Publ. Math. IHES **60** (1984).

[BT2] F. Bruhat et J. Tits, *Groupes algébriques sur un corps local. Chapitre I. Compléments et applications à la cohomologie galoisienne*, J. Fac. Sci. Univ. Tokyo Sect. IA Math. **34** (1987), 671–698.

[C] B. Conrad, *Finiteness theorems for algebraic groups over function fields*, Compos. Math. **148** (2012), 555–639.

[DG] M. Demazure et P. Gabriel, *Groupes algébriques*, Masson (1970).

[EGA4] A. Grothendieck et J. Dieudonné, *Éléments de géométrie algébrique IV.2 et IV.3*, Pub. Math. IHES **24** et **28** (1965).

[GMS] S. Garibaldi, A. Merkurjev, J.-P. Serre, *Cohomological Invariants in Galois Cohomology*, University Lecture Series **28** (2003), American Mathematical Society.

[G] P. Gille, *Torseurs sur la droite affine*, Transform. Groups **7** (2002), 231–245.

[Gi] J. Giraud, *Cohomologie non abélienne*, Grundlehren der mathematischen Wissenschaften **179**, Springer-Verlag (1971).

[H] G. Harder, *Über die Galoiskohomologie halbeinfacher algebraischer Gruppen. III*, J. Reine Angew. Math. **274/275** (1975), 125–138.

[HS] D. Harari et A. N. Skorobogatov, *Non-abelian cohomology and rational points*, Compositio Math. **130** (2002), 241–273.

[J] M. Jarden, *The Cebotarev Density Theorem for Function Fields : An Elementary Approach*, Math. Ann. **261**, 467–475 (1982).

[Kn] M. A. Knus, *Quadratic and hermitian forms over rings*, Grundlehren der mat. Wissenschaften **294** (1991), Springer.

[Ko] R. Kottwitz, *Rational conjugacy classes in reductive groups*, Duke Math. J. **49** (1982), 785-806.

[KS] K. Kato et S. Saito, *Unramified class field theory of arithmetical surfaces*, Ann. of Math. **118** (1983), 241–275.

[M1] J. S. Milne, *Étale cohomology*, Princeton Mathematical Series, **33** (1980), Princeton University Press.

[M2] J. S. Milne, *Arithmetic duality theorems*, Perspectives in Mathematics **1** (1986), Academic Press.

[N] Ye. A. Nisnevich, *Espaces homogènes principaux rationnellement triviaux et arithmétique des schémas en groupes réductifs sur les anneaux de Dedekind*, C.R. Acad Sci. Paris, tome **299** (1984), 5–8.

[O] J. Oesterlé, *Nombres de Tamagawa et groupes unipotents en caractéristique $p > 0$*, Inv. Math. **78** (1984), 13–88.

[P] V. Platonov, *Le problème du genre dans les groupes arithmétiques*, Dokl. Akad. Nauk SSSR **200** (1971), 793–796, traduction anglaise Soviet Math. Dokl. **12** (1971), 1503–1507.

[PR] V. Platonov et A. Rapinchuk, *Algebraic groups and number theory*, Pure and Applied Mathematics 139 (1994), Academic Press.

[Sa] J.-J. Sansuc, *Groupe de Brauer et arithmétique des groupes algébriques linéaires sur un corps de nombres*, J. reine angew. Math. **327** (1981), 12–80.

[S] J.-P. Serre, *Cohomologie galoisienne*, Lecture Notes in Math. 5, 5e édition (1994), Springer-Verlag.

[SGA1] *Séminaire de Géométrie algébrique de l'I.H.E.S., Revêtements étales et groupe fondamental, dirigé par A. Grothendieck*, Lecture Notes in Math. 224. Springer (1971).

[SGA3] *Séminaire de Géométrie algébrique de l'I.H.E.S., 1963-1964, Schémas en groupes, dirigé par M. Demazure et A. Grothendieck*, Lecture Notes in Math. **151–153**, Springer (1970).

[UY] E. Ullmo et A. Yafaev, *Galois orbits and equidistribution of special subvarieties : towards the André-Oort conjecture*, prépublication (2006).

8
Descent theory for open varieties

David Harari
Mathématiques, Bâtiment 425, Université Paris-Sud

Alexei N. Skorobogatov
Department of Mathematics, Imperial College London

Abstract We extend the descent theory of Colliot-Thélène and Sansuc to arbitrary smooth algebraic varieties by removing the condition that every invertible regular function is constant. This links the Brauer–Manin obstruction for integral points on arithmetic schemes to the obstructions defined by torsors under groups of multiplicative type.

Let X be a smooth and geometrically integral variety over a number field k with points everywhere locally. The descent theory of Colliot-Thélène and Sansuc ([8], [29]) describes arithmetic properties of X in terms of X-torsors under k-groups of multiplicative type. It interprets the *Brauer–Manin obstruction* to the existence of a rational point (or to weak approximation) on X in terms of the obstructions defined by torsors.

Let $\bar{k}$ be an algebraic closure of k. Because of its first applications the descent theory was stated in [8] for proper varieties that become rational over $\bar{k}$; in this case it is enough to consider torsors under tori. It was pointed out in [28] that the theory works more generally under the sole assumption that the group $\bar{k}[X]^*$ of invertible regular functions on $\overline{X} := X \times_k \bar{k}$ is the group of constants $\bar{k}^*$. This assumption is satisfied when X is proper, but it often fails for complements to reducible divisors in smooth projective varieties; it also fails for many homogeneous spaces of algebraic groups. In the general case of an arbitrary smooth and geometrically integral variety Colliot-Thélène and Xu Fei have recently introduced a Brauer–Manin obstruction to the existence of integral points [7, Sect. 1]. Descent obstructions to the existence of integral points were briefly considered by Kresch and Tschinkel in [21], Remark 3; see also Section 5.3 of [9]. In the particular case of an open subset of $\mathbf{P}^1_k$, a variant of the main theorem of descent linking the two kinds of obstructions has recently turned up in connection with an old conjecture of Skolem, see [19, Thm. 1].

The goal of this paper is to extend the theory of descent to the general case of a smooth and geometrically integral variety. It turns out that the main results are almost entirely the same. The methods, however, must be completely overhauled. As it frequently happens, one needs to systematically consider Galois hypercohomology of complexes instead of Galois cohomology of individual Galois modules. For principal homogeneous spaces of algebraic groups this approach has already been used in [1], [18] and [12]. But even in the 'classical' case $\bar{k}[X]^* = \bar{k}^*$, working with derived categories and hypercohomology of complexes streamlines the proof of a key result of descent theory ([8], Prop. 3.3.2 and Lemme 3.3.3; [29], Thm. 6.1.2 (a)) by avoiding delicate explicit computations with cocycles (see our Theorem 8.16 and its proof).

Let us now describe the contents of the paper. Let S be a k-group of multiplicative type, that is, a commutative algebraic group whose connected component of the identity is an algebraic torus. In Section 8.1 we define the *extended type* of an X-torsor under S, an invariant that classifies X-torsors up to twists by a k-torsor. When $\bar{k}[X]^* \neq \bar{k}^*$ the extended type defines a stronger equivalence relation on $H^1(X,S)$ than the classical type introduced by Colliot-Thélène and Sansuc in [8].

Let $\mathcal{T}$ be an X-torsor under S. In Section 8.2 we show that if $U \subset X$ is an open set such that the classical type of the torsor $\mathcal{T}_U \to U$ is zero, then $\mathcal{T}_U$ is canonically isomorphic to the fibred product $Z \times_Y U$, where Z and $Y = Z/S$ are k-torsors under groups of multiplicative type, and $U \to Y$ is a certain canonical morphism (Theorem 8.10). This description follows the ideas of Colliot-Thélène and Sansuc [8] who used similar constructions to describe $\mathcal{T}_U$ by explicit equations. Our goal was to obtain a functorial description, so our results are not immediately related to theirs. Corollary 8.11 describes the restriction of torsors of given extended type to sufficiently small open subsets.

In Section 8.3 we prove the main results of our generalised descent theory. The proof of Theorem 8.16 relies on the previous work of T. Szamuely and the first named author [17, 18], in particular, on their version of the Poitou–Tate duality for tori, which was later extended by C. Demarche to the groups of multiplicative type [11].

In Section 8.4 we prove statements about the existence of integral points and strong approximation. As an application we give a short proof of a result by Colliot-Thélène and Xu Fei, generalised by C. Demarche, see Theorem 8.20.

8.1 The extended type of a torsor

Let Z be an integral regular Noetherian scheme, and let $p : X \to Z$ be a faithfully flat morphism of finite type. Let $\mathcal{D}(Z)$ be the derived category of bounded complexes of fppf or étale sheaves on Z. For an object $\mathcal{C}$ of $\mathcal{D}(Z)$, the hypercohomology groups $\mathbb{H}^i(Z, \mathcal{C})$ will be denoted simply by $H^i(Z, \mathcal{C})$. Notation such as $\mathrm{Hom}_Z(A, B)$ or $\mathrm{Ext}^i_Z(A, B)$ will be understood in the category of sheaves on Z, or in $\mathcal{D}(Z)$. The same conventions apply when Z is replaced by X.

Consider the truncated object $\tau_{\leq 1}\mathbf{R}p_*\mathbf{G}_{m,X}$ in $\mathcal{D}(Z)$. Its shift by 1, which has trivial cohomology outside the degrees -1 and 0, is denoted by

$$KD(X) = (\tau_{\leq 1}\mathbf{R}p_*\mathbf{G}_{m,X})[1].$$

There is a canonical morphism $i : \mathbf{G}_{m,Z} \to \tau_{\leq 1}\mathbf{R}p_*\mathbf{G}_{m,X}$, and we define

$$KD'(X) = \mathrm{Coker}\,(i)[1],$$

so that we have an exact triangle

$$\mathbf{G}_{m,Z}[1] \longrightarrow KD(X) \xrightarrow{v} KD'(X) \xrightarrow{w} \mathbf{G}_{m,Z}[2]. \tag{8.1}$$

A group scheme of finite type G over Z is called a Z-group of multiplicative type if locally (for the étale topology) on Z it is isomorphic to a group subscheme of $\mathbf{G}^n_{m,Z}$: this means that for every $z \in Z$, there exists a Zariski open neighborhood U of z and a finite étale morphism $Z' \to U$ such that $G \times_Z Z'$ is isomorphic to a group subscheme of $\mathbf{G}^n_{m,Z'}$. By [14, IX, Prop. 2.1] such a group is affine and faithfully flat over Z. If S is a group of multiplicative type or a finite flat group scheme over Z, we denote by $\widehat{S}$ the *Cartier dual* of S. This is the group scheme over

Z which represents the fppf sheaf $\mathcal{H}om_Z(S, \mathbf{G}_{m,Z})$, see [14, X, Cor. 5.9] when S is of multiplicative type, and [25, Ch. 14] when S is finite flat.

The following proposition is a generalisation of the fundamental exact sequence of Colliot-Thélène and Sansuc (see [29], Thm. 2.3.6 and Cor. 2.3.9).

Proposition 8.1 *Let S be a Z-group scheme. Assume that one of the two following properties is satisfied:*

(a) *S is of multiplicative type;*
(b) *S is finite and flat, and if 2 is a residual characteristic of Z, then the 2-primary torsion subgroup $S\{2\}$ is of multiplicative type (equivalently, the Cartier dual $\widehat{S\{2\}}$ is smooth over Z).*

Then there is an exact sequence

$$H^1(Z,S) \to H^1(X,S) \xrightarrow{\chi} \mathrm{Hom}_Z(\widehat{S}, KD'(X)) \xrightarrow{\partial} H^2(Z,S) \to H^2(X,S). \tag{8.2}$$

To simplify notation, here and elsewhere we write $H^n(X,S)$ for the fppf cohomology group $H^n(X, p^*S)$. If S is smooth, then the fppf topology can be replaced by étale topology.

Proof of Proposition 8.1 We apply the functor $\mathrm{Hom}_k(\widehat{S}, .)$ to the exact triangle (8.1). To identify the terms of the resulting long exact sequence we use the following well known fact: for any scheme X/Z, any Z-group S of multiplicative type and any $n \geq 0$ we have

$$H^n(X,S) = \mathrm{Ext}^n_X(p^*\widehat{S}, \mathbf{G}_{m,X}),$$

see [8], Prop. 1.4.1, or [29], Lemma 2.3.7. Let us recall the argument for the convenience of the reader. One proves first that $\mathcal{E}xt^n_X(p^*\widehat{S}, \mathbf{G}_{m,X}) = 0$ for any $n \geq 1$, and then the local-to-global spectral sequence

$$H^m(X, \mathcal{E}xt^n_X(p^*\widehat{S}, \mathbf{G}_{m,X})) \Rightarrow \mathrm{Ext}^{m+n}_X(p^*\widehat{S}, \mathbf{G}_{m,X})$$

completely degenerates, giving the desired isomorphism.

In case (b) the same argument works for $1 \leq n \leq 3$: indeed, for $\ell \neq 2$ we have $\mathcal{E}xt^n_X(p^*\widehat{S}\{\ell\}, \mathbf{G}_{m,X}) = 0$ by the main result of [6]. (Note that the case $\ell = 2$ is exceptional: for example, $\mathrm{Ext}^2_K(\alpha_2, \mathbf{G}_{m,K}) \neq 0$ if K is a separably closed field of characteristic 2, see [5].)

The functor $\mathbf{R}\mathrm{Hom}_X(p^*\widehat{S}, .)$ from $\mathcal{D}(X)$ to the derived category of abelian groups $\mathcal{D}(\mathrm{Ab})$ is the composition of the functors

$\mathbf{R}p_*(.) : \mathcal{D}(X) \to \mathcal{D}(Z)$ and $\mathbf{R}\mathrm{Hom}_Z(\widehat{S}, .) : \mathcal{D}(Z) \to \mathcal{D}(\mathrm{Ab})$. This formally entails a canonical isomorphism

$$\mathrm{Ext}^n_X(p^*\widehat{S}, \mathbf{G}_{m,X}) = R^n\mathrm{Hom}_Z(\widehat{S}, \mathbf{R}p_*\mathbf{G}_{m,X}).$$

In particular, we have

$$H^n(Z, S) = \mathrm{Ext}^n_Z(\widehat{S}, \mathbf{G}_{m,Z}) = \mathrm{Hom}_Z(\widehat{S}, \mathbf{G}_{m,Z}[n]).$$

Truncation produces an exact triangle

$$\tau_{\leq 1}\mathbf{R}p_*\mathbf{G}_{m,X} \to \mathbf{R}p_*\mathbf{G}_{m,X} \to \tau_{\geq 2}\mathbf{R}p_*\mathbf{G}_{m,X} \to (\tau_{\leq 1}\mathbf{R}p_*\mathbf{G}_{m,X})[1],$$

and here $\tau_{\geq 2}\mathbf{R}p_*\mathbf{G}_{m,X}$ is acyclic in degrees 0 and 1. We deduce canonical isomorphisms

$$R^1\mathrm{Hom}_Z(\widehat{S}, \tau_{\leq 1}\mathbf{R}p_*\mathbf{G}_{m,X}) = R^1\mathrm{Hom}_Z(\widehat{S}, \mathbf{R}p_*\mathbf{G}_{m,X}) = H^1(X, S),$$

and an injection of $R^2\mathrm{Hom}_Z(\widehat{S}, \tau_{\leq 1}\mathbf{R}p_*\mathbf{G}_{m,X})$ into

$$R^2\mathrm{Hom}_Z(\widehat{S}, \mathbf{R}p_*\mathbf{G}_{m,X}) = H^2(X, S).$$

Now (8.2) is obtained by applying $\mathrm{Hom}_Z(\widehat{S}, .)$ to (8.1). □

Remarks

1. Let k be a field of characteristic zero with algebraic closure $\bar{k}$ and Galois group $\Gamma = \mathrm{Gal}(\bar{k}/k)$. In the case when X is smooth over k, $KD(X)$ was introduced in [18] as the following complex of Γ-modules in degrees -1 and 0:

   $$[\bar{k}(X)^* \to \mathrm{Div}(\overline{X})].$$

   Here $\bar{k}(X)$ is the function field of $\overline{X} = X \times_k \bar{k}$, and $\mathrm{Div}(\overline{X})$ is the group of divisors on $\overline{X}$ (see [3], Lemma 2.3 and Remark 2.6). In this case $KD'(X)$ is quasi-isomorphic to the complex of Γ-modules

   $$[\bar{k}(X)^*/\bar{k}^* \to \mathrm{Div}(\overline{X})].$$

   Up to shift, $KD'(X)$ was independently introduced by Borovoi and van Hamel in [3]: in their notation we have $KD'(X) = U\mathrm{Pic}(\overline{X})[1]$. Furthermore, if X^c is a smooth compactification of X, and $\mathrm{Div}_\infty(\overline{X}^c)$ is the group of divisors of $\overline{X}^c$ supported on $\overline{X} - \overline{X}^c$, then, by [18], Lemma 2.2, $KD'(X)$ is quasi-isomorphic to the complex

   $$[\mathrm{Div}_\infty(\overline{X}^c) \to \mathrm{Pic}(\overline{X}^c)].$$

2. In the relative case, when X is smooth over Z, our $KD(X)$ and $KD'(X)$ coincide with analogous objects defined in [18], Remark 2.4 (2). See Appendix A to the present paper for the proof of this fact.
3. In the relative case, when X is proper over Z with geometrically integral fibres, $KD'(X)$ identifies with the sheaf $R^1p_*\mathbf{G}_{m,X}$, the relative Picard functor. When X is also assumed projective over Z, the relative Picard functor is representable by a Z-scheme, separated and locally of finite type, see [4], Ch. 8, Thm. 1 on p. 210.

Let $\mathcal{D}(k)$ be the bounded derived category of the category of continuous discrete Γ-modules.

Definition 8.2 Let X be a smooth and geometrically integral variety over k. Let Y be an X-torsor under a k-group of multiplicative type S, and let $[Y]$ be its class in $H^1(X, S)$. We shall say that the morphism $\chi([Y]) : \widehat{S} \to KD'(X)$ in the derived category $\mathcal{D}(k)$ is the *extended type* of the torsor $Y \to X$.

Remarks

1. There is a canonical morphism from (8.2) to the sequence of Colliot-Thélène and Sansuc ([29], Thm. 2.3.6):

$$\mathrm{Ext}^1_k(\widehat{S}, \bar{k}[X]^*) \to H^1(X, S) \to \mathrm{Hom}_k(\widehat{S}, \mathrm{Pic}(\overline{X})) \to \mathrm{Ext}^2_k(\widehat{S}, \bar{k}[X]^*) \to H^2(X, S). \quad (8.3)$$

Indeed, (8.3) is obtained by applying the functor $\mathrm{Hom}_k(\widehat{S}, .)$ to the exact triangle

$$\bar{k}[X]^*[1] \to KD(X) \to \mathrm{Pic}(\overline{X}) \to \bar{k}[X]^*[2] \quad (8.4)$$

(cf. [29], p. 26), and there is an obvious canonical morphism from (8.1) to (8.4). Recall that if $Y \to X$ is a torsor under S, then the image of the class $[Y] \in H^1(X, S)$ in $\mathrm{Hom}_k(\widehat{S}, \mathrm{Pic}(\overline{X}))$ is called the *type* of $Y \to X$. We see that the notion of extended type defines a stronger equivalence relation on $H^1(X, S)$ than the notion of type. For example two torsors have the same extended type if and only if their classes in $H^1(X, S)$ coincide up to a 'constant element'.
2. If we assume further that $\bar{k}[X]^* = \bar{k}^*$ (e.g. X proper), then $KD'(X)$ is quasi-isomorphic to $[0 \to \mathrm{Pic}(\overline{X})]$, and the exact sequence (8.2) is

just the fundamental exact sequence of Colliot-Thélène and Sansuc ([29], Cor. 2.3.9):

$$H^1(k,S) \to H^1(X,S) \to \mathrm{Hom}_k(\widehat{S}, \mathrm{Pic}(\overline{X})) \to H^2(k,S) \to H^2(X,S).$$

3. The other 'extreme' case is when $\mathrm{Pic}(\overline{X}) = 0$. Then $KD'(X)$ is quasi-isomorphic to $(\bar{k}[X]^*/\bar{k}^*)[1]$, and the extended type is an element of $\mathrm{Ext}^1_k(\widehat{S}, \bar{k}[X]^*/\bar{k}^*)$. One case of interest is when X is a principal homogeneous space of a k-torus T, so that $\mathrm{Pic}(\overline{X}) = \mathrm{Pic}(\overline{T}) = 0$, then the extended type is an element of $\mathrm{Ext}^1_k(\widehat{S}, \widehat{T})$. Suppose that $X = T$, and let $T' \to T$ be a surjective homomorphism of k-tori with kernel S. This is of course a T-torsor under S. We shall show in Remark 2 after Proposition 8.9 below that the extended type of this torsor is given by the natural extension
$$0 \to \widehat{T} \to \widehat{T'} \to \widehat{S} \to 0.$$
This fact was implicitly used in [19, Lemma 2.2].
4. Unlike the classical type, the extended type of a torsor $Y \to X$ is in general not determined by the $\overline{X}$-torsor $\overline{Y}$. For example, if $\mathrm{Pic}(\overline{X}) = 0$ and S is a torus, then $\mathrm{Ext}^1_{\bar{k}}(\widehat{S}, \bar{k}[X]^*/\bar{k}^*) = 0$ because $\widehat{S}$ is a free abelian group.

Proposition 8.3 *Let X be a smooth and geometrically integral variety over k, and let S be a k-group of multiplicative type. If $X(k) \neq \emptyset$, then the map $\chi : H^1(X,S) \to \mathrm{Hom}_k(\widehat{S}, KD'(X))$ is onto. In other words, if $X(k) \neq \emptyset$, then there exist X-torsors of every extended type.*

Proof Since $X(k) \neq \emptyset$, the map $H^2(k,S) \to H^2(X,S)$ has a retraction, hence is injective. Therefore the map ∂ is zero and χ is surjective. □

Let $\mathrm{Br}(X) = H^2(X, \mathbf{G}_{m,X})$ be the cohomological Brauer–Grothendieck group of X. As usual, $\mathrm{Br}_0(X)$ will denote the image of the natural map $\mathrm{Br}(k) \to \mathrm{Br}(X)$, and $\mathrm{Br}_1(X)$ the kernel of the natural map $\mathrm{Br}(X) \to \mathrm{Br}(\overline{X})$.

It is easy to check that $\mathrm{Br}_1(X)$ is canonically isomorphic to $H^1(k, KD(X))$, see [3], Prop. 2.18, or [18], Lemma 2.1. Thus the exact triangle (8.1) induces an exact sequence in Galois hypercohomology

$$\mathrm{Br}(k) \to \mathrm{Br}_1(X) \xrightarrow{r} H^1(k, KD'(X)) \to H^3(k, \bar{k}^*). \tag{8.5}$$

The cup-product in étale cohomology defines the pairing

$$\cup : H^1(k,\widehat{S}) \times H^1(X,S) \to H^1(X,\widehat{S}) \times H^1(X,S) \to \mathrm{Br}(X),$$

whose image visibly belongs to $\mathrm{Br}_1(X)$. The following statement generalises [29, Thm. 4.1.1].

Theorem 8.4 *Let X be a smooth and geometrically integral variety over k, and let $f : Y \to X$ be a torsor under a k-group of multiplicative type S. Let $\lambda : \widehat{S} \to KD'(X)$ be the extended type of this torsor. Then for any $a \in H^1(k, \widehat{S})$ we have*

$$r(a \cup [Y]) = \lambda_*(a),$$

where λ_* is the induced map $H^1(k, \widehat{S}) \to H^1(k, KD'(X))$.

Proof We have canonical isomorphisms

$$H^1(X, S) = \mathrm{Hom}_k(\widehat{S}, \mathbf{R}p_*\mathbf{G}_{m,X}[1]) = \mathrm{Hom}_k(\widehat{S}, KD(X)),$$

cf. the proof of Proposition 8.1. By [23], Prop. V.1.20, these isomorphisms fit into the following commutative diagram of pairings

$$\begin{array}{ccccc}
H^1(k,\widehat{S}) & \times & H^1(X,S) & \to & \mathrm{Br}(X) \\
\| & & \| & & \| \\
H^1(k,\widehat{S}) & \times & \mathrm{Hom}_k(\widehat{S},\mathbf{R}p_*\mathbf{G}_{m,X}[1]) & \to & H^1(k,\mathbf{R}p_*\mathbf{G}_{m,X}[1]) \\
\| & & \| & & \uparrow \\
H^1(k,\widehat{S}) & \times & \mathrm{Hom}_k(\widehat{S},KD(X)) & \to & H^1(k,KD(X)).
\end{array}$$

Here the vertical arrow is induced by the canonical map

$$KD(X) = (\tau_{\leq 1}\mathbf{R}p_*\mathbf{G}_{m,X})[1] \to \mathbf{R}p_*\mathbf{G}_{m,X}[1].$$

Let $u : \widehat{S} \to KD(X)$ be the morphism corresponding to the class $[Y]$. By the commutativity of the diagram we have

$$a \cup [Y] = u_*(a) \in H^1(k, KD(X)) = \mathrm{Br}_1(X).$$

By definition, λ is the composed map

$$\widehat{S} \xrightarrow{u} KD(X) \xrightarrow{v} KD'(X).$$

By construction, the map r from the exact sequence (8.5) is the induced map $v_* : H^1(k, KD(X)) \to H^1(k, KD'(X))$, hence $r(a \cup [Y]) = v_*(u_*(a)) = \lambda_*(a)$. □

8.2 Localisation of torsors

Let U be a smooth and geometrically integral variety over k. The abelian group $\bar{k}[U]^*/\bar{k}^*$ is torsion free, so we can define a k-torus R as the torus whose module of characters $\widehat{R}$ is the Γ-module $\bar{k}[U]^*/\bar{k}^*$. The natural exact sequence of Γ-modules

$$1 \to \bar{k}^* \to \bar{k}[U]^* \to \bar{k}[U]^*/\bar{k}^* \to 1 \qquad (*_U)$$

defines a class

$$[*_U] \in \mathrm{Ext}^1_k(\bar{k}[U]^*/\bar{k}^*, \bar{k}^*) = H^1(k, R).$$

Let Y be the k-torsor under R whose class in $H^1(k, R)$ is $-[*_U]$.

Lemma 8.5 *There exists a morphism $q_U : U \to Y$ such that q_U^* identifies $(*_Y)$ with $(*_U)$. Any morphism from U to a k-torsor under a torus factors through q_U.*

Proof This is Lemma 2.4.4 of [29]. □

To deal with the case of torsors under arbitrary groups of multiplicative type we need to extend this constrution to certain geometrically reducible varieties. However, Lemma 8.5 does not readily generalise, because its essential ingredient is Rosenlicht's lemma which is valid only for connected groups. If S is a torus, it says that the natural map of Γ-modules $\widehat{S} \to \bar{k}[S]^*$ induces an isomorphism $\widehat{S} \cong \bar{k}[S]^*/\bar{k}^*$ (every invertible regular function that takes value 1 at the neutral element of S is a character). This is no longer true if $\widehat{S}$ has non-zero torsion subgroup. For general groups of multiplicative type we propose the following substitute.

Definition 8.6 For a (not necessarily integral) k-variety V with an action of S we define $\bar{k}[V]^*_S$ as the subgroup of $\bar{k}[V]^*$ consisting of the functions $f(x)$ for which there exists a character $\chi \in \widehat{S}$ such that $f(sx) = \chi(s)f(x)$ for any $s \in S(\bar{k})$.

It is easy to see that $\bar{k}[V]^*_S$ is a Γ-submodule of $\bar{k}[V]^*$.

Remark If S is a torus and V is geometrically connected, then $\bar{k}[V]^*_S = \bar{k}[V]^*$, so we are not getting anything new. Indeed, if $x \in V(\bar{k})$ and $f \in \bar{k}[V]^*$, then $f(sx)/f(x)$ is a regular invertible function on $\overline{S}$ with value 1 at the neutral element $e \in S(\bar{k})$. By Rosenlicht's lemma such a function is a character in $\widehat{S}$. We obtain a morphism from a connected variety $\overline{V}$ to a discrete group $\widehat{S}$, which must be a constant map. Hence there exists $\chi \in \widehat{S}$ such that $f(sx) = \chi(s)f(x)$ for any $x \in V(\bar{k})$ and any $s \in S(\bar{k})$.

Proposition 8.7 *Let S be a k-group of multiplicative type. Then the natural map $\widehat{S} \to \bar{k}[S]^*_S$ induces an isomorphism of Γ-modules $\widehat{S} \tilde{\longrightarrow} \bar{k}[S]^*_S/\bar{k}^*$.*

Proof The image of the natural inclusion $\widehat{S} \to \bar{k}[S]^*$ is contained in $\bar{k}[S]^*_S$, so it remains to show that any function from $f(x) \in \bar{k}[S]^*_S$ that takes value 1 at the neutral element e of S is a character. Indeed, for any $s \in S(\bar{k})$ we have $f(sx) = \chi(s)f(x)$, and taking $x = e$ we obtain $f(s) = \chi(s)$. □

Corollary 8.8 *Let V be a k-torsor of S. Then we have an exact sequence of Γ-modules*

$$0 \to \bar{k}^* \to \bar{k}[V]^*_S \to \widehat{S} \to 0. \tag{8.6}$$

The class of extension (8.6) *in* $\mathrm{Ext}^1_k(\widehat{S}, \bar{k}^*) = H^1(k, S)$ *is* $-[V]$.

Proof The action of $S(\bar{k})$ on $\widehat{S} = \bar{k}[S]^*_S/\bar{k}^*$ is trivial, hence the first statement follows from Proposition 8.7 by Galois descent. In the case when S is a torus the last statement is a well known lemma of Sansuc [27], (6.7.3) and (6.7.4), see also Lemma 5.4 of [3]. The same calculation works in the general case. □

We shall need a relative version of Corollary 8.8. Recall that $\pi_*\mathbf{G}_{m,Y}$ is the sheaf on X such that for an étale morphism $U \to X$ we have

$$\pi_*\mathbf{G}_{m,Y}(U) = \mathrm{Mor}_U(Y_U, \mathbf{G}_{m,U}) = \mathrm{Mor}_k(Y_U, \mathbf{G}_{m,k}).$$

Define $(\pi_*\mathbf{G}_{m,Y})_S$ as the subsheaf of $\pi_*\mathbf{G}_{m,Y}$ such that for an étale morphism $U \to X$ the group of sections $(\pi_*\mathbf{G}_{m,Y})_S(U)$ consists of the functions $f(x) \in \mathrm{Mor}_U(Y_U, \mathbf{G}_{m,U})$ for which there exists a group scheme homomorphism $\chi : S_U \to \mathbf{G}_{m,U}$ such that $f(sx) = \chi(s)f(x)$ for any $s \in S_U(\bar{k})$ and any $x \in Y_U(\bar{k})$. If $m : S_U \times_U Y_U = S \times_k Y_U \to Y_U$ is the action of S on Y_U, then the last condition is $m^*f = \chi \cdot f$.

Proposition 8.9 *Let $p : X \to \mathrm{Spec}(k)$ be a smooth and geometrically integral variety, and let $\pi : Y \to X$ be a torsor under S.*

(i) *We have an exact sequence of étale sheaves on X:*

$$0 \to \mathbf{G}_{m,X} \to (\pi_*\mathbf{G}_{m,Y})_S \to p^*\widehat{S} \to 0. \tag{8.7}$$

Applying p_* to (8.7) we obtain an exact sequence of Γ-modules

$$0 \to \bar{k}[X]^* \to \bar{k}[Y]^*_S \to \widehat{S} \to \operatorname{Pic} \overline{X}. \qquad (8.8)$$

(ii) *The class of extension* (8.7) *in* $\operatorname{Ext}^1_X(p^*\widehat{S}, \mathbf{G}_{m,X}) = H^1(X, S)$ *is the class* $[Y/X]$ *of the* X*-torsor* Y *(up to sign).*

(iii) *The last arrow in* (8.8) *is the type of the torsor* $\pi : Y \to X$.

(iv) *When the type of* $\pi : Y \to X$ *is zero, the extension given by the first three non-zero terms of* (8.8) *maps to the class of* (8.7) *by the canonical injective map*

$$0 \to \operatorname{Ext}^1_k(\widehat{S}, \bar{k}[X]^*) \to \operatorname{Ext}^1_X(p^*\widehat{S}, \mathbf{G}_{m,X}) = H^1(X, S).$$

Proof

(i) The maps in this sequence are obvious maps. The exactness can be checked locally, so we can assume that $Y = X \times_k S$, but in this case the exactness is clear. The exact sequence (8.8) follows from (8.7) once we note that the canonical morphism $\widehat{S} \to p_*p^*\widehat{S}$ is an isomorphism since $\overline{X}$ is connected.

(ii) The proof of [8, Prop. 1.4.3] applies as is.

(iii-iv) More generally, let A be a Γ-module, and $\mathcal{F}$ be a sheaf on X. Recall that we have the spectral sequence of the composition of functors $\mathbf{R}p_*$ and $\mathbf{R}\mathrm{Hom}_k(A, \cdot)$:

$$\operatorname{Ext}^m_k(A, H^n(\overline{X}, \mathcal{F})) \Rightarrow \operatorname{Ext}^{m+n}_X(p^*A, \mathcal{F}).$$

It gives rise to the exact sequence

$$0 \to \operatorname{Ext}^1_k(A, p_*\mathcal{F}) \to \operatorname{Ext}^1_X(p^*A, \mathcal{F}) \to \operatorname{Hom}_k(A, R^1p_*\mathcal{F}). \qquad (8.9)$$

The arrows in (8.9) have explicit description. The canonical map $\mathrm{E}^1 \to \mathrm{E}^{0,1}$ sends the class of the extension of sheaves on X

$$0 \to \mathcal{F} \to \mathcal{E} \to p^*A \to 0$$

to the last arrow in

$$0 \to p_*\mathcal{F} \to p_*\mathcal{E} \to p_*p^*A \to R^1p_*\mathcal{F},$$

composed with the canonical map $A \to p_*p^*A$. If the class of the extension $\mathcal{E}$ goes to $0 \in \mathrm{E}^{0,1}$, then this class comes from the extension of Γ-modules

$$0 \to p_*\mathcal{F} \to p_*\mathcal{E} \to \operatorname{Ker}[p_*p^*A \to R^1p_*\mathcal{F}] \to 0$$

pulled back by the same canonical map. See Appendix B to this paper for a proof of these facts. In our case take $A = \widehat{S}$ and $\mathcal{F} = \mathbf{G}_{m,X}$.

□

Remarks

1. The type of the torsor $\pi : Y \to X$, at least up to sign, can also be described explicitly as follows. Let $K = \bar{k}(X)$. The fibre of $p : Y \to X$ over $\mathrm{Spec}(K)$ is a K-torsor Y_K under S. By Corollary 8.8 we can lift any character $\chi \in \widehat{S}$ to a rational function $f \in K[Y_K]^*_S \subset \bar{k}(Y)^*$. By construction f is an invertible regular function on Y_K, hence $\mathrm{div}_{\overline{Y}}(f) = \pi^*(D)$ where D is a divisor on $\overline{X}$. Note that D is uniquely determined by χ up to a principal divisor on $\overline{X}$. It is not hard to check that the class of this divisor in $\mathrm{Pic}\,\overline{X}$ is the image of χ (up to sign). Indeed, by [29], Lemma 2.3.1 (ii), the type associates to χ the subsheaf $\mathcal{O}_\chi$ of χ-semiinvariants of $p_*(\mathcal{O}_Y)$. The function f is a rational section of $\mathcal{O}_\chi$, hence the class $[D]$ represents $\mathcal{O}_\chi \in \mathrm{Pic}\,\overline{X}$. If this description was used as a definition of type, then the exactness of (8.8) is easily checked directly.
2. From (8.8) we obtain the following exact sequence:

$$0 \to \bar{k}[X]^*/\bar{k}^* \to \bar{k}[Y]^*_S/\bar{k}^* \to \widehat{S} \to \mathrm{Pic}\,\overline{X}. \qquad (8.10)$$

   In the same way as in Proposition 8.9 (iii), one shows that when the type of the torsor $\pi : Y \to X$ is zero, the extension given by the first three non-zero terms of (8.10) maps to the extended type of $\pi : Y \to X$ by the canonical injective map

$$0 \to \mathrm{Ext}^1_k(\widehat{S}, \bar{k}[X]^*/\bar{k}^*) \to \mathrm{Hom}_k(\widehat{S}, KD'(X)).$$

   In particular, a surjective homomorphism of k-tori $T_1 \to T_2$ with kernel S is a T_2-torsor under S. The extended type of this torsor comes from the extension

$$0 \to \bar{k}[T_2]^*/\bar{k}^* \to \bar{k}[T_1]^*_S/\bar{k}^* \to \widehat{S} \to 0,$$

   which is precisely the dual exact sequence

$$0 \to \widehat{T_2} \to \widehat{T_1} \to \widehat{S} \to 0.$$

3. For an application of Proposition 8.9 to the computation of $KD'(X)$ when X is a homogeneous space, see [2], Appendix B.

Theorem 8.10 *Let U be a smooth and geometrically integral variety over k, let S be a k-group of multiplicative type, and let $\pi : \mathcal{T} \to U$ be a torsor under S of type zero. Then we have the following statements.*

(i) *There is a natural exact sequence of Γ-modules*

$$0 \to \bar{k}^* \to \bar{k}[\mathcal{T}]^*_S \to \widehat{M} \to 0,$$

which is the definition of the k-group of multiplicative type M.

(ii) *There is a natural exact sequence of Γ-modules*

$$0 \to \bar{k}[U]^* \to \bar{k}[\mathcal{T}]^*_S \to \widehat{S} \to 0. \tag{8.11}$$

Let

$$1 \to S \to M \to R \to 1$$

be the dual exact sequence of k-groups of multiplicative type.

(iii) *We have $\mathcal{T} = Z \times_Y U$, where Z is a k-torsor under M that represents the negative of the class of the extension* (i), *$Z \to Y = Z/S$ is the natural quotient, and $U \to Y$ is the morphism q_U from Lemma* 8.5.

Proof We note that the abelian group $\bar{k}[\mathcal{T}]^*_S/\bar{k}^*$ is finitely generated since the same is true for $\bar{k}[U]^*/\bar{k}^*$ and $\widehat{S}$. Thus we can define M as in (i). The extension (8.11) gives rise to the following commutative diagram:

$$\begin{array}{ccccccccc} & & 0 & & 0 & & & & \\ & & \downarrow & & \downarrow & & & & \\ & & \bar{k}^* & = & \bar{k}^* & & & & \\ & & \downarrow & & \downarrow & & & & \\ 0 & \to & \bar{k}[U]^* & \to & \bar{k}[\mathcal{T}]^*_S & \to & \widehat{S} & \to & 0 \\ & & \downarrow & & \downarrow & & \| & & \\ 0 & \to & \widehat{R} & \to & \widehat{M} & \to & \widehat{S} & \to & 0 \\ & & \downarrow & & \downarrow & & & & \\ & & 0 & & 0. & & & & \end{array} \tag{8.12}$$

Similarly to Lemma 8.5 the extension (i) defines a k-torsor Z under M and a morphism $q : \mathcal{T} \to Z$ which identifies (i) with the extension

$$0 \to \bar{k}^* \to \bar{k}[Z]^*_S \to \widehat{M} \to 0.$$

The functoriality of this construction and the commutativity of (8.12) imply that there is an isomorphism $Z/S \cong Y$ of torsors under R which makes the diagram commute:

$$\begin{array}{ccc} \mathcal{T} & \to & U \\ \downarrow & & \downarrow \\ Z & \to & Y. \end{array}$$

This gives a morphism $\mathcal{T} \to Z \times_Y U$ of U-torsors under S, which, as any such morphism, is an isomorphism. □

Corollary 8.11 *Let X be a smooth geometrically integral variety over k, let S be a k-group of multiplicative type, and let $\lambda \in \mathrm{Hom}_k(\widehat{S}, KD'(X))$. Let U be a dense open set of X such that the induced element $\lambda_U \in \mathrm{Hom}_k(\widehat{S}, KD'(U))$ has trivial image in $\mathrm{Hom}_k(\widehat{S}, \mathrm{Pic}\,\overline{U})$, so that*

$$\lambda_U \in \mathrm{Ext}^1_k(\widehat{S}, \bar{k}[U]^*/\bar{k}^*) = \mathrm{Ext}^1_k(\widehat{S}, \widehat{R}) = \mathrm{Ext}^1_{k-\mathrm{groups}}(R, S),$$

where $\widehat{R} = \bar{k}[U]^*/\bar{k}^*$. Let

$$1 \to S \to M \to R \to 1$$

be an extension representing this class. Then we have the following statements.

(i) *The restriction of an X-torsor of extended type λ to U is isomorphic to $Z \times_Y U$, where Y is a k-torsor under R, $U \to Y$ is the morphism q_U defined in Lemma 8.5, and Z is a k-torsor under M such that $Y = Z/S$.*

(ii) *Conversely, any U-torsor $Z \times_Y U \to U$ extends to an X-torsor under S of extended type λ.*

Proof By Remark 2 after Proposition 8.9 we know that the extension (8.11) represents the class λ_U. Now part (i) follows from Theorem 8.10.

Recall that the embedding $j : U \to X$ gives a natural injective map $\mathbf{G}_{m,X} \to j_*\mathbf{G}_{m,U}$ of étale sheaves on X. On applying $\mathbf{R}p_*$ and the truncation $\tau_{\leq 1}$ we obtain a natural morphism $\tau_{\leq 1}\mathbf{R}p_*\mathbf{G}_{m,X} \to \tau_{\leq 1}\mathbf{R}(pj)_*\mathbf{G}_{m,U}$ in $\mathcal{D}(k)$. It is clear that we have a commutative diagram of exact triangles in $\mathcal{D}(k)$:

$$\begin{array}{ccccc} \bar{k}^* & \to & \tau_{\leq 1}\mathbf{R}p_*\mathbf{G}_{m,X} & \to & KD'(X)[-1] \\ || & & \downarrow & & \downarrow \\ \bar{k}^* & \to & \tau_{\leq 1}\mathbf{R}(pj)_*\mathbf{G}_{m,U} & \to & KD'(U)[-1]. \end{array}$$

It gives rise to the following commutative diagrams of abelian groups:

$$\begin{array}{ccccccc} H^1(k,S) & \to & H^1(X,S) & \to & \mathrm{Hom}_k(\widehat{S},KD'(X)) & \to & H^2(k,S) \\ \| & & \downarrow & & \downarrow & & \| \\ H^1(k,S) & \to & H^1(U,S) & \to & \mathrm{Hom}_k(\widehat{S},KD'(U)) & \to & H^2(k,S). \end{array} \tag{8.13}$$

Now it is easy to complete the proof of the corollary. From Corollary 8.8 and Remark 2 after Proposition 8.9 we see that the extended type of $Z \to Y$ is λ_U. This implies that the extended type of $Z \times_Y U \to U$ is λ_U. Now (ii) is an immediate consequence of (8.13). □

8.3 Descent theory

In this and the following chapters, k is a number field with the ring of integers $\mathcal{O}_k$. Let Ω_k be the set of places of k, and let Ω_∞ (resp. Ω_f) be the set of archimedean (resp. finite) places of k. For $v \in \Omega_k$ we write k_v for the completion of k at v.

For a variety X over k we denote by $X(\mathbf{A}_k)$ the topological space of adelic points of X; it coincides with $\prod_{v\in\Omega_k} X(k_v)$ when X is proper. Recall (cf. [29], Ch. 5) that the Brauer–Manin pairing

$$X(\mathbf{A}_k) \times \mathrm{Br}(X) \to \mathbf{Q}/\mathbf{Z}$$

is defined by the formula

$$((P_v),\alpha) \mapsto \sum_{v\in\Omega_k} j_v(\alpha(P_v)),$$

where $j_v : \mathrm{Br}(k_v) \to \mathbf{Q}/\mathbf{Z}$ is the local invariant in class field theory. By global class field theory we have $((P_v),\alpha) = 0$ for every $\alpha \in \mathrm{Br}_0(X)$. For a subgroup $B \subset \mathrm{Br}(X)$ (or $B \subset \mathrm{Br}(X)/\mathrm{Br}_0(X)$) we denote by $X(\mathbf{A}_k)^B$ the set of those adelic points that are orthogonal to B, and we write $X(\mathbf{A}_k)^{\mathrm{Br}}$ for $X(\mathbf{A}_k)^{\mathrm{Br}(X)}$. By the reciprocity law in global class field theory, we have $X(k) \subset X(\mathbf{A}_k)^{\mathrm{Br}}$.

If $f : Y \to X$ is a torsor under a k-group of multiplicative type S, the *descent set* $X(\mathbf{A}_k)^f$ is defined as the set of adelic points $(P_v) \in X(\mathbf{A}_k)$ such that the family $([Y](P_v))$ is in the image of the diagonal map $H^1(k,G) \to \prod_{v\in\Omega_k} H^1(k_v,G)$, see [29], Section 5.3.

Proposition 8.12 *Let X be a smooth and geometrically integral variety over a number field k, and let S be a k-group of multiplicative type. Keep the notation as in Theorem 8.4. Then an adelic point $(P_v) \in X(\mathbf{A}_k)$*

belongs to the descent set $X(\mathbf{A}_k)^f$ associated to the torsor $f : Y \to X$ under S if and only if (P_v) is orthogonal to the subgroup

$$\mathrm{Br}_\lambda(X) := r^{-1}(\lambda_*(H^1(k, \widehat{S}))) \subset \mathrm{Br}_1(X)$$

with respect to the Brauer–Manin pairing.

Proof The property $(P_v) \in X(\mathbf{A}_k)^f$ means that the family $([Y](P_v))$ is in the image of the diagonal map $H^1(k, S) \to \mathbf{P}^1(S)$, where $\mathbf{P}^1(S)$ is the restricted product of the groups $H^1(k_v, S)$. By the Poitou–Tate exact sequence (see, for example, [11], Thm. 6.3) this is equivalent to the condition

$$\sum_{v \in \Omega_k} j_v((a \cup [Y])(P_v)) = 0$$

for every $a \in H^1(k, \widehat{S})$. On the other hand, by Theorem 8.4 and exact sequence (8.5) every element of $\mathrm{Br}_\lambda(X)$ can be written as $a \cup [Y] + \alpha_0$, where $\alpha_0 \in \mathrm{Br}_0(X)$. The proposition follows. □

What we want now is an 'integral version' of Proposition 8.12. If Σ is a finite set of places of k, we denote by $\mathcal{O}_\Sigma$ the subring of k consisting of the elements integral at the non-archimedean places outside Σ. Then $U = \mathrm{Spec}\,(\mathcal{O}_\Sigma)$ is an open subset of $\mathrm{Spec}\,(\mathcal{O}_k)$. Let us assume that there are

- a faithfully flat and separated U-scheme of finite type $\mathcal{X}$,
- a flat commutative group U-scheme $\mathcal{S}$ of finite type, and
- an fppf $\mathcal{X}$-torsor $\mathcal{Y}$ under $\mathcal{S}$,

such that $X = \mathcal{X} \times_U k$, $S = \mathcal{S} \times_U k$, and $Y = \mathcal{Y} \times_U k$. This assumption can always be satisfied if Σ is large enough.

Let $[\mathcal{Y}]$ be the class of $\mathcal{Y}$ in the fppf cohomology group $H^1(\mathcal{O}_\Sigma, \mathcal{S}) = H^1(U, \mathcal{S})$. For every $\mathcal{O}_\Sigma$-torsor c under $\mathcal{S}$, one defines the *twisted torsor* $\mathcal{Y}^c = (\mathcal{Y} \times_U c)/\mathcal{S}$. This is a U-torsor under $\mathcal{S}$ such that $[\mathcal{Y}^c] = [\mathcal{Y}] - c$ (see [29], Lemma 2.2.3).

Corollary 8.13 *Let $(P_v) \in \prod_{v \in \Sigma} X(k_v) \times \prod_{v \notin \Sigma} \mathcal{X}(\mathcal{O}_v)$. Then the following conditions are equivalent:*

(a) *The adelic point (P_v) is orthogonal to $\mathrm{Br}_\lambda(X)$.*
(b) *There exists a class $[c] \in H^1(\mathcal{O}_\Sigma, \mathcal{S})$ that goes to $([\mathcal{Y}](P_v))$ under the diagonal map*

$$H^1(\mathcal{O}_\Sigma, \mathcal{S}) \to \prod_{v \in \Sigma} H^1(k_v, S) \times \prod_{v \notin \Sigma} H^1(\mathcal{O}_v, \mathcal{S}).$$

(c) *There exists an $\mathcal{O}_\Sigma$-torsor c under $\mathcal{S}$ such that the adelic point (P_v) lifts to an adelic point in $\prod_{v\in\Sigma} Y^c(k_v) \times \prod_{v\notin\Sigma} \mathcal{Y}^c(\mathcal{O}_v)$, where $Y^c = \mathcal{Y}^c \times_{\mathcal{O}_\Sigma} k$ is the generic fibre of the twisted torsor $\mathcal{Y}^c$.*

Proof Let c be an $\mathcal{O}_\Sigma$-torsor under $\mathcal{S}$ with cohomology class $[c] \in H^1(\mathcal{O}_\Sigma, \mathcal{S})$. Then $[c]$ goes to $([\mathcal{Y}(P_v)])$ if and only if $[\mathcal{Y}^c(P_v)] = 0$ for every place v. But this is equivalent to the fact that P_v lifts to a point in $Y^c(k_v)$ if $v \in \Sigma$, and to a point in $\mathcal{Y}^c(\mathcal{O}_v)$ if $v \notin \Sigma$. This proves the equivalence of (b) and (c).

Condition (b) implies that the adelic point (P_v) is in the descent set $X(\mathbf{A}_k)^f$. Hence (a) follows from (b) by Proposition 8.12.

Assume condition (a). By Proposition 8.12, the element

$$([\mathcal{Y}(P_v)]) \in \prod_{v\in\Sigma} H^1(k_v, S) \times \prod_{v\notin\Sigma} H^1(\mathcal{O}_v, \mathcal{S})$$

is in the diagonal image of some $\sigma \in H^1(k, S)$. Since σ is unramified outside Σ, Harder's lemma ([20], Lemma 4.1.3 or [13], Corollary A.8) implies that σ is in the image of the restriction map $H^1(\mathcal{O}_\Sigma, \mathcal{S}) \to H^1(k, S)$. Thus (a) implies (b). □

Remarks

1. If we assume further that $\mathcal{S}$ and $\mathcal{X}$ are smooth over U, then everywhere in the previous corollary we can replace fppf cohomology by étale cohomology. This can be arranged by choosing a sufficiently large set Σ.
2. We refer the reader to [21] and [9] for examples of descent on the torsor $\mathcal{Y} \to \mathcal{X}$ under μ_d, where $\mathcal{X} \subset \mathbf{P}^2_{\mathbf{Z}}$ is the complement to the closed subscheme given by a homogeneous polynomial $f(x, y, z)$ of degree d with integral coefficients, and $\mathcal{Y}$ is given by the equation $u^d = f(x, y, z)$.

Below is a "truncated" variant of Proposition 8.12 where we consider all places of k except finitely many. Keep the notation as above and let Σ_0 be a finite set of places of k. Let $X(\mathbf{A}_k^{\Sigma_0})$ be the topological space of "truncated" adelic points, defined as the restricted product of the spaces $X(k_v)$ for $v \notin \Sigma_0$ with respect to the subsets $\mathcal{X}(\mathcal{O}_v)$, $v \notin \Sigma \cup \Sigma_0$. We define $\mathbf{P}^1_{\Sigma_0}(S)$ as the restricted product of the groups $H^1(k_v, S)$ for $v \notin \Sigma_0$ with respect to the subgroups $H^1(\mathcal{O}_v, \mathcal{S})$, $v \notin \Sigma \cup \Sigma_0$. As in the classical case $\Sigma_0 = \emptyset$, the sets $\mathbf{P}^1_{\Sigma_0}(S)$ and $X(\mathbf{A}_k^{\Sigma_0})$ are independent of the choices of models $\mathcal{S}$ and $\mathcal{X}$. Let $H^1_{\Sigma_0}(k, \widehat{S})$ be the kernel of the restriction map $H^1(k, \widehat{S}) \to \prod_{v \in \Sigma_0} H^1(k_v, \widehat{S})$.

Proposition 8.14 *Let $(P_v)_{v \notin \Sigma_0} \in X(\mathbf{A}_k^{\Sigma_0})$. Then $([Y](P_v))_{v \notin \Sigma_0}$ is in the image of the diagonal map $H^1(k, S) \to \mathbf{P}^1_{\Sigma_0}(S)$ if and only if the "truncated" adelic point $(P_v)_{v \notin \Sigma_0}$ is orthogonal to the subgroup*

$$\mathrm{Br}_{\lambda, \Sigma_0}(X) := r^{-1}(\lambda_*(H^1_{\Sigma_0}(k, \widehat{S}))) \subset \mathrm{Br}_\lambda(X).$$

Proof Using the local Tate duality, we see that the Poitou–Tate exact sequence for S (see [11], Thm. 6.3.) gives rise to the Σ_0-truncated exact sequence

$$H^1(k, S) \to \mathbf{P}^1_{\Sigma_0}(S) \to H^1_{\Sigma_0}(k, \widehat{S})^{\mathrm{D}},$$

where the superscript D denotes the Pontryagin dual $\mathrm{Hom}(\cdot, \mathbf{Q}/\mathbf{Z})$. By this sequence, $(s_v)_{v \notin \Sigma_0} \in \mathbf{P}^1_{\Sigma_0}(S)$ is in the image of $H^1(k, S)$ if and only if

$$\sum_{v \notin \Sigma_0} j_v(a \cup s_v) = 0$$

for every $a \in H^1_{\Sigma_0}(k, \widehat{S})$. The proof finishes in the same way as the proof of Proposition 8.12. □

Taking $\Sigma = \Sigma_0$ we obtain a "truncated" analogue of Corollary 8.13.

Corollary 8.15 *Let $(P_v) \in \prod_{v \notin \Sigma} \mathcal{X}(\mathcal{O}_v)$. Then the following conditions are equivalent.*

(a) *(P_v) is orthogonal to $\mathrm{Br}_{\lambda, \Sigma}(X)$.*
(b) *There exists a class $[c] \in H^1(\mathcal{O}_\Sigma, \mathcal{S})$ that goes to $([\mathcal{Y}](P_v))$ under the diagonal map*

$$H^1(\mathcal{O}_\Sigma, \mathcal{S}) \to \prod_{v \notin \Sigma} H^1(\mathcal{O}_v, \mathcal{S}).$$

(c) *There exists an $\mathcal{O}_\Sigma$-torsor c under $\mathcal{S}$ such that P_v lifts to a point in $\mathcal{Y}^c(\mathcal{O}_v)$ for every $v \notin \Sigma$.*

Thm. 1 of [19] is a particular case of this result.

Let X be a smooth and geometrically integral k-variety. Define

$$Б(X) := \ker[\mathrm{Br}_1(X)/\mathrm{Br}(k) \to \prod_{v \in \Omega_k} \mathrm{Br}_1(X_v)/\mathrm{Br}(k_v)],$$

where $X_v := X \times_k k_v$. For $\alpha \in \text{Б}(X)$ and $(P_v) \in X(\mathbf{A}_k)$ the image α_v of α in $\text{Br}_1(X_v)$ is constant for every place v, hence

$$i(\alpha) = \sum_{v \in \Omega_k} j_v(\alpha(P_v)) \in \mathbf{Q}/\mathbf{Z}$$

is well defined and does not depend on the choice of (P_v). Let us assume that $X(\mathbf{A}_k) \neq \emptyset$. Then we obtain a map $i : \text{Б}(X) \to \mathbf{Q}/\mathbf{Z}$. Note also that this assumption, by global class field theory, implies that the natural map $\text{Br}(k) \to \text{Br}(X)$ is injective. For a number field k we have $H^3(k, \bar{k}^*) = 0$, so we see from (8.5) that the map $r : \text{Br}_1(X) \to H^1(k, KD'(X))$ induces an isomorphism

$$\text{Br}_1(X)/\text{Br}(k) \tilde{\longrightarrow} H^1(k, KD'(X)).$$

If $\mathcal{C}$ is an object of $\mathcal{D}(k)$, and $i > 0$ we define

$$Ш^i(\mathcal{C}) = \ker[H^i(k, \mathcal{C}) \to \prod_{v \in \Omega_k} H^i(k_v, \mathcal{C})].$$

Thus we get an isomorphism $\text{Б}(X) \tilde{\longrightarrow} Ш^1(KD'(X))$, using which we obtain a map $i : Ш^1(KD'(X)) \to \mathbf{Q}/\mathbf{Z}$.

Let S be a k-group of multiplicative type. There is a perfect Poitou–Tate pairing of finite groups (cf. [11], Thm. 5.7)

$$\langle , \rangle_{PT} : Ш^2(S) \times Ш^1(\widehat{S}) \to \mathbf{Q}/\mathbf{Z}$$

defined as follows. Let $a \in Ш^1(\widehat{S})$ and $b \in Ш^2(S)$. By [23], Lemma III.1.16, the group $H^2(k, S)$ is the direct limit of the groups $H^2(U, \mathcal{S})$ where U runs over non-empty open subsets of $\text{Spec}\,(\mathcal{O}_k)$. Note that by taking a smaller U we can assume that S and $\widehat{S}$ extend to smooth U-group schemes $\mathcal{S}$ and $\widehat{\mathcal{S}}$, respectively. For U sufficiently small we can lift b to some $b_U \in H^2(U, \mathcal{S})$, and lift a to some $\tilde{a}_U \in H^1(U, \widehat{\mathcal{S}})$. For any object $\mathcal{C}$ of $\mathcal{D}(U)$ we have the hypercohomology groups with *compact support* $H^i_c(U, \mathcal{C})$, see Section 3 of [17] for definitions. By [22], Prop. II.2.3 (a) (see also [17], Sect. 3), since a is locally trivial everywhere, $\tilde{a}_U$ comes from some $a_U \in H^1_c(U, \widehat{\mathcal{S}})$ under the natural map

$$H^1_c(U, \widehat{\mathcal{S}}) \to H^1(U, \widehat{\mathcal{S}}).$$

Define $\langle b, a \rangle_{PT}$ as the cup-product $b_U \cup a_U \in H^3_c(U, \mathbf{G}_{m,U}) \simeq \mathbf{Q}/\mathbf{Z}$ (the last isomorphism comes from the trace map, see [22], Prop. II.2.6). It is not clear to us whether this definition of the Poitou–Tate pairing coincides with the classical definition in terms of cocycles, but we shall only use the fact that it leads to a perfect pairing.

Recall that the map $\partial : \mathrm{Hom}_k(\widehat{S}, KD'(X)) \to H^2(k, S)$ was defined in the exact sequence (8.2).

Theorem 8.16 *Let X be a smooth and geometrically integral variety over a number field k such that $X(\mathbf{A}_k) \neq \emptyset$. Let S be a k-group of multiplicative type, $\lambda \in \mathrm{Hom}_k(\widehat{S}, KD'(X))$ and $a \in Ш^1(\widehat{S})$. Then $\partial(\lambda) \in Ш^2(S)$, and we have*

$$\langle \partial(\lambda), a \rangle_{PT} = i(\lambda_*(a)).$$

Proof The image of $\partial(\lambda)$ in $H^2(X, S)$ is zero because (8.2) is a complex. The assumption $X(\mathbf{A}_k) \neq \emptyset$ implies that the map $H^2(k_v, S) \to H^2(X_v, S)$ is injective for every place v (cf. also Proposition 8.3). Therefore, we have $\partial(\lambda) \in Ш^2(S)$.

Recall that $w : KD'(X) \to \mathbf{G}_{m,k}[2]$ is the natural map defined in (8.1) for X/k. Since (8.2) is obtained by applying the functor $\mathrm{Hom}_k(\widehat{S}, .)$ to (8.1), under the canonical isomorphism $\mathrm{Hom}_k(\widehat{S}, \mathbf{G}_{m,k}[2]) = H^2(k, S)$ we have the equality $w \circ \lambda = \partial(\lambda)$. Let us write $\alpha = \lambda_*(a) \in Ш^1(KD'(X))$.

Let $U \subset \mathrm{Spec}\,(\mathcal{O}_k)$ be a sufficiently small non-empty open subset such that there exists a smooth U-scheme $\mathcal{X}$ with geometrically integral fibres and the generic fibre $X = \mathcal{X} \times_U k$, and a smooth U-group of multiplicative type $\mathcal{S}$ with the generic fibre $S = \mathcal{S} \times_U k$.

Write $w_U \in \mathrm{Hom}_U(KD'(\mathcal{X}), \mathbf{G}_{m,U}[2])$ for the map in the exact triangle (8.1) for $\mathcal{X}/U$. The passage to the generic point $\mathrm{Spec}\,(k)$ of U defines the restriction map

$$\mathrm{Hom}_U(\widehat{\mathcal{S}}, KD'(\mathcal{X})) \to \mathrm{Hom}_k(\widehat{S}, KD'(X)).$$

Consider the exact sequence (8.2) for $\mathcal{X}_V = \mathcal{X} \times_U V$ and $\mathcal{S}_V = \mathcal{S} \times_U V$, where $V \subset U$ is a non-empty open set, and also for X and S. We obtain a commutative diagram

$$\begin{array}{ccccccc} H^1(\mathcal{X}_V, \mathcal{S}_V) & \longrightarrow & \mathrm{Hom}_V(\widehat{\mathcal{S}}, KD'(\mathcal{X}_V)) & \longrightarrow & H^2(V, \mathcal{S}_V) & \longrightarrow & H^2(\mathcal{X}_V, \mathcal{S}_V) \\ \downarrow & & \downarrow & & \downarrow & & \downarrow \\ H^1(X, S) & \longrightarrow & \mathrm{Hom}_k(\widehat{S}, KD'(X)) & \longrightarrow & H^2(k, S) & \longrightarrow & H^2(X, S). \end{array}$$

Passing to the inductive limit over V and using [23], Lemma III.1.16, we deduce from this diagram a canonical surjective homomorphism

$$\varinjlim_V \mathrm{Hom}_V(\widehat{\mathcal{S}_V}, KD'(\mathcal{X}_V)) \to \mathrm{Hom}_k(\widehat{S}, KD'(X)).$$

Thus, by shrinking U, if necessary, we can lift λ to some $\lambda_U \in \mathrm{Hom}_U(\widehat{\mathcal{S}}, KD'(\mathcal{X}))$. Then

$$w_U \circ \lambda_U \in \mathrm{Hom}_U(\widehat{\mathcal{S}}, \mathbf{G}_{m,U}[2]) = H^2(U, \mathcal{S})$$

(see the proof of Proposition 8.1 for the equality here) goes to $\partial(\lambda)$ under the restriction map to $H^2(k, S)$.

As was explained above, we can lift $a \in Ш^1(\widehat{S})$ to some $a_U \in H^1_c(U, \widehat{\mathcal{S}})$. Write $\alpha_U = \lambda_{U*}(a_U)$. Then α_U is sent to α by the natural map

$$H^1_c(U, KD'(\mathcal{X})) \to H^1(k, KD'(X)).$$

By the remark before Proposition 8.1 we can use [18], Prop. 3.3, which gives

$$\begin{aligned} i(\lambda_*(a)) = i(\alpha) &= w_U \cup \alpha_U = w_{U*}(\alpha_U) = w_{U*}(\lambda_{U*}(a_U)) \\ &= (w_U \circ \lambda_U)_*(a_U) = (w_U \circ \lambda_U) \cup a_U. \end{aligned}$$

The above definition of the Poitou–Tate pairing shows that this equals $\langle \partial(\lambda), a \rangle_{PT}$. □

Remark This proof avoids delicate computations with cocycles as in [29], the proof of Thm. 6.1.2, which follows [8], Prop. 3.3.2.

Corollary 8.17 *Let X be a smooth and geometrically integral variety over a number field k such that $X(\mathbf{A}_k)^{Б(X)} \neq \emptyset$. Then the map*

$$\chi : H^1(X, S) \to \mathrm{Hom}_k(\widehat{S}, KD'(X))$$

is surjective (there exist X-torsors of every extended type). The converse is true when $\mathrm{Pic}(\overline{X})$ is a finitely generated abelian group.

Proof Let $\lambda \in \mathrm{Hom}_k(\widehat{S}, KD'(X))$. Since $X(\mathbf{A}_k)^{Б(X)} \neq \emptyset$, Theorem 8.16 ensures that $\langle \partial(\lambda), a \rangle_{PT} = 0$ for every $a \in Ш^1(\widehat{S})$. The non-degeneracy of the Poitou–Tate pairing implies that $\partial(\lambda) = 0$. By Proposition 8.1 this is equivalent to $\lambda \in \mathrm{Im}(\chi)$.

To prove the converse it is enough to show that $i : Б(X) \to \mathbf{Q}/\mathbf{Z}$ is the zero map. The formation of $Б(X)$ is functorial in X, so there is a natural restriction map $Б(X^c) \to Б(X)$. By [27], formula (6.1.4), this is an isomorphism. The map $i^c : Б(X^c) \to \mathbf{Q}/\mathbf{Z}$ is the composition

$$Б(X^c) \xrightarrow{\sim} Б(X) \xrightarrow{i} \mathbf{Q}/\mathbf{Z},$$

so it is enough to show that i^c is identically zero.

By functoriality of the exact sequence (8.2) we have a commutative diagram with exact rows

$$\begin{array}{ccccccc} H^1(X^c,S) & \xrightarrow{\chi^c} & \mathrm{Hom}_k(\widehat{S},\mathrm{Pic}(\overline{X}^c)) & \xrightarrow{\partial^c} & H^2(k,S) & \longrightarrow & H^2(X^c,S) \\ \downarrow & & \downarrow & & \| & & \downarrow \\ H^1(X,S) & \xrightarrow{\chi} & \mathrm{Hom}_k(\widehat{S},KD'(X)) & \xrightarrow{\partial} & H^2(k,S) & \longrightarrow & H^2(X,S)\,. \end{array}$$

The commutativity of this diagram implies that if χ is surjective, so that ∂ is zero, then ∂^c is also zero, hence χ^c is surjective. The assumption that $\mathrm{Pic}(\overline{X})$ is finitely generated implies that $\mathrm{Pic}(\overline{X}^c)$ is also finitely generated. Using [29], Prop. 6.1.4, we see that $X^c(\mathbf{A}_k)^{\mathrm{Б}(X^c)}$ is not empty, thus i^c is identically zero. □

See [30], Thm. 3.3.1, for miscellaneous characterisations of the property $X(\mathbf{A}_k)^{\mathrm{Б}(X)} \neq \emptyset$ in terms of the so called elementary obstruction and the generic period.

8.4 Application: existence of integral points, obstructions to strong approximation

Recall that for a finite set of places $\Sigma_0 \subset \Omega_k$ we denote by $\mathbf{A}_k^{\Sigma_0}$ the ring of k-adèles without v-components for $v \in \Sigma_0$. Let X be a smooth and geometrically integral k-variety such that $X(\mathbf{A}_k) \neq \emptyset$. There exists a finite set of places Σ containing $\Sigma_0 \cup \Omega_\infty$, and a faithfully flat morphism $\mathcal{X} \to \mathrm{Spec}\,(\mathcal{O}_\Sigma)$ such that $X = \mathcal{X} \times_{\mathcal{O}_\Sigma} k$. We shall say that X satisfies *strong approximation*[1] *outside* Σ_0 if $X(k)$ is dense in the restricted product $X(\mathbf{A}_k^{\Sigma_0})$ of the sets $X(k_v)$ for $v \notin \Sigma_0$ with respect to the subsets $\mathcal{X}(\mathcal{O}_v)$ (defined for $v \notin \Sigma$). The restricted product topology is called the *strong* topology. Explicitly, the base of open subsets of this topology consists of the sets

$$\prod_{v\in T} U_v \times \prod_{v\notin T} \mathcal{X}(\mathcal{O}_v),$$

where T is a finite subset of $\Omega_k \setminus \Sigma_0$ such that $\Sigma \subset T$, and U_v is an open subset of $X(k_v)$ for $v \in T$.

The following theorem gives sufficient conditions for "the Brauer–

[1] We adopt the convention that a variety X such that $X(\mathbf{A}_k) = \emptyset$ satisfies strong approximation outside Σ_0 for every Σ_0.

Manin obstruction to strong approximation outside Σ_0" to be the only obstruction on X.

Theorem 8.18 *Let X be a smooth and geometrically integral k-variety such that $X(\mathbf{A}_k) \neq \emptyset$, and let S be a k-group of multiplicative type. Let Σ_0 be a finite set of places of k. Assume that there exists an X-torsor Y under S with the following property:* For all k-torsors c under S, the twisted torsor Y^c has the strong approximation property outside Σ_0. *If $(P_v) \in X(\mathbf{A}_k)$ is orthogonal to* $\mathrm{Br}_\lambda(X)$, *then $(P_v)_{v \notin \Sigma_0}$ belongs to the closure of $X(k)$ in $X(\mathbf{A}_k^{\Sigma_0})$ for the strong topology.*

Proof Choose a finite set of places Σ containing $\Sigma_0 \cup \Omega_\infty$ such that X is the generic fibre of a flat smooth $\mathcal{O}_\Sigma$-scheme of finite type $\mathcal{X}$. We can also assume that the torsor $Y \to X$ extends to a torsor $\mathcal{Y} \to \mathcal{X}$ under a smooth $\mathcal{O}_\Sigma$-group scheme of multiplicative type $\mathcal{S}$ such that $S = \mathcal{S} \times_{\mathcal{O}_\Sigma} k$. Furthermore, we can assume that the adelic point $(P_v) \in X(\mathbf{A}_k)$ belongs to $\prod_{v \in \Sigma} X(k_v) \times \prod_{v \notin \Sigma} \mathcal{X}(\mathcal{O}_v)$. We want to find a rational point on X very close to P_v for $v \in (\Sigma - \Sigma_0)$ and integral outside Σ.

By Corollary 8.13, the property that (P_v) is orthogonal to $\mathrm{Br}_\lambda(X)$ implies that it can be lifted to an adelic point $(Q_v) \in \prod_{v \in \Sigma} Y^c(k_v) \times \prod_{v \notin \Sigma} \mathcal{Y}^c(\mathcal{O}_v)$ on some twisted torsor Y^c. In particular, $Y^c(\mathbf{A}_k) \neq \emptyset$. Since Y^c satisfies strong approximation outside Σ_0, we can find a rational point $m \in Y^c(k)$ very close to Q_v for $v \in (\Sigma - \Sigma_0)$ and integral outside Σ. Sending m to X produces a rational point $m' \in X(k)$ very close to P_v for $v \in (\Sigma - \Sigma_0)$ and integral outside Σ. □

The following corollary gives sufficient conditions for "the Brauer–Manin obstruction to the integral Hasse principle" to be the only obstruction.

Corollary 8.19 *Let $\mathcal{X}$ be a faithfully flat and separated scheme of finite type over $\mathcal{O}_k$ such that $X = \mathcal{X} \times_{\mathcal{O}_k} k$. Assume that Y^c has the strong approximation property outside Ω_∞ for every k-torsor c under S. If there exists an adelic point $(P_v) \in \prod_{v \in \Omega_k} \mathcal{X}(\mathcal{O}_v)$ orthogonal to* $\mathrm{Br}_\lambda(X)$, *then $\mathcal{X}(\mathcal{O}_k) \neq \emptyset$.*

Proof Theorem 8.18 says that (P_v) can be approximated by a rational point $m \in X(k)$ for the strong topology on $X(\mathbf{A}_k^{\Omega_\infty})$. Since $P_v \in \mathcal{X}(\mathcal{O}_v)$ for $v \in \Omega_f$, this implies that $m \in \mathcal{X}(\mathcal{O}_k)$. □

As an application of Theorem 8.18 we get a short proof of a result that already appeared in C. Demarche's thesis [10, Remark 4.8.2] (see also [7, Thm. 4.5], where there is an additional assumption that the geometric stabiliser $\overline{H}$ is finite).

Theorem 8.20 *Let G be a semi-simple, simply connected linear group over a number field k. Let Σ_0 be a finite set of places of k such that for every almost k-simple factor G_1 of G there exists a place $v \in \Sigma_0$ such that $G_1(k_v)$ is not compact (for example, if k is not totally real we can take $\Sigma_0 = \{v_0\}$, where v_0 is a complex place of k). Let X be a homogeneous space of G such that the geometric stabiliser $\overline{H}$ is a $\bar{k}$-group of multiplicative type. Then for every adelic point $(P_v)_{v \in \Omega_k}$ of X orthogonal to $\mathrm{Br}_1(X)$, the point $(P_v)_{v \notin \Sigma_0}$ is in the closure of $X(k)$ in $X(\mathbf{A}_k^{\Sigma_0})$ for the strong topology.*

In other words: the Brauer–Manin obstruction to strong approximation outside Σ_0 is the only one on X.

Proof Let us assume that G acts on X on the left. Then $\overline{X}$ with the left action of $\overline{G}$ is isomorphic to $\overline{G}/\overline{H}$. Since $\mathrm{Pic}(\overline{G}) = 0$ and $\bar{k}[G]^* = \bar{k}^*$, the abelian group $\mathrm{Pic}(\overline{X})$ is finitely generated, and $\bar{k}[X]^* = \bar{k}^*$. Now the existence of a point $(P_v)_{v \in \Omega_k}$ orthogonal to $\mathrm{Br}_1(X)$ implies that $X(k) \neq \emptyset$ by [29], Prop. 6.1.4, and [16], Prop 3.7 (3) and Example 3.4. Therefore X with the left action of G is isomorphic to $X = G/H$, where H is a k-group of multiplicative type. Taking $Y = G$, we obtain a right torsor $Y \to X$ under H such that for any k-torsor c under H the twist Y^c is a left k-torsor under G.

By the Hasse principle for semi-simple simply connected groups (a theorem of Kneser–Harder–Chernousov), $Y^c(\mathbf{A}_k) \neq \emptyset$ implies $Y^c(k) \neq \emptyset$, hence $Y^c \simeq G$. By the strong approximation theorem (see, for example, [26], Thm. 7.12), G satisfies strong approximation outside Σ_0. It remains to apply Theorem 8.18. □

Remarks

1. It is not clear to us whether Corollary 8.19 still holds if we only assume that all the twists Y^c satisfy the integral Hasse principle: indeed, we do not know in general whether the torsor $Y \to X$ can be extended to an fppf torsor $\mathcal{Y} \to \mathcal{X}$ over $\mathrm{Spec}\,(\mathcal{O}_k)$.
2. The assumptions of Theorem 8.18 and Corollary 8.19 imply that $\bar{k}[X]^* = \bar{k}^*$, that is, we are still in "the classical case" of descent theory. Indeed, if Y satisfies strong approximation outside a finite

set of places, then $\overline{Y}$ is simply connected (this was first observed in [24], Thm. 1, see also [15], Cor. 2.4). This implies $\bar{k}[Y]^* = \bar{k}^*$, and hence $\bar{k}[X]^* = \bar{k}^*$. (Otherwise pick up a function $f \in \bar{k}[Y]^*$ such that the image of f in the free abelian group $\bar{k}[Y]^*/\bar{k}^*$ is not divisible by a prime ℓ. Then the normalisation of $\overline{Y}$ in $\bar{k}(Y)(f^{1/\ell})$ is a connected étale covering of $\overline{Y}$ of degree ℓ.)

Appendix A

Let Z be an integral regular Noetherian scheme, and let $p : X \to Z$ be a smooth morphism of finite type with geometrically integral fibres. The goal of this appendix is to show that the object $\tau_{\leq 1}\mathbf{R}p_*\mathbf{G}_{m,X}$ of the derived category $\mathcal{D}(Z)$ of étale sheaves on Z can be represented by an explicit two-term complex. This links our $KD(X)$ and $KD'(X)$ with their analogues introduced in [18], Remark 2.4 (2).

Let $j : \eta = \mathrm{Spec}(k(X)) \hookrightarrow X$ be the inclusion of the generic point. Since X is regular, there is no difference between Weil and Cartier divisors, so we have the following exact sequence of sheaves on X, see [23], Examples II.3.9 and III.2.22:

$$0 \to \mathbf{G}_{m,X} \to j_*\mathbf{G}_{m,\eta} \to \mathrm{Div}_X \to 0,$$

where Div_X is the sheaf of divisors on X, that is, the sheaf associated to the presheaf such that the group of sections over an étale U/X is the group of divisors on U.

We call an irreducible effective divisor D on X *horizontal* if it is the Zariski closure of a divisor on the generic fibre of $p : X \to Z$. If $D = p^{-1}(D')$ for a divisor D' on Z, we call D *vertical.* The sheaf Div_X is the direct sum of sheaves

$$\mathrm{Div}_X = \mathrm{Div}_{X/Z} \oplus \mathrm{Div}_X^v,$$

where $\mathrm{Div}_{X/Z}$ is the subsheaf of horizontal divisors, and Div_X^v is the subsheaf of vertical divisors.

Define a subsheaf $K_{X/Z}^\times \subset j_*\mathbf{G}_{m,\eta}$ by the condition that the following diagram is commutative and has exact rows and columns:

$$\begin{array}{ccccccccc} & & & & 0 & & 0 & & \\ & & & & \downarrow & & \downarrow & & \\ 0 & \to & \mathbf{G}_{m,X} & \to & K^{\times}_{X/Z} & \to & \mathrm{Div}_{X/Z} & \to & 0 \\ & & \| & & \downarrow & & \downarrow & & \\ 0 & \to & \mathbf{G}_{m,X} & \to & j_*\mathbf{G}_{m,\eta} & \to & \mathrm{Div}_X & \to & 0 \\ & & & & \downarrow & & \downarrow & & \\ & & & & \mathrm{Div}^v_X & = & \mathrm{Div}^v_X & & \\ & & & & \downarrow & & \downarrow & & \\ & & & & 0 & & 0. & & \end{array}$$

The complex of étale sheaves on Z

$$p_*K^{\times}_{X/Z} \to p_* \mathrm{Div}_{X/Z},$$

after the shift by 1 to the left, is the complex $\mathcal{KD}(\mathcal{X})$ defined in [18], Remark 2.4 (2), see also the formulae on the bottom of page 538. There is a natural injective morphism $\mathbf{G}_{m,Z} \to p_*K^{\times}_{X/Z}$; the complex

$$p_*K^{\times}_{X/Z}/\mathbf{G}_{m,Z} \to p_* \mathrm{Div}_{X/Z}$$

was introduced in [18] and denoted there by $\mathcal{KD}'(\mathcal{X})$.

Proposition *The object $\tau_{\leq 1}\mathbf{R}p_*\mathbf{G}_{m,X}$ of the derived category of étale sheaves on Z is represented by the complex $p_*K^{\times}_{X/Z} \to p_* \mathrm{Div}_{X/Z}$.*

Proof The proof of Lemma 2.3 of [3] works in our situation. To complete the proof we only need to show that $R^1p_*K^{\times}_{X/Z} = 0$. Note that the canonical morphism $\mathrm{Div}_Z \to p_* \mathrm{Div}^v_X$ is an isomorphism because p is surjective with geometrically integral fibres. Now the exact sequence of sheaves on X

$$0 \to K^{\times}_{X/Z} \to j_*\mathbf{G}_{m,\eta} \to \mathrm{Div}^v_X \to 0$$

gives rise to the following exact sequence of sheaves on Z:

$$p_*j_*\mathbf{G}_{m,\eta} \to \mathrm{Div}_Z \to R^1p_*(K^{\times}_{X/Z}) \to R^1p_*(j_*\mathbf{G}_{m,\eta}).$$

Using the spectral sequence of the composition of functors $\mathbf{R}p_*$ and $\mathbf{R}j_*$ we see that the sheaf $R^1p_*(j_*\mathbf{G}_{m,\eta})$ has a canonical embedding into $R^1(pj)_*\mathbf{G}_{m,\eta}$. The latter sheaf is zero by Grothendieck's version of Hilbert's Theorem 90.

It remains to prove the surjectivity of $(pj)_*\mathbf{G}_{m,\eta} \to \mathrm{Div}_Z$, which is enough to check at the stalk at any geometric point of Z. But locally every divisor on Z is the divisor of a function, since Z is regular. This completes the proof. □

Remark In this appendix we worked over the small étale site of Z. Applying our arguments to an arbitrary smooth scheme of finite type S/Z one shows that the same results remain true for the smooth site Sm/Z used in [18].

Appendix B

In this appendix k is a field of characteristic zero and $p : X \to \operatorname{Spec} k$ is a geometrically integral k-variety. We consider a Galois module A over $\operatorname{Gal}(\bar{k}/k)$ and an étale sheaf $\mathcal{F}$ on X.

The functor $\mathbf{R}\mathrm{Hom}_X(p^*A, \cdot) : \mathcal{D}(X) \to \mathcal{D}(\mathrm{Ab})$ is the composition of functors $\mathbf{R}p_* : \mathcal{D}(X) \to \mathcal{D}(k)$ and $\mathbf{R}\mathrm{Hom}_k(A, \cdot) : \mathcal{D}(k) \to \mathcal{D}(\mathrm{Ab})$, hence we have

$$\mathbf{R}\mathrm{Hom}_X(p^*A, \mathcal{F}) = \mathbf{R}\mathrm{Hom}_k(A, \mathbf{R}p_*\mathcal{F}).$$

Explicitly, this isomorphism associates to $p^*A \to \mathcal{F}$ the composition

$$A \to \mathbf{R}p_*(p^*A) \to \mathbf{R}p_*\mathcal{F},$$

where the first map is the canonical adjunction morphism. The inverse associates to $A \to \mathbf{R}p_*\mathcal{F}$ the composition

$$p^*A \to p^*(\mathbf{R}p_*\mathcal{F}) \to \mathcal{F},$$

where the last map is the second canonical adjunction morphism.

Let us now complete the proof of Proposition 8.9 (iii). To give an equivalence class of the extension of sheaves on X

$$0 \to \mathcal{F} \to \mathcal{E} \to p^*A \to 0 \tag{8.14}$$

is the same as to give a morphism $p^*A \to \mathcal{F}[1]$ in the derived category $\mathcal{D}(X)$. By the above, to this morphism we associate the composition

$$A \to \mathbf{R}p_*p^*A \to \mathbf{R}p_*\mathcal{F}[1].$$

Since A is a one-term complex concentrated in degree 0 this composition comes from a morphism $\alpha : A \to (\tau_{\leq 1}\mathbf{R}p_*\mathcal{F})[1]$ in $\mathcal{D}(k)$. By taking the 0th cohomology we obtain a homomorphism $\beta : A \to R^1p_*\mathcal{F}$ of discrete Galois modules. Clearly, β is the composition of the canonical map $A \to p_*p^*A$ with the differential in the long exact sequence of cohomology attached to (8.14):

$$0 \to p_*\mathcal{F} \to p_*\mathcal{E} \to p_*p^*A \to R^1p_*\mathcal{F}.$$

To finish the proof of (iii) we need to show that β can also be obtained through the spectral sequence, that is, as the image of the class of (8.14) under the right arrow in (8.9). But (8.9) is obtained by applying $\mathbf{R}\mathrm{Hom}_k(A, \cdot)$ to the exact triangle

$$(p_*\mathcal{F})[1] \to (\tau_{\leq 1}\mathbf{R}p_*\mathcal{F})[1] \to R^1p_*\mathcal{F}. \tag{8.15}$$

By definition, β is the composition of α with the right map in (8.15), so the proof of (iii) is now complete.

Let us complete the proof of Proposition 8.9 (iv). The exact triangle (8.15) gives rise to the exact sequence of abelian groups

$$0 \to \mathrm{Hom}_k(A, (p_*\mathcal{F})[1]) \to \mathrm{Hom}_k(A, (\tau_{\leq 1}\mathbf{R}p_*\mathcal{F})[1]) \to \mathrm{Hom}_k(A, R^1p_*\mathcal{F}),$$

which is the same as (8.9). Since the right arrow here sends α to β, we see that if $\beta = 0$, then α comes from a morphism $A \to (p_*\mathcal{F})[1]$. Hence the class of (8.14) comes from the class of an extension of A by $p_*\mathcal{F}$, say

$$0 \to p_*\mathcal{F} \to B \to A \to 0, \tag{8.16}$$

in the sense that (8.14) is the push-out of

$$0 \to p^*p_*\mathcal{F} \to p^*B \to p^*A \to 0$$

by the adjunction map $p^*p_*\mathcal{F} \to \mathcal{F}$. Therefore, by the description of the adjunction isomorphism and its inverse given above, applying p_* to (8.14), and pulling back the resulting short exact sequence via the adjunction map $A \to \mathrm{Ker}[p_*p^*A \to R^1p_*F]$ (this makes sense when $\beta = 0$) gives back the extension (8.16). □

Acknowledgement We would like to thank the organisers of the workshop on anabelian geometry at the Isaac Newton Institute for Mathematical Sciences in Cambridge (August, 2009) where the work on this paper began.

References

[1] M. Borovoi. Abelianization of the second nonabelian Galois cohomology. *Duke Math. J.* **72** (1993) 217–239.

[2] M. Borovoi, C. Demarche et D. Harari. Complexes de groupes de type multiplicatif et groupe de Brauer nom-ramifié des espaces homogènes. *Ann. Sci. École Norm. Sup.*, to appear. arXiv:1203.5964

[3] M. Borovoi and J. van Hamel. Extended Picard complexes and linear algebraic groups. *J. reine angew. Math.* **627** (2009) 53–82.

[4] S. Bosch, W. Lütkebohmert and M. Raynaud. *Néron models*. Ergebnisse der Mathematik und ihrer Grenzgebiete, Springer-Verlag, 1990.

[5] L. Breen. On a nontrivial higher extension of representable abelian sheaves. *Bull. Amer. Math. Soc.* **75** (1969) 1249–1253.

[6] L. Breen. Un théorème d'annulation pour certains Ext^i de faisceaux abéliens. *Ann. Sci. École Norm. Sup.* **8** (1975) 339–352.

[7] J-L. Colliot-Thélène and Xu Fei. Brauer–Manin obstruction for integral points of homogeneous spaces and representation of integral quadratic forms. *Comp. Math.* **145** (2009) 309–363.

[8] J-L. Colliot-Thélène et J-J. Sansuc. La descente sur les variétés rationnelles, II. *Duke Math. J.* **54** (1987) 375–492.

[9] J-L. Colliot-Thélène et O. Wittenberg. Groupe de Brauer et points entiers de deux familles de surfaces cubiques affines. *Amer. J. Math.* **134** (2012) 1303–1327.

[10] C. Demarche. *Méthodes cohomologiques pour l'étude des points rationnels sur les espaces homogènes*. Thèse de l'Université Paris-Sud, 2009. Available at http://www.math.u-psud.fr/~demarche/thesedemarche.pdf

[11] C. Demarche. Suites de Poitou–Tate pour les complexes de tores à deux termes. *I.M.R.N.* (2011) 135–174.

[12] C. Demarche. Le défaut d'approximation forte dans les groupes linéaires connexes. *Proc. London Math. Soc.* **102** (2011) 563–597.

[13] P. Gille and A. Pianzola. Isotriviality and étale cohomology of Laurent polynomial rings. *J. Pure Appl. Algebra* **212** (2008) 780–800.

[14] A. Grothendieck et al. *Schémas en Groupes* (SGA 3), II, Lecture Notes in Math. **151**, Springer-Verlag, 1970.

[15] D. Harari. Weak approximation and non-abelian fundamental groups. *Ann. Sci. École Norm. Sup.* **33** (2000) 467–484.

[16] D. Harari and A. N. Skorobogatov. Non-abelian cohomology and rational points. *Comp. Math.* **130** (2002) 241–273.

[17] D. Harari and T. Szamuely. Arithmetic duality theorems for 1-motives. *J. reine angew. Math.* **578** (2005) 93–128.

[18] D. Harari and T. Szamuely. Local-global principles for 1-motives. *Duke Math. J.* **143** (2008) 531–557.

[19] D. Harari and J.F. Voloch. The Brauer–Manin obstruction for integral points on curves. *Math. Proc. Cambridge Philos. Soc.* **149** (2010) 413–421.

[20] G. Harder. Halbeinfache Gruppenschemata über Dedekindringen. *Inv. Math.* **4** (1967) 165–191.
[21] A. Kresch and Yu. Tschinkel. Two examples of Brauer–Manin obstruction to integral points. *Bull. London Math. Soc.* **40** (2008) 995–1001.
[22] J.S. Milne. *Arithmetic duality theorems*, Academic Press, 1986.
[23] J.S. Milne. *Étale cohomology*, Princeton University Press, 1980.
[24] Kh. P. Minchev. Strong approximation for varieties over algebraic number fields. *Dokl. Akad. Nauk BSSR* **33** (1989) 5–8. (Russian)
[25] D. Mumford. *Abelian varieties*, Oxford University Press, 1970.
[26] V. Platonov and A. Rapinchuk. *Algebraic groups and number theory*, Academic Press, 1994.
[27] J-J. Sansuc. Groupe de Brauer et arithmétique des groupes algébriques linéaires sur un corps de nombres. *J. reine angew. Math.* **327** (1981) 12–80.
[28] A.N. Skorobogatov. Beyond the Manin obstruction. *Inv. Math.* **135** (1999) 399–424.
[29] A.N. Skorobogatov. *Torsors and rational points*, Cambridge Tracts in Mathematics **144**, Cambridge University Press, 2001.
[30] O. Wittenberg. On Albanese torsors and the elementary obstruction. *Math. Ann.* **340** (2008) 805–838.

9

Homotopy obstructions to rational points

Yonatan Harpaz
Einstein Institute of Mathematics, The Hebrew University of Jerusalem

Tomer M. Schlank
Einstein Institute of Mathematics, The Hebrew University of Jerusalem

Abstract In this paper we propose to use a relative variant of the notion of the étale homotopy type of an algebraic variety in order to study the existence of rational points on it. In particular, we use an appropriate notion of homotopy fixed points in order to construct obstructions to the local-global principle. The main results in this paper are the connections between these obstructions and the classical obstructions, such as the Brauer-Manin, the étale-Brauer and certain descent obstructions. These connections allow one to understand the various classical obstructions in a unified framework.

A note about the structure of this paper. The reader can think of this paper as consisting of three parts:
The **introduction and definitions** are found in sections 9.1 to 9.3. **Galois homotopy fixed points** are the focus of sections 9.4 to 9.7. The remaining sections, 9.8 to 9.12, are devoted to a **comparison of classical and homotopy obstructions**. See below for a more detailed breakdown of the contents.

Chapter contents

9.1 Introduction
9.2 The étale homotopy type and its relative version
9.3 The obstructions
9.4 Homotopy fixed points for pro-finite groups

9.5 The finite pre-image theorem
9.6 Sections and homotopy fixed points
9.7 Homotopy fixed point sets and pro-isomorphisms
9.8 Varieties of dimension zero
9.9 Connection to finite descent
9.10 The equivalence of the homology obstruction and the Brauer-Manin obstruction
9.11 The equivalence of the homotopy obstruction and the étale Brauer obstruction
9.12 Applications

9.1 Introduction

9.1.1 Obstructions to the local global principle - overview

Let X be a smooth variety over a number field K. A prominent problem in arithmetic algebraic geometry is to understand the set $X(K)$. For example, one would like to be able to know whether $X(K) \neq \emptyset$. As a first approximation one can consider the set

$$X(K) \subseteq X(\mathbb{A})$$

where $\mathbb{A}$ is the ring of adeles of K.

It is a classical theorem of Minkowski and Hasse that if $X \subseteq \mathbb{P}^n$ is hypersurface given by one quadratic equation then

$$X(\mathbb{A}) \neq \emptyset \Rightarrow X(K) \neq \emptyset.$$

When a variety X satisfies this property we say that it satisfies the local-global principle. In the 1940's Lind and Reichardt ([Lin40], [Rei42]) gave examples of genus 1 curves that do not satisfy the local-global principle.

More counterexamples to the local-global principle where given throughout the years until in 1971 Manin ([Man70]) described a general obstruction to the local-global principle that explained all the examples that were known to that date. The obstruction (known as the Brauer-Manin obstruction) is defined by considering a set $X(\mathbb{A})^{\mathrm{Br}}$ which satisfies

$$X(K) \subseteq X(\mathbb{A})^{\mathrm{Br}} \subseteq X(\mathbb{A}).$$

If X is a counterexample to the local-global principle we say that it is accounted for or explained by the Brauer-Manin obstruction if

$$\emptyset = X(\mathbb{A})^{\mathrm{Br}} \subseteq X(\mathbb{A}) \neq \emptyset.$$

In 1999 Skorobogatov ([Sko99]) defined a refinement of the Brauer-Manin obstruction (also known as the étale-Brauer-Manin obstruction) and used it to give an example of a variety X for which

$$\emptyset = X(K) \subseteq X(\mathbb{A})^{\mathrm{Br}} \neq \emptyset.$$

More precisely, Skorobogatov described a new intermediate set

$$X(K) \subseteq X(\mathbb{A})^{fin,\mathrm{Br}} \subseteq X(\mathbb{A})^{\mathrm{Br}} \subseteq X(\mathbb{A})$$

and found a variety X such that

$$\emptyset = X(\mathbb{A})^{fin,\mathrm{Br}} \subseteq X(\mathbb{A})^{\mathrm{Br}} \neq \emptyset.$$

In his paper from 2008 (published in 2010), Poonen ([Poo10]) constructed the first and currently only known example of a variety X such that

$$\emptyset = X(K) \subseteq X(\mathbb{A})^{fin,\mathrm{Br}} \neq \emptyset.$$

In 2009 the second author ([Sch09]) showed that in some cases Poonen's counter-example can be explained by showing that some smaller set $X(\mathbb{A})^{fin,\mathrm{Br}\sim D}$ is empty.

A different approach to define obstructions sets of the form

$$X(K) \subseteq X(\mathbb{A})^{obs} \subseteq X(\mathbb{A})$$

is by using descent on torsors over X under linear algebraic groups. This method was studied by Colliot-Thélène and Sansuc ([CTS80] and [CTS87]) for torsors under groups of multiplicative type and by Harari and Skorobogatov ([HSk02]) in the general non-abelian case (see also [Sko01]).

One can define numerous variants of the descent obstruction by considering only torsors under a certain class of groups. We shall denote by $X(\mathbb{A})^{desc}$, $X(\mathbb{A})^{fin}$, $X(\mathbb{A})^{fin-ab}$ and $X(\mathbb{A})^{con}$ the obstructions obtained when considering all, only finite, only finite abelian and only connected linear algebraic groups respectively.

In the case where X is projective Harari ([Har02]) showed that

$$X(\mathbb{A})^{\mathrm{Br}} = X(\mathbb{A})^{con}.$$

Lately, building on this work, Skorobogatov ([Sko09]) and Demarche

([De09a]) showed that in this case one also has

$$X(\mathbb{A})^{fin,\mathrm{Br}} = X(\mathbb{A})^{desc}.$$

9.1.2 Our results

In this paper we use a new method in order to construct natural intermediate sets between $X(K)$ and $X(\mathbb{A})$. This method uses a relative variant of the **étale homotopy type** $\acute{E}t(X)$ of X which was constructed by Artin and Mazur [AMa69].

This variant, denoted by $\acute{E}t_{/K}(X)$, is an inverse system of simplicial sets which carry an action of the absolute Galois group Γ_K of K. We then use an appropriate notion of **homotopy fixed points** to define a (functorial) set $X(hK)$ which serves as a certain homotopical approximation of the set $X(K)$ of rational points. In fact one obtains a natural map

$$h : X(K) \longrightarrow X(hK).$$

In order to apply this idea to the theory of obstructions to the local-global principle one proceeds to construct an adelic analogue, $X(h\mathbb{A})$, which serves as an approximation to the adelic points in X. One then obtains a commutative diagram of sets

$$\begin{array}{ccc} X(K) & \xrightarrow{h} & X(hK) \\ \downarrow{\scriptstyle \mathrm{loc}} & & \downarrow{\scriptstyle \mathrm{loc}_h} \\ X(\mathbb{A}) & \xrightarrow{h} & X(h\mathbb{A}). \end{array}$$

We then define $X(\mathbb{A})^h$ to be the set of adelic points whose corresponding adelic homotopy fixed point is rational, i.e. is in the image of loc_h. This set is intermediate in the sense that

$$X(K) \subseteq X(\mathbb{A})^h \subseteq X(\mathbb{A})$$

and so provides an obstruction to the existence of rational points. By using a close variant of this construction (essentially working with homology instead of homotopy) we define a set $X(\mathbb{A})^{\mathbb{Z}h}$ that satisfies

$$X(K) \subseteq X(\mathbb{A})^h \subseteq X(\mathbb{A})^{\mathbb{Z}h} \subseteq X(\mathbb{A}).$$

A second variant consists of replacing $\acute{E}t_{/K}(X)$ with its n'th Postnikov piece (in the appropriate sense) yielding "bounded" versions of the obstruction above denoted by

$$X(K) \subseteq X(\mathbb{A})^{h,n} \subseteq X(\mathbb{A})^{\mathbb{Z}h,n} \subseteq X(\mathbb{A}).$$

To conclude we get the following diagram of inclusions of obstruction sets:

$$\begin{array}{ccccccccc} & & X(\mathbb{A})^{\mathbb{Z}h} & \hookrightarrow \ldots \hookrightarrow & X(\mathbb{A})^{\mathbb{Z}h,2} & \hookrightarrow & X(\mathbb{A})^{\mathbb{Z}h,1} & \hookrightarrow & X(\mathbb{A}) \\ & & \uparrow & & \uparrow & & \uparrow & & \\ X(K) & \hookrightarrow & X(\mathbb{A})^{h} & \hookrightarrow \ldots \hookrightarrow & X(\mathbb{A})^{h,2} & \hookrightarrow & X(\mathbb{A})^{h,1}. & & \end{array}$$

The main results of this paper (Theorems 9.103, 9.116, 9.136 and Corollary 9.61) describe these obstructions in terms of previously constructed obstructions:

Theorem 9.1 *Let X be smooth geometrically connected variety over a number field K. Then*

$$X(\mathbb{A})^{h} = X(\mathbb{A})^{fin,Br},$$

$$X(\mathbb{A})^{\mathbb{Z}h} = X(\mathbb{A})^{Br},$$

$$X(\mathbb{A})^{h,1} = X(\mathbb{A})^{fin},$$

$$X(\mathbb{A})^{\mathbb{Z}h,1} = X(\mathbb{A})^{fin-ab}.$$

Furthermore, for every $n \geq 2$

$$X(\mathbb{A})^{h,n} = X(\mathbb{A})^{h},$$

$$X(\mathbb{A})^{\mathbb{Z}h,n} = X(\mathbb{A})^{\mathbb{Z}h}.$$

In particular, the diagram above is equal to the diagram

$$\begin{array}{ccccccccc} & & X(\mathbb{A})^{\mathrm{Br}} & \hookrightarrow \ldots \hookrightarrow & X(\mathbb{A})^{\mathrm{Br}} & \hookrightarrow & X(\mathbb{A})^{fin-ab} & \hookrightarrow & X(\mathbb{A}) \\ & & \uparrow & & \uparrow & & \uparrow & & \\ X(K) & \hookrightarrow & X(\mathbb{A})^{fin,\mathrm{Br}} & \hookrightarrow \ldots \hookrightarrow & X(\mathbb{A})^{fin,\mathrm{Br}} & \hookrightarrow & X(\mathbb{A})^{fin}. & & \end{array}$$

This homotopical description of the classical obstructions can be used to relate them in new ways to each other. For example one gets the following consequences:

Corollary 9.2 (Theorem 9.148) *Let K be number field and X a smooth geometrically connected K-variety. Assume further that $\pi_2^{ét}(\overline{X}) = 0$ (which is true, for example, when X is a curve such that $\overline{X} \neq \mathbb{P}^1$). Then*

$$X(\mathbb{A})^{fin} = X(\mathbb{A})^{fin,\mathrm{Br}}.$$

Corollary 9.3 (Theorem 9.147) *Let K be number field and X, Y be two smooth geometrically connected K-varieties, then*

$$(X \times Y)(\mathbb{A})^{fin,\mathrm{Br}} = X(\mathbb{A})^{fin,\mathrm{Br}} \times Y(\mathbb{A})^{fin,\mathrm{Br}}.$$

When studying homotopy fixed points for pro-finite groups we rely heavily on [Goe95] who defines for a profinite group Γ and a simplicial set $\mathbf{X}$ with continuous Γ-action the notion of a homotopy fixed points space $\mathbf{X}^{h\Gamma}$.

Given a field K and a simplicial set $\mathbf{X}$ with continuous Γ_K-action we define a suitable notion of "adelic homotopy fixed points space" $\mathbf{X}^{h\mathbb{A}}$. An additional result we obtain is a generalization of the finiteness of the Tate-Shafarevich groups for finite Galois modules.

Proposition 9.4 *Let K be a number field and let $\mathbf{X}$ be an excellent finite bounded simplicial Γ_K-set. Then the map*

$$\mathrm{loc}_{\mathbf{X}} : \pi_0\left(\mathbf{X}^{h\Gamma_K}\right) \longrightarrow \pi_0\left(\mathbf{X}^{h\mathbb{A}}\right)$$

has finite pre-images, i.e. for every $(x_\nu) \in \mathbf{X}(h\mathbb{A})$ the set $\mathrm{loc}_{\mathbf{X}}{}^{-1}((x_\nu))$ is finite.

Shortly after we published on the ArXiv website a first draft of this paper, Ambrus Pál published a paper with some similar ideas ([Pal10]). In his paper, Pál also uses the étale homotopy type to study rational points on varieties, but takes a slightly different approach.

The paper is divided into three parts. In the first part we give the basic definitions and describe the various obstructions. The first part includes §9.2, where we recall the notion of the étale homotopy type $\acute{E}t(X)$ of X and construct the variations used later for the obstructions, and §9.3 where we recall the definition of the classical obstructions and define our various obstructions $X(\mathbb{A})^{h,n}, X(\mathbb{A})^{\mathbb{Z}h,n}$ using $\acute{E}t^{\natural}_{/K}(X)$ and an appropriate notion of homotopy fixed points.

In the second part of the paper we prove the technical results regarding Galois homotopy fixed points that we will use later to relate different obstructions to each other. As we have already mentioned the fact that we don't use a model structure on pro-spaces has some technical disadvantages that will be addressed in a future work. This part is the one most influenced by this fact. The second part contains 4 sections.

In §9.4 we study methods to work with homotopy fixed points under pro-finite groups. In §9.5 we prove that in order to study the obstructions obtained in this method it is enough to study the obstructions obtained from each space in the diagram $\acute{E}t^{\natural}_{/K}(X)$ separately. This will prove very useful in third part.

In 9.6 we relate the homotopical obstructions we have defined to Grothendieck's section obstruction and to Pál's point of view presented in [Pal10]. In §9.7 we show that our notion of the homotopy fixed points set is stable under a certain type of pro-isomorphisms. This section will be redundant when we work in suitable model structure.

Finally in the third part of paper we relate the homotopy obstructions to classical obstructions. In §9.8 we give basic results for varieties of dimension zero and non-connected varieties as well as analyzing the obstructions $X(\mathbb{A})^{h,0}$ and $X(\mathbb{A})^{\mathbb{Z}h,0}$. In §9.9 the equivalence between finite (finite-abelian) descent and $X(\mathbb{A})^{h,1}$ ($X(\mathbb{A})^{\mathbb{Z}h,1}$) is proven for smooth geometrically connected varieties. In §9.10 we prove (under the same assumptions on X) that the Brauer-Manin obstruction is equivalent to $X(\mathbb{A})^{\mathbb{Z}h}$. In §9.11 we prove that the étale-Brauer-Manin obstruction is equivalent to $X(\mathbb{A})^{h}$ (again for smooth geometrically connected X). Finally in §9.12 we give some applications of the theory developed throughout the paper.

We would like to thank our advisors D. Kazhdan, E. Farjoun and E. De-Shalit for their essential guidance. We would also like to thank J.-L. Colliot-Thélène, A. Skorobogatov, H. Fausk, D. Isaksen and J. Milne for useful discussions. We would also like to thank K. Česnavičius for alerting us to an error in the proof of Lemma 9.40 which we have corrected in the current version.

We would also like to thank the anonymous referee for the very useful comments and detailed corrections, which we found very constructive and helpful to improve our manuscript.

Part of the research presented here was done while the second author was staying at the "Diophantine Equations" trimester program at Hausdorff Institute in Bonn. Another part was done while both authors were visiting MIT. We would like to thank both hosts for their hospitality and excellent working conditions. The second author is supported by the Hoffman Leadership Program.

9.2 The étale homotopy type and its relative version

In this paper we will only be interested in (smooth) **varieties** over fields. By this we mean (smooth) reduced separated schemes of finite type over a field K. In this section we do not assume any restriction on the field K. We denote the category of smooth varieties over K by Var/K. We will also refer to them as smooth K-varieties.

Remark 9.5 In this paper we deal a lot with both algebraic varieties and simplicial sets. In order to distinguish in notation we will use regular font (e.g. X, Y, etc.) to denote algebraic varieties and bold font (e.g. $\mathbf{X}, \mathbf{Y}$, etc.) to denote simplicial sets.

This section is partitioned as follows. In §§9.2.1 we will discuss the notions of skeleton and coskeleton and use them to define the notion of a hypercovering. In §§9.2.2 we will describe the classical construction of the étale homotopy type as well as the construction of a **relative** variant which is the main object of interest in this paper. In §§9.2.4 we will give some computational tools to help understand the homotopy type of the spaces we shall encounter.

9.2.1 Hypercoverings

In this section we shall recall the notion of a **hypercovering** which is used in the definition of the étale homotopy type. For a more detailed treatment of the subject we refer the reader to §8 in [AMa69] or to [Fri82]. We start with a discussion of the notions of skeleton and coskeleton.

Skeleton and coskeleton

For every $n \geq -1$ we will consider the full subcategory $\Delta_{\leq n} \subseteq \Delta$ spanned by the objects $\{[i] \in Ob\Delta | i \leq n\}$. The natural inclusion $\Delta_{\leq n} \hookrightarrow \Delta$ induces a truncation functor

$$\mathrm{tr}_n : \mathrm{Set}^{\Delta^{op}} \longrightarrow \mathrm{Set}^{\Delta^{op}_{\leq n}}$$

that takes a simplicial set and ignores the simplices of degree $> n$.

This functor clearly commutes with both limits and colimits so it has a left adjoint, given by left Kan extension

$$\mathrm{sk}_n : \mathrm{Set}^{\Delta^{op}_{\leq n}} \longrightarrow \mathrm{Set}^{\Delta^{op}}$$

also called the **n-skeleton**, and a right adjoint, given by right Kan extension

$$\mathrm{cosk}_n : \mathrm{Set}^{\Delta^{op}_{\leq n}} \longrightarrow \mathrm{Set}^{\Delta^{op}}$$

called the **n-coskeleton**. To conclude, we have the two adjunctions

$$\mathrm{sk}_n \dashv \mathrm{tr}_n \dashv \mathrm{cosk}_n.$$

The n-skeleton produces a simplicial set that is freely filled with degenerate simplices above degree n.

Definition 9.6 We will use the following notation

$$Q_n = \mathrm{sk}_{n+1} \circ \mathrm{tr}_{n+1} : \mathrm{Set}^{\Delta^{op}} \longrightarrow \mathrm{Set}^{\Delta^{op}}$$

and

$$P_n = \mathrm{cosk}_{n+1} \circ \mathrm{tr}_{n+1} : \mathrm{Set}^{\Delta^{op}} \longrightarrow \mathrm{Set}^{\Delta^{op}}$$

for the composite functors. Now these two functors satisfy the adjunction

$$(Q_n \dashv P_n) : \mathrm{Set}^{\Delta^{op}} \longrightarrow \mathrm{Set}^{\Delta^{op}}.$$

One of the important roles of P_n is that if $\mathbf{X}$ is a Kan simplicial set then $P_n(\mathbf{X})$ is its nth Postnikov piece, i.e. for $k > n$ we have

$$\pi_k(P_n(\mathbf{X})) = 0$$

and for $k \leq n$ the natural map $\mathbf{X} \to P_n(\mathbf{X})$ induces an isomorphism $\pi_k(\mathbf{X}) \longrightarrow \pi_k(P_n(\mathbf{X}))$.

Now suppose we replace the category Set of sets with an arbitrary category C. Then if C has finite colimits one has the functor

$$Q_n : C^{\Delta^{op}} \longrightarrow C^{\Delta^{op}}$$

defined in an analogous way and if C has finite limits then one has the functor

$$P_n : C^{\Delta^{op}} \longrightarrow C^{\Delta^{op}}.$$

The skeleton and coskelaton constructions are very useful and shall be used repeatedly throughout this paper. One useful application is the construction of a contractible space with free group action.

Definition 9.7 Let G be a finite group. We have an action of G on itself by multiplication on the left. Now consider G as a functor:

$$* = \Delta^{op}_{\leq 0} \longrightarrow \mathrm{Set}_G.$$

Define

$$\mathbf{E}G = \text{cosk}(G) \in \text{Set}_G^{\Delta^{op}}.$$

We can write an explicit description of this simplicial G-set by

$$\mathbf{E}G_n = G^{n+1}.$$

Note that the action of G on $\mathbf{E}G$ is free and and that $\mathbf{E}G$ is contractible.

Definition 9.8 Let G be a finite group. Define

$$\mathbf{B}G = \mathbf{E}G/G.$$

Note that $\mathbf{B}G$ is a connected and satisfies

$$\pi_1(\mathbf{B}G) = G.$$

Furthermore it can be checked that the natural map

$$\mathbf{B}G \longrightarrow P_1(\mathbf{B}G)$$

is an isomorphism of simplicial sets.

Hypercoverings

We will apply these concepts for the case of C being the étale site of an algebraic K-variety X.

Definition 9.9 Let X be a K-variety. A **hypercovering** $\mathcal{U}_\bullet \longrightarrow X$ is a simplicial object in the étale site over X satisfying the following conditions:

1. $\mathcal{U}_0 \longrightarrow X$ is a covering in the étale topology.
2. For every n, the canonical map

$$\mathcal{U}_{n+1} \longrightarrow (\text{cosk}_n(\text{tr}_n(\mathcal{U}_\bullet)))_{n+1}$$

   is a covering in the étale topology.

Example 9.10 The most common and simple kind of hypercoverings are those defined through the classical Čech resolution:

Definition 9.11 Let X be a K-variety and $Y \longrightarrow X$ an étale covering of X. Considering Y as a functor from $\Delta^{\leq 0}$ to the étale site of X we will define the Čech hypercovering of Y to be

$$\check{Y}_\bullet = \text{cosk}_0(Y).$$

As above we can write an explicit description of this hypercovering by

$$\check{Y}_n = \overbrace{Y \times_X \ldots \times_X Y}^{n+1}.$$

9.2.2 The étale homotopy type

We shall start our discussion by recalling the definition of the étale homotopy type functor as defined by Artin and Mazur in [AMa69]. We will then describe a variant of this construction which we will use for the rest of the paper.

Both the construction of the étale homotopy type and its variant use categories enriched over simplicial sets. In all cases the enrichment is defined in a very similar way, so it's worthwhile to describe the general pattern.

Suppose that C is an ordinary category which admits finite coproducts. Given a finite set A and an object $X \in C$ we will denote by

$$A \otimes X = \coprod_{a \in A} X$$

the coproduct of copies of P indexed by A. Note that $- \otimes -$ is a functor from $\text{Set} \times C$ to C (where Set is the category of sets) and that $B \otimes (A \otimes X)$ is naturally isomorphic to $(B \times A) \otimes X$.

Let $C^{\Delta^{op}}$ denote the category of simplicial objects in C and $\text{Set}^{\Delta^{op}}$ the category of simplicial sets. Given a simplicial set $\mathbf{S}_\bullet \in \text{Set}^{\Delta^{op}}$ and an object $X_\bullet \in C^{\Delta^{op}}$ we will denote by $\mathbf{S}_\bullet \otimes X_\bullet$ the object given by

$$(\mathbf{S}_\bullet \otimes X_\bullet)_n = \mathbf{S}_n \otimes X_n.$$

Now for every two objects $X_\bullet, Y_\bullet \in C^{\Delta^{op}}$ one can define a mapping simplicial set $\text{Map}(X_\bullet, Y_\bullet) \in \text{Set}^{\Delta^{op}}$ by the formula

$$\text{Map}(X_\bullet, Y_\bullet)_n = \text{Hom}_{C^{\Delta^{op}}}(\Delta^n \otimes X_\bullet, Y_\bullet).$$

There is a natural way to define composition of these mapping simplicial sets (using the natural map $\Delta^n \otimes X \longrightarrow \Delta^n \otimes \Delta^n \otimes X$ induced from the diagonal $\Delta^n \longrightarrow \Delta^n \times \Delta^n$) which is strictly associative. For example, in the case $C = \text{Set}$ we obtain the usual enrichment of $\text{Set}^{\Delta^{op}}$ over itself.

Note that the zero simplices of $\text{Map}(X, Y)$ are just the usual maps in $C^{\Delta^{op}}$ from X to Y. Furthermore if C, D are two such categories and $F : C \longrightarrow D$ is a functor which respects coproducts then it induces a **simplicially enriched** functor

$$F^{\Delta^{op}} : C^{\Delta^{op}} \longrightarrow D^{\Delta^{op}}.$$

Remark 9.12 To avoid confusion, let us emphasize that when we say that $F^{\Delta^{op}}$ is a simplicially enriched functor we mean that for every two objects $X_\bullet, Y_\bullet \in C^{\Delta^{op}}$ we have a map of simplicial sets

$$\mathrm{Map}(X_\bullet, Y_\bullet) \longrightarrow \mathrm{Map}\left(F^{\Delta^{op}}(X_\bullet), F^{\Delta^{op}}(Y_\bullet)\right)$$

which respects the identity and the composition rule.

Now let X be a smooth K-variety and consider the category $\mathrm{Var}_{/X}$ of K-varieties over X. This category admits finite coproducts so one obtains a natural simplicial enrichment of the category $\mathrm{Var}_{/X}^{\Delta^{op}}$ as above. Let $HC(X) \subseteq \mathrm{Var}_{/X}^{\Delta^{op}}$ denote the full (simplicially enriched) subcategory spanned by the hypercoverings with respect to the étale topology. We denote by $I(X) = \mathrm{Ho}(HC(X))$ the homotopy category of $HC(X)$ with respect to this simplicial enrichment.

Now consider the connected component functor (over K):

$$\pi_0 : \mathrm{Var}_{/X} \longrightarrow \mathrm{Set}.$$

This functor preserves finite coproducts and so induces a functor of simplicially enriched categories

$$\pi_0^{\Delta^{op}} : \mathrm{Var}_{/X}^{\Delta^{op}} \longrightarrow \mathrm{Set}^{\Delta^{op}}.$$

Since one wants to think of $\pi_0^{\Delta^{op}}(U_\bullet)$ as a topological space, one can either take the realization of $\pi_0^{\Delta^{op}}(U_\bullet)$ (as is done in [AMa69]), or equivalently, take the Kan replacement $\mathrm{Ex}^\infty(\pi_0^{\Delta^{op}}(U_\bullet))$. The second option is more convenient because it allows one to continue working inside the world of simplicial sets. It is equivalent in the sense that the subcategory

$$\mathrm{Ho}^{\mathrm{Kan}}\left(\mathrm{Set}^{\Delta^{op}}\right) \subseteq \mathrm{Ho}\left(\mathrm{Set}^{\Delta^{op}}\right)$$

consisting of Kan simplicial sets is equivalent to the homotopy category of topological spaces with CW homotopy type.

Now the functor Ex^∞ extends to a simplicially enriched functor: the augmentation map $\Delta^n \longrightarrow \mathrm{Ex}^\infty(\Delta^n)$ induces a natural map on mapping simplicial sets

$$\mathrm{Map}(\mathbf{X}, \mathbf{Y}) \longrightarrow \mathrm{Map}(\mathrm{Ex}^\infty(\mathbf{X}), \mathrm{Ex}^\infty(\mathbf{Y}))$$

which respects composition. Hence the composed functor $\mathrm{Ex}^\infty\left(\pi_0^{\Delta^{op}}(\bullet)\right)$ also extends to a simplicially enriched functor. Restricting $\mathrm{Ex}^\infty\left(\pi_0^{\Delta^{op}}(\bullet)\right)$ to $HC(X)$ and descending to the respective homotopy categories one obtains a functor

$$\acute{E}t(X) : I(X) \longrightarrow \mathrm{Ho}^{\mathrm{Kan}}\left(\mathrm{Set}^{\Delta^{op}}\right).$$

Since the category $I(X)$ is cofiltered (Corollary 8.13 [AMa69]) we can consider $\acute{E}t(X)$ as a **pro-object** in $\mathrm{Ho}^{\mathrm{Kan}}(\mathrm{Set}^{\Delta^{op}})$, i.e. an object in the pro-category of $\mathrm{Ho}^{\mathrm{Kan}}(\mathrm{Set}^{\Delta^{op}})$.

For a category C we will denote the pro-category of C by Pro C. Recall that objects of Pro C are diagrams $\{X_\alpha\}_{\alpha\in I}$ of objects of C indexed by a cofiltered category I and that

$$\mathrm{Hom}_{\mathrm{Pro}\ C}(\{X_\alpha\}_{\alpha\in I}, \{Y_\beta\}_{\beta\in J}) = \lim_{\beta\in J} \operatorname*{colim}_{\alpha\in I} \mathrm{Hom}_C(X_\alpha, Y_\beta).$$

Now if $f : X \longrightarrow Y$ is a map of K-varieties and $U_\bullet \longrightarrow Y$ is a hypercovering of Y we can pull it back to obtain a hypercovering $f^*U_\bullet \longrightarrow X$. One then gets a natural map of simplicial sets

$$\mathrm{Ex}^\infty\left(\pi_0^{\Delta^{op}}(f^*\mathcal{U})\right) \longrightarrow \mathrm{Ex}^\infty\left(\pi_0^{\Delta^{op}}(\mathcal{U})\right).$$

These natural maps fit together to form a map

$$\acute{E}t(X) \longrightarrow \acute{E}t(Y)$$

in $\mathrm{Pro}\,\mathrm{Ho}^{\mathrm{Kan}}(\mathrm{Set}^{\Delta^{op}})$. This exhibits $\acute{E}t$ as a functor

$$\acute{E}t : \mathrm{Var}_{/K} \longrightarrow \mathrm{Pro}\,\mathrm{Ho}^{\mathrm{Kan}}(\mathrm{Set}^{\Delta^{op}}).$$

This is the **étale homotopy type** functor defined in [AMa69].

In [AMa69] Artin and Mazur work with a certain localization of $\mathrm{Pro}\,\mathrm{Ho}^{\mathrm{Kan}}\left(\mathrm{Set}^{\Delta^{op}}\right)$, via Postnikov towers. Postnikov towers are a way of filtering a topological space by higher and higher homotopical information. In order to use Postnikov towers here we need a functorial and simplicial way to describe them. Such a description can found in [GJa99] by using the functor $P_n = \mathrm{cosk}_{n+1} \circ \mathrm{tr}_n$ defined in the previous section.

Note that if $\mathbf{X}$ is a Kan simplicial set then $P_n(\mathbf{X})$ will be a Kan simplicial set as well, i.e. P_n can be considered as a functor from Kan simplicial sets to Kan simplicial sets. Furthermore P_n extends to a functor of simplicially enriched categories. Descending to the homotopy category we get a functor

$$P_n : \mathrm{Ho}^{\mathrm{Kan}}\left(\mathrm{Set}^{\Delta^{op}}\right) \longrightarrow \mathrm{Ho}^{\mathrm{Kan}}\left(\mathrm{Set}^{\Delta^{op}}\right).$$

Then for $\mathbf{X}_I = \{\mathbf{X}_\alpha\}_{\alpha\in I} \in \mathrm{Pro}\,\mathrm{Ho}^{\mathrm{Kan}}(\mathrm{Set}^{\Delta^{op}})$ one defines

$$\mathbf{X}_I^\natural = \{P_n(\mathbf{X}_\alpha)\}_{n,\alpha}.$$

In many ways the object $\mathbf{X}_I^\natural$ is better behaved then $\mathbf{X}_I$. We will imitate this stage as well in our construction of the relative analogue of $\acute{E}t(-)$.

9.2.3 The relative étale homotopy type

We now come to the construction of the relative analogue. If X is a variety over K then it admits a natural structure map $X \longrightarrow \mathrm{Spec}\,(K)$. We wish to replace the functor π_0 from K-varieties to sets with a **relative** version of it. This relative functor should take a variety over K and return a "set over K", i.e. a sheaf of sets on $\mathrm{Spec}\,(K)$ (with respect to the étale topology).

Without getting into the formalities of the theory of sheaves let us take a shortcut and note that to give an étale sheaf of sets on $\mathrm{Spec}\,(K)$ is equivalent to giving a set with a $\Gamma_K = \mathrm{Gal}\left(\overline{K}/K\right)$ action such that each element has an open stabilzer. We will denote such objects by the name Γ_K**-sets** and their category by Set_{Γ_K}.

Remark 9.13 Some authors use the term **discrete** Γ_K-set in order to emphasize the continuous action of Γ_K. Since we will **never** consider non-continuous actions, and in order to improve readability, we chose to omit this adjective.

Note that Set_{Γ_K} admits coproducts and so we have a natural simplicial enrichment of $\mathrm{Set}_{\Gamma_K}^{\Delta^{op}}$. The relative version of π_0 is the functor

$$\pi_{0/K} : \mathrm{Var}/K \longrightarrow \mathrm{Set}_{\Gamma_K}$$

which takes the K-variety X to the Γ_K-set of connected components of

$$\overline{X} = X \otimes_K \overline{K}.$$

This functor preserves coproducts and so induces a functor of simplicially enriched categories

$$\pi_{0/K}^{\Delta^{op}} : \mathrm{Var}_{/K}^{\Delta^{op}} \longrightarrow \mathrm{Set}_{\Gamma_K}^{\Delta^{op}}.$$

As before, we don't really want to work in $\mathrm{Set}_{\Gamma_K}^{\Delta^{op}}$ itself, but with some localization of it. In [Goe95] Goerss considers two simplicial model structures on the simplicially enriched category $\mathrm{Set}_{\Gamma_K}^{\Delta^{op}}$ of simplicial Galois sets. In the first one, which is called the **strict** model structure, the weak equivalences are equivariant maps of $f : \mathbf{X} \longrightarrow \mathbf{Y}$ of simplicial Galois sets such that the induced map

$$f^{\Lambda} : \mathbf{X}^{\Lambda} \longrightarrow \mathbf{Y}^{\Lambda}$$

is a weak equivalence of simplicial sets for every open normal subgroup $\Lambda \lhd \Gamma$ (in what follows we will use the notation $\Lambda \widetilde{\lhd} \Gamma$ to denote an open normal subgroup). In the second model structure, which we will refer to as the **local** model structure (as it is also a particular case of the local model structure developed by Joyal in a well-known unpublished work), weak equivalences are equivariant maps $f : \mathbf{X} \longrightarrow \mathbf{Y}$ of simplicial Galois sets which induce a weak equivalence on the underlying simplicial sets. In both cases the cofibrations are just injective maps (and hence all the objects are cofibrant) and for the strict model structure one also has a concrete description of the fibrations: they are maps $f : \mathbf{X} \longrightarrow \mathbf{Y}$ such that the induced maps $f^\Lambda : \mathbf{X}^\Lambda \longrightarrow \mathbf{Y}^\Lambda$ are Kan fibrations for every $\Lambda \tilde{\triangleleft} \Gamma$.

We will denote by $\mathrm{Ho}\left(\mathrm{Set}_{\Gamma_K}^{\Delta^{op}}\right)$ the homotopy category of $\mathrm{Set}_{\Gamma_K}^{\Delta^{op}}$ with respect to the simplicial enrichment given above. The localized homotopy categories with respect to the strict and local model structures will be denoted by $\mathrm{Ho}^{\mathrm{st}}\left(\mathrm{Set}_{\Gamma_K}^{\Delta^{op}}\right)$ and $\mathrm{Ho}^{\mathrm{lc}}\left(\mathrm{Set}_{\Gamma_K}^{\Delta^{op}}\right)$ respectively. They can be realized as a sequence of full sub-categories

$$\mathrm{Ho}^{\mathrm{lc}}\left(\mathrm{Set}_{\Gamma_K}^{\Delta^{op}}\right) \subseteq \mathrm{Ho}^{\mathrm{st}}\left(\mathrm{Set}_{\Gamma_K}^{\Delta^{op}}\right) \subseteq \mathrm{Ho}\left(\mathrm{Set}_{\Gamma_K}^{\Delta^{op}}\right)$$

obtained by restricting to standardly fibrant and strictly fibrant objects respectively.

Now, similarly to the above, we would like to compose $\pi_0{}_{/K}^{\Delta^{op}}$ with the strict fibrant replacement functor. It is convenient to note that strict fibrant replacement can actually be done using the Kan replacement functor Ex^∞: the functor Ex^∞ also induces a functor $\mathrm{Set}_\Gamma^{\Delta^{op}} \longrightarrow \mathrm{Set}_\Gamma^{\Delta^{op}}$ (which by abuse of notation we will also call Ex^∞), and we have the following observation:

Lemma 9.14 *If* $\mathbf{X} \in \mathrm{Set}_\Gamma^{\Delta^{op}}$ *is a simplicial* Γ*-set then object* $\mathrm{Ex}^\infty(\mathbf{X})$ *is strictly fibrant and the map* $\mathbf{X} \longrightarrow \mathrm{Ex}^\infty(\mathbf{X})$ *is a strict weak equivalence.*

Proof We need to show that for every $\Lambda \tilde{\triangleleft} \Gamma$ the simplicial set $\mathrm{Ex}^\infty(\mathbf{X})^\Lambda$ is Kan and the map $\mathbf{X}^\Lambda \longrightarrow \mathrm{Ex}^\infty(\mathbf{X})^\Lambda$ is a weak equivalence. Both claims follow easily once one shows that

$$\mathrm{Ex}^\infty(\mathbf{X})^\Lambda = \mathrm{Ex}^\infty(\mathbf{X}^\Lambda).$$

This in turn follows from the fact that $\mathrm{Ex}^\infty(\mathbf{X})$ is the colimit of the sequence of **inclusions**:

$$\mathbf{X} \hookrightarrow \mathrm{Ex}(\mathbf{X}) \hookrightarrow \mathrm{Ex}(\mathrm{Ex}(\mathbf{X})) \hookrightarrow \mathrm{Ex}(\mathrm{Ex}(\mathrm{Ex}(\mathbf{X}))) \hookrightarrow \ldots$$

and

$$\mathrm{Ex}(\mathbf{X})^\Lambda = \mathrm{Ex}(\mathbf{X}^\Lambda)$$

since Ex has a left adjoint (given by barycentric subdivision). □

We now take the functor $\mathrm{Ex}^{\infty}\left(\pi^{0}_{/K}(\bullet)\right)$ restricted to $HC(X) \subseteq \mathrm{Var}^{\Delta^{op}}_{/K}$ and descend to the respective homotopy categories. We end up with a functor

$$\acute{E}t_{/K}(X) : I(X) \longrightarrow \mathrm{Pro}\,\mathrm{Ho}^{\mathrm{st}}\left(\mathrm{Set}^{\Delta^{op}}_{\Gamma_K}\right).$$

We consider $\acute{E}t_{/K}(X)$ to be the **relative analogue** of $\acute{E}t(X)$. Now just as for $\acute{E}t$, one can use pullbacks of hypercovering in order to make $\acute{E}t_{/K}$ into a functor:

$$\acute{E}t_{/K} : \mathrm{Var}_{/K} \longrightarrow \mathrm{Pro}\,\mathrm{Ho}^{\mathrm{st}}\left(\mathrm{Set}^{\Delta^{op}}_{\Gamma_K}\right).$$

As in the case of the regular étale homotopy type it will be better to work with an appropriate Postnikov tower $\acute{E}t^{\natural}_{/K}(X)$ of $\acute{E}t_{/K}(X)$, i.e. we want a way to estimate an object by a tower of "bounded" objects. When working with strictly fibrant objects the relevant notion of boundedness is stronger:

Definition 9.15 Let Γ be a pro-finite group and $\mathbf{X}$ a simplicial Γ-set. We will say that $\mathbf{X}$ is **strictly bounded** if it is strictly fibrant and if there exists an N such that for all $\Lambda \widetilde{\triangleleft} \Gamma$ the simplicial set X^{Λ} is N-bounded (i.e. all the homotopy groups $\pi_n\left(X^{\Lambda}\right)$ are trivial for $n > N$. Note in particular that the empty set is considered N-bounded for every $N \geq 0$).

Let $\mathbf{X}$ be a strictly fibrant simplicial Γ_K-set. Recall that the underlying simplicial set of $\mathbf{X}$ is Kan so we can consider

$$P_n(\mathbf{X}) = \mathrm{cosk}_{n+1}(\mathrm{tr}_{n+1}(\mathbf{X})).$$

Since P_n is a functor we have an induced action of Γ_K on $P_n(\mathbf{X})$. Note that under this action the stabilizer of each simplex in $P_n(\mathbf{X})$ will be open and so $P_n(\mathbf{X})$ is a simplicial Γ_K-set. Furthermore since

$$P_n(X)^{\Lambda} = P_n\left(X^{\Lambda}\right)$$

for each $\Lambda \widetilde{\triangleleft} \Gamma_K$, we get that $P_n(X)$ is a strictly bounded simplicial Γ_K-set. Hence we can think of P_n as a simplicially enriched functor from the category of strictly fibrant simplicial Galois sets to itself. Descending to the homotopy category we get a functor

$$P_n : \mathrm{Ho}^{\mathrm{st}}\left(\mathrm{Set}^{\Delta^{op}}_{\Gamma_K}\right) \longrightarrow \mathrm{Ho}^{\mathrm{st}}\left(\mathrm{Set}^{\Delta^{op}}_{\Gamma_K}\right).$$

Then for $\mathbf{X}_I = \{\mathbf{X}_\alpha\}_{\alpha \in I} \in \operatorname{Pro}\operatorname{Ho}^{st}(\operatorname{Set}^{\Delta^{op}})$ we define as above

$$\mathbf{X}_I^\natural = \{P_n(\mathbf{X}_\alpha)\}_{n,\alpha}.$$

It will also be convenient to consider the pro-objects

$$\mathbf{X}_I^n = \{P_k(\mathbf{X}_\alpha)\}_{k \le n,\alpha},$$

where for $n = \infty$ we define:

$$\mathbf{X}_I^\infty = \mathbf{X}_I^\natural.$$

Note also that since $P_n(\bullet)$ is an augmented functor so is $(\bullet)^n$ and so for every $0 \le n \le \infty$ we have a natural map

$$\mathbf{X}_I \longrightarrow \mathbf{X}_I^n.$$

We will use the following notation:

Definition 9.16 Let X be a K-variety and $\mathcal{U} \longrightarrow X$ a hypercovering. We will use the following notations for the relevant simplicial sets constructed from $\mathcal{U}$:

$$\mathbf{N}_\mathcal{U} = \left(\pi_{/K}^{\Delta^{op}}(\mathcal{U})\right),$$

$$\mathbf{X}_\mathcal{U} = \operatorname{Ex}^\infty(\mathbf{N}_\mathcal{U}),$$

$$\mathbf{X}_{\mathcal{U},k} = P_k(\mathbf{X}_\mathcal{U}).$$

Under this notation one has

$$\acute{E}t_{/K}(X) = \{\mathbf{X}_\mathcal{U}\}_{\mathcal{U} \in I(X)},$$

$$\acute{E}t_{/K}^n(X) = \{\mathbf{X}_{\mathcal{U},k}\}_{\mathcal{U} \in I(X), k \le n}.$$

There are several important properties that the underlying simplicial sets in the diagram of $\acute{E}t_{/K}^n(X)$ satisfy. Here are the properties that we will be interested in:

Definition 9.17 Let Γ be a pro-finite group. A simplicial Γ-set $\mathbf{X}$ will be called

1. **Finite** if $\pi_n(\mathbf{X})$ is finite for $n \ge 1$.
2. **Excellent** if the action of Γ on $\mathbf{X}$ factors through a finite quotient of Γ.
3. **Nice** if the action of Γ on every skeleton $\operatorname{sk}_n \mathbf{X}$ factors through a finite quotient of Γ.

Remark 9.18 When we say a finite quotient of a pro-finite group we always mean a **continuous** finite quotient, i.e. a finite quotient in the category of pro-finite groups.

Now let X/K be an algebraic K-variety and $\mathcal{U} \longrightarrow X$ a hypercovering. Then clearly $\pi_{0/K}(\mathcal{U})$ is nice and it is easy to show that $\mathbf{X}_{\mathcal{U}} = \mathrm{Ex}^{\infty}\left(\pi_{0/K}(\mathcal{U})\right)$ is nice as well (this is one of the advantages of using Ex^{∞} as strict fibrant replacement). Then we see that for each $k < \infty$ the simplicial Γ_K-set $\mathbf{X}_{\mathcal{U},k}$ is **excellent** and **strictly bounded**.

It can be shown ([AMa69]) that if X is a smooth K-variety then $\mathbf{X}_{\mathcal{U}}$ is finite. Hence in that case $\mathbf{X}_{\mathcal{U},k}$ is finite as well.

We finish this subsection with a basic comparison result connecting the relative notion $\acute{E}t_{/K}(X)$ and the étale homotopy type $\acute{E}t(\overline{X})$ of $\overline{X} = X \otimes_K \overline{K}$. Note that by forgetting the group action we obtain a forgetful functor from $\mathrm{Ho}(\mathrm{Set}_{\Gamma_K}^{\Delta^{op}})$ to $\mathrm{Ho}(\mathrm{Set}^{\Delta^{op}})$. Prolonging this functor we obtain a forgetful functor

$$F : \mathrm{Pro}\,\mathrm{Ho}^{\mathrm{st}}(\mathrm{Set}_{\Gamma_K}^{\Delta^{op}}) \longrightarrow \mathrm{Pro}\,\mathrm{Ho}^{\mathrm{Kan}}(\mathrm{Set}^{\Delta^{op}}).$$

Proposition 9.19 *Let K be a field and X a K-variety. Then there is an isomorphism in* $\mathrm{Pro}\,\mathrm{Ho}^{\mathrm{Kan}}(\mathrm{Set}^{\Delta^{op}})$*:*

$$f : \acute{E}t^{n}(\overline{X}) \longrightarrow F\left(\acute{E}t^{n}_{/K}(X)\right).$$

Proof Note that the indexing category of $\acute{E}t^{n}_{/K}(X)$ is naturally contained in the indexing category of $\acute{E}t^{n}(\overline{X})$ and this inclusion identifies $F\left(\acute{E}t^{n}_{/K}(X)\right)$ with a sub-diagram of $\acute{E}t^{n}(\overline{X})$. This yields a natural map

$$f : \acute{E}t^{n}(\overline{X}) \longrightarrow F\left(\acute{E}t^{n}_{/K}(X)\right).$$

In order to show that f is an isomorphism one needs to show that this sub-diagram is cofinal. More concretely, one needs to show that for every hypercovering $\mathcal{V}_{\bullet} \longrightarrow \overline{X}$ defined over $\overline{K}$ and every $0 \le k \le n$ there exists a hypercovering $\mathcal{U}_{\bullet} \longrightarrow X$ (defined over K) such that $\mathbf{X}_{\overline{\mathcal{U}},k}$ dominates $\mathbf{X}_{\mathcal{V},k}$.

First, note that it is not restrictive to consider $P_k(\mathcal{V})$ instead of $\mathcal{V}$. Thus we may assume that $\mathcal{V}$ is defined over some finite field extension L/K. Then it is clear that we can take

$$\mathcal{U}_{\bullet} = R^{L/K}_{/X}(\mathcal{V}_{\bullet})$$

where $R^{L/K}_{/X}$ is the relative Weil restriction of scalars functor, i.e.

$$\mathrm{Hom}_{/X}\left(Z, R^{L/K}_{/X}(V_n)\right) = \mathrm{Hom}_{/X\otimes L}(Z \otimes L, V_n).$$

□

9.2.4 The homotopy type of $\mathbf{X}_{\mathcal{U}}$

Let X be a smooth geometrically connected K-variety and $\mathcal{U}_\bullet \longrightarrow X$ a hypercovering. In this section we will give basic results which help to analyze the homotopy type of the spaces $\mathbf{X}_{\mathcal{U}}$ which appear in $\acute{E}t_{/K}(X)$.

Recall that given a map $f : Y \longrightarrow X$ and an étale hypercovering $\mathcal{U}_\bullet \longrightarrow X$ one can levelwise pull $\mathcal{U}$ back to a hypercovering of Y. We then have a natural map

$$\mathbf{X}_f : \mathbf{X}_{f^*\mathcal{U}} \longrightarrow \mathbf{X}_{\mathcal{U}}.$$

Now let X/K be a smooth geometrically connected variety. In this case X has a generic point

$$\xi : \mathrm{Spec}\,(K(X)) \longrightarrow X$$

where $K(X)$ is the function field of X (over K).

Note that $\mathrm{Spec}\,(K(X))$ is not a K-variety. However, the schemes $\mathrm{Spec}\,(K(X))$ and $\mathrm{Spec}\,(K(X)) \otimes \overline{K} \cong \mathrm{Spec}\,(\overline{K}(\overline{X}))$ are **Noetherian**, which means that the functor π_0 is well defined on étale schemes over them. In fact, since $K(X)$ is a field the étale site of $\mathrm{Spec}\,K(X)$ can be identified with the site of finite discrete $\Gamma_{K(X)}$-sets. Under this identification a hypercovering of $\mathrm{Spec}\,(K(X))$ corresponds to a Kan contractible simplicial discrete $\Gamma_{K(X)}$-set which is levelwise finite. We will use the following notation:

Definition 9.20 Let $\mathcal{U} \longrightarrow X$ be a hypercovering. We shall denote by $\widetilde{\mathcal{U}}$ the simplicial $\Gamma_{K(X)}$-set corresponding to the hypercovering $\xi^*\mathcal{U} \longrightarrow \mathrm{Spec}\,(K(X))$. In particular for each n we will denote the $\Gamma_{K(X)}$-set corresponding to $\xi^*\mathcal{U}_n$ by $\widetilde{\mathcal{U}}_n$.

Under the identification above we can describe the functor $\pi_{0/K}$ as

$$\pi_{0/K}(\mathcal{U}_n) \cong \widetilde{\mathcal{U}}_n/\Gamma_{\overline{K}(\overline{X})},$$

where the Γ_K action on the right hand side is induced from the $\Gamma_{K(X)}$ action on $\mathcal{U}$ and the short exact sequence

$$1 \longrightarrow \Gamma_{\overline{K}(\overline{X})} \longrightarrow \Gamma_{K(X)} \longrightarrow \Gamma_K \longrightarrow 1.$$

In particular we get the following expression for $\mathbf{N}_{\mathcal{U}}$ which we frame for future use:

Corollary 9.21 *Let $\mathcal{U} \longrightarrow X$ be a hypercovering. Then we have an isomorphism of simplicial Γ_K-sets*

$$\mathbf{N}_{\mathcal{U}} \cong \widetilde{\mathcal{U}}/\Gamma_{\overline{K(X)}}.$$

This interpretation of $\mathbf{N}_{\mathcal{U}}$ will be a useful later in order to calculate invariants of $\mathbf{X}_{\mathcal{U}} \simeq \mathbf{N}_{\mathcal{U}}$. For example we immediately get the following conclusion:

Corollary 9.22 *Let X/K be a smooth geometrically connected variety and $\mathcal{U}_\bullet \longrightarrow X$ a hypercovering. Then $\mathbf{X}_{\mathcal{U}}$ is connected.*

We can also use this interpretation in order to calculate the fundamental group of $\mathbf{X}_{\mathcal{U}}$. This is done using the following lemma:

Lemma 9.23 *Let G be a group and $\mathbf{X}$ a contractible Kan simplicial G-set. Then, given a base point $\overline{x} \in \mathbf{X}/G$, there is a natural short exact sequence*

$$1 \longrightarrow K \longrightarrow G \longrightarrow \pi_1(\mathbf{X}/G, \overline{x}) \longrightarrow 1$$

where

$$K = \left\langle \bigcup_x \{\mathrm{Stab}_G(x) | x \in \mathbf{X}_0\} \right\rangle.$$

Proof Let

$$\pi : \mathbf{X} \longrightarrow \mathbf{X}/G$$

be the natural quotient map and $a \in \mathbf{X}_0$ a vertex. We will construct a surjective map

$$\phi_a : G \longrightarrow \pi_1(\mathbf{X}/G, \pi(a))$$

such that

$$\ker(\phi_a) = K.$$

Let g be an element of G. Since $\mathbf{X}$ is contractible and Kan there is at least one 1-simplex l_g in $\mathbf{X}$ joining a and ga. We shall take $\phi_a(g)$ to be the element in $\pi_1(\mathbf{X}/G, \pi(a))$ corresponding to $\pi(l_g)$.

We shall first show that ϕ_a is well defined. Assume that l'_g is another 1-simplex connecting a and ga. Since $\mathbf{X}$ is contractible there is an end-points-preserving homotopy between the paths corresponding to l_g and

l'_g in the realization of $\mathbf{X}$. Projecting this homotopy to $\mathbf{X}/G$ we get that the corresponding elements in $\pi_1(\mathbf{X}/G, \pi(a))$ are equal.

We will now prove surjectivity. It is not hard to see that $\mathbf{X}/G$ is Kan in dimension one (see for example Lemma 11.6 in [AMa69]) and therefore every element in $\pi_1(\mathbf{X}/G, \pi(a))$ can be represented by a 1-simplex in $\mathbf{X}/G$ with both end points being $\pi(a)$. We can then lift this 1-simplex to $\mathbf{X}$ (note that the map $\mathbf{X} \longrightarrow \mathbf{X}/G$ is levelwise surjective) obtaining a simplex joining a and ga for some $g \in G$.

It now remains to calculate the kernel of ϕ_a. First we shall show that if b is another point in $\mathbf{X}_0$ then $\ker\phi_b = \ker\phi_a$. Since $\mathbf{X}$ is contractible and Kan there is a 1-simplex in $\mathbf{X}$ going from b to a. We denote this 1-simplex by p_{ba}. Now for every g consider the three 1-simplices p_{ba}, l_g and $g(p_{ba})$ creating together a path between b and gb. Since $\mathbf{X}$ is Kan this path is homotopic to a 1-simplex l_g^b joining b and gb.

We shall use this l_g^b to obtain $\phi_b(g)$. From the discussion above we get that the following diagram commutes

$$\begin{array}{ccc} G & \xrightarrow{\phi_b} & \pi_1(\mathbf{X}/G, \pi(b)) \\ \| & & \downarrow{\scriptstyle c_{\pi(p_{ba})}} \\ G & \xrightarrow{\phi_a} & \pi_1(\mathbf{X}/G, \pi(a)) \end{array}$$

where $c_{\pi(p_{ba})}$ is the natural isomorphism defined by "conjugation" by the path $\pi(p_{ba})$ between $\pi(b)$ and $\pi(a)$. This diagram implies that the kernel $\ker\phi_a$ is independent of the choice of $a \in \mathbf{X}$. Since clearly

$$\mathrm{Stab}_G(a) \subseteq \ker\phi_a,$$

we get that $K \subseteq \ker(\phi_a)$.

To complete the proof of the lemma it suffices to show that $\ker\phi_a \subseteq K$. For every $x \in \mathbf{X}_0$ we have $\mathrm{Stab}_G x \subseteq K$. Thus there is map of G-sets

$$\phi : \mathbf{X}_0 \longrightarrow G/K.$$

Recall that K is normal in G and let $\mathbf{E}(G/K)$ be as in Definition 9.7. $\mathbf{E}(G/K)$ is a Kan contractible simplicial set with the free action of G/K. Pulling this action to G we get a Kan contractible G-simplicial set. The quotient $\mathbf{E}(G/K)/G$ is a weakly equivalent to $\mathbf{B}(G/K)$, whose fundamental group is exactly G/K.

The map ϕ lifts to a unique equivariant map

$$\check{\phi} : \mathbf{X} \longrightarrow \mathbf{E}(G/K),$$

and the following commutative diagram

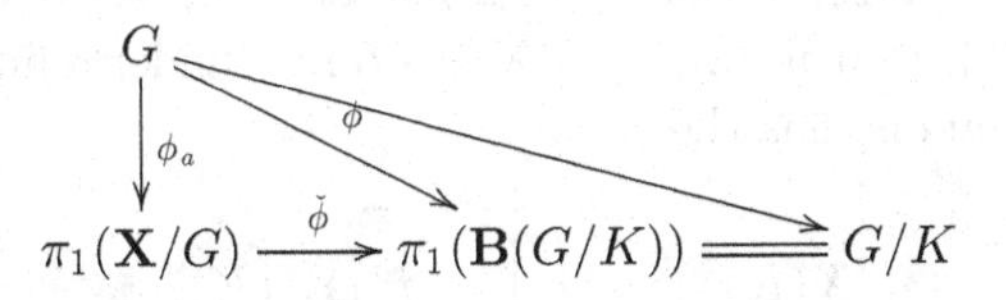

implies that $\ker \phi_a \subseteq K$. □

9.3 The obstructions

9.3.1 The classical obstructions

The Brauer-Manin obstruction

Let K be a number field and X an algebraic variety over K. Given an adelic point $(x_\nu)_\nu \in X(\mathbb{A})$ and an element $u \in H^2_{\acute{e}t}(X, \mathbb{G}_m)$ one can pullback u by each x_ν to obtain an element

$$x_\nu^* u \in H^2_{\acute{e}t}\left(\mathrm{Spec}\left(K_\nu\right), \mathbb{G}_m\right).$$

There is a canonical map

$$\mathrm{inv} : H^2_{\acute{e}t}\left(\mathrm{Spec}\left(K_\nu\right), \mathbb{G}_m\right) \longrightarrow \mathbb{Q}/\mathbb{Z}$$

called the **invariant** which is an isomorphism in non-archimedean places. Summing all the invariants one obtains a pairing

$$X(\mathbb{A}) \times H^2_{\acute{e}t}\left(X, \mathbb{G}_m\right) \longrightarrow \mathbb{Q}/\mathbb{Z}$$

given by

$$((x_\nu)_\nu, u) \mapsto \sum_\nu \mathrm{inv}(x_\nu^* u) \in \mathbb{Q}/\mathbb{Z}.$$

Now by the Hasse-Brauer-Noether Theorem we see that if (x_ν) is actually a rational point then its pairing with any element in $H^2(X, \mathbb{G}_m)$ would be zero. This motivates the definition of the Brauer set

$$X(\mathbb{A})^{\mathrm{Br}} = \left\{(x_\nu)_\nu \in X(\mathbb{A}) | ((x_\nu)_\nu, u) = 0, \forall u \in H^2(X_{\acute{e}t}, \mathbb{G}_m)\right\}$$

and we have

$$X(K) \subseteq X(\mathbb{A})^{\mathrm{Br}} \subseteq X(\mathbb{A}).$$

Descent obstructions

Let X be an algebraic variety over a number field K. It is well-known (see e.g. [Sko01]) that if $f : Y \longrightarrow X$ is a torsor under a linear algebraic K-group G then one has the equality

$$X(K) = \biguplus_{\sigma \in H^1(K,G)} f^\sigma(Y^\sigma(K))$$

and so the set

$$X(\mathbb{A})^f = \bigcup_{\sigma \in H^1(K,G)} f^\sigma(Y^\sigma(\mathbb{A}))$$

has to contain $X(K)$. This motivates the definition

$$X(\mathbb{A})^{desc} = \bigcap_f X(\mathbb{A})^f$$

where f runs over all torsors under linear algebraic K-groups. As before we have

$$X(K) \subseteq X(\mathbb{A})^{desc} \subseteq X(\mathbb{A}).$$

We shall denote by $X(\mathbb{A})^{fin}$, $X(\mathbb{A})^{fin-ab}$ and $X(\mathbb{A})^{con}$ the analogous sets obtained by restricting f to torsors under finite, finite abelian and connected linear algebraic groups respectively.

Applying obstructions to finite torsors

Given a functorial obstruction set

$$X(K) \subseteq X(\mathbb{A})^{obs} \subseteq X(\mathbb{A}),$$

and a torsor under finite K-group

$$f : Y \longrightarrow X,$$

one can always define the set

$$X(\mathbb{A})^{f,obs} = \bigcup_{\sigma \in H^1(K,G)} f^\sigma(Y^\sigma(\mathbb{A})^{obs})$$

satisfying

$$X(K) \subseteq X(\mathbb{A})^{f,obs} \subseteq X(\mathbb{A})^{obs} \subseteq X(\mathbb{A}).$$

Now by going over all such f we get

$$X(\mathbb{A})^{fin,obs} = \bigcap_f X(\mathbb{A})^{f,obs}$$

and

$$X(K) \subseteq X(\mathbb{A})^{fin,obs} \subseteq X(\mathbb{A})^{obs} \subseteq X(\mathbb{A}).$$

In [Sko99] Skorobogatov defines in this way the set

$$X(\mathbb{A})^{fin,\mathrm{Br}}$$

and constructs a variety X such that

$$X(\mathbb{A})^{fin,\mathrm{Br}} = \emptyset$$

but

$$X(\mathbb{A})^{\mathrm{Br}} \neq \emptyset.$$

9.3.2 Homotopy-theoretic obstructions

Homotopy fixed points sets

In this subsection we will explain how the notion of homotopy fixed points by a profinite group Γ fits into the scheme of étale homotopy theory. Although we will be primarily interested in the case where $\Gamma = \Gamma_K$ is the absolute Galois group of a field K, many of the constructions and observations will be valid for a general profinite group, and will hence be presented in the general form. The reader will not lose much by assuming $\Gamma = \Gamma_K$ everywhere.

Recall from the previous section the strict model structure and the local model structure introduced by Goerss ([Goe95]) on the category $\mathrm{Set}_\Gamma^{\Delta^{op}}$ of simplicial (discrete) Γ-sets. We will use the terms strict fibrations/strict weak equivalence for the strict model structure and fibration/weak equivalence for the local model structure. Cofibrations in both case are simply injective maps.

As mentioned in [Goe95] the operation of (local) fibrant replacement can actually be done functorially, yielding a functor

$$(\bullet)^{fib} : \mathrm{Set}_\Gamma^{\Delta^{op}} \longrightarrow \mathrm{Set}_\Gamma^{\Delta^{op}}.$$

One then defines the **homotopy fixed points** of $\mathbf{X}$ to be the fixed points of X^{fib}, i.e.:

$$\mathbf{X}^{h\Gamma} \stackrel{def}{=} \left(\mathbf{X}^{fib}\right)^{\Gamma}.$$

Since fibrant objects are in particular strictly fibrant one gets that $\mathbf{X}^{h\Gamma}$ is always a Kan simplicial set. It can be shown that if two maps

$f, g : \mathbf{X} \longrightarrow \mathbf{Y}$ induce the same map in $\mathrm{Ho}^{\mathrm{lc}}\left(\mathrm{Set}_{\Gamma}^{\Delta^{op}}\right)$ then the induced maps in

$$f_*, g_* : X^{h\Gamma} \longrightarrow Y^{h\Gamma}$$

are simplicially homotopic. In particular this will be true for f, g which are simplicially homotopic.

Since we will be working only with strictly bounded simplicial sets we would like to have a convenient formula for the homotopy fixed points in that case. Note that if S is strictly fibrant then for every open normal subgroup $\Lambda \tilde{\triangleleft} \Gamma_K$ the corresponding fixed points S^{Λ} form a Kan simplicial set. We then have the following formula:

Theorem 9.24 *Let* $\mathbf{Y}$ *be a simplicial* Γ*-set whose underlying simplicial set is Kan. Let* $D^{\Gamma}_{fin} \subseteq \mathrm{Ho}\left(\mathrm{Set}_{\Gamma}^{\Delta^{op}}\right)$ *be the full subcategory spanned by Kan contractible objects which are levelwise finite. Then one has an isomorphism of sets*

$$\pi_0\left((\mathbf{Y})^{h\Gamma}\right) \simeq \operatorname*{colim}_{\mathbf{E} \in D^{\Gamma}_{fin}} [\mathbf{E}, \mathbf{Y}]_{\Gamma}.$$

If in addition $\mathbf{Y}$ *is also* ***strictly bounded*** *then the formula can be refined to*

$$\pi_0\left((\mathbf{Y})^{h\Gamma}\right) \simeq \operatorname*{colim}_{\Lambda \tilde{\triangleleft} \Gamma}[\mathbf{E}(\Gamma/\Lambda), \mathbf{Y}]_{\Gamma} \simeq$$

$$\operatorname*{colim}_{\Lambda \tilde{\triangleleft} \Gamma}\left[\mathbf{E}(\Gamma/\Lambda), \mathbf{Y}^{\Lambda}\right]_{\Gamma} \simeq \operatorname*{colim}_{\Lambda \tilde{\triangleleft} \Gamma} \pi_0\left(\left(\mathbf{Y}^{\Lambda}\right)^{h(\Gamma/\Lambda)}\right).$$

We delay the proof of this formula (which is completely independent of the rest of this chapter) to section §4 (see 9.44).

Now let K be a number field and X an algebraic variety over K. We will denote by Γ_K the absolute Galois group of K and similarly by $\Gamma_L \tilde{\triangleleft} \Gamma_K$ the absolute Galois group of a finite Galois extension L/K.

Every K-rational point in X is a map

$$\mathrm{Spec}\,(K) \longrightarrow X$$

in Var/K. Applying the functor $\acute{E}t_{/K}{}^{\natural}$ we get a map

$$\acute{E}t_{/K}{}^{\natural}(\mathrm{Spec}\,(K)) \longrightarrow \acute{E}t_{/K}{}^{\natural}(X).$$

In order to describe mappings from $\acute{E}t_{/K}{}^{\natural}(\mathrm{Spec}\,(K))$ to some pro-object we need first to understand the pro-object $\acute{E}t_{/K}{}^{\natural}(\mathrm{Spec}\,(K))$ itself.

We start with the simpler task of describing the pro-object

$\acute{E}t_{/K}(\mathrm{Spec}\,(K))$: the site of finite étale varieties over $\mathrm{Spec}\,(K)$ can be identified with the site of finite discrete Γ_K-sets (and surjective maps as coverings) via the fully faithful functor $\pi_{0/K}$. This means that the functor $\pi_{/K}^{\Delta^{op}}$ induces a fully faithful embedding of $I(\mathrm{Spec}\,(K))$ into $\mathrm{Ho}\left(\mathrm{Set}_{\Gamma_K}^{\Delta^{op}}\right)$ whose essential image consists of the full subcategory $D_{fin}^{\Gamma_K} \subseteq \mathrm{Ho}\left(\mathrm{Set}_{\Gamma_K}^{\Delta^{op}}\right)$. In particular we have an equivalence of categories

$$\pi_{0/K} : I(\mathrm{Spec}\,(K)) \xrightarrow{\sim} D_{fin}^{\Gamma_K}.$$

Now for a pro-finite group Γ we will denote by $\mathbf{E}\Gamma$ the pro-object

$$\mathbf{E}\Gamma \stackrel{def}{=} \{\mathrm{Ex}^{\infty}(S)\}_{S\in D_{fin}^{\Gamma}} \in \mathrm{Pro}\,\mathrm{Ho}\left(\mathrm{Set}_{\Gamma}^{\Delta^{op}}\right).$$

We then have an isomorphism of pro-objects

$$\acute{E}t_{/K}(\mathrm{Spec}\,(K)) \simeq \mathbf{E}(\Gamma_K).$$

A simple corollary of this observation plus formula 9.44 is the following:

Corollary 9.25 *Let* $\mathbf{Y}_I = \{\mathbf{Y}_\alpha\}_{\alpha\in I} \in \mathrm{Pro}\,\mathrm{Ho}^{\mathrm{st}}\left(\mathrm{Set}_{\Gamma_K}^{\Delta^{op}}\right)$ *be an object. Then*

$$\mathrm{Hom}_{\mathrm{Pro}\,\mathrm{Ho}^{\mathrm{st}}\left(\mathrm{Set}_{\Gamma_K}^{\Delta^{op}}\right)}\left(\acute{E}t_{/K}(\mathrm{Spec}\,(K)), \mathbf{Y}_I\right) \simeq \lim_{\alpha\in I} \pi_0\left(Y_\alpha^{h\Gamma_K}\right).$$

We now wish to compute $\mathbf{E}\Gamma^{\natural}$ for a pro-finite group Γ. Since the skeletons of the simplicial sets in $\mathbf{E}\Gamma$ are finite the action of Γ on each specific skeleton factors through a finite quotient. This means that the action of Γ on each of the spaces of $\mathbf{E}\Gamma^{\natural}$ factors through a finite quotient $G = \Gamma/\Lambda$ for some $\Lambda \tilde{\triangleleft} \Gamma$.

Now for every such G one has the Kan contractible levelwise finite simplicial Γ-set $\mathbf{E}G = \mathrm{cosk}_0(G)$ with the Γ-action given by pulling the standard G-action. Then $\mathrm{Ex}^{\infty}(\mathbf{E}G)$ appears in the diagram of $\mathbf{E}\Gamma^{\natural}$. Since every $\mathbf{E}G$ is strictly fibrant as a simplicial Γ-set (every fixed point space is either empty or equal to all of $\mathbf{E}G$, which is Kan) the map $\mathbf{E}G \longrightarrow \mathrm{Ex}^{\infty}(\mathbf{E}G)$ admits a simplicial homotopy inverse $\mathrm{Ex}^{\infty}(\mathbf{E}G) \longrightarrow \mathbf{E}G$ which is unique up to simplicial homotopy. This gives a map in $\mathrm{Pro}\,\mathrm{Ho}\left(\mathrm{Set}_{\Gamma}^{\Delta^{op}}\right)$:

$$\varphi : \{\mathbf{E}(\Gamma/\Lambda)\}_{\Lambda\tilde{\triangleleft}\Gamma} \longrightarrow \mathbf{E}\Gamma^{\natural}.$$

We claim that φ is actually an **isomorphism**: every Kan contractible simplicial G-set $\mathbf{X}$ admits a map $\mathbf{E}G \longrightarrow \mathbf{X}$ which is unique up to simplicial homotopy (this can be seen using the projective model structure on simplicial G-sets). These maps fit together to give an inverse to φ. This finishes the computation of $\mathbf{E}\Gamma^{\natural}$.

We wish to compute the set of maps from

$$\acute{E}t^{\natural}(\mathrm{Spec}\,(K)) \cong \mathbf{E}(\Gamma_K)^{\natural} \cong \{\mathbf{E}G_L\}_{L/K}$$

to a pro-object of the form $X_I = \{\mathbf{X}_\alpha\}_{\alpha \in I}$ where each X_α is strictly bounded. By definition this morphism set is the set

$$\lim_{\alpha} \operatorname*{colim}_{L/K} [\mathbf{E}G_L, \mathbf{X}_\alpha]_{\Gamma_K}$$

where L/K runs over all finite Galois extensions, $G_L = \Gamma_K/\Gamma_L$ and $[\mathbf{X}, \mathbf{Y}]_{\Gamma_K}$ denotes simplicial homotopy classes of maps. Since Γ_L stabilizes $\mathbf{E}G_L$ this is the same as

$$\begin{aligned}\lim_{\alpha} \operatorname*{colim}_{L/K} \left[\mathbf{E}G_L, \mathbf{X}_\alpha^{\Gamma_L}\right]_{G_L} &= \lim_{\alpha} \operatorname*{colim}_{L/K} \pi_0\left(\left(\mathbf{X}_\alpha^{\Gamma_L}\right)^{hG_L}\right) \\ &= \lim_{\alpha} \pi_0\left(\mathbf{X}_\alpha^{h\Gamma_K}\right)\end{aligned}$$

where the last equality is obtained by applying formula 9.44 to the strictly bounded simplicial Γ_K-set $\mathbf{X}_\alpha$.

We summarize the above computation in the following definition

Definition 9.26 Let $\mathbf{X}_I = \{\mathbf{X}_\alpha\}_{\alpha \in I} \in \mathrm{Pro}\,\mathrm{Ho}\left(\mathrm{Set}_\Gamma^{\Delta^{op}}\right)$ be an object. We define the **Γ-homotopy fixed points set** of $\mathbf{X}_I$, denoted by $\mathbf{X}_I\left(\mathbf{E}\Gamma^{\natural}\right)$, to be

$$\mathbf{X}_I\left(\mathbf{E}\Gamma^{\natural}\right) = \lim_{\alpha} \pi_0\left(\mathbf{X}_\alpha^{h\Gamma}\right).$$

If $\Gamma = \Gamma_K$ is the absolute Galois group of a field K then we will also denote this set by $\mathbf{X}_I\,(hK)$.

Remark 9.27 Note that if all the simplicial Γ-sets $\mathbf{X}_\alpha$ are strictly bounded then we have the isomorphism of sets

$$\mathbf{X}_I\left(\mathbf{E}\Gamma^{\natural}\right) \simeq \mathrm{Map}_{\mathrm{Pro}\,\mathrm{Ho}\left(\mathrm{Set}_\Gamma^{\Delta^{op}}\right)}\left(\mathbf{E}\Gamma^{\natural}, \mathbf{X}_I\right).$$

Remark 9.28 Usually one would like to get a notion of a homotopy fixed points **space** by taking some kind of a derived equivariant mapping space from a point. Alas our definition **cannot** be obtained this way, and in particular we work with a set which in general is not the set of connected components of an underlying space. This definition is hence not optimal (see for e.g Proposition 9.95). A better definition can be given by using a suitable model structure (see [BSc11]). Other approaches for a "correct" definition of such a notion appear in [Qui09] as well as in [FIs07]. However these definitions are less convenient for our needs.

When we originally wrote this paper we did not have at our disposal the model structure described in [BSc11] and thus we chose to work with the less "correct" notion. We will publish a model structure-based version of the results in this paper in the near future. In any case one should be aware that the model theoretic approach gives rise to **different** objects, but to the **same** obstruction sets (and much more elegant proofs).

When $\mathbf{X}_I$ is the étale homotopy type of an algebraic variety we use the following abbreviations

$$X(hK) = \acute{E}t^{\natural}_{/K}(X)\,(hK) = \lim_{\mathcal{U}\in I(X),k\in\mathbb{N}} \pi_0\left(\mathbf{X}^{h\Gamma_K}_{\mathcal{U},k}\right)$$

and

$$X^n(hK) = \acute{E}t^{n}_{/K}(X)\,(hK) = \lim_{\mathcal{U}\in I(X),k\leq n} \pi_0\left(\mathbf{X}^{h\Gamma_K}_{\mathcal{U},k}\right) = \lim_{\mathcal{U}\in I(X)} \pi_0\left(\mathbf{X}^{h\Gamma_K}_{\mathcal{U},n}\right).$$

Remark 9.29 Let $\mathbf{X}$ be a simplicial Γ-set. By considering $\mathbf{X}$ as a pro-object in $\mathrm{Ho}\left(\mathrm{Set}^{\Delta^{op}}_{\Gamma_K}\right)$ in a trivial way we will write

$$\mathbf{X}\left(\mathbf{E}\Gamma^{\natural}\right) = \pi_0\left(\mathbf{X}^{h\Gamma}_{\alpha}\right).$$

When $\Gamma = \Gamma_K$ is the absolute Galois group of a field we will use the notation

$$\mathbf{X}(hK) = \mathbf{X}\left(\mathbf{E}\Gamma^{\natural}\right).$$

Summarizing the discussion so far we see that for every $0 \leq n \leq \infty$ we get a natural map

$$h_n : X(K) \longrightarrow X^n(hK).$$

It is useful to keep in mind the most trivial example:

Lemma 9.30

$$\mathrm{Spec}\,(K)^n\,(hK) = *\,.$$

Proof We know that $\acute{E}t^{\natural}_{/K}(\mathrm{Spec}\,(K)) \cong \{\mathbf{E}G_L\}_{L/K}$ where L runs over all finite Galois extensions of K and G_L is the Galois group of L over K. Since each $\mathbf{E}G_K$ is Kan contractible and strictly bounded we see that

$$\mathrm{Spec}\,(K)^n\,(hK) = \mathrm{Spec}\,(K)(hK) = \lim_{L/K} \mathbf{E}G_L^{h\Gamma} \cong *\,.$$

□

Remark 9.31 Notions for homotopy fixed points for action of pro-finite groups on pro-spaces were studied by Fausk and Isaksen in [FIs07] and by Quick in [Qui09]. Fausk and Isaksen's approach is to put a model structure on the category of pro-spaces with a pro-finite group action. In Quick's work one replaces pro-spaces by simplicial pro-sets. Both approaches use a model structure in order to produce a homotopy fixed point **space**, and not just a set as we have in Definition 9.26.

In this paper we work with (a relative variation of) the étale homotopy type which is a pro object in the **homotopy** category of spaces, and not of spaces. Hence one cannot apply to it either of the theories above. In order to use Fausk and Isaksen's approach one would need to replace the étale homotopy type with the étale topological type (see Friedlander [Fri82]) which is a pro-space. Alternatively one can convert the étale topological type to a simplicial pro-set and use Quick's theory.

There are some drawbacks for working without a model structure. For example, one gets only a homotopy fixed points set and not a homotopy fixed point space. Furthermore, we only prove that this set is invariant under pro-weak equivalences with heavy assumptions and using very ad-hoc technical machinery.

However, working with the (relative) étale homotopy type has its advantages. In our proofs we always end up analyzing objects with favorable "finiteness" properties (like varieties over number fields). Thus, for our needs in this paper we found the étale homotopy type more suitable than the aforementioned model structures.

Lately, Ilan Barnea and the second author ([BSc11]) defined a new model structure which is more suitable for our needs. A detailed description of how to use this model structure in the context of this work would be published by the authors in another article.

p-adic and adelic homotopy fixed points

Let K be a number field and K_ν a completion of K. The relationship between a number field and its completions will be the basis on which we will construct obstructions. To begin let us denote by $\Gamma_\nu < \Gamma_K$ the decomposition group.

If $\mathbf{X}$ is a simplicial Γ_K-set we can naturally consider it as a Γ_ν set by restricting the action. The group Γ_ν is pro-finite as well and we can apply Goerss theory to it. In particular for every hypercovering $\mathcal{U} \longrightarrow X$ and a natural number k we can consider $\mathbf{X}_{\mathcal{U},k}$ as a simplicial Γ_ν-set and take the corresponding homotopy fixed points

$$\mathbf{X}_{\mathcal{U},k}(hK_\nu) = \pi_0\left(\mathbf{X}_{\mathcal{U},k}^{h\Gamma_\nu}\right)$$

as well as the limit as in Definition 9.26:

$$X^n(hK_\nu) = \lim_{\mathcal{U},k\leq n} \mathbf{X}_{\mathcal{U},k}(hK_\nu).$$

Remark 9.32 Let $\mathbf{X}$ be a simplicial Γ_K-set. It is well known that every subgroup of a pro-finite group is equal to the intersection of all open subgroups containing it. Hence for every $\Lambda \tilde{\triangleleft} \Gamma_\nu$ one gets

$$\mathbf{X}^\Lambda = \operatorname*{colim}_{\Lambda < \Lambda' \tilde{\triangleleft} \Gamma_K} \mathbf{X}^{\Lambda'}.$$

This means that if $f : \mathbf{X} \longrightarrow \mathbf{Y}$ is a strict weak equivalence of simplicial Γ_K-sets then f induces a strict equivalence of simplicial Γ_ν-sets for every ν.

Now for every $0 \leq n \leq \infty$ we get a map

$$h_n : X(K_\nu) \longrightarrow X^n(hK_\nu).$$

Taking into account all the completions of K we get a commutative diagram

$$\begin{array}{ccc} X(K) & \xrightarrow{h_n} & X^n(hK) \\ \Big\downarrow{\scriptstyle \mathrm{loc}} & & \Big\downarrow{\scriptstyle \mathrm{loc}_{h,n}} \\ \prod_\nu X(K_\nu) & \xrightarrow{h_n} & \prod_\nu X^n(hK_\nu). \end{array}$$

Note that we abuse notation and use h_n for the rational case, the p-adic case and the product of p-adics case.

Now consider the set

$$\left(\prod_\nu X(K_\nu)\right)^{h,n} \subseteq \prod_\nu X(K_\nu)$$

given by all elements $(x_\nu) \in \prod_\nu X(K_\nu)$ such that $h_n((x_\nu)) \in \mathrm{im}(\mathrm{loc}_{\mathrm{h,n}})$.

Note that

$$X(K) \subseteq \left(\prod_\nu X(K_\nu)\right)^{h,n}.$$

Thus the comparison of local and global homotopy fixed points can be used to define obstructions sets. However, when studying the local global principle on a variety X one sees that in general it is better to work with the set of adelic points on $X(\mathbb{A})$ rather then the entire product $\prod_{\nu} X(K_\nu)$. Similarly we would like to replace the set $\prod_{\nu} X^n(hK_\nu)$ with an analogous set $X^n(h\mathbb{A})$.

Before defining such a notion for an object in $\mathrm{Pro}\,\mathrm{Ho}^{st}(\mathrm{Set}_{\Gamma_K}^{\Delta^{op}})$ we shall define it for the more simple case of a simplicial Γ_K-set.

Definition 9.33 Let K_ν be a non-archimedean local field, $I_\nu \lhd \Gamma_\nu$ the inertia group and

$$\Gamma_\nu^{ur} = \Gamma_\nu / I_\nu$$

the unramified Galois group. Let $\mathbf{X}$ be a simplicial Γ_ν-set. We define the **unramified** Γ_{K_ν}-homotopy fixed points to be the simplicial set

$$\mathbf{X}^{h^{ur}\Gamma_\nu} = \left(\mathbf{X}^{I_\nu}\right)^{h\Gamma_\nu^{ur}}.$$

Remark 9.34 In light of remark 9.32 we see that if $f : \mathbf{X} \longrightarrow \mathbf{Y}$ is a strict weak equivalence of simplicial Γ_K-sets then f induces a weak equivalence of simplicial sets

$$\mathbf{X}^{I_\nu} \xrightarrow{\sim} \mathbf{Y}^{I_\nu}$$

and so a weak equivalence

$$\mathbf{X}^{h^{ur}\Gamma_\nu} \xrightarrow{\sim} \mathbf{Y}^{h^{ur}\Gamma_\nu}.$$

We will not need this result in this paper.

Definition 9.35 Let K be a number field, S a set of places of K and $\mathbf{X}$ a simplicial Γ_K-set. We define the restricted product of S-homotopy fixed points space to be

$$\mathbf{X}^{h\mathbb{A}_S} = \operatorname*{hocolim}_{T} \prod_{\nu \in T} \mathbf{X}^{h\Gamma_\nu} \times \prod_{\nu \in S \setminus T} \mathbf{X}^{h^{ur}\Gamma_\nu}$$

where T runs over all the finite subsets of S. We also denote

$$\mathbf{X}(h\mathbb{A}_s) = \pi_0\left(\mathbf{X}^{h\mathbb{A}_S}\right).$$

Remark 9.36 Note that as a restricted product of discrete sets $\mathbf{X}(h\mathbb{A}_s)$ carries the restricted product topology.

Since homotopy groups commute with products and directed colimits we get that

$$\pi_n\left(\mathbf{X}^{h\mathbb{A}_S}\right) = \prod_{\nu\in S}{}' \pi_n\left(\mathbf{X}^{h\Gamma_\nu}\right)$$

when the restricted product is taken with respect to the subsets

$$\mathrm{im}\left[\pi_n\left(\mathbf{X}^{h^{ur}\Gamma_\nu}\right) \longrightarrow \pi_n\left(\mathbf{X}^{h\Gamma_\nu}\right)\right].$$

When S is the set of all places we denote $\mathbf{X}(h\mathbb{A}_s)$ and $\mathbf{X}^{h\mathbb{A}_S}$ by $\mathbf{X}(h\mathbb{A})$ and $\mathbf{X}^{h\mathbb{A}}$ respectively. Similarly when S is the set of all finite places we denote $\mathbf{X}(h\mathbb{A}_s)$ and $\mathbf{X}^{h\mathbb{A}_S}$ by $\mathbf{X}(h\mathbb{A}_f)$ and $\mathbf{X}^{h\mathbb{A}_f}$ respectively.

Definition 9.37 Let K be a number field. Let $\mathbf{X}_I = \{\mathbf{X}_\alpha\}_{\alpha\in I}$ be a pro-Γ_K-simplicial set. We define

$$\mathbf{X}_I(h\mathbb{A}_s) = \lim_{\alpha\in I} \mathbf{X}_\alpha(h\mathbb{A}_s).$$

In the case where we are dealing with the étale homotopy type of an algebraic variety X over K we abbreviate

$$X^n(h\mathbb{A}) = \acute{E}t^n_{/K}(X)(h\mathbb{A})$$

and

$$X(h\mathbb{A}) = X^\infty(h\mathbb{A}).$$

Lemma 9.38 *Let K be number field and $\mathbf{X}$ a strictly bounded Γ_K-simplicial set. Then the natural map*

$$\mathrm{loc} : \mathbf{X}(hK) \longrightarrow \prod_\nu \mathbf{X}(hK_\nu)$$

factors through a natural map

$$f_0 : \mathbf{X}(hK) \longrightarrow \mathbf{X}(h\mathbb{A}).$$

Proof Let L/K be finite extension and T_L is the set of places of K that ramify L. Since for $\nu \notin T_L$ we have $I_\nu < \Gamma_L$ there is a natural map:

$$f_L : \mathbf{X}^{\Gamma_L} \longrightarrow \prod_{\nu\in T_L} \mathbf{X} \times \prod_{\nu\notin T_L} \mathbf{X}^{I_\nu}.$$

Now this map induces a map

$$f : (\mathbf{X}^{\Gamma_L})^{hG_L} \longrightarrow \prod_{\nu\in T_L} \mathbf{X}^{h\Gamma_\nu} \times \prod_{\nu\notin T_L} (\mathbf{X}^{I_\nu})^{h\Gamma_\nu^{ur}}.$$

By passing to homotopy colimit on L we get a map

$$f : \mathbf{X}^{h\Gamma_K} \longrightarrow \mathbf{X}^{h\mathbb{A}}$$

and we can choose $f_0 = \pi_0(f)$. □

Lemma 9.39 *Let K_ν be a local field, let X be a variety over K_ν and let $\mathcal{U}_\bullet \longrightarrow X$ be an étale hypercovering. Then for every $n \geq 0$ the map*

$$h : X(K_\nu) \longrightarrow \mathbf{X}_{\mathcal{U},n}\,(hK_\nu)$$

is continuous (where $X(K_\nu)$ inherits a natural topology from the topology of K_ν and $\mathbf{X}_{\mathcal{U},n}\,(hK_\nu)$ is discrete).

Proof Let $x \in X(K_\nu)$ be a point. It is enough to find a neighborhood V of x in $X(K_\nu)$ (with respect to the K_ν-topology) such that for every $y \in V$ we have $h(x) = h(y)$. Consider x as a map

$$x : \operatorname{Spec} K_\nu \longrightarrow X$$

and let $x^*(\mathcal{U})$ be the hypercovering of $\operatorname{Spec} K_\nu$ which is the pullback of $\mathcal{U}$ by x. The underlying simplicial set of $\mathbf{X}_{x^*(\mathcal{U}),n}$ is contractible and hence $\mathbf{X}^{h\Gamma_v}_{x^*(\mathcal{U}),n}$ is also contractible. The map x gives a map

$$\tilde{x} : \mathbf{X}_{x^*(\mathcal{U}),n} \longrightarrow \mathbf{X}_{\mathcal{U},n}$$

such that the induced map

$$\tilde{x}^{h\Gamma_v} : \mathbf{X}^{h\Gamma_v}_{x^*(\mathcal{U}),n} \longrightarrow \mathbf{X}^{h\Gamma_v}_{\mathcal{U},n}$$

sends $\mathbf{X}^{h\Gamma_v}_{x^*(\mathcal{U}),n}$ to the connected component $h(x)$.

Now given any étale map $f : U \longrightarrow X$ the Inverse Function Theorem insures that there is an open neighborhood V of x in the K_ν-topology such that for every $y \in V$ there is a natural Galois equivariant identification of the fibers

$$F_y : f^{-1}(x) \longrightarrow f^{-1}(y).$$

This identification takes any point in $f^{-1}(x)$ to the unique point in $f^{-1}(y)$ sitting with it in the same connected component of $f^{-1}(V)$.

This means that there is an open neighborhood $V_{\mathcal{U},n}$ of x in the K_ν-topology such that for every $y \in V_{\mathcal{U},n}$ there is a natural Galois equivariant map

$$F_{\mathcal{U},n} : \operatorname{sk}_n(x^*(\mathcal{U})) \longrightarrow \operatorname{sk}_n(y^*(\mathcal{U})).$$

Now since for every $y \in V_{\mathcal{U},n}$ the map $\tilde{x}$ factors through the map $\tilde{y}$ we see that $\tilde{x}^{h\Gamma_v}$ and $\tilde{y}^{h\Gamma_v}$ must land in the same connected component and so $h(y) = h(x)$. □

Lemma 9.40 *Let X be an algebraic variety over a number field K and let $\mathcal{U}_\bullet \longrightarrow X$ be a hypercovering. Then the natural continuous map*

$$X(\mathbb{A}) \longrightarrow \prod_{\nu} \mathbf{X}_{\mathcal{U},n}\left(hK_\nu\right)$$

factors through a natural continuous map

$$h_{\mathcal{U},n} : X(\mathbb{A}) \longrightarrow \mathbf{X}_{\mathcal{U},n}\left(h\mathbb{A}\right).$$

Proof First choose some model for $\mathcal{U}_\bullet \longrightarrow X$ over Spec $(\mathbb{Z})$,

$$(\mathcal{U}_\mathbb{Z})_\bullet \longrightarrow X_\mathbb{Z},$$

then there is a finite set of primes T_1 such that the maps $(\mathcal{U}_\mathbb{Z})_m \longrightarrow X_\mathbb{Z}$ are étale outside T_1 for $m \leq n+1$. We will show that

$$h(X_\mathbb{Z}(O_\nu)) \subseteq \operatorname{im}\left[\pi_0\left(\mathbf{X}_{\mathcal{U},\mathrm{n}}^{\mathrm{h^{ur}}\Gamma_\nu}\right) \longrightarrow \mathbf{X}_{\mathcal{U},\mathrm{n}}\left(\mathrm{hK}_\nu\right)\right]$$

for almost all ν. Choose some finite extension L/K such that Γ_L fixes $\mathrm{sk}_n\left(\mathbf{X}_\mathcal{U}\right)$ and denote by T_0 the set of ramified places in L.

Consider now a place $\nu \notin T_0 \cup T_1$. Since $\nu \notin T_0$ we have

$$\mathbf{X}_{\mathcal{U},n}^{I_\nu} = \mathbf{X}_{\mathcal{U},n}.$$

Hence all we need to show is that the homotopy fixed point $h_n(x)$ comes from a homotopy fixed point of the quotient group $\Gamma_\nu^{ur} = \Gamma_\nu / I_\nu$.

Since $\nu \notin T_1$ pulling back $\mathcal{U}_\mathbb{Z}$ to a point $x \in X_\mathbb{Z}(O_\nu)$ will yield a (contractible) simplicial set E that is stabilized up to level $n+1$ by some unramified extension of K_ν. Then $h_n(x)$ is the image of the unique point $\pi_0(E^{h\Gamma_\nu}) = *$ in $\mathbf{X}_{\mathcal{U},n}$. Since E is stabilized by I_ν this homotopy fixed point comes from Γ_ν^{ur} and we are done.

This means that in order to prove Lemma 9.40 it will be enough to prove the following general lemma on the behavior of maps between restricted products of topological spaces:

Lemma 9.41 *Let*

$$\{f_\lambda : X_\lambda \longrightarrow Y_\lambda\}_{\lambda \in \Lambda}$$

be a family of continuous maps of topological spaces and let

$$\{O_\lambda, U_\lambda\}_{\lambda \in \Lambda}$$

be a family of open subsets $O_\lambda \subseteq X_\lambda, U_\lambda \subseteq Y_\lambda$.

Assume that $f(O_\lambda) \subseteq U_\lambda$ for almost all λ. Then the map

$$F_\Lambda = \prod_{\lambda \in \Lambda} f_\lambda : \prod_{\lambda \in \Lambda} X_\lambda \longrightarrow \prod_{\lambda \in \Lambda} Y_\lambda$$

induces a continuous map

$$F_\Lambda : \prod_{\lambda\in\Lambda}' X_\lambda \longrightarrow \prod_{\lambda\in\Lambda}' Y_\lambda$$

where the restricted product is taken with respect to O_λ, U_λ respectively.

Proof It is clear that

$$F_\Lambda\left(\prod_{\lambda\in\Lambda}' X_\lambda\right) \subseteq \prod_{\lambda\in\Lambda}' Y_\lambda.$$

Hence it is enough to show that F_Λ is continuous.

We need to show that if $S \subseteq \Lambda$ is a finite set and $\{A_\lambda \subseteq Y_\lambda\}_{\lambda\in S}$ are open then the set

$$(F_\Lambda)^{-1}\left(\prod_{\lambda\in S} A_\lambda \times \prod_{\lambda\in\Lambda\setminus S} X_\lambda\right)$$

is open in $\prod_{\lambda\in\Lambda}' X_\lambda$. Let

$$(x_\lambda)_\lambda \in (F_\Lambda)^{-1}\left(\prod_{\lambda\in S} A_\lambda \times \prod_{\lambda\in\Lambda\setminus S} U_\lambda\right)$$

and let $T \subseteq \Lambda$ be a finite set containing S such that

$$x_\lambda \in O_\lambda, f(O_\lambda) \subseteq U_\lambda$$

for every $\lambda \in \Lambda \setminus T$. Note that such a set exists due the assumptions of the lemma. Consider the set

$$N_x = \prod_{\lambda\in S} f_\lambda^{-1}(A_\lambda) \times \prod_{\lambda\in T\setminus S} f_\lambda^{-1}(U_\lambda) \times \prod_{\lambda\in\Lambda\setminus T} O_\lambda.$$

It is clear that N_x is an open neighborhood of x and that

$$N_x \subseteq (F_\Lambda)^{-1}\left(\prod_{\lambda\in S} A_\lambda \times \prod_{\lambda\in\Lambda\setminus S} X_\lambda\right).$$

□

This completes the proof of Lemma 9.40. □

The étale homotopy obstructions

By Lemma 9.40 and Lemma 9.38 we have a commutative diagram

$$\begin{array}{ccc} X(K) & \xrightarrow{h_n} & X^n(hK) \\ \downarrow{\scriptstyle \mathrm{loc}} & & \downarrow{\scriptstyle \mathrm{loc}_{h,n}} \\ X(\mathbb{A}) & \xrightarrow{h_n} & X^n(h\mathbb{A})\,. \end{array}$$

Note that we abuse notation and use h_n for both the rational and adelic cases.

We denote by $X(\mathbb{A})^{h,n} \subseteq X(\mathbb{A})$ the set of adelic points whose image in $X^n(h\mathbb{A})$ lies in the image of $\mathrm{loc}_{h,n}$. Note that

$$X(\mathbb{A})^{h,n} = \left(\prod_{\nu} X(K_\nu)\right)^{h,n} \cap X(\mathbb{A}) \subseteq X(\mathbb{A})$$

and also

$$X(K) \subseteq X(\mathbb{A})^h \subseteq \cdots \subseteq X(\mathbb{A})^{h,2} \subseteq \cdots \subseteq X(\mathbb{A})^{h,1} \subseteq X(\mathbb{A}).$$

We denote $X(\mathbb{A})^{h,\infty}$ simply by $X(\mathbb{A})^h$. We call the elements of $X(\mathbb{A})^h$ the set of **homotopically rational points**.

Definition 9.42 We say that the lack of K-rational points in X is explained by the **étale homotopy obstruction** if the set $X(\mathbb{A})^h$ is empty.

The étale homology obstructions

Let Mod_Γ be the category of discrete Γ-modules. Consider the augmented functor

$$\mathbb{Z} : \mathrm{Set}_\Gamma \longrightarrow \mathrm{Mod}_\Gamma$$

which associates to the Γ-set A the free abelian group $\mathbb{Z}A$ generated from A with the induced Galois action. The terminal map $A \longrightarrow \{*\}$ defines a map $\mathbb{Z}A \longrightarrow \mathbb{Z}$ which we will call the **degree** map. Note that the image of the augmentation map $A \longrightarrow \mathbb{Z}A$ lies in the subset of elements of degree 1.

For a simplicial Γ-set $\mathbf{X}$ we will denote by $\mathbb{Z}\mathbf{X}$ the simplicial Γ-module obtained by applying the $\mathbb{Z}$ functor levelwise, i.e.

$$(\mathbb{Z}\mathbf{X})_n = \mathbb{Z}(\mathbf{X}_n).$$

The terminal map $\mathbf{X} \longrightarrow *$ induces a map from $\mathbb{Z}\mathbf{X}$ to the constant simplicial Γ-module $\mathbb{Z}$ (this is the discrete simplicial Γ-module with the trivial action). Note that again the elements in the image of the augmentation map have degree 1.

The homotopy groups of $\mathbb{Z}\mathbf{X}$ can be naturally identified with the homology of $\mathbf{X}$ and the augmentation map induces the Hurewicz map

$$\pi_*(\mathbf{X}) \longrightarrow \pi_*(\mathbb{Z}\mathbf{X}) = H_*(\mathbf{X}).$$

We shall refer to the augmentation map as the Hurewicz map as well.

We can now consider the $\mathbb{Z}$-variant of the functor $\acute{E}t_{/K}$ applying the $\mathbb{Z}$ functor on each simplicial set in the diagram:

$$\mathbb{Z}\acute{E}t_{/K} = \{\mathbb{Z}\mathbf{X}_{\mathcal{U}}\}_{\mathcal{U}\in I(X)}.$$

As before we prefer to work with bounded simplicial Γ-sets and so we replace this object by its Postnikov tower

$$(\mathbb{Z}\acute{E}t_{/K})^n = \{P_k(\mathbb{Z}\mathbf{X}_{\mathcal{U}})\}_{\mathcal{U}\in I(X),k\le n}$$

as well as the full Postnikov tower

$$(\mathbb{Z}\acute{E}t_{/K})^\natural = (\mathbb{Z}\acute{E}t_{/K})^\infty = \{P_k(\mathbb{Z}\mathbf{X}_{\mathcal{U}})\}_{\mathcal{U}\in I(X),k\in\mathbb{N}}.$$

For every $0 \le n \le \infty$ we have a natural transformation

$$\acute{E}t^n_{/K}(X) \longrightarrow (\mathbb{Z}\acute{E}t_{/K})^n(X)$$

and so we can consider the commutative diagram

$$\begin{array}{ccccc}
X(K) & \xrightarrow{h_n} & X^n(hK) & \longrightarrow & X^{\mathbb{Z},n}(hK) \\
\downarrow & & \downarrow \text{loc}_{h,n} & & \downarrow \text{loc}_{\mathbb{Z}h,n} \\
X(\mathbb{A}) & \xrightarrow{h_n} & X^n(h\mathbb{A}) & \longrightarrow & X^{\mathbb{Z},n}(h\mathbb{A})
\end{array}$$

where

$$X^{\mathbb{Z},n}(hK) = ((\mathbb{Z}\acute{E}t_{/K})^n(X))(hK)$$

and

$$X^{\mathbb{Z},n}(h\mathbb{A}) = ((\mathbb{Z}\acute{E}t_{/K})^n(X))(h\mathbb{A}).$$

We say that an adelic point

$$(x_\nu) \in X(\mathbb{A}) = \prod_\nu{}' X(K_\nu)$$

is **n-homologically rational** if its image in $X^{\mathbb{Z},n}(h\mathbb{A})$ is rational, i.e.

is in the image of $\mathrm{loc}_{\mathbb{Z}h,n}$. We denote by $X(\mathbb{A})^{\mathbb{Z}h,n} \subseteq X(\mathbb{A})$ the set of n-homologically rational points. We also denote

$$X(\mathbb{A})^{\mathbb{Z}h} = X(\mathbb{A})^{\mathbb{Z}h,\infty}.$$

Definition 9.43 We say that the lack of K-rational points in X is explained by the **étale homology obstruction** if the set $X(\mathbb{A})^{\mathbb{Z}h}$ is empty.

From the above discussion we immediately see that we have the following diagram of inclusions

$$\begin{array}{ccccccccc} & & X(\mathbb{A})^{\mathbb{Z}h} & \hookrightarrow \cdots \hookrightarrow & X(\mathbb{A})^{\mathbb{Z}h,2} & \hookrightarrow & X(\mathbb{A})^{\mathbb{Z}h,1} & \hookrightarrow & X(\mathbb{A}) \\ & & \uparrow & & \uparrow & & \uparrow & & \\ X(K) & \hookrightarrow & X(\mathbb{A})^{h} & \hookrightarrow \cdots \hookrightarrow & X(\mathbb{A})^{h,2} & \hookrightarrow & X(\mathbb{A})^{h,1} & & \end{array}$$

and so the étale n-homology obstruction is in general weaker then the étale n-homotopy obstruction.

The single hypercovering version

It will sometimes be convenient to consider the information obtained from a single hypercovering. Let X be an algebraic variety over K and

$$\acute{E}t^{\natural}_{/K}(X) = \{\mathbf{X}_{\mathcal{U},n}\}_{\mathcal{U}\in I(X),n\in\mathbb{N}}$$

For each hypercovering $\mathcal{U} \in I(X)$ and $n \in \mathbb{N}$ we can consider the diagram

$$\begin{array}{ccc} X(K) & \xrightarrow{h_{\mathcal{U},n}} & \mathbf{X}_{\mathcal{U},n}(hK) \\ \downarrow & & \downarrow \mathrm{loc}_{\mathcal{U},n} \\ X(\mathbb{A}) & \xrightarrow{h_{\mathcal{U},n}} & \mathbf{X}_{\mathcal{U},n}(h\mathbb{A}) . \end{array}$$

We denote by $X(\mathbb{A})^{\mathcal{U},n}$ the set of adelic points $(x_\nu) \in X(\mathbb{A})$ whose image in $\mathbf{X}_{\mathcal{U},n}(h\mathbb{A})$ is rational (i.e. is in the image of $\mathrm{loc}_{\mathcal{U},n}$).

Similarly we can consider the diagrams

$$\begin{array}{ccc} X(K) & \xrightarrow{h_{\mathbb{Z}\mathcal{U},n}} & \mathbb{Z}\mathbf{X}_{\mathcal{U},n}(hK) \\ \downarrow & & \downarrow \mathrm{loc}_{\mathbb{Z}\mathcal{U},n} \\ X(\mathbb{A}) & \xrightarrow{h_{\mathbb{Z}\mathcal{U},n}} & \mathbb{Z}\mathbf{X}_{\mathcal{U},n}(h\mathbb{A}) . \end{array}$$

We denote by $X(\mathbb{A})^{\mathbb{Z}\mathcal{U},n}$ the set of adelic points $(x_\nu) \in X(\mathbb{A})$ whose image in $\mathbb{Z}\mathbf{X}_{\mathcal{U},n}(h\mathbb{A})$ is rational (i.e. is in the image of $\mathrm{loc}_{\mathbb{Z}\mathcal{U},n}$).

Note that for every object $\{\mathbf{X}_\alpha\}_{\alpha \in I} \in \mathrm{Pro}\,\mathrm{Ho}\left(\mathrm{Set}_\Gamma^{\Delta^{op}}\right)$ and every $\alpha_0 \in I$ we have a natural map

$$\{\mathbf{X}_\alpha\}_{\alpha \in I} \longrightarrow \mathbf{X}_{\alpha_0}.$$

Thus

$$X(\mathbb{A})^{h,n} \subseteq X(\mathbb{A})^{\mathcal{U},n}$$

and

$$X(\mathbb{A})^{\mathbb{Z}h,n} \subseteq X(\mathbb{A})^{\mathbb{Z}\mathcal{U},n}$$

for every hypercovering $\mathcal{U}_\bullet \longrightarrow X$.

9.3.3 Summary of notation

The following is a summary of the notation we've used so far for maps from points to corresponding homotopy fixed points.

$$h : X(K) \longrightarrow X(hK),$$

$$h_n : X(K) \longrightarrow X^n(hK),$$

$$h_{\mathcal{U}} : X(K) \longrightarrow \mathbf{X}_{\mathcal{U}}(hK),$$

$$h_{\mathcal{U},n} : X(K) \longrightarrow \mathbf{X}_{\mathcal{U},n}(hK),$$

$$h_{\mathbb{Z}} : X(K) \longrightarrow X^{\mathbb{Z}}(hK),$$

$$h_{\mathbb{Z},n} : X(K) \longrightarrow X^{\mathbb{Z},n}(hK),$$

$$h_{\mathbb{Z}\mathcal{U}} : X(K) \longrightarrow \mathbb{Z}\mathbf{X}_{\mathcal{U}}(hK),$$

$$h_{\mathbb{Z}\mathcal{U},n} : X(K) \longrightarrow P_n(\mathbb{Z}\mathbf{X}_{\mathcal{U}})(hK).$$

We will abuse notation and use the exact same notation for the adelic versions of all of these maps.

9.4 Homotopy fixed points for pro-finite groups

Let Γ be a pro-finite group. We start with the basic calculative theorem regarding homotopy fixed points of simplicial Γ-sets in Goerss' model category:

Theorem 9.44 *Let* $\mathbf{Y}$ *be a simplicial* Γ*-set whose underlying simplicial set is Kan. Let* $D_{fin} \subseteq \mathrm{Ho}\left(\mathrm{Set}_{\Gamma}^{\Delta^{op}}\right)$ *be the full subcategory spanned by Kan contractible objects which are levelwise finite. Then one has an isomorphism of sets*

$$\pi_0\left((\mathbf{Y})^{h\Gamma}\right) \simeq \operatorname*{colim}_{\mathbf{E}\in D_{fin}} [\mathbf{E},\mathbf{Y}]_{\Gamma}.$$

If in addition $\mathbf{Y}$ *is also* ***strictly bounded*** *then the formula can be refined to*

$$\pi_0\left((\mathbf{Y})^{h\Gamma}\right) \simeq \operatorname*{colim}_{\Lambda \trianglelefteq \Gamma}[\mathbf{E}(\Gamma/\Lambda),\mathbf{Y}]_{\Gamma} \simeq$$
$$\operatorname*{colim}_{\Lambda \trianglelefteq \Gamma}\left[\mathbf{E}(\Gamma/\Lambda),\mathbf{Y}^{\Lambda}\right]_{\Gamma} \simeq \operatorname*{colim}_{\Lambda \trianglelefteq \Gamma}\pi_0\left(\left(\mathbf{Y}^{\Lambda}\right)^{h(\Gamma/\Lambda)}\right).$$

Proof We use a formalism developed by Brown in [Bro73] called a **category of fibrant objects**. This is a notion of a category with weak equivalences and fibrations satisfying certain properties (see [Bro73] pages 420-421).

We will apply this formalism to an example analogous to one appearing in [Bro73] itself: let $C \subseteq \mathrm{Set}_{\Gamma}^{\Delta^{op}}$ be the full subcategory consisting of simplicial Γ-sets whose underlying simplicial set is Kan. We will declare a morphism in C to be a fibration if it induces a Kan fibration on the underlying simplicial sets and a weak equivalence if it induces a weak equivalence on the underlying simplicial set (so in particular weak equivalences in C coincide with those of the local model structure). It can be shown that these choices endow C with the structure of a category with fibrant objects.

Now let $\mathbf{X},\mathbf{Y}$ be two simplicial Γ-sets and let $\varphi : \mathbf{X}' \xrightarrow{\sim} \mathbf{X}$ be a weak equivalence (with respect to the local model structure). Let $g : \mathbf{X}' \longrightarrow \mathbf{Y}$ be a map. Then there exists a unique map $h : \mathbf{X} \longrightarrow \mathbf{Y}^{fib}$ such that the square

$$\begin{array}{ccc} \mathbf{X}' & \xrightarrow{g} & \mathbf{Y} \\ \downarrow{\scriptstyle \varphi} & & \downarrow{\scriptstyle f} \\ \mathbf{X} & \xrightarrow{h} & \mathbf{Y}^{fib} \end{array}$$

commutes up to simplicial homotopy. This gives us a map of sets

$$[\mathbf{X}', \mathbf{Y}]_\Gamma \longrightarrow \left[\mathbf{X}, \mathbf{Y}^{fib}\right]_\Gamma$$

for every weak equivalence $\varphi : \mathbf{X}' \xrightarrow{\simeq} \mathbf{X}$. Now Theorem 1 in [Bro73] applied to C (taking into account remark 5 on page 427 of [Bro73]) yields the following: if $\mathbf{X}$ and $\mathbf{Y}$ are in C, and if we take the colimit over all weak equivalences $\varphi : \mathbf{X}' \xrightarrow{\simeq} \mathbf{X}$ in C, then the resulting map

$$\operatorname*{colim}_{\varphi:\mathbf{X}' \xrightarrow{\simeq} \mathbf{X}} [\mathbf{X}', \mathbf{Y}]_\Gamma \longrightarrow \left[\mathbf{X}, \mathbf{Y}^{fib}\right]_\Gamma$$

is actually an isomorphism of sets.

In particular if we denote by $D \subseteq \mathrm{Ho}\left(\mathrm{Set}_\Gamma^\Delta\right)$ the full subcategory of Kan contractible simplicial Γ-sets then we get an isomorphism of sets

$$\operatorname*{colim}_{\mathbf{E}\in D} [\mathbf{E}, \mathbf{Y}]_\Gamma \xrightarrow{\simeq} \left[*, \mathbf{Y}^{fib}\right]_\Gamma = \pi_0\left(\left(\mathbf{Y}^{fib}\right)^\Gamma\right) = \pi_0\left(\mathbf{Y}^{h\Gamma}\right).$$

This colimit is indexed by a category which is a bit too big for our purposes, but this problem can easily be mended:

Lemma 9.45 *The subcategory $D_{fin} \subseteq D$ consisting of objects which are levelwise finite is cofinal.*

Proof We need to show that every object $\mathbf{E} \in D$ admits a map $\mathbf{E}' \longrightarrow \mathbf{E}$ where $\mathbf{E}' \in D$ is a levelwise finite Kan contractible simplicial Γ-set. Construct $\mathbf{E}'$ inductively as follows: let $E'_{-1} = \emptyset$ and $f_{-1} : \mathbf{E}'_{-1} \longrightarrow \mathbf{E}$ the unique map. We will extend $\{\mathbf{E}'_{-1}\}$ to an increasing family of simplicial Γ-sets

$$\mathbf{E}'_{-1} \subseteq \mathbf{E}'_0 \subseteq ... \subseteq \mathbf{E}'_n \subseteq ...$$

such that $\mathbf{E}'_n$ is an n-dimensional levelwise finite simplicial Γ-set and the map $\mathbf{E}'_n \hookrightarrow \mathbf{E}'_{n+1}$ is the inclusion of the n-skeleton. Furthermore the simplicial set E'_n will be $(n-1)$-connected in the following sense: every map $\partial\Delta^m \longrightarrow E'_n$ with $m \leq n$ extends to Δ^m. This will guarantee that

$$\mathbf{E}' = \bigcup_n \mathbf{E}'_n$$

is a Kan contractible levelwise finite simplicial Γ-set. Furthermore we will construct a compatible family of equivariant maps

$$f_n : \mathbf{E}'_n \longrightarrow \mathbf{E}$$

which will induce one big equivariant map $\mathbf{E}' \longrightarrow \mathbf{E}$.

Now let $n \geq -1$ and suppose $f_n : E'_n \longrightarrow E$ as above has already been constructed. We will describe the construction of $\mathbf{E}'_{n+1}$. First for a **finite** simplicial set $\mathbf{X}$ and a simplicial Γ-set $\mathbf{Y}$ we will denote by

$$\mathbf{Y}^{\mathbf{X}} = \mathrm{Hom}_{\mathrm{Set}^{\Delta}}(\mathbf{X}, \mathbf{Y})$$

the **set** of maps of simplicial sets from $\mathbf{X}$ to $\mathbf{Y}$. The action of Γ on $\mathbf{Y}$ induces an action of Γ on $\mathbf{Y}^{\mathbf{X}}$ rendering it a Γ-set (i.e. all the stabilizers are open because $\mathbf{X}$ is finite).

Now consider the Γ-set

$$A = \mathbf{E}_n'^{\partial\Delta^{n+1}} \times_{\mathbf{E}^{\partial\Delta^{n+1}}} \mathbf{E}^{\Delta^{n+1}}.$$

This set parameterizes commutative diagrams in the category of simplicial sets of the form

$$\begin{array}{ccc} \partial\Delta^{n+1} & \longrightarrow & \mathbf{E}'_n \\ \downarrow & & \downarrow f_n \\ \Delta^{n+1} & \longrightarrow & \mathbf{E}. \end{array}$$

Since $\mathbf{E}$ is Kan contractible the map $A \longrightarrow \mathbf{E}_n'^{\partial\Delta^{n+1}}$ is **surjective**. Since $\mathbf{E}'_n$ is levelwise finite the set $\mathbf{E}_n'^{\partial\Delta^{n+1}}$ is finite, so we can choose a finite subset $A' \subseteq A$ such that the restricted map

$$A' \longrightarrow \mathbf{E}_n'^{\partial\Delta^{n+1}}$$

is still surjective (recall that all the orbits in A are finite). Then we get one big commutative diagram of simplicial Γ-sets and **equivariant** maps

$$\begin{array}{ccc} \partial\Delta^{n+1} \times A' & \longrightarrow & \mathbf{E}'_n \\ \downarrow & & \downarrow f_n \\ \Delta^{n+1} \times A' & \longrightarrow & \mathbf{E} \end{array}$$

and we define $\mathbf{E}'_{n+1}$ to be the pushout of the diagram

$$\begin{array}{ccc} \partial\Delta^{n+1} \times A' & \longrightarrow & \mathbf{E}'_n \\ \downarrow & & \\ \Delta^{n+1} \times A' & & \end{array}$$

which admits a natural equivariant extension

$$f_{n+1} : \mathbf{E}'_{n+1} \longrightarrow \mathbf{E}$$

of f_n. Since A' is finite $\mathbf{E}'_{n+1}$ is still levelwise finite. Furthermore since the map $A' \longrightarrow \mathbf{E}_n'^{\partial\Delta^{n+1}}$ is surjective we see that every map $\partial\Delta^{n+1} \longrightarrow \mathbf{E}'_{n+1}$ extends to all of Δ^{n+1}. This finishes the proof of the lemma. □

We now get the first desired formula:

$$\operatorname*{colim}_{\mathbf{E}\in D_{fin}} [\mathbf{E}, \mathbf{Y}]_\Gamma \cong \pi_0\left(\mathbf{Y}^{h\Gamma}\right).$$

Now for every finite quotient $G = \Gamma/\Lambda$ for $\Lambda \tilde{\triangleleft} \Gamma$ we can consider $\mathbf{E}G$ as a simplicial Γ-set with the action induced from the action of G. Then $\mathbf{E}G$ is Kan contractible and levelwise finite, i.e. $\mathbf{E}G \in D_{fin}$. We then have a map of sets

$$F_{\mathbf{Y}} : \operatorname*{colim}_{\Lambda \tilde{\triangleleft} \Gamma} \pi_0\left(\left(\mathbf{Y}^\Lambda\right)^{h(\Gamma/\Lambda)}\right) \cong \operatorname*{colim}_{\Lambda \tilde{\triangleleft} \Gamma} [\mathbf{E}(\Gamma/\Lambda), \mathbf{Y}]_\Gamma \longrightarrow$$
$$\operatorname*{colim}_{\mathbf{E}\in D_{fin}} [\mathbf{E}, \mathbf{Y}]_\Gamma \cong \pi_0\left(\mathbf{Y}^{h\Gamma}\right).$$

We will finish the proof by showing that if $\mathbf{Y}$ is nice and strictly bounded then $F_{\mathbf{Y}}$ is an isomorphism of sets.

Let n be big enough so that $\pi_n\left(\mathbf{Y}^\Lambda\right) = 0$ for all $\Lambda \tilde{\triangleleft} \Gamma$. Then the map

$$\mathbf{Y} \longrightarrow P_n(\mathbf{Y})$$

is a strict weak equivalence. Note that both the domain and range of $F_{\mathbf{Y}}$ are invariant under strict weak equivalence in $\mathbf{Y}$, so it will be enough to prove the theorem for $P_n(\mathbf{Y})$. We will start by showing that $F_{P_n(\mathbf{Y})}$ is surjective.

Let $g : \mathbf{E} \longrightarrow P_n(\mathbf{Y})$ be a map. Then g factors

$$\mathbf{E} \longrightarrow P_n(\mathbf{E}) \xrightarrow{g'} P_n(\mathbf{Y}).$$

Since $\mathbf{E} \in D_{fin}$, it is in particular nice and so $P_n(\mathbf{E})$ is excellent, i.e. the action of Γ on $P_n(\mathbf{E})$ factors through a finite quotient $G = \Gamma/\Lambda$. Furthermore $P_n(\mathbf{E})$ is also Kan contractible so it admits a G-homotopy fixed point, i.e. a map

$$h : \mathbf{E}G \longrightarrow P_n(\mathbf{E}).$$

The fact that such a map exists **simplicially** can be seen by using the projective model structure on simplicial G-sets. Now the composition

$$\mathbf{E}G \xrightarrow{h} P_n(\mathbf{E}) \xrightarrow{g'} P_n(\mathbf{Y})$$

and

$$g : \mathbf{E} \longrightarrow P_n(\mathbf{Y})$$

both factor through g' and so represent the same element in

$$\operatorname*{colim}_{\mathbf{E}\in D_{fin}} [\mathbf{E}, \mathbf{Y}]_\Gamma.$$

This means that $F_{P_n(\mathbf{Y})}$ is surjective. Now consider a diagram

$$\begin{array}{ccc} \mathbf{E} & \xrightarrow{p_1} & \mathbf{E}(\Gamma/\Lambda_1) \\ \downarrow{\scriptstyle p_2} & & \downarrow{\scriptstyle f_1} \\ \mathbf{E}(\Gamma/\Lambda_2) & \xrightarrow{f_2} & P_n(\mathbf{Y}) \end{array}$$

which commutes up to simplicial homotopy, i.e. f_1 and f_2 represent the same element in $\operatorname{colim}_{\mathbf{E}\in D_{fin}} [\mathbf{E}, \mathbf{Y}]_\Gamma$. Since $P_n(\mathbf{E}(\Gamma/\Lambda_i)) = \mathbf{E}(\Gamma/\Lambda_i)$ we get that this diagram factors through a diagram

$$\begin{array}{ccc} P_n(\mathbf{E}) & \xrightarrow{p_1'} & \mathbf{E}(\Gamma/\Lambda_1) \\ \downarrow{\scriptstyle p_2'} & & \downarrow{\scriptstyle f_1} \\ \mathbf{E}(\Gamma/\Lambda_2) & \xrightarrow{f_2} & P_n(\mathbf{Y}) \end{array}$$

which also commutes up to simplicial homotopy since we have a map

$$P_n(\mathbf{E}) \times I \xrightarrow{\sim} P_n(\mathbf{E}) \times P_n(I) \cong P_n(\mathbf{E} \times I).$$

Now there exists a $\Lambda_3 \subseteq \Lambda_1 \cap \Lambda_2$ such that the action of Γ on $P_n(\mathbf{E})$ factors through Γ/Λ_3. Since $P_n(\mathbf{E})$ is Kan contractible it admits a map $h : \mathbf{E}(\Gamma/\Lambda_3) \longrightarrow P_n(\mathbf{E})$. Pulling the diagram by h we obtain a new diagram

$$\begin{array}{ccc} \mathbf{E}(\Gamma/\Lambda_3) & \xrightarrow{p_1'} & \mathbf{E}(\Gamma/\Lambda_1) \\ \downarrow{\scriptstyle p_2'} & & \downarrow{\scriptstyle f_1} \\ \mathbf{E}(\Gamma/\Lambda_2) & \xrightarrow{f_2} & P_n(\mathbf{Y}) \end{array}$$

which commutes up to simplicial homotopy. This shows that g_1, g_2 represent the same element in

$$\operatorname*{colim}_{\Lambda \trianglelefteq \Gamma} [\mathbf{E}(\Gamma/\Lambda), \mathbf{Y}]_\Gamma$$

and so $F_{P_n(\mathbf{Y})}$ is injective. This finishes the proof of the theorem. □

Aside for having an explicit formula we would also like to have a concrete computation aid, in the form of a spectral sequence. The following theorem appears in the paper of Goerss [Goe95]:

Theorem 9.46 *Let* $\mathbf{X}$ *be a bounded simplicial* Γ*-set and let* $x \in \mathbf{X}$ *be a fixed point of* Γ. *Then there exists a spectral sequence of pointed sets*

$$E^r_{s,t} \Rightarrow \pi_{t-s}\left(\mathbf{X}^{h\Gamma}, x\right)$$

such that

$$E^2_{s,t} \cong H^s(\Gamma, \pi_t(\mathbf{X}, x)).$$

Remark 9.47 When we write $H^*(\Gamma, A)$ for Γ profinite we always mean **Galois** cohomology. In [Goe95] Goerss uses the notation $H^*_{\mathrm{Gal}}(\Gamma, A)$ for this notion. For simplicity of notation we chose to omit the Gal subscript. Note that when Γ is a finite group, Galois cohomology coincides with regular group cohomology.

Remark 9.48 The above spectral sequence is of the form used to compute homotopy groups of homotopy limits. It is concentrated in the domain $t \geq s - 1$ and its differential $d^r_{s,t}$ goes from $E^r_{s,t}$ to $E^r_{s+r,t+r-1}$. We call such spectral sequences HL-spectral sequences.

We wish to drop the assumption that $\mathbf{X}$ has an actual fixed point and replace it by the assumption that $\mathbf{X}$ is Kan and admits a **homotopy fixed point**. From Theorem 9.44 this implies the existence of an equivariant map

$$f : \mathbf{E} \longrightarrow \mathbf{X}$$

for some Kan contractible simplicial Γ-set $\mathbf{E}$. We can then take the cofiber C_f and extend the action of Γ to it.

Since $\mathbf{E}$ is contractible the map $\mathbf{X} \longrightarrow C_f$ induces a homotopy equivalence of the underlying simplicial sets and so is a weak equivalence. This means that f induces a weak equivalence $\mathbf{X}^{h\Gamma} \longrightarrow C_f^{h\Gamma}$. But C_f has an actual fixed point and so we can use Goerss' Theorem on it and obtain the desired spectral sequence.

Another aspect of homotopy fixed points for finite groups is that of **obstruction theory**. Let G be a finite group acting on a simplicial set $\mathbf{X}$. We want to know whether there exists a homotopy fixed point.

Suppose for simplicity that $\mathbf{X}$ is a bounded Kan simplicial set and let n be such that $\mathbf{X} \simeq P_n(\mathbf{X})$. Then we can reduce the question of whether $\mathbf{X}$ has a homotopy fixed point to the question of whether $P_n(\mathbf{X})$ has a homotopy fixed point. We then consider the sequence of simplicial G-sets and equivariant maps

$$P_n(\mathbf{X}) \longrightarrow P_{n-1}(\mathbf{X}) \longrightarrow \ldots \longrightarrow P_0(\mathbf{X}).$$

We can then break the non-emptiness question of $\mathbf{X}^{hG}$ into a finite number of stages: for every $i = 0, ..., n$ we can ask whether $P_i(\mathbf{X})^{hG}$ is non-empty. Note that $P_i(\mathbf{X})^{hG}$ is non-empty if and only if $\pi_0\left(P_i(\mathbf{X})^{hG}\right)$ is non-empty, so we can work with sets instead of spaces.

For $i = 0$ we have $P_0(\mathbf{X}) = \pi_0(\mathbf{X})$ and so

$$\pi_0\left(P_0(\mathbf{X})^{hG}\right) = P_0(\mathbf{X})^{hG} = \pi_0(\mathbf{X})^G$$

is just the set of G-invariant connected components. Now given a G-invariant connected component $x_0 \in \pi_0\left(P_0(\mathbf{X})^{hG}\right)$ one can ask if it lifts to an element $x_1 \in \pi_0\left(P_1(\mathbf{X})^{hG}\right)$. This results in a short exact sequence

$$1 \longrightarrow \pi_1(X, x_0) \longrightarrow H \longrightarrow G \longrightarrow 1$$

where $\pi_1(X, x_0)$ here denotes the fundamental group of the component of X corresponding to x_0 (choosing different base points in the same connected component will lead to isomorphic short exact sequences). Obstruction theory then tells us that this sequence splits if and only if x_0 lifts to $\pi_0\left(P_1(\mathbf{X})^{hG}\right)$.

Now suppose we have an element $x_1 \in \pi_0\left(P_1(\mathbf{X})^{hG}\right)$ and let $p_1 \in x_1$ be a point. Note that p_1 is a homotopy fixed point, and not an actual point. However, since it encodes a map from a contractible space to $P_1(X)$ we can still use it as if it were a base point for purposes of homotopy groups, i.e. we can write

$$\pi_n(X, p_1).$$

Furthermore p_1 behaves like a G-invariant base point and so we can use it to define an action of Γ on each $\pi_n(X, p_1)$.

Obstruction theory then proceeds as follows: let $x_{i-1} \in \pi_0\left(P_{i-1}(\mathbf{X})^{hG}\right)$ be an element, let $x_1 \in \pi_0\left(P_1(\mathbf{X})^{hG}\right)$ be its image and let $p_1 \in x_1$ be a point. Then one obtains an obstruction element

$$o_{x_{i-1}} \in H^{i+1}(G, \pi_i(\mathbf{X}, p_1))$$

which is trivial if and only if x_{i-1} lifts to $\pi_0\left(P_i(\mathbf{X})^{hG}\right)$.

Now consider the case of a pro-finite group Γ. Let $\mathbf{X}$ be a bounded Kan simplicial Γ-set. Then there is an n such that the map $\mathbf{X} \longrightarrow P_n(\mathbf{X})$ is a weak equivalence. Hence as above we reduce the question of emptiness of $\mathbf{X}^{h\Gamma}$ to that of $P_n(\mathbf{X})^{h\Gamma}$ which in turn leads to a sequence of lifting problems via the sequence of simplicial Γ-sets

$$P_n(\mathbf{X}) \longrightarrow P_{n-1}(\mathbf{X}) \longrightarrow ... \longrightarrow P_0(\mathbf{X}).$$

We now claim that the obstruction theory above generalizes to the pro-finite case by replacing the group cohomology

$$H^{i+1}(G, \pi_i(\mathbf{X}, p_1))$$

with **Galois cohomology**

$$H^{i+1}(\Gamma, \pi_i(\mathbf{X}, p_1)).$$

We will prove this here for the case $i \geq 2$. Note that in this case a homotopy fixed point $x \in P_{i-1}(\mathbf{X})^{h\Gamma}$ gives an element in $P_1(\mathbf{X})^{h\Gamma}$ which as above can act as a Γ-invariant base point and determine an action of Γ on all the homotopy groups of $\mathbf{X}$. The case $i = 1$ will be dealt with in subsection §§9.6.2 (see Theorem 9.78).

Proposition 9.49 *Let $\mathbf{X}$ be a bounded excellent strictly fibrant simplicial Γ-set. Let $x_{i-1} \in \pi_0\left(P_{i-1}(\mathbf{X})^{h\Gamma}\right)$ $(i \geq 2)$ be a homotopy fixed point component, $x_1 \in \pi_0\left(P_1(\mathbf{X})^{h\Gamma}\right)$ its image and $p_1 \in x_1$ a point. Then there exists an obstruction in the Galois cohomology group*

$$o(x_{i-1}) \in H^{i+1}(\Gamma, \pi_i(\mathbf{X}, p_1))$$

which vanish if and only x_{i-1} lifts to an element $x_i \in \pi_0\left(P_i(\mathbf{X})^{h\Gamma}\right)$.

Proof Since $\mathbf{X}$ is excellent there exists a $\Lambda \tilde{\triangleleft} \Gamma$ such that the action of Γ on $\mathbf{X}$ factors through Γ/Λ. Since $\mathbf{X}$ is bounded $P_{i-1}(\mathbf{X})$ and $P_i(\mathbf{X})$ are strictly bounded and so we can apply formula 9.44 to get

$$\begin{aligned}\pi_0\left(P_{i-1}(\mathbf{X})^{h\Gamma}\right) &= \operatorname*{colim}_{\Lambda' \tilde{\triangleleft} \Gamma} \pi_0\left(\left(P_{i-1}(\mathbf{X})^{\Lambda'}\right)^{h(\Gamma/\Lambda')}\right) \\ &= \operatorname*{colim}_{\Lambda' \subseteq \Lambda, \Lambda' \tilde{\triangleleft} \Gamma} \pi_0\left(P_{i-1}(\mathbf{X})^{h(\Gamma/\Lambda')}\right)\end{aligned}$$

and similarly

$$\pi_0\left(P_i(\mathbf{X})^{h\Gamma}\right) = \operatorname*{colim}_{\Lambda' \subseteq \Lambda, \Lambda' \tilde{\triangleleft} \Gamma} \pi_0\left(P_i(\mathbf{X})^{h(\Gamma/\Lambda')}\right).$$

Let $\Lambda_1 \subseteq \Lambda$ be such that x_{i-1} comes from $\pi_0\left(P_{i-1}(\mathbf{X})^{h(\Gamma/\Lambda_1)}\right)$. Then we want to know if there exists a $\Lambda_2 \subseteq \Lambda_1$ such that the image of x_{i-1} in $\pi_0\left(P_{i-1}^{h(\Gamma/\Lambda_2)}\right)$ lifts to $\pi_0\left(P_i^{h(\Gamma/\Lambda_2)}\right)$. From the obstruction theory in the finite case we know that this is equivalent to the vanishing of a certain obstruction element

$$o_{\Lambda_2}(x_{i-1}) \in H^{i+1}(\Gamma/\Lambda_2, \pi_i(\mathbf{X}, p_1)).$$

Hence we see that x_{i-1} lifts to $\pi_0\left(P_i^{h\Gamma}\right)$ if and only if $o_{\Lambda_2} = 0$ for **some** Λ_2.

Now each such $o_{\Lambda_2}(x_{i-1})$ defines (the same) element

$$o(x_{i-1}) \in H^{i+1}(\Gamma, \pi_i(\mathbf{X}, p_1)) = \operatorname*{colim}_{\Lambda_2 \subseteq \Lambda, \Lambda_2 \trianglelefteq \Gamma} H^{i+1}(\Gamma/\Lambda_2, \pi_i(\mathbf{X}, p_1)^{\Lambda_2})$$
$$= \operatorname*{colim}_{\Lambda_2 \subseteq \Lambda, \Lambda_2 \trianglelefteq \Gamma} H^{i+1}(\Gamma/\Lambda_2, \pi_i(\mathbf{X}, p_1))$$

and $o(x_{i-1})$ vanishes if and only if o_{Λ_2} vanishes for some Λ_2, so we are done. □

Remark 9.50 It is not hard to show that under the same assumptions for every $x_{i-1} \in P_{i-1}(\mathbf{X})(h\mathbb{A})$ there exists an obstruction

$$o(x_{i-1}) \in H^{i+1}(\mathbb{A}, \pi_i(\mathbf{X}, p_1)),$$

which vanishes if and only if x_{i-1} lifts to $P_i(\mathbf{X})(h\mathbb{A})$, where $H^{i+1}(\mathbb{A}, -)$ is the suitable notion of restricted product of local cohomologies. We shall give an exact definition of a generalization of this notion in §9.10.

Note that if

$$f : \mathbf{X}_1 \longrightarrow \mathbf{X}_2$$

is a weak equivalence of simplicial Γ_K-sets then the induced map

$$f^{h\Gamma_K} : \mathbf{X}_1^{h\Gamma_K} \longrightarrow \mathbf{X}_2^{h\Gamma_K}$$

is a weak equivalence. We want a similar property to hold for adelic homotopy fixed points. This will require the additional assumption that the spaces are nice:

Theorem 9.51 *Let*

$$f : \mathbf{X}_1 \longrightarrow \mathbf{X}_2$$

be a weak equivalence of nice and bounded simplicial Γ_K-sets. Then the induced map

$$f^{h\mathbb{A}} : \mathbf{X}_1^{h\mathbb{A}} \longrightarrow \mathbf{X}_2^{h\mathbb{A}}$$

is a weak equivalence.

Proof We start with two lemmas which give a connection between the connectivity of f and the corresponding connectivity of $f^{h\Gamma_K}$ and $f^{h\mathbb{A}}$.

Lemma 9.52 *Assume that Γ_K is a pro-finite group of finite strict (non-strict) cohomological dimension d and let*

$$f : \mathbf{X}_1 \longrightarrow \mathbf{X}_2$$

be an n-connected map of (finite) nice bounded simplicial Γ_K-sets. Then the induced map

$$f^{h\Gamma_K} : \mathbf{X}_1^{h\Gamma_K} \longrightarrow \mathbf{X}_2^{h\Gamma_K}$$

is $(n-d)$-connected.

Proof Consider the corresponding spectral sequences $E^r_{s,t}, F^r_{s,t}$. From our assumption the map

$$E^2_{s,t} \longrightarrow F^2_{s,t}$$

is an isomorphism for $t \leq n$. Since the differential $d^r_{s,t}$ goes from (s,t) to $(s+r, t+r-1)$ we see that if $t - s \leq n - d$ then the map

$$E^r_{s,t} \longrightarrow F^r_{s,t}$$

remains an isomorphism for all r. □

Lemma 9.53 *Let K be number field S a set of places of K that does not contain the real places. Let*

$$f : \mathbf{X}_1 \longrightarrow \mathbf{X}_2$$

be an n-connected map of nice bounded simplicial Γ_K-sets. Then the induced map

$$f : \mathbf{X}_1^{h\mathbb{A}_S} \longrightarrow \mathbf{X}_2^{h\mathbb{A}_S}$$

is $(n-3)$-connected.

Proof Since $\mathbf{X}_1, \mathbf{X}_2$ are nice there exists a finite Galois extension L/K such that Γ_L stabilizes the $(n+1)$-skeleton of both $\mathbf{X}_1$ and $\mathbf{X}_2$. Let T_0 be the finite set of places the ramifies in L. Then if $\nu \notin T_0$ we get that I_ν stabilizes the $(n+1)$-skeleton of both $\mathbf{X}_1$ and $\mathbf{X}_2$. Thus for every finite set of places $T_0 \subseteq T \subseteq S$ we have that the maps

$$f_\nu : \mathbf{X}_1 \longrightarrow \mathbf{X}_2, \quad \nu \in T$$

and

$$f_\nu : \mathbf{X}_1^{I_\nu} \longrightarrow \mathbf{X}_2^{I_\nu}, \quad \nu \notin T$$

are n-connected. Thus by Lemma 9.52 we get that the maps

$$f_\nu : \mathbf{X}_1^{h\Gamma_\nu} \longrightarrow \mathbf{X}_2^{h\Gamma_\nu}, \quad \nu \in T$$

and

$$f_\nu : (\mathbf{X}_1^{I_\nu})^{h\Gamma_\nu^{ur}} \longrightarrow (\mathbf{X}_2^{I_\nu})^{h\Gamma_\nu^{ur}}, \quad \nu \notin T$$

are $(n-3)$-connected. Therefore the map

$$\prod_{\nu} f_\nu : \operatorname*{hocolim}_{T_0 \subseteq T \subseteq S} \prod_{\nu \in T} \mathbf{X}_1^{h\Gamma_\nu} \times \prod_{\nu \in S \backslash T} \mathbf{X}_1^{h^{ur}\Gamma_\nu} \longrightarrow$$

$$\operatorname*{hocolim}_{T_0 \subseteq T \subseteq S} \prod_{\nu \in T} \mathbf{X}_2^{h\Gamma_\nu} \times \prod_{\nu \in S \backslash T} \mathbf{X}_2^{h^{ur}\Gamma_\nu}$$

is $(n-3)$-connected. □

We now complete the proof of the theorem. Denote by S_∞ the finite set of archimedean places in S and let

$$S_f = S \backslash S_\infty .$$

We get

$$\mathbf{X}_i^{h\mathbb{A}_S} = \mathbf{X}_i^{h\mathbb{A}_{S_\infty}} \times \mathbf{X}_i^{h\mathbb{A}_{S_f}}$$

and since S_∞ is finite,

$$\mathbf{X}_i^{h\mathbb{A}_{S_\infty}} = \prod_{\nu \in S_\infty} \mathbf{X}_i^{h\Gamma_\nu} .$$

To conclude we have that

$$f : \mathbf{X}_1^{h\mathbb{A}_{S_\infty}} \longrightarrow \mathbf{X}_2^{h\mathbb{A}_{S_\infty}}$$

is a weak equivalence and by Lemma 9.53

$$f : \mathbf{X}_1^{h\mathbb{A}_{S_f}} \longrightarrow \mathbf{X}_2^{h\mathbb{A}_{S_f}}$$

is a weak equivalence. This means that

$$f : \mathbf{X}_1^{h\mathbb{A}_S} \longrightarrow \mathbf{X}_2^{h\mathbb{A}_S}$$

is a weak equivalence. □

Corollary 9.54 *In Lemma 9.38 we may replace the assumption that* $\mathbf{X}$ *is strictly bounded with the assumption that* $\mathbf{X}$ *is bounded, i.e if* K *is number field and* $\mathbf{X}$ *a bounded* Γ_K*-simplicial set then the natural map*

$$\text{loc} : \mathbf{X}(hK) \longrightarrow \prod_{\nu} \mathbf{X}(hK_\nu)$$

factors through a natural map

$$f_0 : \mathbf{X}(hK) \longrightarrow \mathbf{X}(h\mathbb{A}) .$$

Proof In light of Theorem 9.51 and Lemma 9.38 it is enough to show that every bounded Γ_K-simplicial set is weakly equivalent strictly to a bounded Γ_K-simplicial set. Indeed, if $\mathbf{X}$ is bounded then for large enough n the map

$$\mathbf{X} \to P_n(\mathbf{X})$$

is a weak equivalence, and $P_n(\mathbf{X})$ is always strictly bounded. □

Lemma 9.55 *Let Γ be a pro-finite group and let*

$$\mathbf{X}_1 \longrightarrow \mathbf{X}_2 \longrightarrow \mathbf{X}_3$$

be a homotopy fibration sequence of simplicial Γ-sets. Then

$$\mathbf{X}_1^{h\Gamma} \longrightarrow \mathbf{X}_2^{h\Gamma} \longrightarrow \mathbf{X}_3^{h\Gamma}$$

is a homotopy fibration sequence of simplicial sets.

Proof First note that we can change the map $\mathbf{X}_2 \longrightarrow \mathbf{X}_3$ to a fibration $\widetilde{\mathbf{X}}_2 \longrightarrow \mathbf{X}_3$ in the local model structure without changing the homotopy types. Now any fibration in the local model structure is also a Kan fibration and thus the fibre of the map $\widetilde{\mathbf{X}}_2 \longrightarrow \mathbf{X}_3$ is standardly equivalent to $\mathbf{X}_1$. Hence we may assume that

$$\mathbf{X}_1 \longrightarrow \mathbf{X}_2 \longrightarrow \mathbf{X}_3$$

is a fibration sequence of simplicial Γ-sets in the local model structure and the lemma follows from the fact that $\mathbf{X}^{h\Gamma}$ is the derived mapping space from the terminal object to $\mathbf{X}$. □

Corollary 9.56 *Let K be a number field and let*

$$\mathbf{X}_1 \longrightarrow \mathbf{X}_2 \longrightarrow \mathbf{X}_3$$

be a homotopy fibration sequence of nice simplicial Γ_K-sets. Then

$$\mathbf{X}_1^{h\mathbb{A}} \longrightarrow \mathbf{X}_2^{h\mathbb{A}} \longrightarrow \mathbf{X}_3^{h\mathbb{A}}$$

is a homotopy fibration sequence of simplicial sets.

Proof By applying P_n for large enough n it is enough to prove this when the $\mathbf{X}_i$ are excellent. Now take S to be a finite set of places such that all the $\mathbf{X}_i$'s are unramified outside S. Now if T is any finite set of places such that $S \subseteq T$ then by Lemma 9.55

$$\prod_{\nu \in T} \mathbf{X}_1^{h\Gamma_\nu} \times \prod_{\nu \notin T} \mathbf{X}_1^{h^{ur}\Gamma_\nu} \longrightarrow \prod_{\nu \in T} \mathbf{X}_2^{h\Gamma_\nu} \times \prod_{\nu \notin T} \mathbf{X}_2^{h^{ur}\Gamma_\nu} \longrightarrow \prod_{\nu \in T} \mathbf{X}_3^{h\Gamma_\nu} \times \prod_{\nu \notin T} \mathbf{X}_3^{h^{ur}\Gamma_\nu}$$

is a homotopy fibration sequence. Now by passing to the limit and using the fact that direct homotopy colimits preserve homotopy fibration sequences (as they commute with finite homotopy limits) we get that

$$\mathbf{X}_1^{h\mathbb{A}} \longrightarrow \mathbf{X}_2^{h\mathbb{A}} \longrightarrow \mathbf{X}_3^{h\mathbb{A}}$$

is a homotopy fibration sequence. □

Definition 9.57 We shall say that a commutative diagram

$$\begin{array}{ccc} A & \longrightarrow & B \\ \downarrow & & \downarrow \\ C & \longrightarrow & D \end{array}$$

in the category of sets is **semi-Cartesian**, if the map

$$A \longrightarrow B \times_D C$$

is onto.

Proposition 9.58 *Let K be a number field and let $\mathbf{X}$ be an excellent bounded simplicial Γ_K-set. Then the commutative diagram*

$$\begin{array}{ccc} \mathbf{X}(hK) & \longrightarrow & P_2(\mathbf{X})(hK) \\ \downarrow & & \downarrow \\ \mathbf{X}(h\mathbb{A}) & \longrightarrow & P_2(\mathbf{X})(h\mathbb{A}) \end{array}$$

is semi-Cartesian.

Proof We shall first prove the proposition for the case that $\mathbf{X}$ is 2-connected. In that case $P_2(\mathbf{X})$ is contractible and thus the claim is reduced to the following lemma:

Lemma 9.59 *Let K be number field and $\mathbf{X}$ be a 2-connected excellent bounded simplicial Γ_K-set. Then the map* $\mathrm{loc} : \mathbf{X}(hK) \longrightarrow \mathbf{X}(h\mathbb{A})$ *is surjective.*

Proof First we will show that if $\mathbf{X}(hK) = \emptyset$ then $\mathbf{X}(h\mathbb{A}) = \emptyset$ as well. Since $\mathbf{X}$ is excellent and bounded we can use the obstruction theory described above (see 9.49).

Since $\mathbf{X}$ is 2-connected the obstructions fall in the groups

$$H^{i+1}(K, \pi_i(\mathbf{X})),$$

$$H^{i+1}(\mathbb{A}, \pi_i(\mathbf{X})),$$

for $i \geq 3$ (note that since $P_1(\mathbf{X})$ is contractible we can suppress the base point). Since the map

$$H^{i+1}(K, A) \longrightarrow H^{i+1}(\mathbb{A}, A)$$

is an isomorphism for $i \geq 2$ and every finite module A (see [Mil06], Theorem 4.10 (c)) we get that if $\mathbf{X}(hK)$ is empty then so is $\mathbf{X}(h\mathbb{A})$.

Now assume that $\mathbf{X}(hK) \neq \emptyset$. We shall prove that the map

$$\text{loc} : \mathbf{X}(hK) \longrightarrow \prod \mathbf{X}(hK_\nu)$$

(and thus the map loc : $\mathbf{X}(hK) \longrightarrow \mathbf{X}(h\mathbb{A})$) is surjective. Since $\mathbf{X}(hK) \neq \emptyset$ we have $\prod \mathbf{X}(hK_\nu) \neq \emptyset$ and so we can use the spectral sequence from Theorem 9.46. Let $p \in \mathbf{X}^{h\Gamma_K}$ be a chosen base homotopy fixed point.

The lemma will follow by carefully investigating these spectral sequences. We shall denote by $E^r_{s,t}(K)$ the spectral sequence that converges to $\pi_{t-s}\left(\mathbf{X}^{h\Gamma_K}, p\right)$ and by $E^r_{s,t}(K_\nu)$ the spectral sequence that converges to $\pi_{t-s}\left(\mathbf{X}^{h\Gamma_\nu}, p_\nu\right)$. We also denote by $\prod_\nu E^r_{s,t}(K_\nu)$ the product spectral sequence (since $\mathbf{X}$ is bounded the spectral sequence collapses after a finite number of pages. This fact together with the exactness of products insures that $\prod_\nu E^r_{s,t}(K_\nu)$ is indeed a spectral sequence and that it converges to the product $\prod_\nu \pi_{t-s}\left(\mathbf{X}^{h\Gamma_K}, p_\nu\right)$). Consider the map of spectral sequences

$$\text{loc}\,^r_{s,t} : E^r_{s,t}(K) \longrightarrow \prod E^r_{s,t}(K_\nu).$$

This map converges to

$$\text{loc}\,_{t-s} : \pi_{t-s}\left(\mathbf{X}^{h\Gamma_\nu}, p\right) \longrightarrow \prod_\nu \pi_{t-s}\left(\mathbf{X}^{h\Gamma_\nu}, p_\nu\right).$$

Now since for a finite module M and $k \geq 3$ one has

$$H^k(K, M) \cong \prod_\nu H^k(K_\nu, M)$$

we see that $\text{loc}\,^2_{s,t}$ is an isomorphism for $t \geq 3$. Since $\mathbf{X}$ is 2-connected, $\text{loc}\,^2_{s,t}$ is also an isomorphism for $0 \leq t \leq 2$. Hence in particular it is an isomorphism on the lines $t - s = 0$ and $t - s = -1$. We shall use the following lemma:

Lemma 9.60 *If $f^r_{s,t} : E^r_{s,t} \longrightarrow F^r_{s,t}$ is map of HL-spectral sequences such that $f^2_{s,t}$ is injective on the line $t - s = d$ and surjective on the line $t - s = d + 1$ then the same is true for all $f^r_{s,t}$.*

Proof By induction on r and a simple diagram chase. □

Now since $\mathbf{X}$ is bounded the spectral sequences $E^r_{s,t}(K), \prod E^r_{s,t}(K_\nu)$ collapse in some page r and we get that the map

$$\mathrm{loc}^{\infty}_{t,t} : E^{\infty}_{t,t}(K) \longrightarrow \prod E^{\infty}_{t,t}(K_\nu)$$

is surjective for all $t \geq 0$. Hence

$$\mathrm{loc}_0 : \pi_0\left(\mathbf{X}^{h\Gamma_K}, p\right) \longrightarrow \prod_\nu \pi_0\left(\mathbf{X}^{h\Gamma_K}, p_\nu\right)$$

is also surjective. □

We shall now prove the claim in the case of a general $\mathbf{X}$. Let

$$((a_\nu), x_0) \in \mathbf{X}(h\mathbb{A}) \times_{P_2(\mathbf{X})(h\mathbb{A})} P_2(\mathbf{X})(hK)$$

be a general point. Let $\Lambda \widetilde{\triangleleft} \Gamma_K$ be such that x_0 can be represented by a Γ_K equivariant map $\mathbf{E}(\Gamma_K/\Lambda) \longrightarrow P_2(\mathbf{X})$. We get a diagram of excellent strictly fibrant simplicial Γ_K-sets:

$$\begin{array}{ccc} & & \mathbf{E}(\Gamma_K/\Lambda) \\ & & \downarrow \\ \mathbf{X} & \longrightarrow & P_2(\mathbf{X})\,. \end{array}$$

We denote the homotopy pullback of this diagram by $\mathbf{X}\langle 2\rangle$. Note that $\mathbf{X}\langle 2\rangle$ is 2-connected and excellent and that the sequence

$$\mathbf{X}\langle 2\rangle \longrightarrow \mathbf{X} \longrightarrow P_2(\mathbf{X})$$

is a homotopy fibration sequence of excellent strictly fibrant simplicial Γ_K-sets. Hence by applying Lemma 9.55 and Corollary 9.56 we get the following commutative diagram with exact rows

$$\begin{array}{ccccc} \mathbf{X}\langle 2\rangle(hK) & \xrightarrow{p} & \mathbf{X}(hK) & \xrightarrow{c_2} & (P_2(\mathbf{X})(hK), x_0) \\ \downarrow{\scriptstyle \mathrm{loc}_1} & & \downarrow{\scriptstyle \mathrm{loc}_2} & & \downarrow{\scriptstyle \mathrm{loc}_3} \\ \mathbf{X}\langle 2\rangle(h\mathbb{A}) & \xrightarrow{p} & \mathbf{X}(h\mathbb{A}) & \xrightarrow{c_2} & (P_2(\mathbf{X})(h\mathbb{A}), \mathrm{loc}_3(x_0)) \end{array}$$

where the notation (A, a) means that the element $a \in A$ is the neutral element in the pointed set A. Now since

$$c_2((a_\nu)) = \mathrm{loc}_3(x_0)$$

there is an element $(b_\nu) \in \mathbf{X}\langle 2\rangle(h\mathbb{A})$ such that $p((b_\nu)) = (a_\nu)$. Now by

Lemma 9.59 there is an element $k_0 \in \mathbf{X}\langle 2\rangle(hK)$ such that $\mathrm{loc}_1(k_0) = (b_\nu)$. We denote $k_1 = p(k_0)$ and get

$$c_2(k_1) = c_2(p(k_0)) = x_0$$

and

$$\mathrm{loc}_2(k_1) = \mathrm{loc}_2(p(k_0)) = p(\mathrm{loc}_1(k_0)) = p((b_\nu)) = (a_\nu).$$

□

Now from Proposition 9.58 we immediately get the following two corollaries.

Corollary 9.61 *Let K be a number field, X/K a smooth variety and $\mathcal{U} \longrightarrow X$ an hypercovering. Then for every $n \geq 2$ we have*

$$X(\mathbb{A})^{\mathcal{U},n} = X(\mathbb{A})^{\mathcal{U},2},$$

$$X(\mathbb{A})^{\mathbb{Z}\mathcal{U},n} = X(\mathbb{A})^{\mathbb{Z}\mathcal{U},2}.$$

Corollary 9.62 *Let K be a number field, X/K a smooth variety and $\mathcal{U} \longrightarrow X$ a hypercovering such that $\mathbf{X}_{\mathcal{U}}$ is simply connected. Then*

$$X(\mathbb{A})^{\mathcal{U},n} = X(\mathbb{A})^{\mathbb{Z}\mathcal{U},n}$$

for every $n \geq 0$.

9.5 The finite pre-image theorem

In this section we will prove a theorem that will be very helpful in analyzing our various obstructions.

Theorem 9.63 *Let K be a number field, $\mathbf{X}_I = \{\mathbf{X}_\alpha\}_{\alpha\in I} \in \mathrm{Pro\,Ho}\left(\mathrm{Set}_{\Gamma_K}^{\Delta^{op}}\right)$ such that each $\mathbf{X}_\alpha$ is finite, bounded and excellent. Let $(x_\nu) \in \mathbf{X}_I(h\mathbb{A})$ be an adelic homotopy fixed point. Then (x_ν) is rational if and only if its image in each $\mathbf{X}_\alpha(h\mathbb{A})$ is rational.*

Proof Since an inverse system of non-empty finite sets has a non-empty inverse limit (see Lemma 9.76) it is enough to show that:

Proposition 9.64 *Let K be a number field and let $\mathbf{X}$ be an excellent finite bounded simplicial Γ_K-set. Then the map*

$$\mathrm{loc}_{\mathbf{X}} : \mathbf{X}(hK) \longrightarrow \mathbf{X}(h\mathbb{A})$$

has finite pre-images. i.e. for every $(x_\nu) \in \mathbf{X}(h\mathbb{A})$ *the set* $\mathrm{loc}_{\mathbf{X}}{}^{-1}((x_\nu))$ *is finite.*

Proof Since

$$\mathbf{X}(h\mathbb{A}) \subseteq \prod_\nu \mathbf{X}(hK_\nu)$$

we can work with this product instead of $\mathbf{X}(h\mathbb{A})$.

First note that the theorem is trivial if either of the sets is empty. If both of them are non-empty we can use the spectral sequence of Theorem 9.46 in order to compute them. Let $p \in \mathbf{X}^{h\Gamma_K}$ be a base homotopy fixed point.

The proposition will follow by carefully investigating these spectral sequences. We shall denote by $E^r_{s,t}(K)$ the spectral sequence that converges to $\pi_{t-s}\left(\mathbf{X}^{h\Gamma_K}, p\right)$ and by $E^r_{s,t}(K_\nu)$ the spectral sequence that converges to $\pi_{t-s}\left(\mathbf{X}^{h\Gamma_\nu}, p_\nu\right)$. Consider the map of spectral sequences

$$\mathrm{loc}^r : E^r_{s,t}(K) \longrightarrow \prod E^r_{s,t}(K_\nu).$$

This map converges to

$$\mathrm{loc}_{t-s} : \pi_{t-s}\left(\mathbf{X}^{h\Gamma_K}, p\right) \longrightarrow \prod_\nu \pi_{t-s}\left(\mathbf{X}^{h\Gamma_K}, p_\nu\right).$$

We are interested in the components which contribute to π_0 so we would like to understand the pre-images of the maps

$$\mathrm{loc}^\infty_t : E^\infty_{t,t}(K) \longrightarrow \prod E^\infty_{t,t}(K_\nu).$$

Since $\mathbf{X}$ is bounded these groups/pointed sets are trivial for large enough t. For the rest of the t's we will prove the following:

Proposition 9.65 *The maps*

$$\mathrm{loc}^\infty_t : E^\infty_{t,t}(K) \longrightarrow \prod_\nu E^\infty_{t,t}(K_\nu)$$

have finite pre-images for all $t \geq 0$.

Before we begin the proof let us explain how this proves that the pre-image of (x_ν) is finite, i.e. Proposition 9.64.

Note the $E^\infty_{t,t}$ terms are pointed sets which filter the set $\pi_0\left(\mathbf{X}^{h\Gamma_K}\right)$ in a way which we describe below. The idea is to use this filtration on the pre-image $\mathrm{loc}_{\mathbf{X}}{}^{-1}((x_\nu))$ of some $((x_\nu)) \in \mathbf{X}(h\mathbb{A})$. We will assume that $\mathrm{loc}_{\mathbf{X}}{}^{-1}((x_\nu))$ is infinite and get a contradiction using this filtration.

Recall that in order to construct the spectral sequence we had to choose some homotopy fixed point $p \in \mathbf{X}^{h\Gamma_K}$, which we call the **base**

point of the spectral sequence. We pick it so that its connected component $x \in \pi_0\left(\mathbf{X}^{h\Gamma_K}\right)$ is in ${\operatorname{loc}_{\mathbf{X}}}^{-1}((x_\nu))$.

The first filtration map is the map

$$f_0 : \pi_0\left(\mathbf{X}^{h\Gamma_K}\right) \longrightarrow E_{0,0}^{\infty} \subseteq H^0(\Gamma_K, \pi_0(\mathbf{X}))$$

which associates to every homotopy fixed point the invariant connected component of $\mathbf{X}$ which it lies in. Note that all of ${\operatorname{loc}_{\mathbf{X}}}^{-1}((x_\nu))$ is mapped to the connected component of $\mathbf{X}$ which $((x_\nu))$ maps to, and so this filtration step is trivial when restricted to ${\operatorname{loc}_{\mathbf{X}}}^{-1}((x_\nu))$.

Now for those homotopy fixed points which are mapped to the same connected component as x we get the next filtration map

$$f_1 : f_0^{-1}(f_0(x)) \longrightarrow E_{1,1}^{\infty} \subseteq H^1(\Gamma, \pi_1(\mathbf{X})).$$

Now if we restrict this filtration map to ${\operatorname{loc}_{\mathbf{X}}}^{-1}((x_\nu))$ we get that their image under f_1 lies in the appropriate pre-image of the map

$$\operatorname{loc}_1^{\infty} : E_{1,1}^{\infty}(K) \longrightarrow \prod E_{1,1}^{\infty}(K_\nu).$$

From Proposition 9.65 such a pre-image must be finite. Hence if ${\operatorname{loc}_{\mathbf{X}}}^{-1}((x_\nu))$ is infinite then there exists a fiber of f_1 which has an infinite intersection with ${\operatorname{loc}_{\mathbf{X}}}^{-1}((x_\nu))$. Let $F \subseteq {\operatorname{loc}_{\mathbf{X}}}^{-1}((x_\nu))$ be this fiber. Then we can assume without loss of generality that $x \in F$.

Now since all the elements in F agree on the first two filtration steps it follows from the general construction of the spectral sequence that if we change the base point p to any other base point in F then the spectral sequence will be isomorphic. This makes the remainder of the steps of the filtration independent of p.

Continuing on, we have filtration maps

$$f_t : f_{t-1}^{-1}(f_{t-1}(x)) \longrightarrow E_{t,t}^{\infty}.$$

For $t \geq 2$ these are abelian groups and elements of ${\operatorname{loc}_{\mathbf{X}}}^{-1}((x_\nu))$ are mapped to the kernel of the map $\operatorname{loc}_t^{\infty}$, which is finite by Proposition 9.65.

Hence we can continue the process of choosing the infinite fiber each time and assume that x is in that infinite fiber. Since $\mathbf{X}$ is bounded there will be only a finite number of filtration steps and since the spectral sequence will no longer change when we change x within the infinite fiber we get from Proposition 9.65 the desired contradiction.

We now finish the proof of Theorem 9.63 by proving Proposition 9.65.

Proof of Proposition 9.65 For $t = 0$ note that the map

$$E^2_{0,0}(K) \longrightarrow \prod E^2_{0,0}(K_\nu)$$

is injective and thus so is the map

$$E^\infty_{0,0}(K) \longrightarrow \prod E^\infty_{0,0}(K_\nu).$$

For $t > 2$ the set $E^2_{t,t}(K) = H^t(K, \pi_t(\mathbf{X}))$ is finite and therefore the set $E^\infty_{t,t}(K)$ is finite. For $t = 1$ we have that $E^\infty_{1,1} \subseteq E^2_{1,1}$ so it is enough to show that the map

$$\mathrm{loc}^2 : E^2_{1,1}(K) \longrightarrow \prod E^2_{1,1}(K_\nu)$$

has finite pre-images. This is the map

$$\mathrm{loc}_{\pi_1} : H^1(K, \pi_1(\mathbf{X})) \longrightarrow \prod_\nu H^1(K_\nu, \pi_1(\mathbf{X})).$$

The fact that this map has finite pre-images appears for example in Borel-Serre [BSe64] §7.

We shall now prove the result for $t = 2$. We have

$$E^\infty_{2,2} \subseteq E^2_{2,2}/d^2\left(E^2_{1,0}\right).$$

Consider first the map

$$\mathrm{loc}^2_{2,2} : E^2_{2,2}(K) \longrightarrow \prod E^2_{2,2}(K_\nu)$$

which is the map of abelian groups

$$\mathrm{loc}_{\pi_2} : H^2(K, \pi_2(\mathbf{X})) \longrightarrow \prod_\nu H^2(K_\nu, \pi_2(\mathbf{X}))$$

and the kernel of this map is $\text{Ш}^2(\pi_2(\mathbf{X}))$ which is finite since $\pi_2(\mathbf{X})$ is finite (see Milne [Mil06] Theorem 4.10).

Hence in order to show that $\mathrm{loc}^\infty_{2,2}$ has finite kernel it is enough to show that

$$d^2 : \prod E^2_{1,0}(K_\nu) \longrightarrow \prod E^2_{2,2}(K_\nu)$$

has finite image. Now for each ν the group

$$E^2_{1,0}(K_\nu) = H^0(\Gamma_\nu, \pi_1(\mathbf{X}))$$

is finite because $\pi_1(\mathbf{X})$ is finite and so it is enough to show that for almost all ν the map

$$d^2 : E^2_{1,0}(K_\nu) \longrightarrow E^2_{2,2}(K_\nu)$$

is the zero map. This is achieved in the following lemma:

Lemma 9.66 *Let $\mathbf{X}$ be a Γ_K-simplicial set such that the 3-skeleton $\mathbf{X}_3$ is stabilized by some open subgroup $\Gamma_L \subseteq \Gamma_K$ where L/K is a finite Galois extension. Then if ν is a place of K which is non-ramified in L, the differential*

$$d^2 : E^2_{1,0}(K_\nu) \longrightarrow E^2_{2,2}(K_\nu)$$

is zero.

Proof The action of the group Γ_ν on $\mathbf{X}_3$ factors through the group Γ_ν^{ur}. Now consider the natural maps

$$\mathbf{X}_3^{h\Gamma_\nu^{ur}} \xrightarrow{f_1} \mathbf{X}_3^{h\Gamma_\nu} \xrightarrow{f_2} \mathbf{X}^{h\Gamma_\nu}.$$

Let

$$F^r_{s,t}(K_\nu^{ur}),\ F^r_{s,t}(K_\nu),\ E^r_{s,t}(K_\nu)$$

be the spectral sequences converging to $\pi_{t-s}\left(\mathbf{X}_3^{h\Gamma_\nu^{ur}}\right), \pi_{t-s}\left(\mathbf{X}_3^{h\Gamma_\nu}\right)$ and $\pi_{t-s}\left(\mathbf{X}^{h\Gamma_\nu}\right)$ respectively. Then we have corresponding maps of spectral sequences

$$F^r_{s,t}(K_\nu^{ur}) \xrightarrow{f_1^r} F^r_{s,t}(K_\nu) \xrightarrow{f_2^r} E^r_{s,t}(K_\nu).$$

We denote by $f_3^2 = f_2^2 \circ f_1^2$. Now let

$$a \in E^2_{1,0}(K_\nu) = H^0(\Gamma_\nu, \pi_1(\mathbf{X})).$$

Since the map

$$\pi_1(\mathbf{X}_3) \longrightarrow \pi_1(\mathbf{X})$$

is an isomorphism and the action on $\pi_1(\mathbf{X}_3)$ factors through Γ_ν^{ur} we see that

$$f_3^2 : H^0(\Gamma_\nu^{ur}, \pi_1(\mathbf{X}_3)) \longrightarrow H^0(\Gamma_\nu, \pi_1(\mathbf{X}))$$

is an isomorphism. This means that there exists a

$$b \in H^0(\Gamma_\nu^{ur}, \pi_1(\mathbf{X}_3)) = F^2_{1,0}(K_\nu^{ur})$$

such that $f_3^2(b) = a$.

Now since Γ_ν^{ur} has cohomological dimension 1 and $\pi_2(\mathbf{X}_3) \cong \pi_2(\mathbf{X})$ is finite we have

$$F^2_{2,2}(K_\nu^{ur}) = H^2(\Gamma_\nu^{ur}, \pi_2(\mathbf{X}_3)) = 0$$

and therefore $d^2(b) = 0$. Hence

$$d^2(a) = d^2(f_3^2(b)) = f_3^2(d^2(b)) = f_3^2(0) = 0$$

as required. □

This completes the proof of Proposition 9.65, which also completes the proof of Proposition 9.64. By the above considerations, this concludes the proof of Theorem 9.63. □

Corollary 9.67 *Let X be a smooth algebraic variety over a number field K. Then for every $0 \leq n \leq \infty$ one has*

$$X(\mathbb{A})^{h,n} = \bigcap_{\mathcal{U},k\leq n} X(\mathbb{A})^{\mathcal{U},k},$$

$$X(\mathbb{A})^{\mathbb{Z}h,n} = \bigcap_{\mathcal{U},k\leq n} X(\mathbb{A})^{\mathbb{Z}\mathcal{U},k}.$$

Using Corollary 9.61 we then get the following important conclusion:

Corollary 9.68 *Let K be a number field and X/K a smooth variety. Then for every $2 \leq n \leq \infty$ we have*

$$X(\mathbb{A})^{h,n} = X(\mathbb{A})^{h,2},$$

$$X(\mathbb{A})^{\mathbb{Z}h,n} = X(\mathbb{A})^{\mathbb{Z}h,2}.$$

In particular, the homotopy and homology obstructions **depend only on the 2-truncation of the étale homotopy type.**

9.6 Sections and homotopy fixed points

Let G be a finite group acting on a space $\mathbf{X}$. Then it is known that the space of homotopy fixed points $\mathbf{X}^{hG} = [\mathbf{E}G, X]_G$ is naturally equivalent to the space of sections of the classifying map

$$p : \mathbf{X}_{hG} \longrightarrow \mathbf{B}G$$

where

$$\mathbf{X}_{hG} = (X \times \mathbf{E}G)/G$$

is the homotopy quotient. In particular the question of existence of a homotopy fixed point can be translated to the question of existence of a section to p.

In this section we will discuss the generalization of this alternative approach to the case of pro-spaces and pro-homotopy types. A similar approach is taken by Pàl in his paper [Pal10].

9.6.1 The pro fundamental group

In order to study the notion of fundamental groups one needs to be able to work with base points. Since the étale homotopy type is not naturally a pro object in the homotopy category of pointed spaces one needs to make some choices in order to identify base points.

One way to tackle this issue is to lift pro-homotopy-types to pro-spaces. We will use the following observation:

Lemma 9.69 1. *Let* $\mathbf{X}_I = \{\mathbf{X}_\alpha\}_{\alpha \in I} \in \mathrm{Pro}\,\mathrm{Ho}\left(\mathrm{Set}_{\Gamma_K}^{\Delta^{op}}\right)$ *be an object such that* I *is* ***countable****. Then there exists an object* $\widetilde{\mathbf{X}}_{I'} \in \mathrm{ProSet}_{\Gamma_K}^{\Delta^{op}s}$ *whose image in* $\mathrm{Pro}\,\mathrm{Ho}\left(\mathrm{Set}_{\Gamma_K}^{\Delta^{op}}\right)$ *is isomorphic to* $\mathbf{X}_I$*. Furthermore one can choose* I' *to be the poset* $\mathbb{N}$ *of natural numbers such that* $\widetilde{\mathbf{X}}_0$ *is fibrant and all the maps* $\widetilde{\mathbf{X}}_n \longrightarrow \widetilde{\mathbf{X}}_{n-1}$ *are fibrations in the local model structure. We will refer to such pro-objects as* ***fibration towers****.*

2. *Let* $f : \mathbf{X}_I \longrightarrow \mathbf{Y}_J$ *be a map in* $\mathrm{Pro}\,\mathrm{Ho}\left(\mathrm{Set}_{\Gamma_K}^{\Delta^{op}}\right)$ *and assume that both* I, J *are countable. Then one can lift* f *to a map*

$$\widetilde{f} : \widetilde{\mathbf{X}}_{I'} \longrightarrow \widetilde{\mathbf{Y}}_{J'}$$

in $\mathrm{Pro}\,\mathrm{Set}_{\Gamma_K}^{\Delta^{op}}$*. Furthermore one can choose* I' *and* J' *to be* $\mathbb{N}$ *and* $\widetilde{f}$ *to be levelwise.*

Proof 1. Since I is countable it contains the poset $\mathbb{N} \subseteq I$ as cofinal subcategory. One then constructs the lifts $\widetilde{\mathbf{X}}_n$ by induction on n by each time representing the homotopy class $\mathbf{X}_n \longrightarrow \mathbf{X}_{n-1}$ by a fibration.

2. By choosing $\widetilde{\mathbf{X}}_{I'}$ and $\widetilde{\mathbf{Y}}_{J'}$ to be towers of fibrations one can lift f to $\widetilde{f}$ using standard lifting properties of fibrations. One can then replace $\widetilde{\mathbf{X}}_{I'}$ with a sub-tower in order to make f into a levelwise map.

□

Remark 9.70 The lift described above is unique up to homotopy but is **not functorial**. There is a way to lift the étale homotopy type functorially to a pro-space using Friedlander's construction of the étale topological type ([Fri82]). However we will not make use of this construction in this paper.

Remark 9.71 Given a K-variety X, the category $I(X)$ is in general not countable. However, if K is a countable field (e.g. a number field) then the category of K-varieties is essentially countable (every K-variety can be described by a finite set of equations and inequalities over K).

This means that the category of **truncated** hypercoverings (i.e. hypercoverings $\mathcal{U}_\bullet \longrightarrow X$ which satisfy $\mathcal{U}_\bullet \cong \mathrm{cosk}_n(\mathrm{tr}_n(U_\bullet))$ for some n) is essentially countable as well. Now since all the simplicial sets in $\acute{E}t^\natural_{/K}(X)$ were explicitly truncated we see that it is isomorphic to its sub-diagram indexed by truncated hypercoverings.

Definition 9.72 Let $\{\mathbf{X}_\alpha\} \in \mathrm{Pro\ Set}^{\Delta^{op}}$ be an object and $\{x_\alpha\} \in \lim_\alpha \mathbf{X}_\alpha$ a base point. We will define the nth pro-homotopy group to be the pro-group

$$\pi_n(\mathbf{X}_{\mathbb{N}}, \{x_\alpha\}) \stackrel{def}{=} \{\pi_n(\mathbf{X}_\alpha, x_\alpha)\}.$$

Remark 9.73 Note that if $\{\mathbf{X}_\alpha\}$ is a pro-simplicial Γ-set we will define pro-homotopy groups by forgetting the group action. In particular we will not require base points to be Γ-invariant.

We would like to restrict our selves to the following nice class of objects:

Definition 9.74 An object $\mathbf{X}_I \in \mathrm{Pro\ Set}_\Gamma^{\Delta^{op}}$ will be called **pro-finite** if each $\mathbf{X}_\alpha$ is an excellent, strictly bounded and finite simplicial Γ-set (see Definition 9.17).

The issue of uniqueness of the base-point is dealt with in the following lemma:

Lemma 9.75 *Let $\mathbf{X}_{\mathbb{N}}$ be a tower of Kan fibrations such that each $\mathbf{X}_n$ is finite, non-empty and connected. Then $\lim_n \mathbf{X}_n$ is non-empty and connected.*

Proof Since $\mathbf{X}_{\mathbb{N}}$ is a tower of Kan fibrations the non-emptiness of each $\mathbf{X}_n$ implies the non-emptiness of $\lim_n \mathbf{X}_n$.

Now let $\{x_n\}, \{x'_n\} \in \lim_n \mathbf{X}_n$ be two points. For each n, let $P(\mathbf{X}_n, x_n, x'_n)$ denote the space of paths from x_n to x'_n in $\mathbf{X}_n$. Then $\{P(\mathbf{X}_n, x_n, x'_n)\}$ is a tower of Kan fibrations as well and so it is enough to show that

$$\lim_n \pi_0(P(\mathbf{X}_n, x_n, x'_n)) \neq \emptyset.$$

But this now follows from the following standard lemma (since by our assumptions each $\pi_0(P(\mathbf{X}_n, x_n, x'_n))$ is non-empty and finite):

Lemma 9.76 *Let $\{A_\alpha\}_{\alpha \in I}$ be an inverse system of finite non-empty sets. Then*

$$\lim_{\overleftarrow{\alpha}} A_\alpha \neq \emptyset.$$

Proof A standard compactness argument. □

This completes the proof of Lemma 9.75. □

It will be useful to keep in mind the following classical observation:

Lemma 9.77 *The pro-category of finite groups is equivalent to the category of pro-finite groups (and continuous homomorphisms). The equivalence is given by $\{G_\alpha\} \mapsto \lim_\alpha G_\alpha$ with the natural pro-finite topology.*

In particular we think of the pro-homotopy groups of pro-finite objects $\mathbf{X}_I \in \text{Pro Set}^{\Delta^{op}}$ as pro-finite groups. From now on we will not distinguish between pro-finite groups and pro-objects in the category of finite groups.

9.6.2 The pro homotopy quotient

Recall that the categorical product of $\{\mathbf{X}_\alpha\}, \{\mathbf{Y}_\beta\}$ in the category $\text{Pro Set}_{\Gamma_K}^{\Delta^{op}}$ is given by

$$\{\mathbf{X}_\alpha\} \times \{\mathbf{Y}_\beta\} \stackrel{def}{=} \{\mathbf{X}_\alpha \times \mathbf{Y}_\beta\}_{\alpha,\beta}.$$

Let $\widetilde{\mathbf{E}\Gamma^\natural} \in \text{Pro Set}_\Gamma^{\Delta^{op}}$ be the lift of $\mathbf{E}\Gamma^\natural$ given by

$$\widetilde{\mathbf{E}\Gamma^\natural} = \{\mathbf{E}(\Gamma/\Lambda)\}_{\Lambda \trianglelefteq \Gamma}.$$

We will now define the pro-homotopy quotient. Let $\mathbf{X}_\mathbb{N} \in \text{Pro Set}_\Gamma^{\Delta^{op}}$ be a pro-finite fibration tower. We define its **pro-homotopy quotient** to be the levelwise quotient (which is also the categorical quotient, see 9.93) of $\mathbf{X}_\mathbb{N} \times \widetilde{\mathbf{E}\Gamma^\natural}$ by Γ, which we write as

$$(\mathbf{X}_\mathbb{N})_{h\Gamma} = \left(\mathbf{X}_\mathbb{N} \times \widetilde{\mathbf{E}\Gamma^\natural}\right)/\Gamma = \{\mathbf{X}_{n,\Lambda}\}_{n,\Lambda \trianglelefteq \Gamma}$$

where

$$\mathbf{X}_{n,\Lambda} \stackrel{def}{=} (\mathbf{X}_n \times \mathbf{E}(\Gamma/\Lambda))/\Gamma.$$

Note that whenever Λ fixes $\mathbf{X}_n$ we get that $\mathbf{X}_{n,\Lambda}$ is bounded and finite. Since the pairs (n, Λ) for which Λ fixes $\mathbf{X}_n$ are a cofinal subfamily we get that $(\mathbf{X}_\mathbb{N})_{h\Gamma}$ is isomorphic to a pro-finite object. In particular we can consider its pro-fundamental group as a pro-finite group. This pro-simplicial Γ-set fits into a short sequence

$$\mathbf{X}_{\mathbb{N}} \longrightarrow (\mathbf{X}_{\mathbb{N}})_{h\Gamma} \longrightarrow \{\mathbf{B}(\Gamma/\Lambda)\}_{\Lambda \trianglelefteq \Gamma}.$$

Choosing a base point $\{x_n\} \in \lim_n \mathbf{X}_n$ we get an induced base point in $(\mathbf{X}_{\mathbb{N}})_{h\Gamma}$ (which we will denote $\{x_n\}$ as well) which is mapped to the unique base point of $\{\mathbf{B}(\Gamma/\Lambda)\}$ yielding a short sequence of pro-finite groups:

$$1 \longrightarrow \pi_1(\mathbf{X}_{\mathbb{N}}, \{x_n\}) \longrightarrow \pi_1((\mathbf{X}_{\mathbb{N}})_{h\Gamma}, \{x_n\}) \longrightarrow \Gamma \longrightarrow 1$$

which is exact because whenever Λ fixes $\mathbf{X}_n$ the short sequence of groups

$$1 \longrightarrow \pi_1(\mathbf{X}_n, x_n) \longrightarrow \pi_1(\mathbf{X}_{n,\Lambda}, x_n) \longrightarrow \Gamma/\Lambda \longrightarrow 1$$

is exact.

Our main claim of this section is the following:

Theorem 9.78 *Let $\mathbf{X}_{\mathbb{N}}$ a pro-finite fibration tower and*

$$\mathbf{X}^1_{\mathbb{N}} = \{P_1(\mathbf{X}_n)\}$$

its 1-truncation. Then the sequence

$$1 \longrightarrow \pi_1(\mathbf{X}_{\mathbb{N}}, \{x_n\}) \longrightarrow \pi_1((\mathbf{X}_{\mathbb{N}})_{h\Gamma}, \{x_n\}) \longrightarrow \Gamma \longrightarrow 1$$

splits if and only if $\mathbf{X}^1_{\mathbb{N}}(E\Gamma^{\natural}) \neq \emptyset$.

Proof Since each $\mathbf{X}^1_n$ is strictly bounded the non-emptyness of $\mathbf{X}^1_{\mathbb{N}}(E\Gamma^{\natural})$ is equivalent to the existence of a map

$$\widetilde{\mathbf{E}\Gamma^{\natural}} \longrightarrow \mathbf{X}^1_{\mathbb{N}}$$

in Pro $\mathrm{Set}_{\Gamma}^{\Delta^{op}}$ (we can lift maps into $\mathbf{X}^1_{\mathbb{N}}$ from Pro Ho $\left(\mathrm{Set}_{\Gamma}^{\Delta^{op}}\right)$ to Pro $\mathrm{Set}_{\Gamma}^{\Delta^{op}}$ because $\mathbf{X}^1_{\mathbb{N}}$ is a tower of fibrations).

Now given a map

$$h : \widetilde{\mathbf{E}\Gamma^{\natural}} \longrightarrow \mathbf{X}^1_{\mathbb{N}}$$

we can take the map

$$\widetilde{h} \times Id : \widetilde{\mathbf{E}\Gamma^{\natural}} \longrightarrow \mathbf{X}^1_{\mathbb{N}} \times \widetilde{\mathbf{E}\Gamma^{\natural}}$$

and descend it to the Γ-quotients

$$s : \{\mathbf{B}(\Gamma/\Lambda)\} = \widetilde{\mathbf{E}\Gamma^{\natural}}/\Gamma \longrightarrow (\mathbf{X}^1_{\mathbb{N}})_{h\Gamma}$$

obtaining a section of the natural map

$$(\mathbf{X}^1_{\mathbb{N}})_{h\Gamma} \longrightarrow \{\mathbf{B}(\Gamma/\Lambda)\}.$$

Lemma 9.79 *Every section*

$$s : \{\mathbf{B}(\Gamma/\Lambda)\} \longrightarrow (\mathbf{X}^1_{\mathbb{N}})_{h\Gamma}$$

is induced by a map

$$h : \widetilde{\mathbf{E}\Gamma^{\natural}} \longrightarrow \mathbf{X}^1_{\mathbb{N}}$$

in this way.

Proof The map s can be described by the following information: for each n and Λ that stabilizes $\mathbf{X}_n$ we are given a normal subgroup $\Lambda' \widetilde{\triangleleft} \Gamma$ which is contained in Λ and a map

$$s_{n,\Lambda} : \mathbf{B}(\Gamma/\Lambda') \longrightarrow \mathbf{X}^1_{n,\Lambda} = (\mathbf{X}^1_n \times \mathbf{E}(\Gamma/\Lambda))/\Gamma$$

such that the composition

$$\mathbf{B}(\Gamma/\Lambda') \longrightarrow \mathbf{X}^1_{n,\Lambda} \longrightarrow \mathbf{B}(\Gamma/\Lambda)$$

is equal to the map $q_* : \mathbf{B}(\Gamma/\Lambda') \longrightarrow \mathbf{B}(\Gamma/\Lambda)$ induced by the natural projection $q : \Gamma/\Lambda' \longrightarrow \Gamma/\Lambda$. Since Λ stabilizes $\mathbf{X}^1_n$ the action Γ on $\mathbf{X}^1_n \times \mathbf{E}(\Gamma/\Lambda)$ factors through a free action of Γ/Λ. Hence we have a pullback square

$$\begin{array}{ccc} \mathbf{X}^1_n \times \mathbf{E}(\Gamma/\Lambda) & \longrightarrow & \mathbf{E}(\Gamma/\Lambda) \\ \downarrow & & \downarrow \\ \mathbf{X}^1_{n,\Lambda} & \longrightarrow & \mathbf{B}(\Gamma/\Lambda). \end{array}$$

The maps $q_* : \mathbf{E}(\Gamma/\Lambda') \longrightarrow \mathbf{E}(\Gamma/\Lambda)$ and the composition

$$\mathbf{E}(\Gamma/\Lambda') \longrightarrow \mathbf{B}(\Gamma/\Lambda') \xrightarrow{s_{i,j}} \mathbf{X}^1_{n,\Lambda}$$

combine to form a map

$$\widetilde{s}_{n,\Lambda} : \mathbf{E}(\Gamma/\Lambda') \longrightarrow \mathbf{X}^1_{n,\Lambda}$$

which lifts $s_{n,\Lambda}$.

It is left to show that the maps $\widetilde{s}_{n,\Lambda}$ can be chosen in a compatible way. Since the map $\mathbf{X}^1_n \times E(\Gamma/\Lambda) \longrightarrow \mathbf{X}^1_{n,\Lambda}$ is a covering map with fiber of size $|\Gamma/\Lambda|$ we see that there are no more then $|\Gamma/\Lambda|$ equivariant maps $\mathbf{E}(\Gamma/\Lambda') \longrightarrow \mathbf{X}^1_{n,\Lambda}$ which lift $s_{n,\Lambda}$. Since a filtered colimit of sets of size $\leq |\Gamma/\Lambda|$ is of size $\leq |\Gamma/\Lambda|$ the result will now follow from Lemma 9.76. □

To summarize so far, we see that the non-emptiness of $\mathbf{X}^1_{\mathbb{N}}(E\Gamma^{\natural})$ is equivalent to the existence of the section

$$s : \{\mathbf{B}(\Gamma/\Lambda)\} \longrightarrow (\mathbf{X}^1_{\mathbb{N}})_{h\Gamma}$$

to the natural map

$$(\mathbf{X}^1_{\mathbb{N}})_{h\Gamma} \longrightarrow \{\mathbf{B}(\Gamma/\Lambda)\}.$$

It is left to show that the existence of such a section is equivalent to a section in the respective fundamental groups. Since $\mathbf{X}^1_{n,\Lambda}$ is 1-bounded whenever Λ fixes n we see that a section of the pro-fundamental groups induces a section

$$s : \{\mathbf{B}(\Gamma/\Lambda)\} \longrightarrow (\mathbf{X}^1_{\mathbb{N}})_{h\Gamma}.$$

In the other direction let

$$s : \{\mathbf{B}(\Gamma/\Lambda)\} \longrightarrow (\mathbf{X}^1_{\mathbb{N}})_{h\Gamma}$$

be a section. Note that s might send the base point of $\{B(\Gamma/\Lambda)\}$ to a point other then our chosen base point $\{x_n\}$. However from 9.75 the simplicial set $\lim_n \mathbf{X}^1_n$ is connected and so we will know how to translate s into a section in the fundamental groups. □

We finish by the following useful criterion:

Lemma 9.80 *The map $\mathbf{X}^1_{\mathbb{N}h\Gamma} \longrightarrow \{B(\Gamma/\Lambda)\}$ has a section if and only if the induced map in* Pro Ho $\left(\mathrm{Set}^{\Delta^{op}}\right)$ *has a section.*

Proof We will use the following lemma whose proof is easy and classical:

Lemma 9.81 *Let H, G be two groups then*

$$\mathrm{Hom}_{\mathrm{Set}^{\Delta^{op}}}(\mathbf{B}H, \mathbf{B}G) \cong \mathrm{Hom}_{Grp}(H, G),$$

$$\mathrm{Hom}_{\mathrm{Ho}(\mathrm{Set}^{\Delta^{op}})}(\mathbf{B}H, \mathbf{B}G) \cong \mathrm{Hom}_{Grp}(H, G)/\sim.$$

where for $p_1, p_2 : H \to G$ we have $p_1 \sim p_2$ if there exist $g \in G$ such that $p_1(\bullet) = gp_2(\bullet)g^{-1}$.

Now let

$$s : \{\mathbf{B}(\Gamma/\Lambda)\} \longrightarrow (\mathbf{X}^1_{\mathbb{N}})_{h\Gamma}$$

be a section in Pro Ho $\left(\mathrm{Set}^{\Delta^{op}}\right)$. As above the map s can be described by the following information: for each n and Λ that stabilizes $\mathbf{X}_n$ we are given a normal subgroup $\Lambda' \trianglelefteq \Gamma$ which is contained in Λ and a map

$$s_{n,\Lambda} : \mathbf{B}(\Gamma/\Lambda') \longrightarrow \mathbf{X}^1_{n,\Lambda} = (\mathbf{X}^1_n \times \mathbf{E}(\Gamma/\Lambda))/\Gamma$$

such that the composition

$$\mathbf{B}(\Gamma/\Lambda') \longrightarrow \mathbf{X}^1_{n,\Lambda} \longrightarrow \mathbf{B}(\Gamma/\Lambda)$$

is **homotopic** to the map $q_* : \mathbf{B}(\Gamma/\Lambda') \longrightarrow \mathbf{B}(\Gamma/\Lambda)$ induced by the natural projection $q : \Gamma/\Lambda' \longrightarrow \Gamma/\Lambda$.

By Lemma 9.81 the two maps differ by a conjugation by some element of Γ/Λ. Lifting this element to Γ/Λ' we can find a map $s'_{n,\Lambda}$ homotopic to $s_{n,\Lambda}$ such that the composition

$$\mathbf{B}(\Gamma/\Lambda') \longrightarrow \mathbf{X}_i/(\Gamma/\Lambda) \longrightarrow \mathbf{B}(\Gamma/\Lambda)$$

is exactly the natural projection map $q_* : \mathbf{B}(\Gamma/\Lambda') \longrightarrow \mathbf{B}(\Gamma/\Lambda)$. Now similarly to the proof of Lemma 9.79 there are only finitely many such possible maps so by Lemma 9.76 we have a section in Pro $\mathrm{Set}^{\Delta^{op}}$ and we are done. □

9.6.3 The étale fundamental group

In this section we will connect the notions described in the previous section to the étale fundamental groups.

Proposition 9.82 *Let K be a field and X a K-variety. Then*

$$\left(\acute{E}t^n_{/K}(X)\right)_{h\Gamma_K} \cong \acute{E}t^n(X)$$

in $\mathrm{Pro}\,\mathrm{Ho}\left(\mathrm{Set}^{\Delta^{op}}\right)$ *for every $n \le \infty$.*

Proof We start by showing that

$$\acute{E}t^n_{/K}(X) \times \mathbf{E}\Gamma^\natural_K \cong \acute{E}t^n_{/K}(X)$$

for every $n \le \infty$. Since P_k commutes with products and the simplicial Γ_K-sets in $\mathbf{E}\Gamma^\natural_K$ are 0-bounded it will be enough to prove that

$$\acute{E}t_{/K}(X) \times \mathbf{E}\Gamma^\natural_K \cong \acute{E}t_{/K}(X).$$

Now we have a natural projection map

$$p : \acute{E}t_{/K}(X) \times \mathbf{E}\Gamma^\natural_K \cong \acute{E}t_{/K}(X)$$

and we will show that it is in fact an isomorphism. Let $\mathcal{U} \longrightarrow X$ be a hypercovering, n a natural number and $\Lambda \tilde{\lhd} \Gamma$ an open normal subgroup. Let L/K be the finite Galois extension corresponding to it. We will use the following construction:

Definition 9.83 Let L/K be a finite Galois extension. Let X_L be the restriction of scalars of $X \otimes_K L$ from L to K. Note that there is a natural map

$$X_L \longrightarrow X$$

which is an étale cover. We then denote by

$$\check{X}_L \longrightarrow X$$

the hypercovering obtained by the Čech construction (see §9.2.1, Definition 9.11).

The connected components of $X_L \otimes_K \overline{K}$ (each of which is isomorphic to X) can be identified (not uniquely) with $G_L = \mathrm{Gal}(L/K)$ where the Galois action of G_L on itself is by left translations. Such an identification induces an isomorphism of simplicial Γ_K-sets

$$\mathbf{X}_{\check{X}_L} \longrightarrow \mathbf{E}G_L.$$

In particular, the action of Γ_K on $\mathbf{X}_{\check{X}_L}$ factors through a **free** action of G_L.

Now let

$$\mathcal{U}_L = \mathcal{U} \times_X \check{X}_L.$$

Then it is easy to verify that the natural map

$$\varphi_{\mathcal{U},L} : \mathbf{X}_{\mathcal{U}_L} \longrightarrow \mathbf{X}_{\mathcal{U}} \times \mathbf{X}_{\check{X}_L} = \mathbf{X}_{\mathcal{U}} \times \mathbf{E}G_L$$

is an isomorphism of simplicial Γ_K-set. The maps $\varphi_{\mathcal{U},L}$ fit together to form a map

$$\acute{E}t_{/K}(X) \longrightarrow \acute{E}t_{/K}(X) \times \mathbf{E}\Gamma^{\natural}$$

which is an inverse to p. We leave it to the reader to verify that this is indeed an inverse to p. The proof is analogous to the case appearing in the proof of Theorem 9.131.

In view of Corollary 9.94 (whose proof is independent of the rest of the paper) we can finish the proof by showing that

$$\acute{E}t^n(X) \cong \acute{E}t^n_{/K}(X)/\Gamma = \{\mathbf{X}_{\mathcal{U},k}/\Gamma\}_{\mathcal{U} \in I(K), k \le n}.$$

Now by definition we have that

$$\acute{E}t(X) \cong \acute{E}t_{/K}(X)/\Gamma.$$

Hence in order to get our result we need to verify that taking the quotient commutes with truncation in this case. This will be done using the following two lemmas:

Lemma 9.84 *Let K be a field and X a K-variety. Then there is a cofinal subcategory $J(X) \subseteq I(X) \times \mathbb{N}$ such that for each $(\mathcal{U}, n) \in J(X)$ the action of Γ_K on $\mathbf{X}_{\mathcal{U},n}$ factors through a* ***free*** *action of a finite quotient of Γ_K.*

Proof Let $\mathcal{U} \longrightarrow X$ be a hypercovering and $n \in \mathbb{N}$ a number. Let L/K be a finite Galois extension such that the action of Γ_K on the $(n+1)$-skeleton of $\mathcal{U}$ factors through $G_L = \Gamma_K/\Gamma_L$.

Consider as above

$$\mathcal{U}_L = \mathcal{U} \times_X \check{X}_L.$$

Then the action of Γ_K on the $(n+1)$-skeleton of $\mathbf{X}_{\mathcal{U}_L}$ factors through G_L. Since we have a map

$$\mathbf{X}_{\mathcal{U}_L} \longrightarrow \mathbf{X}_{\check{X}_L}$$

we see that the action of G_L on the $(n+1)$-skeleton of $\mathbf{X}_{\mathcal{U}_L,n}$ is free. We also have a map of hypercoverings $\mathcal{U}_L \longrightarrow U$. Hence the full subcategory on

$$J(X) = \{(\mathcal{U}_L, n)\} \subseteq I(X) \times \mathbb{N}$$

is cofinal and we are done. □

Lemma 9.85 *Let $\mathbf{X}$ be a free simplicial G-set and $n \geq 1$ be a number. Then G acts freely on $P_n(\mathbf{X})$ and*

$$P_n(\mathbf{X}/G) \cong P_n(\mathbf{X})/G.$$

Proof Let G be finite group, $\mathbf{X}$ a simplicial G-set and $\mathbf{Y}$ a simplicial set considered as a simplicial G-set with trivial action. Let $f : \mathbf{X} \longrightarrow \mathbf{Y}$ be a map of simplicial G-sets. We say that **$\mathbf{Y}$ is a free G-quotient by** f if G acts freely on X and f induces an isomorphism as simplicial sets

$$\widetilde{f} : \mathbf{X}/G \xrightarrow{\simeq} \mathbf{Y}.$$

Hence, we want to show that if $\mathbf{Y}$ is a free G-quotient by

$$f : \mathbf{X} \longrightarrow \mathbf{Y}$$

then $P_n(\mathbf{Y})$ is a free G quotient by

$$P_n(f) : P_n(\mathbf{X}) \longrightarrow P_n(\mathbf{Y}).$$

Now it is easy to see that $\mathbf{Y}$ is a free G-quotient by f if and only if f

fits in a pullback diagram of the form

$$\begin{array}{ccc} \mathbf{X} & \longrightarrow & \mathbf{E}G \\ \downarrow f & & \downarrow \\ \mathbf{Y} & \longrightarrow & \mathbf{B}G. \end{array}$$

Since P_n has a left adjoint, it commutes with pullbacks and we can apply it to the above diagram and get another pullback square

$$\begin{array}{ccc} P_n(\mathbf{X}) & \longrightarrow & P_n(\mathbf{E}G) \\ \downarrow f & & \downarrow \\ P_n(\mathbf{X}/G) & \longrightarrow & P_n(\mathbf{B}G). \end{array}$$

However, both $\mathbf{E}G$ and $\mathbf{B}G$ are nerves of categories and thus since $n \geq 1$ we have

$$P_n(\mathbf{E}G) = \mathbf{E}G$$

and

$$P_n(\mathbf{B}G) = \mathbf{B}G.$$

Thus we get that pull back diagram:

$$\begin{array}{ccc} P_n(\mathbf{X}) & \longrightarrow & \mathbf{E}G \\ \downarrow f & & \downarrow \\ P_n(\mathbf{X}/G) & \longrightarrow & \mathbf{B}G \end{array}$$

and so $P_n(\mathbf{Y})$ is indeed a free G-quotient by $P_n(f)$. □

We now proceed with the proof of Proposition 9.82. From Lemma 9.84 we may replace $\{\mathbf{X}_{\mathcal{U},n}\}_{\mathcal{U}\in I(X),n\in\mathbb{N}}$ with $\{\mathbf{X}_{\mathcal{U},n}\}_{(\mathcal{U},n)\in J(X)}$. Now the action of Γ_K on the $(n+1)$-skeleton $\mathbf{X}_{\mathcal{U}}$ factors through a free action of a finite quotient G of Γ_K. Then we have a free action of G on $Q_n(\mathbf{X}_{\mathcal{U}})$ (see Definition 9.6). Applying Lemma 9.85 to $Q_n(\mathbf{X}_{\mathcal{U}})$ we get that

$$P_n(\mathbf{X}_{\mathcal{U}}/\Gamma_K) = P_n(Q_n(\mathbf{X}_{\mathcal{U}}/\Gamma_K)) \cong P_n(Q_n(\mathbf{X}_{\mathcal{U}})/G) \cong$$

$$P_n(Q_n(\mathbf{X}_{\mathcal{U}}))/\Gamma_K = P_n(\mathbf{X}_{\mathcal{U}}) = \mathbf{X}_{\mathcal{U},n}/\Gamma_K$$

for every $n \geq 1$. This means that

$$\acute{E}t^n(X) = \acute{E}t^n_{/K}(X)/\Gamma_K$$

and we are done. □

9.6.4 A generalized version of Grothendieck's obstruction

Going back to the map

$$h_n : X(K) \longrightarrow X^n(hK)$$

We see that

$$X^n(hK) = \emptyset \Rightarrow X(K) = \emptyset$$

and so one can consider the emptiness of each $X^n(hK)$ as an obstruction to existence of a rational point. The following lemma shows that for $n = 1$ this obstruction is actually Grothendieck's section obstruction. See also Pál [Pal10] or Quick [Qui09] for a similar discussion.

Keeping in mind Remark 9.71 and the fact that the pro-fundamental group of (a pointed lift of) $\acute{E}t(X)$ can be identified with the étale fundamental group of X we can use Theorem 9.78, Proposition 9.82 and Lemma 9.80 to obtain the following corollary:

Corollary 9.86 *Let K be a field and X a geometrically connected smooth variety. Then the following conditions are equivalent:*

1. *The set $X^1(hK)$ is nonempty.*
2. *The sequence of pro-finite groups*

$$1 \longrightarrow \pi_1(\overline{X}) \rightarrow \pi_1(X) \longrightarrow \Gamma_K \longrightarrow 1$$

   admits a continuous section.

3. *The map*

$$\acute{E}t^1(X) \longrightarrow \acute{E}t^1(\mathrm{Spec}\,(K))$$

   admits a section in $\mathrm{Pro\,Ho}\left(\mathrm{Set}_{\Gamma_K}^{\Delta^{op}}\right)$.

9.7 Homotopy fixed point sets and pro-isomorphisms

In this section we assume that K is a number field. Let X be a K-variety. Let

$$f : \acute{E}t_{/K}(X) \longrightarrow \mathbf{Y}_I$$

be map in $\mathrm{Pro\,Ho}\left(\mathrm{Set}_{\Gamma_K}^{\Delta^{op}}\right)$. One can then consider the subset of adelic points

$$X(\mathbb{A})^{\mathbf{Y}_I} \subseteq X(\mathbb{A})$$

containing the points (x_ν) whose corresponding homotopy fixed point

$$f_*(h((x_\nu))) \in \mathbf{Y}_I(h\mathbb{A})$$

is rational, i.e. comes from $\mathbf{Y}_I(hK)$. In general this obstruction can only be weaker then the étale homotopy obstruction, i.e. $X^h(\mathbb{A}) \subseteq X(\mathbb{A})^{\mathbf{Y}_I}$. However the freedom to replace $\acute{E}t_{/K}(X)$ with $\mathbf{Y}_I$ can be useful.

In this section we will prove the following theorem:

Theorem 9.87 *Let K be a number field and X a K-variety. Let $\mathbf{X}_I, \mathbf{Y}_I \in \operatorname{Pro}\operatorname{Ho}\left(\operatorname{Set}_{\Gamma_K}^{\Delta^{op}}\right)$ be two objects such that I and J are countable and all the $\mathbf{X}_\alpha$'s and $\mathbf{Y}_\alpha$'s are finite, excellent and connected. Let*

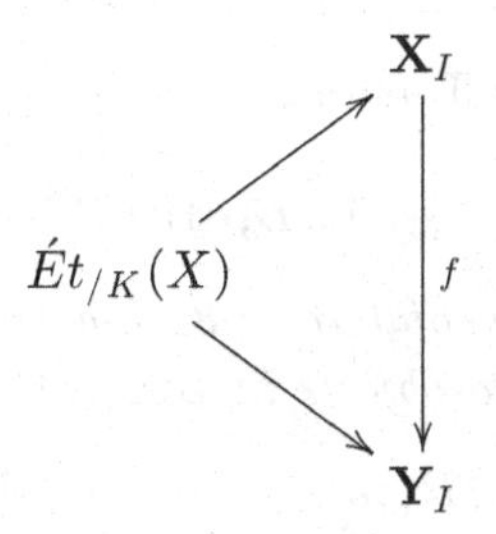

be a commutative triangle such that f induces an isomorphism

$$\overline{f} : \overline{\mathbf{X}}_I^\natural \longrightarrow \overline{\mathbf{Y}}_I^\natural$$

where

$$\overline{(\bullet)} : \operatorname{Pro}\operatorname{Ho}\left(\operatorname{Set}_{\Gamma_K}^{\Delta^{op}}\right) \longrightarrow \operatorname{Pro}\operatorname{Ho}\left(\operatorname{Set}^{\Delta^{op}}\right)$$

is the forgetful functor. Then

$$X(\mathbb{A})^{\mathbf{X}_I} = X(\mathbb{A})^{\mathbf{Y}_I}.$$

Now before we come to the proof of 9.87 we will need to develop some terminology regarding torsors of pointed sets.

Definition 9.88 Let A_* be a pointed set. An A_***-Torsor** is a **non-empty** set B together with an map

$$a : A_* \times B \longrightarrow B$$

such that

1. $a(*, b) = b$ for all $b \in B$.
2. The map

$$A_* \times B \overset{a \times p_2}{\longrightarrow} B \times B$$

is an isomorphism of sets.

We call the map a the "action" of A_* on B.

We denote by Tors(Set$_*$, Set) the category whose objects are triples (A_*, B, a) such that B is an A_*-torsor with action a and whose morphisms are maps of pairs $(A_*, B) \longrightarrow (A'_*, B')$ which commute with the respective actions.

We have two natural functors:

$$\mathfrak{A}_* : \mathrm{Tors}(\mathrm{Set}_*, \mathrm{Set}) \longrightarrow \mathrm{Set}_*,$$

$$\mathfrak{A}_*((A_*, B, a)) = A_*,$$

and

$$\mathfrak{B} : \mathrm{Tors}(\mathrm{Set}_*, \mathrm{Set}) \longrightarrow \mathrm{Set},$$

$$\mathfrak{B}((A_*, B, a)) = B.$$

Lemma 9.89 *Let I be a cofiltered indexing poset and $\{(A_\alpha, B_\alpha, a_\alpha)\}_{\alpha \in I}$, $\{(A'_\beta, B'_\beta, a'_\beta)\}_{\beta \in I}$ two pro-objects in* Tors(Set$_*$, Set). *Let*

$$F : \{(A_\alpha, B_\alpha, a_\alpha)\} \longrightarrow \{(A'_\alpha, B'_\alpha, a'_\alpha)\}$$

be a levelwise map such that $\mathfrak{A}(F)$ is an isomorphism in Pro Set$_*$. *Then $\mathfrak{B}(F)$ is an isomorphism in* Pro Set.

Proof We can consider F as a compatible family of commutative diagrams

$$\begin{array}{ccc} A_\alpha \times B_\alpha & \xrightarrow{a_\alpha} & B_\alpha \\ \Big\downarrow{\scriptstyle f_\alpha \times g_\alpha} & & \Big\downarrow{\scriptstyle g_\alpha} \\ A'_\alpha \times B'_\alpha & \xrightarrow{a'_\alpha} & B'_\alpha . \end{array}$$

Now the map $\mathfrak{A}_*(F)$ in Pro(Set$_*$) is represented by the levelwise map f_α. Since $f_I = \{f_\alpha\}_{\alpha \in I}$ is an isomorphism we have a map

$$f_I^{-1} : \{A'_\alpha\}_{\alpha \in I} \longrightarrow \{A_\alpha\}_{\alpha \in I}$$

which is the inverse of f_I. Note that f_I^{-1} can be represented by a map $\theta : I \longrightarrow I$ and a collection of maps $f_\alpha^{-1} : A'_{\theta(\alpha)} \to A_\alpha$ satisfying some natural compatibility conditions (See [EHa76]). Without loss of generality we may assume that $\theta(\alpha) \geq \alpha$ for all α.

We would like to construct an inverse to the map $\mathfrak{B}(F)$ in Pro(Set) which is represented by the levelwise map g_α. We denote this inverse by

$$g_I^{-1} = \{B'_\alpha\}_{\alpha \in I} \longrightarrow \{B_\alpha\}_{\alpha \in I}.$$

To describe g_I^{-1} we will give a collection of maps

$$g_\alpha^{-1} : B'_{\theta(\alpha)} \longrightarrow B_\alpha$$

and we leave it to the reader to verify the easy diagram chasing that is required to check that the maps g_α^{-1} are indeed defining a pro-map which is an inverse to $g_I = \mathfrak{B}(F)$.

Now let $\alpha \in I$ be an index. Choose an arbitrary element $b_0 \in B_{\theta(\alpha)}$. We shall denote by $\tilde{b_0} \in B_\alpha$ the image of b_0 by the structure map $B_{\theta(\alpha)} \longrightarrow B_\alpha$ and by

$$b'_0 = g_{\theta(\alpha)}(b_0) \in B'_{\theta(\alpha)}$$

the respective image in $B'_{\theta(\alpha)}$.

Now for every $b' \in B'_{\theta(\alpha)}$ we can apply the inverse of the map

$$A'_{\theta(\alpha)} \times B'_{\theta(\alpha)} \longrightarrow B'_{\theta(\alpha)} \times B'_{\theta(\alpha)}$$

to the tuple (b', b'_0) and get an element $(c', b'_0) \in A'_{\theta(\alpha)} \times B'_{\theta(\alpha)}$. We then define

$$g_\alpha^{-1}(b') = a(f_\alpha^{-1}(c'), \tilde{b_0}) \in B_\alpha.$$

□

Now in the proof of Theorem 9.87 we will need to work with pro-objects of various kinds which carry actions of pro-objects in the category of finite groups. In order to work well with such objects it will be useful to recall first that pro objects in the category of finite groups are the same as pro-finite groups in the usual sense (see 9.77). We will be interested in the following kind of actions:

Definition 9.90 Let C be a category, G a pro-finite group and $\{\mathbf{X}_\alpha\}$ an object in Pro C. An **excellent** action of G on $\{\mathbf{X}_\alpha\}$ is an action of G on $\{\mathbf{X}_\alpha\}$ via levelwise maps such that the induced action on each $\mathbf{X}_\alpha$ factors through a finite (continuous) quotient of G. In this case we will say $\{\mathbf{X}_\alpha\}$ is an excellent G-pro-object.

We say that a map in Pro C between two excellent G-pro-objects

$$f : \{\mathbf{X}_\alpha\} \longrightarrow \{\mathbf{Y}_\beta\}$$

is **equivariant** if $g \circ f \circ g^{-1} = f$ for every $g \in G$.

We then have the following basic lemma:

Lemma 9.91 *Let C be a category, G a pro-finite group and let*

$$f : \{\mathbf{X}_\alpha\} \longrightarrow \{\mathbf{Y}_\beta\}$$

be an equivariant map between two excellent G-objects. Then f can be represented by a compatible family of maps

$$f_\beta : \mathbf{X}_{\alpha(\beta)} \longrightarrow \mathbf{Y}_\beta$$

such that each f_β is equivariant.

Proof It is enough to prove for $\{\mathbf{Y}_\beta\}$ which is a single space $\mathbf{Y}$.

Start with any map $f' : \mathbf{X}_\alpha \longrightarrow \mathbf{Y}$ representing f and consider the **finite** orbit $\{f' \circ g\}_{g \in G}$. Since $f \circ g = f$ for every $g \in G$ we get that all the maps $f' \circ g$ represent the same map f. Thus there exist some $\alpha \leq \beta \in I$ such that the images of all the $f' \circ g$ in $\mathrm{Hom}(\mathbf{X}_\beta, \mathbf{Y})$ are the same. We will denote this unified image by

$$f_\beta : \mathbf{X}_\beta \longrightarrow \mathbf{Y}.$$

Then we see that f_β is an equivariant map representing f and we are done. □

In particular we see that any equivariant map

$$f : \{\mathbf{X}_\alpha\} \longrightarrow \{\mathbf{Y}_\beta\}$$

between two excellent G-objects induces a map

$$f' : \{\mathbf{X}_\alpha\} \xrightarrow{\simeq} \{\mathbf{Y}_\beta\}$$

in the pro-category of G-objects in C.

This simple observation has several useful corollaries:

Corollary 9.92 *Let*

$$f : \{\mathbf{X}_\alpha\} \longrightarrow \{\mathbf{Y}_\beta\}$$

be an equivariant map between two excellent G-objects such that the underlying map of pro-objects is an isomorphism. Then f induces an isomorphism in the pro-category of G-objects.

Corollary 9.93 *Let $\mathbf{X}_I = \{\mathbf{X}_\alpha\}$ be an excellent G-object. Then the levelwise quotient*

$$\mathbf{X}_I/\Gamma = \{\mathbf{X}_\alpha/\Gamma\}$$

coincides with the categorical quotient (i.e. with the corresponding colimit) in Pro C.

Corollary 9.94 *Let*

$$f : \{\mathbf{X}_\alpha\} \longrightarrow \{\mathbf{Y}_\beta\}$$

be an equivariant map between two excellent G-objects such that the underlying map of pro-objects is an isomorphism. Then f induces an isomorphism in Pro C

$$\{\mathbf{X}_\alpha/G\} \xrightarrow{\simeq} \{\mathbf{Y}_\beta/G\}.$$

We are now ready to prove the main ingredient of the proof:

Proposition 9.95 *Let* Γ *be a pro-finite group and* $\mathbf{X}_I = \{\mathbf{X}_\alpha\}, \mathbf{Y}_J = \{\mathbf{Y}_\beta\}$ *be two objects in* $\operatorname{Pro}\operatorname{Ho}\left(\operatorname{Set}_\Gamma^{\Delta^{op}}\right)$ *such that* I, J *are countable and all the* $\mathbf{X}_\alpha$*'s and* $\mathbf{Y}_\alpha$*'s are excellent, finite, connected and 2-bounded. Let* $f : \mathbf{X}_I \longrightarrow \mathbf{Y}_J$ *be a map such that*

$$\overline{f} : \overline{\mathbf{X}}_I \longrightarrow \overline{\mathbf{Y}}_I$$

is an isomorphism in $\operatorname{Pro}\operatorname{Ho}\left(\operatorname{Set}^{\Delta^{op}}\right)$. *Then f induces an isomorphism of sets*

$$\mathbf{X}_I\left(\mathbf{E}\Gamma^{\natural}\right) \simeq \mathbf{Y}_I\left(\mathbf{E}\Gamma^{\natural}\right).$$

Remark 9.96 Once again we pay a price for not having a suitable model structure on $\operatorname{Pro}\left(\operatorname{Set}^{\Delta^{op}}\right)$. Of course in such a model structure this statement would be immediate and without the heavy assumptions. As mentioned before, this better approach will be taken in a future paper based on the model structure of [BSc11].

Proof Recalling the discussion in the beginning of subsection §§9.6.1 we start by lifting f to a levelwise map

$$\widetilde{f} : \widetilde{\mathbf{X}}_{\mathbb{N}} \longrightarrow \widetilde{\mathbf{Y}}_{\mathbb{N}}$$

in $\operatorname{ProSet}_\Gamma^{\Delta^{op}}$ between two pro-finite fibration towers.

Let $\widetilde{\mathbf{X}}_{\mathbb{N}}^1 = \left\{P_1\left(\widetilde{\mathbf{X}}_n\right)\right\}$ and $\widetilde{\mathbf{Y}}_{\mathbb{N}}^1 = \left\{P_1\left(\widetilde{\mathbf{Y}}_n\right)\right\}$. We will start by showing that

$$\widetilde{\mathbf{X}}_{\mathbb{N}}^1\left(\mathbf{E}\Gamma^{\natural}\right) \simeq \widetilde{\mathbf{Y}}_{\mathbb{N}}^1\left(\mathbf{E}\Gamma^{\natural}\right).$$

Choose a compatible set of (not-necessarily-Γ-invariant) base points $x_n \in \mathbf{X}_n$ and let $y_n = f_n(x_n) \in \mathbf{Y}_n$. Let $G_i = \pi_1\left(\widetilde{\mathbf{X}}_n, x_n\right)$ and $H_n = \pi_1\left(\widetilde{\mathbf{Y}}_n, y_n\right)$.

Now the map f induces an isomorphism of short exact sequences (note that any bijective map of pro-finite groups is an isomorphism because the topology on both is Hausdorff-compact):

$$\begin{array}{ccccccccc}
1 & \longrightarrow & \pi_1(\mathbf{X}_{\mathbb{N}}, \{x_n\}) & \longrightarrow & \pi_1((\mathbf{X}_{\mathbb{N}})_{h\Gamma}, \{x_n\}) & \longrightarrow & \Gamma & \longrightarrow & 1 \\
 & & \downarrow & & \downarrow & & \| & & \\
1 & \longrightarrow & \pi_1(\mathbf{Y}_{\mathbb{N}}, \{x_n\}) & \longrightarrow & \pi_1((\mathbf{Y}_{\mathbb{N}})_{h\Gamma}, \{x_n\}) & \longrightarrow & \Gamma & \longrightarrow & 1
\end{array}$$

and so the first sequence splits if and only if the second does. Hence from Theorem 9.78 we see that

$$\widetilde{\mathbf{X}}^1_{\mathbb{N}}\left(\mathbf{E}\Gamma^{\natural}\right) \neq \emptyset$$

if and only if

$$\widetilde{\mathbf{Y}}^1_{\mathbb{N}}\left(\mathbf{E}\Gamma^{\natural}\right) \neq \emptyset.$$

Hence it is enough to deal with the case that arises when both of them are non-empty. Since $\widetilde{\mathbf{X}}^1_n$ is a fibration tower we get that $\{(\widetilde{\mathbf{X}}^1_n)^{h\Gamma}\}$ is a tower of Kan fibrations and so one can choose a compatible family $\widetilde{h}_n \in (\widetilde{\mathbf{X}}^1_n)^{h\Gamma}$.

Now from Corollary 9.92 we get that the equivariant isomorphism of pro-finite groups

$$\left\{\pi_1\left(\widetilde{\mathbf{X}}^1_n, \widetilde{h}_n\right)\right\} \longrightarrow \left\{\pi_1\left(\widetilde{\mathbf{Y}}^1_n, f_n(\widetilde{h}_n)\right)\right\}$$

together with the spectral sequence of 9.46 induce an isomorphism

$$\widetilde{\mathbf{X}}^1_I\left(\mathbf{E}\Gamma^{\natural}\right) \simeq \lim_n \widetilde{\mathbf{X}}^1_n(hK) = \lim_n H^1\left(\Gamma, \pi_1\left(\widetilde{\mathbf{X}}^1_n, \widetilde{h}_n\right)\right) \cong$$
$$\lim_n H^1\left(\Gamma, \pi_1\left(\widetilde{\mathbf{Y}}^1_n, f_n(\widetilde{h}_n)\right)\right) \cong \lim_n \widetilde{\mathbf{Y}}^1_n(hK) \cong \widetilde{\mathbf{Y}}^1_{\mathbb{N}}\left(\mathbf{E}\Gamma^{\natural}\right).$$

Now consider the commutative square

$$\begin{array}{ccc}
\mathbf{X}_{\mathbb{N}}\left(\mathbf{E}\Gamma^{\natural}\right) & \xrightarrow{f_*} & \mathbf{Y}_{\mathbb{N}}\left(\mathbf{E}\Gamma^{\natural}\right) \\
\downarrow & & \downarrow \\
\mathbf{X}^1_{\mathbb{N}}\left(\mathbf{E}\Gamma^{\natural}\right) & \longrightarrow & \mathbf{Y}^1_{\mathbb{N}}\left(\mathbf{E}\Gamma^{\natural}\right).
\end{array}$$

In order to finish the proof we will show that for every $\{h_n\} \in \mathbf{X}^1_{\mathbb{N}}\left(\mathbf{E}\Gamma^{\natural}\right)$ the map f_* maps the fiber over $\{h_n\}$ isomorphically to the fiber over $\{f_n(h_n)\}$.

As above, choose a compatible family $\widetilde{h}_n \in (\widetilde{\mathbf{X}}^1_n)^{h\Gamma}$ such that h_n is the connected component of $\widetilde{h}_n$.

First of all we need to address the possibility that this fiber above

$\{h_n\}$ is empty. By Proposition 9.49 we see that for every n we have an obstruction element

$$o_n \in H^3\left(\Gamma, \pi_2\left(\widetilde{\mathbf{X}}_n, \widetilde{h}_n\right)\right)$$

which is trivial if and only if h_n lifts to $\pi_0\left(\widetilde{\mathbf{X}}_n^{h\Gamma}\right) \cong \pi_0\left(P_2\left(\widetilde{\mathbf{X}}_n\right)^{h\Gamma}\right)$. From Corollary 9.92 the equivariant isomorphism of pro-finite groups

$$\left\{\pi_2\left(\widetilde{\mathbf{X}}_n, \widetilde{h}_n\right)\right\} \longrightarrow \left\{\pi_2\left(\widetilde{\mathbf{Y}}_n, f_n(\widetilde{h}_n)\right)\right\}$$

induces an isomorphism of pro-finite groups

$$\left\{H^3\left(\Gamma, \pi_2\left(\widetilde{\mathbf{X}}_n, \widetilde{h}_n\right)\right)\right\} \longrightarrow \left\{H^3\left(\Gamma, \pi_2\left(\widetilde{\mathbf{X}}_n, \widetilde{h}_n\right)\right)\right\}$$

and so the element

$$\{o_n\} \in \lim_n H^3\left(\Gamma, \pi_2\left(\widetilde{\mathbf{X}}_n, \widetilde{h}_n\right)\right)$$

is zero if and only if the element

$$\{f_n(o_n)\} \in \lim_n H^3\left(\Gamma, \pi_2\left(\widetilde{\mathbf{Y}}_n, f_n\left(\widetilde{h}_n\right)\right)\right)$$

is zero. If both of them are non-zero then both fibers are empty and we are done. Hence we can assume that both are zero.

Let $A_n \subseteq \pi_0\left(\widetilde{\mathbf{X}}_n^{h\Gamma}\right)$ be the (non-empty) pre-image of h_n and $B_n \subseteq \pi_0\left(\widetilde{\mathbf{Y}}_n^{h\Gamma}\right)$ the pre-image of $f_n(h_n)$. We need to show that the induced map

$$f_* : \lim A_n \longrightarrow \lim B_n$$

is an isomorphism.

Now given any homotopy fixed point $x \in \widetilde{\mathbf{X}}_n^{h\Gamma}$ we can run the spectral sequence of 9.46 to compute $\pi_0\left(\widetilde{\mathbf{X}}_n^{h\Gamma}\right)$. Since $\widetilde{\mathbf{X}}_n$ is 2-bounded this spectral sequence will degenerate in the third page and hence we see that in order to compute the cell $E_{2,2}^{\infty} = E_{2,2}^{3}$ it is enough to choose the image of x in $P_1(\widetilde{\mathbf{X}}_n)^{h\Gamma}$. In particular we can calculate this cell using the homotopy fixed point $\widetilde{h}_n$.

Now by analyzing the spectral sequence we see that the pointed set $E_{2,2}^3(\widetilde{h}_n)$ is obtained from the group $H^2\left(\Gamma, \pi_2\left(\widetilde{\mathbf{X}}_n, \widetilde{h}_n\right)\right)$ by taking the quotient **pointed set** under the action of $H^0\left(\Gamma, \pi_1\left(\widetilde{\mathbf{X}}_n, \widetilde{h}_n\right)\right)$ which is induced by the action of $H^0\left(\Gamma, \pi_1\left(\widetilde{\mathbf{X}}_n, \widetilde{h}_n\right)\right)$ on $\pi_2\left(\widetilde{\mathbf{X}}_n, \widetilde{h}_n\right)$.

Now the equivariant isomorphism of pro-groups

$$\left\{\pi_2\left(\widetilde{\mathbf{X}}_n,\widetilde{h}_n\right)\right\}\longrightarrow\left\{\pi_2\left(\widetilde{\mathbf{Y}}_n,f_n(\widetilde{h}_n)\right)\right\}$$

induces an isomorphism of pro-groups

$$\left\{H^2\left(\Gamma,\pi_2\left(\widetilde{\mathbf{X}}_n,\widetilde{h}_n\right)\right)\right\}\longrightarrow\left\{H^2\left(\Gamma,\pi_2\left(\widetilde{\mathbf{Y}}_n,f_n\left(\widetilde{h}_n\right)\right)\right)\right\}.$$

From Corollary 9.92 we get that the equivariant isomorphism of pro-finite groups

$$\left\{\pi_1\left(\widetilde{\mathbf{X}}_n,\widetilde{h}_n\right)\right\}\longrightarrow\left\{\pi_1\left(\widetilde{\mathbf{Y}}_n,f_n(\widetilde{h}_n)\right)\right\}$$

induces an isomorphism of pro-finite groups

$$\left\{H^0\left(\Gamma,\pi_1\left(\widetilde{\mathbf{X}}_n,\widetilde{h}_n\right)\right)\right\}\longrightarrow\left\{H^0\left(\Gamma,\pi_1\left(\widetilde{\mathbf{Y}}_n,f_n(\widetilde{h}_n)\right)\right)\right\}.$$

Hence from Corollary 9.94 we get that the induced map of pro-pointed-sets

$$\left\{E_{2,2}^{\infty}\left(\widetilde{h}_n\right)\right\}\longrightarrow\left\{E_{2,2}^{\infty}\left(f_n(\widetilde{h}_n\right)\right\}$$

is an isomorphism. Now from the Bausfield-Kan type spectral sequence of 9.46 the set A_n admits a natural torsor structure under $E_{2,2}^{\infty}\left(\widetilde{h}_n\right)$ and similarly B_n under $E_{2,2}^{\infty}\left(f_n(\widetilde{h}_n)\right)$. From Lemma 9.89 we then get that the induced map of pro-sets

$$\{A_n\}\longrightarrow\{B_n\}$$

is an isomorphism and induces an isomorphism

$$\lim A_n\xrightarrow{\simeq}\lim_n B_n.$$

□

We now come to the proof of Theorem 9.87. From Theorem 9.63 and Proposition 9.58 we can assume that all the spaces in the diagrams of $\mathbf{X}_I$ and $\mathbf{Y}_I$ are 2-bounded. The theorem will then follow from the following corollary:

Corollary 9.97 *Let K be a number field and $\mathbf{X}_I=\{\mathbf{X}_\alpha\},\mathbf{Y}_J=\{\mathbf{Y}_n\}$ two objects in* $\operatorname{Pro}\operatorname{Ho}\left(\operatorname{Set}_{\Gamma_K}^{\Delta^{op}}\right)$ *such that I and J are countable and all the $\mathbf{X}_\alpha$'s and $\mathbf{Y}_\alpha$'s are finite, excellent, connected and 2-bounded. Let $f:\mathbf{X}_I\longrightarrow\mathbf{Y}_J$ be a map such that*

$$\overline{f}:\overline{\mathbf{X}}_I\longrightarrow\overline{\mathbf{Y}}_I$$

is an isomorphism. Assume that $\mathbf{X}_I(h\mathbb{A}) \neq \emptyset$. *Then an element* $(x_\nu) \in \mathbf{X}_I(h\mathbb{A})$ *is rational (i.e. is the image of am element in* $\mathbf{X}_I(hK)$*) if and only if* $f_*(x_\nu) \in \mathbf{Y}_I(\mathbb{A})$ *is rational.*

Proof From Proposition 9.95 we get that f induces isomorphisms of sets:

$$\mathbf{X}_I(hK) \xrightarrow{\simeq} \mathbf{Y}_I(hK)$$

and

$$\prod_\nu \mathbf{X}_I(hK_\nu) \xrightarrow{\simeq} \prod_\nu \mathbf{Y}_I(hK_\nu).$$

Now since $\mathbf{X}_I(h\mathbb{A}) \subseteq \prod_\nu \mathbf{X}_I(hK_\nu)$ the result follows. □

In subsections 9.3.2 and 9.3.2 we have defined a family of obstruction sets that fit together into the diagram

$$\begin{array}{ccccccccc}
 & & X(\mathbb{A})^{\mathbb{Z}h} & \hookrightarrow \cdots \hookrightarrow & X(\mathbb{A})^{\mathbb{Z}h,1} & \hookrightarrow & X(\mathbb{A})^{\mathbb{Z}h,0} & \hookrightarrow & X(\mathbb{A}) \\
 & & \uparrow & & \uparrow & & \uparrow & & \\
X(K) & \hookrightarrow & X(\mathbb{A})^{h} & \hookrightarrow \cdots \hookrightarrow & X(\mathbb{A})^{h,1} & \hookrightarrow & X(\mathbb{A})^{h,0}. & &
\end{array}$$

In section 9.8 we will consider zero-dimensional varieties, and show that in that case even the weakest obstruction set $X(\mathbb{A})^{\mathbb{Z},0}$ is equal to $X(K)$. This can be used to show that for general X the non-emptiness of $X(\mathbb{A})^{\mathbb{Z},0}$ is equivalent to the existence of a Galois invariant geometric connected component.

In the next three sections we will relate all the rest of the above obstruction sets to the classical ones.

9.8 Varieties of dimension zero

Proposition 9.98 *Let X be a zero dimensional variety. Then for every* $0 \leq n \leq \infty$ *one has*

$$X(\mathbb{A})^{h,n} = X(\mathbb{A})^{\mathbb{Z}h,n} = X(K).$$

Proof Consider the hypercovering $X_\bullet \longrightarrow X$ where $X_\bullet$ is the constant simplicial variety $X_n = X$. Then $P_0(\mathbb{Z}\mathbf{X}_{X_\bullet})$ is a discrete simplicial set which is the Γ_K module freely generated by $X(\overline{K})$. From the discreteness of $P_0(\mathbb{Z}\mathbf{X}_{X_\bullet})$ and the fact that

$$X(\overline{K}) \subseteq P_0(\mathbb{Z}\mathbf{X}_{X_\bullet}).$$

we see that

$$X(\mathbb{A})^{\mathbb{Z}X_\bullet,0} = X(K).$$

Now since

$$X(K) \subseteq X(\mathbb{A})^{h,n} \subseteq X(\mathbb{A})^{\mathbb{Z}h,n} \subseteq X(\mathbb{A})^{\mathbb{Z}X_\bullet,0} = X(K)$$

we get that

$$X(K) = X(\mathbb{A})^{\mathbb{Z}h,n} = X(\mathbb{A})^{h,n}$$

as required. □

Corollary 9.99 *Let X be a zero dimensional variety. Then*

$$X(\mathbb{A})^{fin,h} = X(\mathbb{A})^{fin,\mathbb{Z}h} = X(K).$$

Proof Immediate from Proposition 9.98 and the fact that

$$X(\mathbb{A})^{fin,h} \subseteq X(\mathbb{A})^{fin,\mathbb{Z}h} \subseteq X(\mathbb{A})^{\mathbb{Z}h}.$$

□

Corollary 9.100 *If $X = X_1 \coprod X_2$ then*

$$X(\mathbb{A})^{h,n} = X_1(\mathbb{A})^{h,n} \coprod X_2(\mathbb{A})^{h,n},$$

$$X(\mathbb{A})^{\mathbb{Z}h,n} = X_1(\mathbb{A})^{\mathbb{Z}h,n} \coprod X_2(\mathbb{A})^{\mathbb{Z}h,n},$$

$$X(\mathbb{A})^{fin,h} = X_1(\mathbb{A})^{fin,h} \coprod X_2(\mathbb{A})^{fin,h},$$

$$X(\mathbb{A})^{fin,\mathbb{Z}h} = X_1(\mathbb{A})^{fin,\mathbb{Z}h} \coprod X_2(\mathbb{A})^{fin,\mathbb{Z}h}.$$

Proof Consider the obvious map

$$X \longrightarrow \mathrm{Spec}\,(K) \coprod \mathrm{Spec}\,(K)$$

and use functoriality together with Proposition 9.98 and Corollary 9.99. □

Corollary 9.101 *Let X be a K-variety. Then both $X(\mathbb{A})^{h,0}$ and $X(\mathbb{A})^{\mathbb{Z}h,0}$ contains exactly the adelic points which sit in a Γ_K invariant connected component.*

Proof Again consider the map $X \longrightarrow \pi_0(X)$ where $\pi_0(X)$ is considered as a zero dimensional K-variety. From Proposition 9.98 we see that if $(x_\nu) \in X(\mathbb{A})^{\mathbb{Z}h,0}$ then (x_ν) sits on a Γ_K invariant connected component. Hence it will be enough to show that if X is connected over K then $X(\mathbb{A})^{h,0} = X(\mathbb{A})$. But this follows from the fact that when X is connected over K then all the spaces $\mathbf{X}_\mathcal{U}$ are connected and so

$$\mathbf{X}_{\mathcal{U},0} = * .$$

This means that in that case,

$$\lim_{\mathcal{U}} \mathbf{X}_{\mathcal{U},0}(hK) = *$$

and so all adelic points are in $X(\mathbb{A})^{h,0}$. □

Corollary 9.102 *If $\overline{X}$ does not have a Γ_K-invariant connected component then*

$$X(\mathbb{A})^{fin,h} = X(\mathbb{A})^{fin,\mathbb{Z}h} = X(\mathbb{A})^{h,n} = X(\mathbb{A})^{\mathbb{Z}h,n} = \emptyset.$$

Proof Immediate from Corollary 9.101 and the fact that

$$X(\mathbb{A})^{h,n} \subseteq X(\mathbb{A})^{0,h}.$$

□

9.9 Connection to finite descent

Let K be a number field and X a K-variety. Recall sections §9.3.2 and 9.3.2 in which we have defined for each n two obstruction sets

$$X(K) \subseteq X(\mathbb{A})^{h,n} \subseteq X(\mathbb{A})^{\mathbb{Z}h,n} \subseteq X(\mathbb{A}).$$

In section §9.5 we have proved Corollary 9.68 which states that for each $n \geq 2$ one has

$$X(\mathbb{A})^{h,n} = X(\mathbb{A})^{h,2}$$

and

$$X(\mathbb{A})^{\mathbb{Z}h,n} = X(\mathbb{A})^{\mathbb{Z}h,2}.$$

In Corollary 9.101, we have determined $X(\mathbb{A})^{h,0}$ and $X(\mathbb{A})^{\mathbb{Z}h,0}$, and we saw that they contain only information coming from the zero dimensional scheme of connected components of X. It is hence left to analyze the sets $X(\mathbb{A})^{h,n}$ and $X(\mathbb{A})^{\mathbb{Z}h,n}$ for $n = 1, 2$. In this section we will deal with the $n = 1$ case and prove the following theorem, connecting the corresponding obstruction sets to the obstruction sets obtained by descent over certain finite groups:

Theorem 9.103 *Let K be field and X a smooth connected K-variety. Then*

$$X(\mathbb{A})^{h,1} = X(\mathbb{A})^{fin},$$

$$X(\mathbb{A})^{\mathbb{Z}h,1} = X(\mathbb{A})^{fin-ab}.$$

The rest of this section will be devoted to the proof of Theorem 9.103. Let us begin by establishing some notation. In order to understand the finite descent obstructions we will be considering torsors

$$f : Y \longrightarrow X$$

under finite K-groups G. We will always assume that Y is connected over K (i.e. that its geometric connected components form a single Galois orbit), as this is no loss of generality for descent obstruction. When we don't wish to specify the K-group G we will simply refer to $f : Y \longrightarrow X$ as a **finite torsor**. We will denote by

$$X^f(\mathbb{A}) \subseteq X(\mathbb{A})$$

(see §§§9.3.1) the set of adelic points which survive descent by Y, i.e. the set of adelic points which lift to some K-twist of Y. In this notation $X(\mathbb{A})^{fin}$ is obtained by intersecting $X^f(\mathbb{A})$ over all finite torsors and X^{fin-ab} is obtained by intersecting $X(\mathbb{A})^f$ over all finite abelian torsors.

The essence of the connection between torsors under finite K-groups and the relative étale homotopy type is given by applying the Čech construction to the étale (covering) map $Y \longrightarrow X$. One then obtains a hypercovering $\check{Y}_\bullet$ given by

$$\check{Y}_n = \overbrace{Y \times_X ... \times_X Y}^{n+1}.$$

Let us now analyze the simplicial Γ_K-set $\mathbf{N}_{\check{Y}}$ (see Definition 9.16). We have a natural isomorphism (over K)

$$\overbrace{Y \times_X ... \times_X Y}^{n+1} \cong Y \times G^n$$

and so

$$(\mathbf{N}_{\check{Y}})_n = \pi_{0/K}(\check{Y}_n) \cong \pi_{0/K}(Y) \times G^n.$$

We then observe that we can identify $\mathbf{N}_{\check{Y}}$ with the nerve of a certain groupoid $\mathcal{Y}$:

Definition 9.104 Let G be a finite K-group and $f : Y \longrightarrow X$ a torsor under G. We define the groupoid $\mathcal{Y}$ to be the groupoid whose objects set is

$$\mathrm{Ob}(\mathcal{Y}) = \pi_{0/K}(Y)$$

and whose morphism sets are

$$\mathrm{Hom}(C_1, C_2) = \{g \in G | gC_1 = C_2\}$$

for each $C_1, C_2 \in \pi_{0/K}(Y)$.

It is easy to verify that:

Lemma 9.105

$$\mathbf{N}_{\check{Y}} = N(\mathcal{Y}).$$

Now recalling again Definition 9.16 we can write

$$\mathbf{X}_{\check{Y},1} = P_1(\mathrm{Ex}^{\infty}(\mathbf{N}_{\check{Y}})) = P_1(\mathrm{Ex}^{\infty}(N(\mathcal{Y})).$$

Now it is a well known result in homotopy theory that since $\mathcal{Y}$ is a connected groupoid the nerve $N(\mathcal{G})$ is a connected Kan simplicial set and for each $C \in \mathrm{Ob}(\mathcal{Y})$ one has

$$\pi_n(N(\mathcal{Y}), C) = \begin{cases} \mathrm{Hom}_{\mathcal{Y}}(C, C) & n = 1 \\ 0 & n > 1. \end{cases} \qquad (*)$$

In particular this means that the map

$$\mathbf{N}_{\check{Y}} \longrightarrow P_1(\mathrm{Ex}^{\infty}(\mathbf{N}_{\check{Y}})) = \mathbf{X}_{\check{Y},1}$$

is a local weak equivalence of simplicial Γ_K-sets, so we can consider $\mathbf{N}_{\check{Y}}$ as a model for $\mathbf{X}_{\check{Y},1}$. In particular we have isomorphisms of sets

$$\begin{aligned} \mathbf{N}_{\check{Y}}(hK) &\xrightarrow{\simeq} \mathbf{X}_{\check{Y},1}(hK), \\ \mathbf{N}_{\check{Y}}(h\mathbb{A}) &\xrightarrow{\simeq} \mathbf{X}_{\check{Y},1}(h\mathbb{A}). \end{aligned} \qquad (**)$$

We consider the K-group G as the finite group $G(\overline{K})$ with an action of Γ_K on it. We then consider the standard model for $\mathbf{B}G$ as the nerve of the groupoid $\mathcal{B}G$ with one object and morphism set G. We have a Γ_K action on this groupoid and so an action of Γ_K on our model for $\mathbf{B}G$.

We have an equivariant map of groupoids

$$\mathcal{Y} \longrightarrow \mathcal{B}G$$

induced by the natural inclusion $\mathrm{Hom}(C_1, C_2) \subseteq G$. This groupoid map induces a map

$$c_Y : \mathbf{N}_{\check{Y}} \longrightarrow \mathbf{B}G.$$

The computation $(*)$ has the following corollary describing the behavior of the map c_Y:

Corollary 9.106 *Let G be a finite K-group and $f : Y \longrightarrow X$ a torsor under G. If Y is* ***geometrically connected*** *then the map*

$$c_Y : \mathbf{N}_{\check{Y}} \longrightarrow \mathbf{B}G$$

is an ***isomorphism*** *and induces a natural identification*

$$\pi_1\left(\mathbf{N}_{\check{Y}}, *\right) \cong G.$$

If Y is not geometrically connected then the map c_Y will still induce an embedding on the level of fundamental groups.

We now come to the main lemma which connects the descent obstruction given by Y and the homotopy obstruction given by the hypercovering $\check{Y}$:

Proposition 9.107 *Let X be a K-variety and $f : Y \longrightarrow X$ a torsor under a finite K-group G. Then*

$$X(\mathbb{A})^{\check{Y},1} \subseteq X(\mathbb{A})^f.$$

If, furthermore, Y is connected over $\overline{K}$ then

$$X(\mathbb{A})^{\check{Y},1} = X(\mathbb{A})^f.$$

Proof In this model $\mathbf{B}G$ has a single vertex giving it a natural base point preserved by Γ_K. Since $\mathbf{B}G$ is also bounded we can use this base point to compute

$$\mathbf{B}G(hK) = \pi_0\left(\mathbf{B}G^{h\Gamma_K}\right)$$

via the spectral sequence of 9.46. Since $\mathbf{B}G$ is connected and π_1 is its only non-trivial homotopy group the spectral sequence collapses and we obtain an isomorphism

$$\mathbf{B}G(hK) \xrightarrow{\simeq} H^1(K, G).$$

Let us try to make this identification more explicit. Let $\Lambda \tilde{\triangleleft} \Gamma_K$ be an open normal subgroup and define $H = \Gamma_K/\Lambda$. Then the simplicial set $\mathbf{E}H$ can be realized as the nerve of the groupoid $\mathcal{E}H$ whose objects are

H and whose morphism sets are all singletons. Now suppose that our homotopy fixed point is given by a map

$$F : \mathbf{E}H \longrightarrow \mathbf{B}G$$

which is induced by a map of groupoids

$$\mathcal{F} : \mathcal{E}H \longrightarrow \mathcal{B}G.$$

We can identify H with the union of singletons

$$\cup_{\sigma \in H} \mathrm{Hom}_{\mathcal{E}H}(1, \sigma).$$

Then $\mathcal{F}$ maps this set to the morphism set from the single object of $\mathcal{B}G$ to itself which can be identified with G. The obtained map $u : H \longrightarrow G$ gives the desired 1-cocycle in $H^1(K, G)$.

Note that since $\mathcal{F}$ is H-equivariant it is completely determined by u. If one uses this determination in order to write the condition that $\mathcal{F}$ respects composition in terms of u one will find exactly the familiar 1-cocycle condition.

Now recall the map c_Y defined above. By (**) we get a map (which we denote by the same name)

$$c_Y : \mathbf{X}_{\check{Y},1}(hK) \longrightarrow \mathbf{B}G(hK) = H^1(K, G).$$

With some further abuse of notation, we will call the adelic version of this map by the same name:

$$c_Y : \mathbf{X}_{\check{Y},1}(h\mathbb{A}) \longrightarrow \mathbf{B}G(h\mathbb{A}) = H^1(\mathbb{A}, G).$$

Lemma 9.108 *Let $p \in X(K)$ be a point and*

$$h(p) \in X(hK)$$

the corresponding homotopy fixed point. Then

$$c_Y(h(p)) \in H^1(K, G)$$

is the element which classifies the G-torsor $Y_p = f^{-1}(p)$. Similarly if $p \in X(\mathbb{A})$ is an adelic point then $c_Y(h(p))$ classifies the adelic G-torsor $f^{-1}(p)$.

Proof We will prove the lemma for K. The proof of the adelic version is completely analogous. From functoriality it is enough to prove the claim for $X = p$ and $Y = Y_p$.

In this case $\mathcal{Y}$ is the groupoid whose objects set is a principle homogenous G-set classified by an element $\alpha \in H^1(K, G)$ and there is a unique morphism between any two objects.

Let $y_0 \in Y$ be an arbitrary base point and let H be a big enough finite quotient of Γ_K such that the action of Γ_K on both G and Y, as well as the element α, factor through H. In that case we can represent α by a 1-cocycle $u : H \longrightarrow G$ (denoted by $\sigma \longrightarrow u_\sigma$) such that

$$\sigma(y_0) = u_\sigma(y_0)$$

for every $\sigma \in H$.

Now since the elements in $\mathcal{Y}$ don't have automorphisms we see that $\mathbf{N}_{\check{Y}}$ and hence $\mathbf{X}_{\check{Y},1}$ are contractible and so $\mathbf{X}_{\check{Y},1}(hK) = *$. This means that if we construct any concrete element in $\mathbf{X}_{\check{Y}}(hK)$ (or $\mathbf{N}_{\check{Y}}(hK)$) then it has to coincide with $h(p)$.

The homotopy fixed point which we will construct is the one given by the equivariant map of groupoids

$$\varphi : \mathcal{E}H \longrightarrow \mathcal{Y}$$

which sends the object $\sigma \in H$ to the object $\sigma y_0 \in \mathcal{Y}$ (since all the morphism sets are singletons there is no problem to construct such a map). We then need to compose the map φ with the groupoid map $\mathcal{Y} \longrightarrow \mathcal{B}G$ and consider the obtained map

$$\psi : \mathcal{E}H \longrightarrow \mathcal{B}G.$$

In order to find the corresponding element in $H^1(K, G)$ we need to check to which morphism ψ sends the unique morphism from 1 to $\sigma \in H$. Now φ sends this morphism to the unique morphism from y_0 to σy_0. Since

$$\sigma y_0 = u_\sigma(y_0),$$

we see that ψ sends this morphism to the morphism $u_\sigma \in G$. This finishes the proof of the lemma. □

We are now ready to finish the proof of Proposition 9.107. Let $(x_\nu) \in X(\mathbb{A})^{\check{Y}}$ be an adelic point. Then the homotopy fixed point $h((x_\nu)) \in \mathbf{X}_{\check{Y},1}(h\mathbb{A})$ is rational which means that the adelic cohomology class

$$c_Y(h(x_\nu)) \in H^1(\mathbb{A}, G)$$

is rational as well. By Lemma 9.108 the torsor type of the fiber over (x_ν) is rational and so there exists a K-twist of Y such that (x_ν) lifts to it. This means that $(x_\nu) \in X(\mathbb{A})^f$ and so

$$X(\mathbb{A})^{\check{Y},1} \subseteq X(\mathbb{A})^f.$$

Now assume Y is connected over $\overline{K}$ and let $(x_\nu) \in X(\mathbb{A})^f$ be a point surviving descent by Y. In that case the groupoid $\mathcal{Y}$ has only one object and so the map

$$\mathcal{Y} \longrightarrow \mathcal{B}G$$

is an **isomorphism**. This means that the map

$$\mathbf{N}_{\check{Y}} \longrightarrow \mathbf{B}G$$

is an isomorphism and by (**), the maps

$$\mathbf{X}_{\check{Y},1}(hK) \longrightarrow \mathbf{B}G(hK)$$

and

$$\mathbf{X}_{\check{Y},1}(h\mathbb{A}) \longrightarrow \mathbf{B}G(h\mathbb{A})$$

are bijections of sets. By Lemma 9.108 we get that $c_Y(h(x_\nu)) \in \mathbf{B}G(h\mathbb{A})$ is rational and so $h(x_\nu) \in \mathbf{X}_{\check{Y},1}(h\mathbb{A})$ is rational as well. Hence in this case

$$X(\mathbb{A})^{\check{Y},1} = X(\mathbb{A})^f$$

and we are done. □

We are now ready to prove Theorem 9.103.

Proof of 9.103 Since $X(\mathbb{A})^{h,1} \subseteq X^{\check{Y},1}(\mathbb{A})$ for every finite torsor Y we immediately obtain from Proposition 9.107 that

$$X^{h,1}(\mathbb{A}) \subseteq X(\mathbb{A})^{fin}.$$

A similar arguments works for $X^{\mathbb{Z}h,1}$: if $f : Y \longrightarrow X$ is a finite **abelian** torsor then the map

$$\mathbf{N}_{\check{Y}} \longrightarrow \mathbb{Z}\mathbf{N}_{\check{Y}}$$

induces a weak equivalence

$$\mathbf{N}_{\check{Y}} \longrightarrow P_1((\mathbb{Z}\mathbf{N}_{\check{Y}})^1)$$

where $(\mathbb{Z}\mathbf{N}_{\check{Y}})^1$ denotes the connected component of $\mathbb{Z}\mathbf{N}_{\check{Y}}$ given by

$$1 \in \pi_0(\mathbb{Z}\mathbf{N}_{\check{Y}}) \cong H_0(\mathbf{N}_{\check{Y}}) \cong \mathbb{Z}.$$

The last isomorphism here is canonical. This implies that in that case

$$X(\mathbb{A})^{\check{Y},1} = X(\mathbb{A})^{\mathbb{Z}\check{Y},1} \qquad (***)$$

and since $X(\mathbb{A})^{\mathbb{Z}h,1} \subseteq X(\mathbb{A})^{\mathbb{Z}\check{Y},1}$ we get that

$$X(\mathbb{A})^{\mathbb{Z}h,1} \subseteq X(\mathbb{A})^{fin-ab}.$$

It hence remains to prove the inverse inclusion, i.e. to show that

$$X(\mathbb{A})^{fin} \subseteq X^{h,1}(\mathbb{A}),$$

$$X(\mathbb{A})^{fin-ab} \subseteq X(\mathbb{A})^{\mathbb{Z}h,1}.$$

Let $(x_\nu) \in X(\mathbb{A})^{fin}$ be a point. We need to show that $(x_\nu) \in X(\mathbb{A})^{h,1}$. Our first step will be to show that $(x_\nu) \in X(\mathbb{A})^{\check{Y},1}$ for each finite torsor $f : Y \longrightarrow X$.

Proposition 9.109 1. *Let $f : Y \longrightarrow X$ be a torsor under a finite K-group G. Then*

$$X^{fin}(\mathbb{A}) \subseteq X(\mathbb{A})^{\check{Y},1}.$$

2. *Let $f : Y \longrightarrow X$ be a torsor under a finite abelian K-group G. Then*

$$X^{fin-ab}(\mathbb{A}) \subseteq X(\mathbb{A})^{\check{Y},1}.$$

Proof We start with the following lemma:

Lemma 9.110 *Let X, K be as above and assume that $X(\mathbb{A})^{fin} \neq \emptyset$. Let*

$$f : Y \longrightarrow X$$

be a torsor under a finite K-group G. Then there is a twist

$$f^\alpha : Y^\alpha \longrightarrow X$$

such that $\overline{Y}$ has a connected component defined over K. If one assumes instead that $X(\mathbb{A})^{fin-ab} \neq \emptyset$ then the result is true for abelian G.

Proof Note that over $\overline{K}$ there is an inclusion

$$(\overline{Y}_0, \overline{G_0}) \longrightarrow (\overline{Y}, \overline{G})$$

which is a map of torsors over $\overline{X}$, where $\overline{Y}_0$ is any connected component of $\overline{Y}$. Since $X(\mathbb{A})^{fin} \neq \emptyset$ we have from Lemma 5.7 of [Sto07] that there exists a map

$$(\overline{Y'}, \overline{G'}) \longrightarrow (\overline{Y}_0, \overline{G_0})$$

where (Y', G') is a geometrically connected torsor over X. By composing we get a map

$$(\overline{Y'}, \overline{G'}) \longrightarrow (\overline{Y}, \overline{G}).$$

Now by Lemma 5.6 of [Sto07] there exists an $\alpha \in H^1(K, G)$ for which there is a map

$$(Y', G') \longrightarrow (Y^\alpha, G^\alpha).$$

Since (Y', G') is geometrically connected and defined over K the same is true for its image. The abelian version is similar. □

Now assume that $(x_\nu) \in X(\mathbb{A})^{fin}$ and let $f : Y \longrightarrow X$ be a torsor under a finite K-group G. Since $X(\mathbb{A})^{fin} \neq \emptyset$ we get from Lemma 9.110 that there exists a twist

$$f^\alpha : Y^\alpha \longrightarrow X$$

such that $\overline{Y^\alpha}$ has a Γ_K-invariant connected component. Let $Y_0^\alpha \subseteq Y^\alpha$ be that connected component and consider it as torsor over X under its stablizer $G_0^\alpha \subseteq G^\alpha$. Now by calculation $(*)$ above the equivariant map

$$\mathbf{X}_{\check{Y}_0^\alpha,1} \longrightarrow \mathbf{X}_{\check{Y}^\alpha,1}$$

is a weak equivalence. Applying Proposition 9.107 to the torsor $f_0^\alpha : Y_0^\alpha \longrightarrow X$ we then get

$$X^{\check{Y}^\alpha,1}(\mathbb{A}) = X^{\check{Y}_0^\alpha,1}(\mathbb{A}) = X^{f_0^\alpha}(\mathbb{A})$$

and since $(x_\nu) \in X(\mathbb{A})^{fin} \subseteq X^{f_0^\alpha}(\mathbb{A})$ we get that

$$(x_\nu) \in X^{\check{Y}^\alpha,1}(\mathbb{A}).$$

We need to show that $(x_\nu) \in X^{\check{Y},1}(\mathbb{A})$, so we will need to be able to "undo" the twist.

Let H be a big enough finite quotient of Γ_K such that the action of Γ_K on G as well as the element α factor through H. In that case we can represent α by a 1-cocycle $u : H \longrightarrow G$ (denoted by $\sigma \longrightarrow u_\sigma$) satisfying the cocycle condition

$$u_{\sigma\tau} = \sigma(u_\tau)u_\sigma.$$

Let L/K be a finite Galois extension corresponding to H. Let $f^\alpha : Y^\alpha \longrightarrow X$ be the corresponding twist by α which is a finite torsor under G^α. Let

$$X_L \longrightarrow X$$

be as in Definition 9.83. Note that this is a H-torsor.

Now consider the torsors (over X)

$$Y_L = Y \times_X X_L,$$

$$Y_L^\alpha = Y^\alpha \times_X X_L,$$

and the natural projections

$$Y_L \longrightarrow Y,$$

$$Y_L^\alpha \longrightarrow Y^\alpha.$$

Note that these maps induce $\overline{K}$-isomorphism from each $\overline{K}$-connected component in their domain to its image. Hence from $(*)$ above we see that they induce weak equivalences

$$\mathbf{N}_{\check{Y}_L} \longrightarrow \mathbf{N}_{\check{Y}},$$

$$\mathbf{N}_{\check{Y}_L^\alpha} \longrightarrow \mathbf{N}_{\check{Y}^\alpha}.$$

Lemma 9.111 *There exists an isomorphism of étale coverings over* X

$$T_\alpha : Y_L \longrightarrow Y_L^\alpha.$$

Proof After choosing an identification of $\pi_0(\overline{X_L})$ with H it is enough to construct an isomorphism

$$T_\alpha : H \times \overline{Y} \longrightarrow H \times \overline{Y} = H \times \overline{Y^\alpha}$$

which commutes with the action of Γ_K (note that the action is different on both sides). We shall define

$$T_\alpha(\tau, y) = \left(\tau, u_\tau^{-1} y\right).$$

Now note that indeed

$$\begin{aligned} \sigma(T_\alpha(\tau, y)) = \sigma\left(\tau, u_\tau^{-1} y\right) = \left(\sigma\tau, \sigma\left(u_\tau^{-1} y\right)\right) = \\ \left(\sigma\tau, u_\sigma^{-1} \sigma\left(u_\tau^{-1}\right) \sigma(y)\right) = \left(\sigma\tau, u_{\sigma\tau}^{-1} \sigma(y)\right) = \\ T_\alpha(\sigma\tau, \sigma(y)) = T_\alpha(\sigma(\tau, y)). \end{aligned}$$

This finishes the proof of the lemma. □

Now from Lemma 9.111 it follows that

$$X^{\check{Y},1}(\mathbb{A}) = X^{\check{Y}_L,1}(\mathbb{A}) = X^{\check{Y}_L^\alpha,1}(\mathbb{A}) = X^{\check{Y}^\alpha,1}(\mathbb{A})$$

and so $(x_\nu) \in X^{\check{Y},1}(\mathbb{A})$ as required. □

Let us now return to the proof of Theorem 9.103. Let $\mathcal{U} \longrightarrow X$ be a hypercovering. We need to show that $(x_\nu) \in X(\mathbb{A})^{\mathcal{U},1}$. We will require the following definition:

Definition 9.112 Let $\overline{X} = X \otimes_K \overline{K}$ and let $f : \Upsilon \longrightarrow \overline{X}$ be a torsor under a finite group G defined over $\overline{K}$. We say that (f, Υ, G) is **Galois invariant** if for each $\sigma \in \Gamma_K$ the G-torsor $f^\sigma : \Upsilon^\sigma \longrightarrow \overline{X}$ is isomorphic to the G-torsor $f : \Upsilon \longrightarrow \overline{X}$ over $\overline{K}$.

The following lemma will make the crucial step enabling us to apply information on torsors to the general hypercoverings:

Lemma 9.113 *Let K be a field and X a geometrically connected K-variety. Let $\mathcal{U}_\bullet \longrightarrow X$ be a hypercovering and $G = \pi_1(\mathbf{X}_{\mathcal{U}})$. We will denote by G^{ab} the* ***abelianization*** *of G. Then*

1. *There exists a geometrically connected G-torsor*
$$f : \Upsilon \longrightarrow \overline{X}$$
and a map of hypercoverings
$$\gamma : \mathcal{U}_\bullet \longrightarrow \check{Y}_\bullet$$
which induces the identity
$$G = \pi_1\left(\mathbf{X}_{\mathcal{U},1}\right) \xrightarrow{\simeq} \pi_1\left(\mathbf{X}_{\check{\Upsilon},1}\right) = G$$
where the last identification is by Corollary 9.106.
2. *There exist a exists a geometrically connected G^{ab}-torsor*
$$f : \Upsilon \longrightarrow \overline{X}$$
and a map of hypercoverings
$$\gamma : \mathcal{U}_\bullet \longrightarrow \check{Y}_\bullet$$
which induces the abelianization
$$G = \pi_1\left(\mathbf{X}_{\mathcal{U},1}\right) \longrightarrow \pi_1\left(\mathbf{X}_{\check{\Upsilon},1}\right) = G^{ab}.$$

Furthermore in both cases the torsor $f : \Upsilon \longrightarrow \overline{X}$ can be taken to be ***Galois invariant.***

Proof By the discussion in subsection §§9.2.4 we can naturally consider the pullback of $\mathcal{U}_n$ to the generic point of X as a finite $\Gamma_{K(X)}$-set which we denote by $\widetilde{\mathcal{U}}_n$ (see Definition 9.20). By restriction we will also consider $\widetilde{\mathcal{U}}_n$ as a $\Gamma_{\overline{K}(\overline{X})}$-set.

We denote by
$$\mathrm{Stab}_{\Gamma_{\overline{K}(\overline{X})}}\left(\widetilde{\mathcal{U}}_0\right) = \{H_1, ..., H_t\}$$
the set of stabilizers in $\Gamma_{\overline{K}(\overline{X})}$ of the points of $\widetilde{\mathcal{U}}_0$.

In order to prove part 1, consider the group
$$E = \langle H_1, ..., H_t \rangle,$$
i.e E is the group generated inside $\Gamma_{\overline{K}(\overline{X})}$ by all the groups in

$\mathrm{Stab}_{\Gamma_{\overline{K}(\overline{X})}}\left(\widetilde{\mathcal{U}}_0\right)$. Since the set $\mathrm{Stab}_{\Gamma_{\overline{K}(\overline{X})}}\left(\widetilde{\mathcal{U}}_0\right)$ is invariant under conjugation by elements from $\Gamma_{K(X)}$ we see that E is actually normal in $\Gamma_{K(X)}$ (and so in particular normal in $\Gamma_{\overline{K}(\overline{X})}$).

Since $\widetilde{\mathcal{U}}_0$ is a finite set each H_i has finite index in $\Gamma_{\overline{K}(\overline{X})}$ which means that E has finite index in $\Gamma_{\overline{K}(\overline{X})}$ as well. Let $F/\overline{K}(\overline{X})$ be the finite field extension that corresponds to E.

Now for each i let $\overline{K}(\overline{X}) \subseteq F_i$ be the field extension corresponding to H_i. Then F_i is unramified over the image of the component of $\widetilde{\mathcal{U}}_0$ corresponding to the $\Gamma_{\overline{K}(\overline{X})}$-orbit containing i. Since $\mathcal{U}_0 \longrightarrow \overline{X}$ is surjective and since $F \subseteq F_i$ for every i we see that F is unramified on $\overline{X}$. Thus E contains the kernel of the surjective map

$$\rho : \Gamma_{\overline{K}(\overline{X})} \longrightarrow \pi_1(\overline{X})$$

and the normal subgroup $\rho(E) \triangleleft \pi_1(\overline{X})$ corresponds to a geometrically connected torsor

$$\phi : \Upsilon \longrightarrow \overline{X}$$

under the group $\pi_1(\overline{X})/\rho(E)$. Now from Corollary 9.21 and Lemma 9.23 we see that $\pi_1(\overline{X})/\rho(E) \cong G$. By construction, we have a map $\Gamma_{\overline{K}(\overline{X})}$ sets

$$\gamma : \widetilde{\mathcal{U}}_0 \longrightarrow \widetilde{\Upsilon},$$

which induces a map

$$\mathcal{U}_0 \longrightarrow \Upsilon,$$

and so a map

$$\mathcal{U}_{\bullet} \longrightarrow \check{\Upsilon}$$

induces the identification

$$G = \pi_1\left(\mathbf{X}_{\mathcal{U}}\right) \longrightarrow \pi_1\left(\mathbf{X}_{\check{Y}}\right) = G.$$

In order to prove part 2 consider the group $E \subseteq E' \subseteq \mathrm{Stab}_{\Gamma_{\overline{K}(\overline{X})}}\left(\widetilde{\mathcal{U}}_0\right)$ generated by E and the commutator subgroup of $\mathrm{Stab}_{\Gamma_{\overline{K}(\overline{X})}}\left(\widetilde{\mathcal{U}}_0\right)$. Then E' is again normal in $\Gamma_{K(X)}$. The image $\rho(E') \subseteq \pi_1(\overline{X})$ corresponds to a finite Galois cover

$$\phi' : \Upsilon' \longrightarrow \overline{X}$$

under the group $\pi_1(\overline{X})/\rho(E') \cong G^{ab}$. As above we get a map of hypercoverings

$$\mathcal{U}_\bullet \longrightarrow \check{\Upsilon}$$

which this time induces the abelianization

$$G = \pi_1(\mathbf{N}_{\mathcal{U}}) \longrightarrow \pi_1(\mathbf{N}_{\check{Y}}) = G^{ab}.$$

Note that since both $\rho(E), \rho(E') \subseteq \pi_1(\overline{X})$ are invariant to conjugation by elements of $\pi_1(X)$ we get that in both cases Υ is Galois invariant. $\square$

Now let $\mathcal{U} \longrightarrow X$ be a hypercovering with $\pi_1(\mathcal{U}) = G$ and let $f : \Upsilon \longrightarrow \overline{X}$ be a geometrically connected G-torsor which admits a map

$$\mathcal{U} \longrightarrow \check{\Upsilon}$$

inducing the identity on fundamental groups.

Now there exists a large enough finite Galois extension L/K such that we can put an L-group structure on G and an L-variety structure on Υ making $f : \Upsilon \longrightarrow X$ into a G-torsor defined over L. Since Υ was Galois invariant we may assume that L is big enough so that Υ is isomorphic over L to all its Galois conjugates.

By "averaging" over $\mathrm{Gal}(L/K) = \{\sigma_1, ..., \sigma_n\}$ (where $n = |\mathrm{Gal}(L/K)|$) we obtain a map of hypercoverings over K given by

$$\mathcal{U} \longrightarrow \Upsilon^{\sigma_1} \times_X ... \times_X \Upsilon^{\sigma_n} = R^{L/K}_{/X}(\Upsilon)$$

where $R^{L/K}_{/X}$ is the Weil restriction of scalars over X. Let us denote $Y = R^{L/K}_{/X}(\Upsilon)$. Now Y is a torsor under the finite K-group

$$R^{L/K}(G) = G^{\sigma_1} \times ... \times G^{\sigma_n}.$$

Furthermore since Υ was Galois invariant we get that

$$Y \otimes_K L \cong \overbrace{\Upsilon \times_X ... \times_X \Upsilon}^{n} \cong \Upsilon \times G^{n-1}$$

where all the isomorphisms are over L. In particular we have a diagonal embedding of Υ in $Y \otimes_K L$ whose image is a geometric connected component (defined over L). The stabilzer of this connected component is the diagonal subgroup of G^n, which is isomorphic to G (as L-groups). By our analysis above we get that the map

$$\pi_{0/L}(\check{\Upsilon}) \longrightarrow \pi_{0/L}(\check{Y})$$

is a weak equivalence. This means that the map

$$\mathbf{X}_{\mathcal{U},1} \longrightarrow \mathbf{X}_{\check{Y},1}$$

is a weak equivalence, and so

$$X(\mathbb{A})^{\mathcal{U},1} = X(\mathbb{A})^{\check{Y},1}.$$

Since $(x_\nu) \in X(\mathbb{A})^{\check{Y},1}$ by Proposition 9.109 we get that $(x_\nu) \in X(\mathbb{A})^{\mathcal{U},1}$ and we are done.

This finishes part 1 of 9.103. For part 2 we need to show that if $(x_\nu) \in X(\mathbb{A})^{fin-ab}$ then $(x_\nu) \in X(\mathbb{A})^{\mathbb{Z}\mathcal{U},1}$. Using the second part of Lemma 9.113 there exists a normal field extension $K \subseteq L$ and a G^{ab}-torsor $f : \Upsilon \longrightarrow \overline{X}$ which admits a map

$$\mathcal{U} \longrightarrow \check{\Upsilon}$$

inducing the abelianization on fundamental groups. Again we can choose a large enough finite Galois extension L/K such that G, Υ and the map $\mathcal{U} \longrightarrow \check{\Upsilon}$ are all defined over L . Let $Y = R^{L/K}_{/X}(\Upsilon)$. Then as above we get a map defined over K

$$\mathcal{U} \longrightarrow \check{Y}$$

and we know that the diagonal map induces a natural isomorphism

$$\pi_1(\mathbf{N}_{\check{\Upsilon}}) \xrightarrow{\simeq} \pi_1(\mathbf{N}_{\check{Y}}).$$

Hence the map

$$\mathbf{N}_{\mathcal{U}} \longrightarrow \mathbf{N}_{\check{Y}}$$

induces the abelianization on the level of fundamental groups and hence an isomorphism on first homology groups. This means that the map

$$P_1(\mathbb{Z}\mathbf{N}_{\mathcal{U}}) \longrightarrow P_1(\mathbb{Z}\mathbf{N}_{\check{Y}})$$

is a weak equivalence and so

$$X(\mathbb{A})^{\mathbb{Z}\mathcal{U},1} = X(\mathbb{A})^{\mathbb{Z}\check{Y},1}.$$

By $(***)$ above we see that

$$X(\mathbb{A})^{\mathbb{Z}\check{Y},1} = X(\mathbb{A})^{\check{Y},1}$$

and since $(x_\nu) \in X(\mathbb{A})^{\check{Y},1}$ by 9.109 we get that

$$(x_\nu) \in X(\mathbb{A})^{\mathbb{Z}\mathcal{U},1}$$

as needed. This finishes the proof of Theorem 9.103. □

Remark 9.114 One might wonder what the role of connectivity is in the proof of Theorem 9.103. To see this one should consider the case

$$X = \operatorname{Spec}(K) \coprod \operatorname{Spec}(K).$$

By Proposition 9.98 we know that

$$X(\mathbb{A})^{\mathbb{Z}h,1} = X(\mathbb{A})^{h,1} = X(K).$$

On the other hand, when K has a complex place then

$$X(\mathbb{A})^{fin-ab} = X(\mathbb{A})^{fin} \neq X(K).$$

Indeed if $(x_\nu), (x'_\nu) \in X(\mathbb{A})$ are two adelic points that differ only in a complex coordinate then $(x_\nu) \in X(\mathbb{A})^{fin}$ if and only if $(x'_\nu) \in X(\mathbb{A})^{fin}$, and similarly for $X(\mathbb{A})^{fin-ab}$.

This is, however, the only fault in the desired equality of $X(\mathbb{A})^{fin}$ and $X(K)$. In fact, in [Sto07] Stoll gives a proof of this equality (Lemma 5.10 in [Sto07]). However this proof is erroneous and the error becomes a problem exactly in the complex places.

To conclude, when one ignores the complex places (or if K is totally real) one can write

$$X(\mathbb{A})^{h,1} = X(\mathbb{A})^{fin}$$

and

$$X(\mathbb{A})^{\mathbb{Z}h,1} = X(\mathbb{A})^{fin-ab}$$

even when X is not connected.

The following corollary is originally due to Harari and Stix (unpublished work) and is closely related to Lemma 5.7 in [Sto07]:

Corollary 9.115 *Let K and X a smooth geometrically connected variety over K. If*

$$X(\mathbb{A})^{fin} \neq \emptyset$$

then the short exact sequence

$$1 \longrightarrow \pi_1(\overline{X}) \longrightarrow \pi_1(X) \longrightarrow \Gamma_K \longrightarrow 1$$

splits.

Proof Since

$$X(\mathbb{A})^{h,1} = X(\mathbb{A})^{fin} \neq \emptyset$$

we know that $\acute{E}t_{/K}^{\,1}(X)(hK)$ is non-empty thus we get the claim from Lemma 9.86. □

9.10 The equivalence of the homology obstruction and the Brauer-Manin obstruction

In this section we shall prove to following theorem:

Theorem 9.116 *Let $2 \leq n \leq \infty$ and let X be a smooth algebraic variety over a number field K. Then*

$$X(\mathbb{A})^{\mathbb{Z}h,n} \subseteq X(\mathbb{A})^{\mathrm{Br}}.$$

If X is also connected then

$$X(\mathbb{A})^{\mathbb{Z}h,n} = X(\mathbb{A})^{\mathrm{Br}}.$$

In light of Corollary 9.68 it is enough to prove Theorem 9.116 for any specific $2 \leq n \leq \infty$. In particular we can assume that n is some fixed number $3 < n < \infty$. The reader can replace n with 4 everywhere if he or she so pleases.

In light of Corollary 9.67 we see that in order to understand

$$X(\mathbb{A})^{\mathbb{Z}h,n}$$

we first need to understand the sets $P_n(\mathbb{Z}\mathbf{X}_{\mathcal{U}})(h\mathbb{A})$ and $P_n(\mathbb{Z}\mathbf{X}_{\mathcal{U}})(hK)$ for each hypercovering $\mathcal{U}$. Note that $P_n(\mathbb{Z}\mathbf{X}_{\mathcal{U}})$ is not only a simplicial Γ_K-set but also a simplicial Γ_K-module i.e. a simplicial object in the category Mod_{Γ_K} of Γ_K modules.

The category Mod_{Γ_K} is abelian. An important tool of homotopy theory is the Dold-Kan correspondence which allows one to reduce the study of the homotopy theory of simplicial objects in an abelian category $\mathcal{A}$ to homological algebra of complexes over $\mathcal{A}$.

9.10.1 The Dold-Kan correspondence

Definition 9.117 Let $\mathcal{A}$ be an abelian category. We denote by $\mathcal{C}\mathcal{A}$ the category of complexes over A with a differential of degree -1. We denote by $\mathcal{C}^{\geq 0}\mathcal{A}$ the category of complexes which are bounded below by dimension 0 (i.e. complexes $C_\bullet$ such that $C_i = 0$ for $i < 0$).

Definition 9.118 Let $\mathcal{A}$ be an abelian category. If $\mathbf{X}$ is a simplicial object in $\mathcal{A}$ we denote by $\underline{\mathbf{X}} \in \mathcal{C}^{\geq 0}\mathcal{A}$ the complex over $\mathcal{A}$ given by

$$\underline{\mathbf{X}}_n = \mathbf{X}_n$$

with the differential $d = \sum_i (-1)^i d_i$.

Proposition 9.119 *(The Dold-Kan Correspondence) Let $\mathcal{A}$ be an abelian category. The category $\mathcal{A}^{\Delta^{op}}$ of simplicial objects in $\mathcal{A}$ is equivalent to the category $\mathcal{C}^{\geq 0}\mathcal{A}$. Moreover for every two simplicial $\mathcal{A}$ objects one has*

$$[\mathbf{X}, \mathbf{Y}] \cong [\underline{\mathbf{X}}, \underline{\mathbf{Y}}]$$

where the first denotes simplicial homotopy and the second chain homotopy.

Proof See [Dol58]. The proof there uses a slightly different functor called the Moore complex, but this functor is naturally chain equivalent to the functor $\underline{\bullet}$ (called the unnormalized complex in [Dol58]). □

Remark 9.120 The functor $\underline{\bullet}$ admits a right adjoint

$$\overline{\bullet} : \mathcal{C}^{\geq 0}\mathcal{A} \longrightarrow \mathcal{A}^{\Delta^{op}}$$

which is its "homotopy inverse" i.e.

$$\mathbf{X} \xrightarrow{\simeq} \overline{(\underline{\mathbf{X}})}, \quad \forall\, \mathbf{X} \in \mathcal{A}^{\Delta^{op}},$$

$$\underline{(\overline{\mathbf{Y}})} \xrightarrow{\simeq} \mathbf{Y}, \quad \forall\, \mathbf{Y} \in \mathcal{C}^{\geq 0}\mathcal{A}.$$

In particular for every $\mathbf{X} \in \mathcal{C}\mathcal{A}, \mathbf{Y} \in \mathcal{A}^{\Delta^{op}}$ one has an isomorphism of sets

$$[\mathbf{X}, \overline{\mathbf{Y}}] \cong [\underline{\mathbf{X}}, \mathbf{Y}]$$

where the first denotes simplicial homotopy and the second chain homotopy.

Remark 9.121 In the case where $\mathcal{A}$ is the category of abelian groups then the Dold-Kan correspondence replaces homotopy groups with homology groups, i.e

$$H_n(\underline{\mathbf{X}}) \cong \pi_n(\mathbf{X}), \quad \forall\, \mathbf{X} \in \mathcal{A}^{\Delta^{op}}.$$

9.10.2 Postnikov towers For complexes

In order to continue along this line we would like to use a Postnikov-like construction for complexes which allows us to truncate homology groups. This construction is actually quite simpler than the Postinikov piece functor for spaces.

Let $C \in \mathcal{C}\mathcal{A}$ be a complex. We denote by $P_n^+(C)$ the complex

$$\ldots \longrightarrow 0 \longrightarrow C_n/d(C_{n+1}) \longrightarrow C_{n-1} \longrightarrow C_{n-2} \longrightarrow \ldots$$

Note that $H_i(P_i^+(C)) = 0$ for $i > n$ and there is a natural map $C \longrightarrow P_n^+(C)$ which induces isomorphisms

$$H_i(C) \xrightarrow{\simeq} H_i(P_n^+(C))$$

for $i \leq n$. The functor $C \mapsto P_n^+(C)$ from general complexes to complexes bounded from above by dimension n is left adjoint to the inclusion functor.

Definition 9.122 Analogously to Definition 9.17 we shall define

1. A complex $C \in \mathcal{C}\mathrm{Mod}_\Gamma$ will be called **finite** if $H_i(C)$ is finite for all $i \in \mathbb{Z}$.
2. A complex $C \in \mathcal{C}\mathrm{Mod}_\Gamma$ will be called **bounded** if there exists an $N > 0$ such that
$$C_i = 0$$
for $|i| > N$.
3. A complex $C \in \mathcal{C}\mathrm{Mod}_\Gamma$ will be called **excellent** if the action of Γ on C factors through a finite quotient.

The Kan-Dold correspondence allows us to study the complex $\underline{P_n(\mathbb{Z}\mathbf{X}_{\mathcal{U}})}$ instead of the space $P_n(\mathbb{Z}\mathbf{X}_{\mathcal{U}})$. Note that since $P_n(\mathbb{Z}\mathbf{X}_{\mathcal{U}})$ is bounded the homology groups of $\underline{P_n(\mathbb{Z}\mathbf{X}_{\mathcal{U}})}$ are almost always zero. However the complex $\underline{P_n(\mathbb{Z}\mathbf{X}_{\mathcal{U}})}$ itself will not necessarily be bounded.

Since it will be convenient for us to work only with bounded complexes, we would like to find a bounded complex C such that $P_n(\mathbb{Z}\mathbf{X}_{\mathcal{U}})$ is naturally weakly equivalent to $\overline{C}$ in the local model structure. The complex C can be taken to be $P_n^+(\underline{\mathbb{Z}\mathbf{N}_{\mathcal{U}}})$ (see Definition 9.16).

Lemma 9.123 *let $\mathcal{U} \longrightarrow X$ be a hypercover and $n \in \mathbb{N}$, then $\overline{P_n^+(\underline{\mathbb{Z}\mathbf{N}_{\mathcal{U}}})}$ is naturally weakly equivalent to $P_n(\mathbb{Z}\mathbf{X}_{\mathcal{U}})$ in the local model structure on $\mathrm{Set}_{\Gamma_K}^{\Delta^{op}}$.*

Proof We prove this theorem by writing a natural zig-zag of weak-equivalence

$$P_n(\mathbb{Z}\mathbf{X}_{\mathcal{U}}) \longrightarrow \overline{(\underline{P_n(\mathbb{Z}\mathbf{X}_{\mathcal{U}})})} \longrightarrow \overline{P_n^+(\underline{P_n(\mathbb{Z}\mathbf{X}_{\mathcal{U}})})} \longleftarrow \overline{P_n^+(\underline{\mathbb{Z}\mathbf{X}_{\mathcal{U}}})} \longleftarrow \overline{P_n^+(\underline{\mathbb{Z}\mathbf{N}_{\mathcal{U}}})}.$$

□

Note that in view of Lemma 9.123 we get natural bijections of sets.

$$P_n(\mathbb{Z}\mathbf{X}_{\mathcal{U}})(hK) = \overline{P_n^+(\underline{\mathbb{Z}\mathbf{N}_{\mathcal{U}}})}(hK),$$

$$P_n(\mathbb{Z}\mathbf{X}_{\mathcal{U}})(h\mathbb{A}) = \overline{P_n^+(\underline{\mathbb{Z}\mathbf{N}_{\mathcal{U}}})}(h\mathbb{A}).$$

9.10.3 Hypercohomology and homotopy fixed points

In view of our discussion so far we see that in order to understand the homological obstruction we should understand the relation between the set $\overline{P_n^+(\mathbb{Z}\mathbf{N}_{\mathcal{U}})}(hK)$ and the set $\overline{P_n^+(\mathbb{Z}\mathbf{N}_{\mathcal{U}})}(h\mathbb{A})$.

Let $C \in \mathcal{C}^b \operatorname{Mod}_{\Gamma_K}$ be a bounded complex. Then $\overline{C}$ is a strictly bounded simplicial Γ_K-set. By Theorem 9.44 we then get

$$\begin{aligned}\overline{C}(hK) &= \pi_0\left(\overline{C}^{h\Gamma_K}\right) = \operatorname*{colim}_{\Lambda \tilde{\lhd} \Gamma_K} \left[\mathbf{E}(\Gamma_K/\Lambda), \overline{C}\right]_{\mathrm{Set}_{\Gamma_K}^{\Delta^{op}}} \\ &= \operatorname*{colim}_{\Lambda \tilde{\lhd} \Gamma_K} \left[\mathbb{Z}\mathbf{E}(\Gamma_K/\Lambda), \overline{C}\right]_{\mathrm{Mod}_{\Gamma_K}^{\Delta^{op}}} \\ &= \operatorname*{colim}_{\Lambda \tilde{\lhd} \Gamma_K} \left[\underline{\mathbb{Z}\mathbf{E}(\Gamma_K/\Lambda)}, C\right]_{\mathcal{C}^{\geq 0}\operatorname{Mod}_{\Gamma_K}} \\ &= \mathbb{H}^0(\Gamma_K, C) = \mathbb{H}^0(K, C),\end{aligned} \qquad (\star)$$

where

$$\mathbb{H}^i(K, C) = \operatorname*{colim}_{\Lambda \tilde{\lhd} \Gamma_K} \left[\underline{\mathbb{Z}\mathbf{E}(\Gamma_K/\Lambda)}, C\right]^i_{\mathcal{C}^{\geq 0}\operatorname{Mod}_{\Gamma_K}}$$

is the Galois hypercohomology of the C.

We would like to have an analogous hypercohomology counterpart for the adelic homolotopy fixed points. This can be obtained via the following natural definition:

Definition 9.124 Let C be a bounded complex of Γ_K-modules. We define the **adelic hypercohomology** of C to be

$$\mathbb{H}^i(\mathbb{A}, C) = \lim_{\overrightarrow{T}} \prod_{\nu \in T} \mathbb{H}^i(K_\nu, C) \times \prod_{\nu \notin T} \mathbb{H}^i\left(\Gamma_\nu^{ur}, C^{I_\nu}\right)$$

where T runs over all finite subsets of S.

Using the notion of adelic hypercohomology we get the following analogue to equality $(\star)$:

$$\overline{C}(h\mathbb{A}) = H^0(\mathbb{A}, C).$$

Note that adelic hypercohomology, like Galois hypercohomology, transforms short exact sequences to long exact sequences:

Lemma 9.125 *Let*

$$0 \longrightarrow A \longrightarrow B \longrightarrow C \longrightarrow 0$$

be a short exact sequence of excellent bounded complexes in $\mathcal{C}^b \operatorname{Mod}_{\Gamma_K}$.

Then there is a natural long exact sequence

$$\ldots \longrightarrow \mathbb{H}^{i-1}(\mathbb{A},C) \longrightarrow \mathbb{H}^{i}(\mathbb{A},A) \longrightarrow \mathbb{H}^{i}(\mathbb{A},B) \longrightarrow \mathbb{H}^{i}(\mathbb{A},C) \longrightarrow \mathbb{H}^{i+1}(\mathbb{A},A) \longrightarrow \ldots$$

Proof Products and filtered direct limits are exact. □

9.10.4 Proof of the main theorem

This section will be devoted to the proof of Theorem 9.116. As mentioned above we fix an $3 < n < \infty$ and prove the theorem for n. Let $(x_\nu) \in X(\mathbb{A})$ be an adelic point. By the discussion above we see that $(x_\nu) \in X(\mathbb{A})^{\mathbb{Z}h,n}$ if and only if for every hypercovering $\mathcal{U} \longrightarrow X$ the corresponding adelic hypercohomology class

$$h((x_\nu)) \in P_n(\underline{\mathbb{Z}\mathbf{X}_{\mathcal{U}}})(h\mathbb{A}) = \overline{P_n^+(\underline{\mathbb{Z}\mathbf{N}_{\mathcal{U}}})(h\mathbb{A})} = \mathbb{H}^0(\mathbb{A}, P_n^+(\underline{\mathbb{Z}\mathbf{N}_{\mathcal{U}}}))$$

is rational. i.e. lies in the image of the map

$$\mathrm{loc} : \mathbb{H}^0(K, P_n^+(\underline{\mathbb{Z}\mathbf{N}_{\mathcal{U}}})) \longrightarrow \mathbb{H}^0(\mathbb{A}, P_n^+(\underline{\mathbb{Z}\mathbf{N}_{\mathcal{U}}})).$$

On the other hand, the Brauer set is defined via pairings with elements in $H^2_{\acute{e}t}(X, \mathbb{G}_m)$. The main ingredient of the proof of 9.116 will be to show that the Brauer pairing actually factors through the map

$$X(\mathbb{A}) \longrightarrow \varinjlim_{\mathcal{U}} \mathbb{H}^0(\mathbb{A}, P_n^+(\underline{\mathbb{Z}\mathbf{N}_{\mathcal{U}}}))$$

i.e, depends only on the colimit of adelic cohomology classes of (x_ν). First we will need some terminology.

Let X/K be an algebraic variety over K with $t : X \longrightarrow \mathrm{Spec}\, K$ the structure map. Given a Galois module A we can consider it as an étale sheaf over Spec (K). We denote by t^*A the pullback of this sheaf from Spec (K) to X.

The sheaf t^*A can be described more concretely as follows: it associates to an étale map $V \longrightarrow X$ the group of Γ_K equivariant maps from the Γ_K-set $\pi_{0/K}(V)$ to A. We refer to t^*A as the sheaf of locally constant maps to A.

Our first step towards proving Theorem 9.116 will be to describe the Brauer set via pairing with elements in $H^2(X, t^*\mathbb{G}_m)$ rather than

$H^2_{ét}(X, \mathbb{G}_m)$. In order to do this we consider the commutative diagram

$$\begin{array}{ccccc} X(\mathbb{A}) & \times & H^2_{ét}(X, \mathbb{G}_m) & \longrightarrow & \mathbb{Q}/\mathbb{Z} \\ \| & & \uparrow{\scriptstyle i_*} & & \| \\ X(\mathbb{A}) & \times & H^2_{ét}(X, t^*\mathbb{G}_m) & \longrightarrow & \mathbb{Q}/\mathbb{Z} \end{array}$$

where in both cases the pairing is defined by

$$((x_\nu), u) \mapsto \sum_\nu \mathrm{inv}(x_\nu^* u)$$

with

$$\mathrm{inv} : H^2_{ét}(\mathrm{Spec}\,(K_\nu), x_\nu^* t^*\mathbb{G}_m) = H^2_{ét}(\mathrm{Spec}\,(K_\nu), \mathbb{G}_m) \xrightarrow{\simeq} \mathbb{Q}/\mathbb{Z}$$

being the natural isomorphism of class field theory. We refer to the pairing of the second row as the locally-constant Brauer pairing. We claim that the left kernel of the Brauer pairing is equal to the left kernel of the locally-constant Brauer pairing. This will follow immediately from the following lemma:

Lemma 9.126 *Let X be a smooth variety. Then the map*

$$i_* : H^2_{ét}(X, t^*\mathbb{G}_m) \longrightarrow H^2_{ét}(X, \mathbb{G}_m),$$

is surjective.

Proof Since X is a smooth variety $H^2_{ét}(X, \mathbb{G}_m)$ is torsion group (e.g. by [Lie08] Corollary 3.1.3.4). Hence it will be enough to prove surjectivity on torsion elements. Let $1 \leq k \in \mathbb{N}$ be a natural number and consider the map of short Kummer sequences

$$\begin{array}{ccccccccc} 0 & \longrightarrow & t^*\mu_k & \longrightarrow & t^*\mathbb{G}_m & \xrightarrow{\times k} & t^*\mathbb{G}_m & \longrightarrow & 0 \\ & & \| & & \downarrow & & \downarrow & & \\ 0 & \longrightarrow & t^*\mu_k & \longrightarrow & \mathbb{G}_m & \xrightarrow{\times k} & \mathbb{G}_m & \longrightarrow & 0. \end{array}$$

We get an induced map of long exact sequences

$$\begin{array}{ccccccccc} \ldots & \longrightarrow & H^2_{ét}(X, t^*\mu_k) & \longrightarrow & H^2_{ét}(X, t^*\mathbb{G}_m) & \xrightarrow{\times k} & H^2_{ét}(X, t^*\mathbb{G}_m) & \longrightarrow & \ldots \\ & & \| & & \downarrow & & \downarrow & & \\ \ldots & \longrightarrow & H^2_{ét}(X, t^*\mu_k) & \longrightarrow & H^2_{ét}(X, \mathbb{G}_m) & \xrightarrow{\times k} & H^2_{ét}(X, \mathbb{G}_m) & \longrightarrow & \ldots \end{array}$$

inducing a commutative square

$$\begin{array}{ccc} H^2_{\acute{e}t}(X, t^*\mu_k) & \twoheadrightarrow & H^2_{\acute{e}t}(X, t^*\mathbb{G}_m)[k] \\ \| & & \downarrow i_* \\ H^2_{\acute{e}t}(X, t^*\mu_k) & \twoheadrightarrow & H^2_{\acute{e}t}(X, \mathbb{G}_m)[k] \end{array}$$

with the two horizontal maps being surjective. This means that the map

$$i_* : H^2_{\acute{e}t}(X, t^*\mathbb{G}_m)[k] \longrightarrow H^2_{\acute{e}t}(X, \mathbb{G}_m)[k]$$

is surjective as well. This finishes the proof of the lemma. □

From Lemma 9.126 we see that we can describe the Brauer set equivalently as the left kernel of the locally-constant Brauer pairing, i.e. as the set of all adelic points which are orthogonal to all the elements in $H^2_{\acute{e}t}(X, t^*\mathbb{G}_m)$.

We now wish to show that the locally-constant Brauer pairing depends only on (the colimit of) the adelic cohomology classes of (x_ν). For this we will need the following two standard definitions:

Definition 9.127 Let $C_\bullet$ and $D_\bullet$ be two complexes of Γ_K-modules. We define the complex of Γ_K-modules $\mathrm{Hom}(C_\bullet, D_\bullet)$ as follows:

$$\mathrm{Hom}(C_\bullet, D_\bullet)_n = \prod_{j-i=n} \mathrm{Hom}_{\mathrm{Ab}}(C_i, D_j)$$

with the differential given by

$$\partial f = \partial_D \circ f + (-1)^n f \circ \partial_C.$$

The Γ_K action on $\mathrm{Hom}_{\mathrm{Ab}}(C_i, D_j)$ is given by

$$\sigma(f)(c) = \sigma(f(\sigma^{-1}(c))).$$

Definition 9.128 Let K be a field and $\Gamma = \Gamma_K$. Consider $\mathbb{G}_m$ as a complex concentrated in degree zero such that $\mathbb{G}_{m\,0} = \mathbb{G}_m(\overline{K}) = \overline{K}^*$. We denote by $\widehat{C}$ the mapping complex $\underline{\mathrm{Hom}}(C, \mathbb{G}_m)$ and refer to it as the **dual complex**. In particular one has

$$\widehat{C}_n = \mathrm{Hom}_{\mathrm{Ab}}\left(C_{-n}, \overline{K}^*\right).$$

Now the key element in the proof of Theorem 9.116 will be to describe the locally-constant Brauer pairing via the adelic cohomology classes of the adelic point. Note that we have a natural pairing

$$\mathbb{H}^0(\mathbb{A}, P_n^+(\underline{\mathbb{Z}\mathbf{N}_{\mathcal{U}}})) \quad \times \quad \mathbb{H}^2\left(K, \widehat{P_n^+(\underline{\mathbb{Z}\mathbf{N}_{\mathcal{U}}})}\right) \longrightarrow \mathbb{Q}/\mathbb{Z}$$

which is defined as follows. Given

$$x \in \mathbb{H}^0(\mathbb{A}, P_n^+(\underline{\mathbb{Z}\mathbf{N}_{\mathcal{U}}})), \quad y \in \mathbb{H}^2\left(K, \widehat{P_n^+(\underline{\mathbb{Z}\mathbf{N}_{\mathcal{U}}})}\right),$$

we go over all places ν of K, project x to $\mathbb{H}^0(K_\nu, P_n^+(\underline{\mathbb{Z}\mathbf{N}_{\mathcal{U}}}))$ and restrict y to $\mathbb{H}^2\left(K_\nu, \widehat{P_n^+(\underline{\mathbb{Z}\mathbf{N}_{\mathcal{U}}})}\right)$. We then pair using the cup product (note that all complexes here are bounded) to get an element in

$$\mathbb{H}^2(K_\nu, \mathbb{G}_m) = H^2(K_\nu, \mathbb{G}_m) \stackrel{\text{inv}}{\cong} \mathbb{Q}/\mathbb{Z}$$

then as in the Brauer-Manin pairing we sum up the invariants in all the places to get the pairing $(x, y) \in \mathbb{Q}/\mathbb{Z}$. Note that since the cohomology class y is rational its restrictions to $\mathbb{H}^2\left(K_\nu, \widehat{P_n^+(\underline{\mathbb{Z}\mathbf{N}_{\mathcal{U}}})}\right)$ are almost always unramified, hence almost always zero, so the summation is well-defined. We will refer to this pairing as the cup-product pairing. The following theorem connects this cup product pairing with the locally constant Brauer-Manin pairing:

Proposition 9.129 *Let $\mathcal{U} \longrightarrow X$ be a hypercovering. Then there exists a commutative diagram*

$$\begin{array}{ccccc} X(\mathbb{A}) & \times & H^2_{\acute{e}t}(X, t^*\mathbb{G}_m) & \longrightarrow & \mathbb{Q}/\mathbb{Z} \\ \downarrow h & & \uparrow \psi_{\mathcal{U}} & & \| \\ \mathbb{H}^0(\mathbb{A}, P_n^+(\underline{\mathbb{Z}\mathbf{N}_{\mathcal{U}}})) & \times & \mathbb{H}^2\left(K, \widehat{P_n^+(\underline{\mathbb{Z}\mathbf{N}_{\mathcal{U}}})}\right) & \longrightarrow & \mathbb{Q}/\mathbb{Z} \end{array}$$

natural in $\mathcal{U}$, where the upper row is the locally constant Brauer-Manin pairing and the lower row is the cup-product pairing. Further more the maps $\Psi_{\mathcal{U}}$ induce an ***isomorphism:***

$$\varinjlim_{\mathcal{U}} \mathbb{H}^2\left(\Gamma_K, \widehat{P_n^+(\underline{\mathbb{Z}\mathbf{N}_{\mathcal{U}}})}\right) \longrightarrow H^2_{\acute{e}t}(X, t^*\mathbb{G}_m).$$

Before we come to the proof of Proposition 9.129 let us explain how it implies Theorem 9.116. First of all we see that Proposition 9.129 implies that an adelic point is in the Brauer set if and only if its image in $\mathbb{H}^0(\mathbb{A}, P_n^+(\underline{\mathbb{Z}\mathbf{N}_{\mathcal{U}}}))$ is orthogonal to all the elements in $\mathbb{H}^2\left(\Gamma_K, \widehat{P_n^+(\underline{\mathbb{Z}\mathbf{N}_{\mathcal{U}}})}\right)$.

From the Hasse-Brauer-Noether Theorem in class field theory it follows that if an element in $\mathbb{H}^0(\mathbb{A}, P_n^+(\underline{\mathbb{Z}\mathbf{N}_{\mathcal{U}}}))$ is rational then it is orthogonal to every element in $\mathbb{H}^2(K, \widehat{P_n^+(\underline{\mathbb{Z}\mathbf{N}_{\mathcal{U}}})})$. This gives us the first part of the Theorem 9.116, i.e. that

$$X(\mathbb{A})^{\mathbb{Z}h} \subseteq X(\mathbb{A})^{\mathrm{Br}}.$$

Now assume that our variety X is connected. Then $C = P_n^+(\underline{\mathbb{Z}\mathbf{N}_{\mathcal{U}}})$ satisfies the following properties:

1. It is excellent.
2. It is bounded.
3. It is bounded below by dimension 0.
4. $H_0(C) = H_0(\mathbf{N}_{\mathcal{U}}) = \mathbb{Z}$.
5. $H_i(C)$ is finite for $i > 0$.

From property 3 we have a natural map $C \longrightarrow \mathbb{Z}$ which induces a map

$$\pi : \mathbb{H}^0(\mathbb{A}, C) \longrightarrow \mathbb{H}^0(\mathbb{A}, \mathbb{Z}).$$

Then for every $(x_\nu) \in X(\mathbb{A})$ we have that

$$h_{\mathbb{Z}\mathcal{U},n}((x_\nu)) \in \mathbb{H}^0(\mathbb{A}, C)$$

is an element which is mapped by π to the element

$$(1, 1, ..., 1) \in \prod_\nu \mathbb{Z} \cong \mathbb{H}^0(\mathbb{A}, \mathbb{Z})$$

which is clearly rational. Since $H^3\left(K, \widehat{Z}\right) = H^3\left(K, \overline{K}^*\right) = 0$ we get from Theorem 9.135 that $h_{\mathbb{Z}\mathcal{U},n}((x_\nu))$ is rational if and only if it is orthogonal to all the elements in $\mathbb{H}^2\left(\Gamma_K, \widehat{P_n^+(\underline{\mathbb{Z}\mathbf{N}_{\mathcal{U}}})}\right)$.

Remark 9.130 One might wonder what is the role of the connectivity of X in the proof of Theorem 9.116. The situation here is very similar to that of Remark 9.114. Again by Proposition 9.98 we recall that if $X = \mathrm{Spec}(K) \coprod \mathrm{Spec}(K)$ then

$$X(\mathbb{A})^{\mathbb{Z}h} = X(K).$$

On the other hand if K is not totally real then

$$X(\mathbb{A})^{\mathrm{Br}} \neq X(K).$$

As in the case of Remark 9.114, the fault lies in the behavior of the complex places. Indeed since $\mathrm{Br}\,\mathbb{C} = 0$ the pairing

$$X(\mathbb{A}) \times \mathrm{Br}\, X \longrightarrow \mathbb{Q}/\mathbb{Z}$$

is not effected by the complex coordinate. Similarly to the case in Remark 9.114 this is the only fault, i.e. if X is not geometrically connected but we ignore the complex places (or if K is totally real) we can still apply Theorem 9.135 (since $M = H_0(\mathbf{N}_{\mathcal{U}})$ is always a permutation module we have $H^3\left(K, \widehat{M}\right) = 0$) and get that

$$X(\mathbb{A})^{\mathbb{Z}h} = X(\mathbb{A})^{\mathrm{Br}}.$$

We will now finish this section by proving Proposition 9.129:

Proof of Proposition 9.129 Let $\Lambda \triangleleft \Gamma_K$ be an open normal subgroup. In the course of the proof we shall consider the Kan contractible simplicial Γ_K-set $\mathbf{E}(\Gamma_K/\Lambda)$. Note that $\mathbf{E}(\Gamma_K/\Lambda)$ can be considered as a hypercovering in the étale site of $\operatorname{Spec} K$.

We will then consider the pullback of $\mathbf{E}(\Gamma_K/\Lambda)$ from $\operatorname{Spec}(K)$ to X via $t : X \longrightarrow \operatorname{Spec}(K)$ (see §§9.2.4 for the definition of pullback of hypercoverings). Note that if $\mathbf{E}$ is a hypercovering of $\operatorname{Spec} K$ then there is always a natural map $\mathbf{N}_{t^*\mathbf{E}} \longrightarrow \mathbf{E}$.

We start by constructing the maps $\Psi_{\mathcal{U}}$ and showing that they induce an isomorphism:

Theorem 9.131 *Let K be an arbitrary field with absolute Galois group Γ_K and X an algebraic variety over K. Let A be a Γ_K-module and consider the étale sheaf t^*A on X of locally constant maps into A (see definition above) and let $n > i+1$. Then there exist natural maps*

$$\Psi_{\mathcal{U}} : \mathbb{H}^i(K, \underline{\mathrm{Hom}}(P_n^+(\underline{\mathbb{Z}\mathbf{N}_{\mathcal{U}}}), A)) \longrightarrow H^i_{\acute{e}t}(X, t^*A)$$

which induce an isomorphism

$$\varinjlim_{\mathcal{U}} \mathbb{H}^i(K, \underline{\mathrm{Hom}}(P_n^+(\underline{\mathbb{Z}\mathbf{N}_{\mathcal{U}}}), A)) \xrightarrow{\simeq} H^i_{\acute{e}t}(X, t^*A)$$

where A is considered as a complex concentrated at degree 0.

Proof To compute étale cohomology we note that

$$H^i_{\acute{e}t}(X, t^*A) = \mathrm{Ext}^i_{X,\acute{e}t}(t^*\mathbb{Z}, t^*A).$$

We start by computing the left hand side. The hypercovering $\mathcal{U}$ can be used to construct the sheaves

$$\mathcal{P}_n(\mathcal{V}) = t^*\mathbb{Z}(\mathcal{V} \times_X \mathcal{U}_n)$$

which fit in a resolution

$$\ldots \longrightarrow \mathcal{P}_2 \longrightarrow \mathcal{P}_1 \longrightarrow \mathcal{P}_0 \longrightarrow t^*\mathbb{Z}$$

of $t^*\mathbb{Z}$. This is unfortunately not a projective resolution and so the resulting cohomology groups

$$H^i(\Gamma(\mathcal{P}_\bullet)) = H^i\left(\mathrm{Hom}_{\Gamma_K}(\underline{\mathbb{Z}\mathbf{N}_{\mathcal{U}_\bullet}}, A)\right) = \left[\underline{\mathbb{Z}\mathbf{N}_{\mathcal{U}}}, A\right]^i_{\Gamma_K}$$

are not equal to the étale cohomology groups. We do, however, get a map

$$\Phi_{\mathcal{U}} : \left[\underline{\mathbb{Z}\mathbf{N}_{\mathcal{U}}}, A\right]^i_{\Gamma_K} = H^i(\Gamma(\mathcal{P}_\bullet)) \longrightarrow H^i_{\acute{e}t}(X, t^*A)$$

and we know by Verdier's hypercovering theorem for étale cohomology ([SGA4, Exposé V, 7.4.1(4)]) that by taking the direct limit over all étale hypercoverings $\mathcal{U}$ we get an isomorphism

$$\Phi : \varinjlim_{\mathcal{U}} [\underline{\mathbb{Z}\mathbf{N}_{\mathcal{U}}}, A]^i_{\Gamma_K} \xrightarrow{\simeq} H^i_{\acute{e}t}(X, t^*A).$$

Since $n > i + 1$ we have

$$[\underline{\mathbb{Z}\mathbf{N}_{\mathcal{U}}}, A]^i_{\Gamma_K} = [P_n^+(\underline{\mathbb{Z}\mathbf{N}_{\mathcal{U}}}), A]^i_{\Gamma_K}$$

so we get a system of maps

$$\Phi_{\mathcal{U}} : [P_n^+(\underline{\mathbb{Z}\mathbf{N}_{\mathcal{U}}}), A]^i_{\Gamma_K} \longrightarrow H^i_{\acute{e}t}(X, t^*A)$$

and in the limit the isomorphism

$$\Phi : \varinjlim_{\mathcal{U}} [P_n^+(\underline{\mathbb{Z}\mathbf{N}_{\mathcal{U}}}), A]^i_{\Gamma_K} \xrightarrow{\simeq} H^i_{\acute{e}t}(X, t^*A).$$

Since $\underline{\mathrm{Hom}}(P_n^+(\underline{\mathbb{Z}\mathbf{N}_{\mathcal{U}}}), A)$ is bounded we can compute its hypercohomology and get:

$$\begin{aligned}
\mathbb{H}^i(K, \underline{\mathrm{Hom}}(P_n^+(\underline{\mathbb{Z}\mathbf{N}_{\mathcal{U}}}), A)) &= \varinjlim_{\Lambda \tilde{\triangleleft} \Gamma_K} \left[\underline{\mathbb{Z}\mathbf{E}(\Gamma_K/\Lambda)}, \underline{\mathrm{Hom}}(P_n^+(\underline{\mathbb{Z}\mathbf{N}_{\mathcal{U}}}), A)\right]^i_{\Gamma_K} \\
&= \varinjlim_{\Lambda \tilde{\triangleleft} \Gamma_K} \left[\underline{\mathbb{Z}\mathbf{E}(\Gamma_K/\Lambda)} \otimes P_n^+(\underline{\mathbb{Z}\mathbf{N}_{\mathcal{U}}}), A\right]^i_{\Gamma_K} \\
&= \varinjlim_{\Lambda \tilde{\triangleleft} \Gamma_K} \left[\underline{\mathbb{Z}\mathbf{E}(\Gamma_K/\Lambda)} \otimes \underline{\mathbb{Z}\mathbf{N}_{\mathcal{U}}}, A\right]^i_{\Gamma_K}.
\end{aligned}$$

Thus we get in the limit

$$\varinjlim_{\mathcal{U}} \mathbb{H}^i(K, \underline{\mathrm{Hom}}(P_n^+(\underline{\mathbb{Z}\mathbf{N}_{\mathcal{U}}}), A)) = \varinjlim_{\mathcal{U},\Lambda} \left[\underline{\mathbb{Z}\mathbf{E}(\Gamma_K/\Lambda)} \otimes \underline{\mathbb{Z}\mathbf{N}_{\mathcal{U}}}, A\right]^i_{\Gamma_K}.$$

So in order to prove the theorem we will construct a natural equivalence between the functors

$$X \mapsto \{\underline{\mathbb{Z}\mathbf{N}_{\mathcal{U}}}\}_{\mathcal{U}}$$

and

$$X \mapsto \left\{ \underline{\mathbb{Z}\mathbf{E}(\Gamma_K/\Lambda)} \otimes \underline{\mathbb{Z}\mathbf{N}_{\mathcal{U}}} \right\}_{\mathcal{U},\Lambda}$$

as functors from algebraic varieties over K to the pro-category of $\mathrm{Ho}(\mathcal{C}\,\mathrm{Mod}_{\Gamma_K})$.

In order to construct a transformation

$$F : \left\{ \underline{\mathbb{Z}\mathbf{E}(\Gamma_K/\Lambda)} \otimes \underline{\mathbb{Z}\mathbf{N}_{\mathcal{U}}} \right\}_{\mathcal{U},\Lambda} \longrightarrow \{\underline{\mathbb{Z}\mathbf{N}_{\mathcal{U}}}\}_{\mathcal{U}}$$

we need to pick (compatibly) for each étale hypercovering $\mathcal{U}$ an étale hypercovering of X $\mathcal{U}'$, an open normal subgroup $\Lambda \tilde{\triangleleft} \Gamma_K$ and a map

$$F_{\mathcal{U}} : \underline{\mathbb{Z}\mathbf{E}(\Gamma_K/\Lambda)} \otimes \underline{\mathbb{Z}\mathbf{N}_{\mathcal{U}'}} \longrightarrow \underline{\mathbb{Z}\mathbf{N}_{\mathcal{U}}}.$$

Our choice here is simple. Take $\mathcal{U}' = \mathcal{U}$ and $\Lambda = \Gamma_K, E(\Gamma_K/\Gamma_K) = *$. Then choose $F_{\mathcal{U}}$ to be the natural map

$$F_{\mathcal{U}} : \underline{\mathbb{Z}*} \otimes \underline{\mathbb{Z}\mathbf{N}_{\mathcal{U}}} \longrightarrow \underline{\mathbb{Z}\mathbf{N}_{\mathcal{U}}}.$$

The other direction is more tricky. In order to construct a transformation

$$G : \{\underline{\mathbb{Z}\mathbf{N}_{\mathcal{U}}}\}_{\mathcal{U}} \longrightarrow \left\{ \underline{\mathbb{Z}\mathbf{E}(\Gamma_K/\Lambda)} \otimes \underline{\mathbb{Z}\mathbf{N}_{\mathcal{U}}} \right\}_{\mathcal{U},\Lambda}$$

we need to pick (compatibly) for each étale hypercovering $\mathcal{U}$ and an open normal subgroup $\Lambda \tilde{\triangleleft} \Gamma_K$ an étale hypercovering $\mathcal{U}'$ and a map

$$G_{\mathcal{U},\Lambda} : \underline{\mathbb{Z}\mathbf{N}_{\mathcal{U}'}} \longrightarrow \underline{\mathbb{Z}\mathbf{E}(\Gamma_K/\Lambda)} \otimes \underline{\mathbb{Z}\mathbf{N}_{\mathcal{U}}}.$$

Recall that we denote by $t : X \to \operatorname{Spec} X$ the structure map. We choose

$$\mathcal{U}' = \mathcal{U}_\Lambda := t^*(\mathbf{E}(\Gamma_K/\Lambda)) \times_X \mathcal{U}.$$

There are natural maps

$$\mathbf{N}_{\mathcal{U}_\Lambda} \longrightarrow \mathbf{N}_{t^*(\mathbf{E}(\Gamma_K/\Lambda))} \times \mathbf{N}_{\mathcal{U}} \longrightarrow \mathbf{E}(\Gamma_K/\Lambda) \times \mathbf{N}_{\mathcal{U}}$$

which give a map

$$\underline{\mathbb{Z}\mathbf{N}_{\mathcal{U}_\Lambda}} \longrightarrow \underline{\mathbb{Z}(\mathbf{E}(\Gamma_K/\Lambda) \times \mathbf{N}_{\mathcal{U}})}.$$

By composing with the Alexander-Whitney map (which is a homotopy equivalence of complexes by the Künneth Theorem):

$$\underline{\mathbb{Z}(\mathbf{E}(\Gamma_K/\Lambda) \times \mathbf{N}_{\mathcal{U}})} \longrightarrow \underline{\mathbb{Z}\mathbf{E}(\Gamma_K/\Lambda)} \otimes \underline{\mathbb{Z}\mathbf{N}_{\mathcal{U}}},$$

we construct our map

$$G_{\mathcal{U},\Lambda} : \underline{\mathbb{Z}\mathbf{N}_{\mathcal{U}_\Lambda}} \longrightarrow \underline{\mathbb{Z}\mathbf{E}(\Gamma_K/\Lambda)} \otimes \underline{\mathbb{Z}\mathbf{N}_{\mathcal{U}}}.$$

The map $F \circ G$ is clearly the identity (in the category $\mathrm{Pro}\,\mathrm{Ho}(\mathcal{C}\,\mathrm{Mod}_{\Gamma_K})$). Now consider

$$G \circ F : \left\{\underline{\mathbb{Z}\mathbf{E}(\Gamma_K/\Lambda)} \otimes \underline{\mathbb{Z}\mathbf{N}_{\mathcal{U}}}\right\}_{\mathcal{U},\Lambda} \longrightarrow \left\{\underline{\mathbb{Z}\mathbf{E}(\Gamma_K/\Lambda)} \otimes \underline{\mathbb{Z}\mathbf{N}_{\mathcal{U}}}\right\}_{\mathcal{U},\Lambda}.$$

This is the pro-map that for each $\mathcal{U}, \Lambda$ chooses $\mathcal{U}' = \mathcal{U}_\Lambda$, $\Lambda' = \Gamma_K$ and the map

$$\underline{\mathbb{Z}*} \otimes \underline{\mathbb{Z}\mathbf{N}_{\mathcal{U}_\Lambda}} \longrightarrow \underline{\mathbb{Z}\mathbf{E}(\Gamma_K/\Lambda)} \otimes \underline{\mathbb{Z}\mathbf{N}_{\mathcal{U}}}$$

obtained as above.

In order to show that this pro-map represents the identity in $\mathrm{Pro}\,\mathrm{Ho}(\mathcal{C}\,\mathrm{Mod}_{\Gamma_K})$ we will show that the following diagram commutes in $\mathrm{Ho}(\mathcal{C}\,\mathrm{Mod}_{\Gamma_K})$:

$$\begin{array}{ccc} & & \underline{\mathbb{Z}\mathbf{E}(\Gamma_K/\Lambda)} \otimes \underline{\mathbb{Z}\mathbf{N}_{\mathcal{U}_\Lambda}} \\ & \swarrow^{r_1} & \downarrow^{r_2} \\ \underline{\mathbb{Z}*} \otimes \underline{\mathbb{Z}\mathbf{N}_{\mathcal{U}_\Lambda}} & \xrightarrow{G \circ F} & \underline{\mathbb{Z}\mathbf{E}(\Gamma_K/\Lambda)} \otimes \underline{\mathbb{Z}\mathbf{N}_{\mathcal{U}}} \end{array}$$

where r_1, r_2 are refinement maps which are the structure maps of our pro-object. The two maps $G \circ F \circ r_1$ and r_2 both factor through $\underline{\mathbb{Z}\mathbf{E}(\Gamma_K/\Lambda)} \otimes \underline{\mathbb{Z}\mathbf{E}(\Gamma_K/\Lambda)} \otimes \underline{\mathbb{Z}\mathbf{N}_{\mathcal{U}}}$ as

$$G \circ F \circ r_1 = f_1 \circ (Id \otimes G)$$

and

$$r_2 = f_2 \circ (Id \otimes G)$$

where

$$f_1 = p \otimes Id \otimes Id : \underline{\mathbb{Z}\mathbf{E}(\Gamma_K/\Lambda)} \otimes \underline{\mathbb{Z}\mathbf{E}(\Gamma_K/\Lambda)} \otimes \underline{\mathbb{Z}\mathbf{N}_{\mathcal{U}}} \longrightarrow \underline{\mathbb{Z}\mathbf{E}(\Gamma_K/\Lambda)} \otimes \underline{\mathbb{Z}\mathbf{N}_{\mathcal{U}}},$$

$$f_2 = Id \otimes p \otimes Id : \underline{\mathbb{Z}\mathbf{E}(\Gamma_K/\Lambda)} \otimes \underline{\mathbb{Z}\mathbf{E}(\Gamma_K/\Lambda)} \otimes \underline{\mathbb{Z}\mathbf{N}_{\mathcal{U}}} \longrightarrow \underline{\mathbb{Z}\mathbf{E}(\Gamma_K/\Lambda)} \otimes \underline{\mathbb{Z}\mathbf{N}_{\mathcal{U}}},$$

and $p : \underline{\mathbb{Z}\mathbf{E}(\Gamma_K/\Lambda)} \longrightarrow \underline{\mathbb{Z}}$ is the natural projection.

Hence it is enough to prove that the square

$$\begin{array}{ccc} \underline{\mathbb{Z}\mathbf{E}(\Gamma_K/\Lambda)} \otimes \underline{\mathbb{Z}\mathbf{N}_{\mathcal{U}_\Lambda}} & \xrightarrow{Id \otimes G} & \underline{\mathbb{Z}\mathbf{E}(\Gamma_K/\Lambda)} \otimes \underline{\mathbb{Z}\mathbf{E}(\Gamma_K/\Lambda)} \otimes \underline{\mathbb{Z}\mathbf{N}_{\mathcal{U}}} \\ \downarrow{\scriptstyle Id \otimes G} & & \downarrow{\scriptstyle f_2} \\ \underline{\mathbb{Z}\mathbf{E}(\Gamma_K/\Lambda)} \otimes \underline{\mathbb{Z}\mathbf{E}(\Gamma_K/\Lambda)} \otimes \underline{\mathbb{Z}\mathbf{N}_{\mathcal{U}}} & \xrightarrow{f_1} & \underline{\mathbb{Z}\mathbf{E}G_L} \otimes \underline{\mathbb{Z}\mathbf{N}_{\mathcal{U}}} \end{array}$$

commutes up to equivariant homotopy, or simply that f_1 and f_2 are equivariantly homotopic as chain maps.

Let $H = \Gamma_K/\Lambda$. Since the action of Γ_K on $\mathbb{Z}\mathbf{E}(\Gamma_K/\Lambda)$ factors through H it is enough to show that

$$p \otimes Id, \quad Id \otimes p : \underline{\mathbb{Z}\mathbf{E}H} \otimes \underline{\mathbb{Z}\mathbf{E}H} \longrightarrow \underline{\mathbb{Z}\mathbf{E}H}$$

are H-equivariantly homotopic as chain maps. Recall that $\underline{\mathbb{Z}\mathbf{E}H} \otimes \underline{\mathbb{Z}\mathbf{E}H}$ is equivariantly homotopy equivalent to $\underline{\mathbb{Z}(\mathbf{E}H \times \mathbf{E}H)}$. Hence it is enough to show that the two projections

$$p_1, p_2 : \mathbf{E}H \times \mathbf{E}H \longrightarrow \mathbf{E}H$$

are equivariantly homotopic. Note that both $\mathbf{E}H$ and $\mathbf{E}H \times \mathbf{E}H$ are contractible free H-spaces and so the equivariant mapping space from $\mathbf{E}H \times \mathbf{E}H$ to $\mathbf{E}H$ is homotopy equivalent to $\mathbf{E}H^{hH}$ which is contractible. This means that p_1 and p_2 are H-equivariantly homotopic and we are done (the fact that this homotopy can be done simplicially can be seen using the projective model structure on simplicial H-sets).

Now the maps

$$\Psi_{\mathcal{U}} : \mathbb{H}^2\left(\Gamma_K, \widehat{P_n^+(\underline{\mathbb{Z}\mathbf{N}_{\mathcal{U}}})}\right) \longrightarrow H^2_{\acute{e}t}(X, t^*\mathbb{G}_m)$$

are obtained as the composition

$$\begin{aligned} \mathbb{H}^2\left(\Gamma_K, \widehat{P_n^+(\underline{\mathbb{Z}\mathbf{N}_{\mathcal{U}}})}\right) &= \varinjlim_{\Lambda \triangleleft \Gamma_K} \left[\underline{\mathbb{Z}\mathbf{E}(\Gamma_K/\Lambda)} \otimes \underline{\mathbb{Z}\mathbf{N}_{\mathcal{U}}}, \mathbb{G}_m\right]^2_{\Gamma_K} \\ &\xrightarrow{G^*_{\mathcal{U}}} \varinjlim_{\Lambda \triangleleft \Gamma} \left[\underline{\mathbb{Z}\mathbf{N}_{\mathcal{U}_\Lambda}}, \mathbb{G}_m\right]^2_{\Gamma_K} \xrightarrow{\Phi_{\mathcal{U}*}} H^2_{\acute{e}t}(X, t^*\mathbb{G}_m). \end{aligned} \qquad (\clubsuit)$$

This concludes the proof of 9.131. □

Remark 9.132 Note that for $X = \mathrm{Spec}\,(K)$ and every hypercovering $\mathcal{U} \longrightarrow \mathrm{Spec}\,(K)$ we have a quasi-isomorphism

$$\underline{\mathrm{Hom}}\left(P_n^+(\underline{\mathbb{Z}\mathbf{N}_{\mathcal{U}}}), A\right) \simeq \underline{\mathrm{Hom}}\left(P_n^+(\underline{\mathbb{Z}\mathbf{N}_X}), A\right) = \underline{\mathrm{Hom}}\,(\underline{\mathbb{Z}}, \underline{A}) = \underline{A}$$

where $X \longrightarrow X$ is the trivial hypercovering. Substituting $\mathcal{U} = X$ in ($\clubsuit$) one sees that $G^*_{\mathcal{U}}$ becomes the identity and $\Psi_{\mathcal{U}}$ becomes the standard identification

$$\mathbb{H}^i(K, A) = H^i(K, A) \xrightarrow{\simeq} H^i_{\acute{e}t}(\mathrm{Spec}\,(K), A).$$

We shall now finish the proof of Proposition 9.129 by showing that the map $\Psi_{\mathcal{U}}$ renders the diagram

$$\begin{array}{ccccc} X(\mathbb{A}) & \times & H^2_{\acute{e}t}(X, t^*\mathbb{G}_m) & \longrightarrow & \mathbb{Q}/\mathbb{Z} \\ \downarrow & & \uparrow{\scriptstyle \Psi_{\mathcal{U}}} & & \| \\ \mathbb{H}^0(\mathbb{A}, P_n^+(\underline{\mathbb{Z}\mathbf{N}_{\mathcal{U}}})) & \times & \mathbb{H}^2\left(K, \widehat{P_n^+(\underline{\mathbb{Z}\mathbf{N}_{\mathcal{U}}})}\right) & \longrightarrow & \mathbb{Q}/\mathbb{Z} \end{array}$$

commutative.

Since both pairings are defined by summing the local contributions it is enough to show that the diagram

$$\begin{array}{ccccc} X(K_\nu) & \times & H^2_{\acute{e}t}(X_\nu, t^*\mathbb{G}_m) & \longrightarrow & \mathbb{Q}/\mathbb{Z} \\ \downarrow & & \uparrow{\scriptstyle \Psi_{\mathcal{U}_\nu, n}} & & \| \\ \mathbb{H}^0\left(K_\nu, P_n^+(\underline{\mathbb{Z}\mathbf{N}_{\mathcal{U}_\nu}})\right) & \times & \mathbb{H}^2\left(K_\nu, \widehat{P_n^+(\underline{\mathbb{Z}\mathbf{N}_{\mathcal{U}_\nu}})}\right) & \longrightarrow & \mathbb{Q}/\mathbb{Z} \end{array}$$

is commutative, where $X_\nu, \mathcal{U}_\nu$ are the base changes of X and $\mathcal{U}$ respectively from K to K_ν (note that this base change doesn't change geometric connected components so $\mathbf{N}_{\mathcal{U}_\nu}$ is actually the same simplicial set as $\mathbf{N}_{\mathcal{U}}$).

Since both rows of the diagram are functorial in X_ν it is enough to prove the commutativity for the case $X_\nu = \mathrm{Spec}\,(K_\nu)$. In that case $P_n^+(\underline{\mathbb{Z}\mathbf{N}_{\mathcal{U}_\nu}})$ is quasi-isomorphic to $\mathbb{Z}$ considered as a complex concentrated at degree 0 (for all hypercoverings $\mathcal{U}_\nu \longrightarrow \mathrm{Spec}\,(K_\nu)$) and we get the diagram

$$\begin{array}{ccccc} \{\bullet\} & \times & H^2_{\acute{e}t}(\mathrm{Spec}\,(K_\nu), t^*\mathbb{G}_m) & \longrightarrow & \mathbb{Q}/\mathbb{Z} \\ \downarrow & & \uparrow{\scriptstyle \Psi_{\mathcal{U}}} & & \| \\ \mathbb{Z} & \times & \mathbb{H}^2(K_\nu, \mathbb{G}_m) & \longrightarrow & \mathbb{Q}/\mathbb{Z} \end{array}$$

where the point $\bullet$ is mapped to $1 \in \mathbb{Z}$. But this pairing diagram is compatible in view of Remark 9.132 so we are done. This finishes the proof of Proposition 9.129. □

9.10.5 Proof of arithmetic duality results

In this section we will prove the main auxiliary result (Theorem 9.135) which we need in order to prove the equivalence of the homological and Brauer-Manin obstructions (Theorem 9.116). We will need various generalizations of results from the theory of arithmetic duality of Galois modules (See [Mil06]) to Galois complexes. Similar and related results appear in [HSz05], [De09b] and in [Jos09].

Let K be a number field with absolute Galois group Γ. Consider the Galois module

$$\mathfrak{J} = \varinjlim_{L/K} \varinjlim_{T} \prod_{\omega \in T} L_\omega^* \times \prod_{\omega \notin T} O_\omega^*$$

where L runs over finite extensions of K, T over finite sets of places of L and $O_\omega^* \subseteq L_\omega^*$ is the group of O_ω-units. There is a natural inclusion

$$\overline{K}^* \subseteq \mathfrak{J}$$

and the quotient is denoted by

$$\mathfrak{C} = \mathfrak{J}/\overline{K}^*.$$

The module $\mathfrak{C}$ is called a **class formation**. Using the Yoneda product one obtains for every Γ-module M a pairing

$$H^{2-r}(\Gamma, M) \times \mathrm{Ext}_\Gamma^r(M, \mathfrak{C}) \longrightarrow H^2(\Gamma, \mathfrak{C}) \xrightarrow{\simeq} \mathbb{Q}/\mathbb{Z}$$

which is the basis for all arithmetic duality results.

We would like to generalize these notions from Γ-modules to Γ-complexes. By replacing Ext with its extension $\mathbb{E}\mathrm{xt}$ to the category of bounded Γ_K-complexes and cohomology with hypercohomology one obtains analogous pairings

$$\mathbb{H}^{2-r}(\Gamma, C) \times \mathbb{E}\mathrm{xt}_\Gamma^r(C, \mathfrak{C}) \longrightarrow \mathbb{H}^2(\Gamma, \mathfrak{C}) \xrightarrow{\simeq} \mathbb{Q}/\mathbb{Z}$$

for every bounded Γ_K-complex C (where $\mathfrak{C}$ is considered as a complex concentrated in degree 0). These pairings induce maps:

$$\alpha_C^r : \mathbb{E}\mathrm{xt}_\Gamma^r(C, \mathfrak{C}) \longrightarrow \left(\mathbb{H}^{2-r}(\Gamma, C)\right)^* = \left(\mathbb{H}^{2-r}(C)\right)^*.$$

The following two lemmas generalize two of the main aspects of the theory from modules to complexes under various finiteness conditions.

Lemma 9.133 *Let K be a field of characteristic zero and let C be an excellent bounded complex such that $H_n(C)$ is finitely generated for every n. Then there are natural isomorphisms*

$$\mathbb{H}^i(K, C) \cong \mathbb{E}\mathrm{xt}^i\left(\widehat{C}, \overline{K}^*\right),$$

$$\mathbb{H}^i(\mathbb{A}, C) \cong \mathbb{E}\mathrm{xt}^i\left(\widehat{C}, \mathfrak{J}\right).$$

*where $\widehat{C}$ denotes the **dual complex** of C, see Definition 9.128.*

Proof For C which is concentrated at degree 0 (i.e. is actually a Γ-module) this claim is proved in the [Mil06] Lemmas 4.12 and 14.3. One can then proceed by induction on the length of C using the fact that for bounded complexes both $\mathbb{H}^i(\Gamma, \bullet)$ and $\mathbb{H}^i(\mathbb{A}, \bullet)$ transform short exact sequences to long exact sequences (Lemma 9.125) and applying the five lemma. □

Lemma 9.134 *Let K be a number field and Let C be an excellent complex bounded below by dimension 0 such that*

1. *$H_n(C)$ is finite for all $n > 0$.*
2. *$H_0(C) = 0$.*

Then the map

$$\alpha^r_{\widehat{C}} : \mathbb{E}\mathrm{xt}^r\left(\widehat{C}, \mathfrak{C}\right) \longrightarrow \mathbb{H}^{2-r}\left(K, \widehat{C}\right)^*$$

is an isomorphism for every $r \geq 0$ and is surjective for $r = -1$.

Proof We will say that a complex $C \in \mathcal{C}^{\geq 0}\,\mathrm{Mod}_{\Gamma_K}$ is n-bounded if $C_i = 0$ for $i > n$. We will prove the lemma for n-bounded complexes by induction on n.

For $n = 1$ we get that $\widehat{C}$ is quasi-isomorphic to a complex of the form ΣM for some finite Galois module M. The claim then is just Theorem 4.6 in [Mil06] after the suitable dimension shift (note that both sides are invariant to quasi-isomorphisms between bounded complexes).

Now assume that the lemma is true for n-bounded complexes ($n \geq 1$) and let C be an $(n+1)$-bounded complex satisfying the assumption of the lemma. Consider the short exact sequence

$$0 \longrightarrow C\langle 1\rangle \longrightarrow C \longrightarrow P_1^+(C) \longrightarrow 0.$$

Since

$$\mathbb{H}^r\left(K, \widehat{C\langle 1\rangle}\right) = \mathbb{H}^{r-1}\left(K, \widehat{\Sigma^{-1}C\langle 1\rangle}\right),$$

$$\mathbb{E}\mathrm{xt}^r\left(\widehat{C\langle 1\rangle}, \mathfrak{C}\right) = \mathbb{E}\mathrm{xt}^{r+1}\left(\widehat{\Sigma^{-1}C\langle 1\rangle}, \mathfrak{C}\right),$$

and since $\Sigma^{-1}C\langle 1\rangle$ is n-bounded we get that the lemma is true for $\alpha^r_{C\langle 1\rangle}$. Since $P_1^+(C)$ is 1-bounded the lemma is true for $\alpha^r_{C\langle 1\rangle}$ as well. We then observe the map of long exact sequences:

$$\begin{array}{ccccccc}
\to \mathbb{E}\mathrm{xt}^{r-1}\left(\widehat{P_1^+(C)}, \mathfrak{C}\right) & \longrightarrow & \mathbb{E}\mathrm{xt}^{r}\left(\widehat{C\langle 1\rangle}, \mathfrak{C}\right) & \longrightarrow & \mathbb{E}\mathrm{xt}^{r}\left(\widehat{C}, \mathfrak{C}\right) & \longrightarrow \cdots \\
\downarrow \alpha^{r-1}_{P_1^+(C)} & & \downarrow \alpha^{r}_{C\langle 1\rangle} & & \downarrow \alpha^{r}_{C} & \\
\to \left(\mathbb{H}^{3-r}(\widehat{P_1^+(C)})\right)^* & \longrightarrow & \left(\mathbb{H}^{2-r}\left(\widehat{C\langle 1\rangle}\right)\right)^* & \longrightarrow & \left(\mathbb{H}^{2-r}\left(\widehat{C}\right)\right)^* & \longrightarrow \cdots
\end{array}$$

$$\begin{array}{ccccc}
\cdots \longrightarrow \mathbb{E}\mathrm{xt}^{r}\left(\widehat{P_1^+(C)}, \mathfrak{C}\right) & \longrightarrow & \mathbb{E}\mathrm{xt}^{r+1}\left(\widehat{C\langle 1\rangle}, \mathfrak{C}\right) & \to \\
\downarrow \alpha^{r}_{P_1^+(C)} & & \downarrow \alpha^{r+1}_{C\langle 1\rangle} & \\
\cdots \longrightarrow \left(\mathbb{H}^{2-r}\left(\widehat{P_1^+(C)}\right)\right)^* & \longrightarrow & \left(\mathbb{H}^{1-r}\left(\widehat{C\langle 1\rangle}\right)\right)^* & \to
\end{array}$$

For $r \geq 0$ we get from the five lemma that α^r_C is an isomorphism (recall that in the five lemma it is enough to assume the left most map is surjective). If $r = -1$ then an analogous diagram chase shows that α^r_C is surjective. □

Now let C be an excellent bounded Γ_K-complex. Let

$$p : \mathbb{H}^0(\mathbb{A}, C) \longrightarrow \mathbb{H}^2\left(K, \widehat{C}\right)^*$$

be the composition

$$\mathbb{H}^0(\mathbb{A}, C) \xrightarrow{\cong} \mathbb{E}\mathrm{xt}^0\left(\widehat{C}, \mathfrak{J}\right) \longrightarrow \mathbb{E}\mathrm{xt}^0\left(\widehat{C}, \mathfrak{C}\right) \xrightarrow{\alpha^0_{\widehat{C}}} \mathbb{H}^2\left(K, \widehat{C}\right)^*$$

where the first isomorphism is the one given in Lemma 9.133. Then p induces a pairing

$$\mathbb{H}^0(\mathbb{A}, C) \times \mathbb{H}^2\left(K, \widehat{C}\right) \longrightarrow \mathbb{Q}/\mathbb{Z}.$$

Unwinding the definitions one can see that this pairing is given by using the cup product

$$\mathbb{H}^0(K_\nu, C) \times \mathbb{H}^2\left(K_\nu, \widehat{C}\right) \longrightarrow \mathbb{H}^2(K_\nu, \mathbb{G}_m) \cong \mathbb{Q}/\mathbb{Z}$$

for each ν separately and summing the results.

For our purposes we need to understand the connection between the left kernel of the cup-product pairing (♣) and the subgroup of $\mathbb{H}^0(\mathbb{A}, C)$ coming from $\mathbb{H}^0(K, C)$. This is done in the following lemma, under some assumptions, which are slightly weaker than those of Lemma 9.134 (these assumptions will be satisfied in the cases we are interested in):

Theorem 9.135 *Let C be an excellent bounded Γ_K-complex, bounded below by dimension 0, such that:*

1. *$H_n(C)$ is finite for all $n > 0$.*
2. *The Galois module $M = H_0(C)$ satisfies the property that $H^3(K, \widehat{M}) = 0$.*

Denote by $\pi : C \longrightarrow M$ the natural map. Consider the (non-exact) sequence

$$\mathbb{H}^0(K, C) \xrightarrow{\text{loc}} \mathbb{H}^0(\mathbb{A}, C) \xrightarrow{p} \mathbb{H}^2\left(K, \widehat{C}\right)^*.$$

Let $(x_\nu) \in \mathbb{H}^0(\mathbb{A}, C)$ be such that $\pi_*((x_\nu))$ is rational and $p((x_\nu)) = 0$. Then (x_ν) is rational, i.e. is in the image of loc.

Proof Consider the following diagram with the two middle rows exact

$$\begin{array}{ccccc}
\mathbb{H}^0(K, \widetilde{C}) & \xrightarrow{\text{loc}} & \mathbb{H}^0(\mathbb{A}, \widetilde{C}) & & \\
\downarrow \cong & & \downarrow \cong & & \\
\mathbb{E}\text{xt}^0\left(\widehat{C}, \overline{K}^*\right) & \xrightarrow{\text{loc}} & \mathbb{E}\text{xt}^0\left(\widehat{C}, \mathfrak{J}\right) & \longrightarrow & \mathbb{E}\text{xt}^0\left(\widehat{C}, \mathfrak{C}\right) \\
\downarrow & & \downarrow & & \downarrow \widehat{\pi}_* \\
\mathbb{E}\text{xt}^0\left(\widehat{M}, \overline{K}^*\right) & \xrightarrow{\text{loc}} & \mathbb{E}\text{xt}^0\left(\widehat{M}, \mathfrak{J}\right) & \longrightarrow & \mathbb{E}\text{xt}^0\left(\widehat{M}, \mathfrak{C}\right) \\
\downarrow \cong & & \downarrow \cong & & \\
\mathbb{H}^0(K, M) & \xrightarrow{\text{loc}} & \mathbb{H}^0(\mathbb{A}, M) & &
\end{array}$$

where the isomorphisms in the diagram are those of Lemma 9.133. We can hence consider (x_ν) as an element of $\mathbb{E}\text{xt}^0\left(\widehat{C}, \mathfrak{J}\right)$ whose image in $\mathbb{E}\text{xt}^0\left(\widehat{M}, \mathfrak{J}\right)$ comes from $\mathbb{E}\text{xt}^0\left(\widehat{M}, \overline{K}^*\right)$. Let β be the image of (x_ν) in $\mathbb{E}\text{xt}^0\left(\widehat{C}, \mathfrak{C}\right)$. From the exactness of the middle rows we see that it is enough to show $\beta = 0$.

From commutativity we get that

$$\widehat{\pi}_*(\beta) = 0 \in \mathbb{E}\text{xt}^0\left(\widehat{M}, \mathfrak{C}\right).$$

Let $\widetilde{C} = \ker(\pi)$ and consider the exact sequence of Γ_K-complexes

$$0 \longrightarrow \widehat{M} \longrightarrow \widehat{C} \longrightarrow \widehat{\widetilde{C}} \longrightarrow 0.$$

Consider the following commutative diagram with exact columns

$$\begin{array}{ccccc}
 & & & & \mathbb{H}^3\left(K,\widehat{M}\right)^* \\
 & & & & \downarrow \\
\mathbb{E}\mathrm{xt}^0\left(\widehat{\widetilde{C}},\mathfrak{J}\right) & \longrightarrow & \mathbb{E}\mathrm{xt}^0\left(\widehat{\widetilde{C}},\mathfrak{C}\right) & \xrightarrow{\alpha^0_{\widetilde{C}}} & \mathbb{H}^2\left(K,\widehat{\widetilde{C}}\right)^* \\
\downarrow & & \downarrow & & \downarrow \\
\mathbb{E}\mathrm{xt}^0\left(\widehat{C},\mathfrak{J}\right) & \longrightarrow & \mathbb{E}\mathrm{xt}^0\left(\widehat{C},\mathfrak{C}\right) & \xrightarrow{\alpha^0_{C}} & \mathbb{H}^2\left(K,\widehat{C}\right)^* \\
\downarrow & & \downarrow{\scriptstyle \widehat{\pi}_*} & & \\
\mathbb{E}\mathrm{xt}^0\left(\widehat{M},\mathfrak{J}\right) & \longrightarrow & \mathbb{E}\mathrm{xt}^0\left(\widehat{M},\mathfrak{C}\right). & &
\end{array}$$

Since $\widehat{\pi}_*(\beta) = 0$ we get that there is a $\gamma \in \mathbb{E}\mathrm{xt}^0\left(\widehat{\widetilde{C}},\mathfrak{C}\right)$ mapping to β. The fact that $p(x_\nu) = 0$ means that $\alpha^0_C(\beta) = 0$. Since we assume $\mathbb{H}^3\left(K,\widehat{M}\right) = 0$ we get that $\alpha^0_{\widetilde{C}}(\gamma) = 0$. But by Lemma 9.134 the map $\alpha^0_{\widetilde{C}}$ is an isomorphism and so $\gamma = 0$. This means that $\beta = 0$ as well and we are done. □

9.11 The equivalence of the homotopy obstruction and the étale Brauer obstruction

The main result of this section is the following theorem:

Theorem 9.136 *Let K be a number field and X a smooth geometrically connected variety over K. Then the étale homotopy obstruction is equivalent to the étale-Brauer obstruction (see §9.3), i.e.*

$$X(\mathbb{A})^{fin,\mathrm{Br}} = X(\mathbb{A})^h.$$

Remark 9.137 If one wants to omit the condition that X is geometrically connected one again faces a small problem in the complex places of K (the situation here is completely analogous to the one in Remarks 9.114 and 9.130).

In the proof we will use the idea of applying obstructions to finite torsors (see 9.3.1 for the definition). In particular we will apply the

homology obstruction to finite torsors. This will plug into the proof by breaking it into two propositions:

1. Proposition 9.139 will show that for any smooth variety over K we have
$$X(\mathbb{A})^{fin,\mathbb{Z}h} \subseteq X(\mathbb{A})^h.$$
2. Proposition 9.141 will show that for any smooth variety over K we have
$$X(\mathbb{A})^h \subseteq X(\mathbb{A})^{fin,h}.$$

We then get

$$X(\mathbb{A})^{fin,\mathbb{Z}h} \subseteq X(\mathbb{A})^h \subseteq X(\mathbb{A})^{fin,h} \subseteq X(\mathbb{A})^{fin,\mathbb{Z}h}$$

which means that
$$X(\mathbb{A})^{fin,\mathbb{Z}h} = X(\mathbb{A})^h.$$

The theorem will then follow from the following lemma:

Lemma 9.138 *Let K be a number field and X a geometrically connected smooth variety over K. Then*
$$X(\mathbb{A})^{fin,\mathrm{Br}} = X(\mathbb{A})^{fin,\mathbb{Z}h}.$$

Proof By Theorem 9.116 this claim is almost immediate. The only point one would verify is the geometric connectivity issue (note that even though X is geometrically connected, this is not true for every Y which is a torsor over X). By Remarks 9.114, 9.130 and 9.137 the only problem lies in the complex places and this can be easily treated. □

Proposition 9.139 *Let K be a number field and X a smooth variety over K. Then*
$$X(\mathbb{A})^{fin,\mathbb{Z}h} \subseteq X(\mathbb{A})^h.$$

Proof From Corollary 9.100 we can assume without loss of generality that X is connected over K. But in this case both sets are empty unless X is geometrically connected (Corollary 9.102). Hence we can assume that X is geometrically connected.

We shall prove the proposition by using the following lemma

Lemma 9.140 *Let K be a field and X a smooth geometrically connected variety over K such that the short exact sequence*
$$1 \longrightarrow \pi_1(\overline{X}) \longrightarrow \pi_1(X) \longrightarrow \Gamma_K \longrightarrow 1$$

admits a continuous section. Then for every hypercovering $\mathcal{U} \longrightarrow X$ there exists a geometrically connected torsor $f : Y \longrightarrow X$ under a finite K-group G such that $\mathbf{X}_{f^*(\mathcal{U})}$ is simply connected.

Before proving Lemma 9.140 we shall explain why it implies the proposition. Let

$$(x_\nu) \in X(\mathbb{A})^{fin,\mathbb{Z}h}$$

be an adelic point. By Corollary 9.67 it is enough to prove that for every $\mathcal{U} \longrightarrow X$ and every n we have

$$(x_\nu) \in X(\mathbb{A})^{\mathcal{U},n}.$$

Now since

$$(x_\nu) \in X(\mathbb{A})^{fin,\mathbb{Z}h} \subseteq X(\mathbb{A})^{fin}$$

we have that

$$X(\mathbb{A})^{fin} \neq \emptyset$$

and by Corollary 9.115 the sequence

$$1 \longrightarrow \pi_1(\overline{X}) \longrightarrow \pi_1(X) \longrightarrow \Gamma_K \longrightarrow 1$$

admits a continuous section. Now by Lemma 9.140 there exists a geometrically connected torsor $f : Y \longrightarrow X$ under a finite K-group G such that $\mathbf{X}_{f^*(\mathcal{U})}$ is simply connected. Since

$$(x_\nu) \in X(\mathbb{A})^{fin,\mathbb{Z}h}$$

there exists some twist $f^\alpha : Y^\alpha \longrightarrow X^\alpha$ of f and a point $(y_\nu) \in Y^\alpha(\mathbb{A})^{\mathbb{Z}h}$ such that

$$(x_\nu) = f^\alpha((y_\nu)).$$

Now we have

$$(y_\nu) \in Y^\alpha(\mathbb{A})^{\mathbb{Z}h} \subseteq Y^\alpha(\mathbb{A})^{\mathbb{Z}(f^\alpha)^*(\mathcal{U}),n},$$

but since $\mathbf{X}_{(f^\alpha)^*(\mathcal{U})}$ (which is isomorphic to $\mathbf{X}_{f^*(\mathcal{U})}$ as a simplicial set) is simply connected we get from Corollary 9.62 that

$$(y_\nu) \in Y^\alpha(\mathbb{A})^{(f^\alpha)^*(\mathcal{U}),n}$$

as well.

Now by considering the commutative diagram

$$\begin{array}{ccccc} Y^{\alpha}(\mathbb{A}) & \xrightarrow{h} & \mathbf{X}_{(f^{\alpha})^*(\mathcal{U}),n}(h\mathbb{A}) & \xleftarrow[\text{loc}]{} & \mathbf{X}_{(f^{\alpha})^*(\mathcal{U}),n}(hK) \\ \downarrow f^{\alpha} & & \downarrow & & \downarrow \\ X(\mathbb{A}) & \xrightarrow{h} & \mathbf{X}_{\mathcal{U},n}(h\mathbb{A}) & \xleftarrow[\text{loc}]{} & \mathbf{X}_{\mathcal{U},n}(hK) \end{array}$$

one sees that

$$(x_{\nu}) = f^{\alpha}((y_{\nu})) \in X(\mathbb{A})^{\mathcal{U},n}$$

as needed.

We shall now prove Lemma 9.140.

Proof Let $\mathcal{U} \longrightarrow X$ be a hypercovering. By the discussion in subsection §§9.2.4 we can naturally consider the pullback of $\mathcal{U}_n$ to the generic point of X as a finite $\Gamma_{K(X)}$-set which we denote by $\widetilde{\mathcal{U}}_n$ (see Definition 9.20). By restriction we will also consider $\widetilde{\mathcal{U}}_n$ as a $\Gamma_{\overline{K}(\overline{X})}$-set.

We denote by

$$\mathrm{Stab}_{\Gamma_{\overline{K}(\overline{X})}}(\mathcal{U}_0) = \{H_1, ..., H_t\}$$

the set of stabilizers in $\Gamma_{\overline{K}(\overline{X})}$ of the points of $\mathcal{U}_0$. Consider the group

$$E = \langle H_1, ..., H_t \rangle$$

generated by all the stabilizers. Let $F/\overline{K}(\overline{X})$ be the finite field extension that corresponds to E.

Observing Lemma 9.113 and its proof we see that there exists a G-torsor

$$f : \Upsilon \longrightarrow \overline{X}$$

such that $\overline{K}(\Upsilon) = F$.

Now consider $f^*\mathcal{U}_0$. By pulling back to the generic point of Υ we get a set $\widetilde{f^*\mathcal{U}_0}$ with an action of $\Gamma_{\overline{K}(\Upsilon)} = E$. Note that pulling back an étale covering from the generic point of X to the generic point of Υ corresponds to restricting the Galois action. In other words the set $\widetilde{f^*\mathcal{U}_0}$ can be naturally identified with $\widetilde{\mathcal{U}}_0$ and the action of E is obtained by restricting the action of $\Gamma_{\overline{K}(\overline{X})}$.

This means that the set of stabilizers of the points of $\widetilde{f^*\mathcal{U}_0}$ is

$$\{H_1 \cap E, ..., H_t \cap E\} = \{H_1, ..., H_n\}$$

as E contains all the H_i's. In particular the stabilzers of $\widetilde{f^*\mathcal{U}_0}$ in E

generate E. Now from Corollary 9.21 and Lemma 9.23 we get that $\mathbf{X}_{f^*\mathcal{U}}$ is **simply connected.**

To conclude, it is enough to show that we can find a K-form of Υ. Since E is normal in $\Gamma_{K(X)}$ we get that $\rho(E)$ is normal in $\pi_1(X)$. Hence the covering

$$f : \Upsilon \longrightarrow \overline{X}$$

is isomorphic over $\overline{K}$ to all its twists. By standard arguments we get that there exists a K-form for f if and only if the short exact sequence

$$1 \longrightarrow \pi_1(\overline{X})/\rho(E) \longrightarrow \pi_1(X)/\rho(E) \longrightarrow \Gamma_K \longrightarrow 1$$

has a continuous section. But this is true because

$$1 \longrightarrow \pi_1(\overline{X}) \longrightarrow \pi_1(X) \longrightarrow \Gamma_K \longrightarrow 1$$

has a continuous section. □

This finishes the proof of 9.139. □

We now come to the second part of the proof of Theorem 9.136, namely:

Proposition 9.141 *Let K be a number field and X a smooth K-variety over then*

$$X(\mathbb{A})^h \subseteq X(\mathbb{A})^{fin,h}.$$

Proof Like in the proof of Proposition 9.139 we can assume without loss of generality that X is geometrically connected.

Let $(x_\nu) \in X(\mathbb{A})^h$ be a point and $f : Y \longrightarrow X$ a K-connected torsor under a finite K-group G. We need to show that there exists an $\alpha \in H^1(K, G)$ such that the corresponding twist

$$f^\alpha : Y^\alpha \longrightarrow X$$

satisfies

$$(f^\alpha)^{-1}((x_\nu)) \cap Y^\alpha(\mathbb{A})^h \neq \emptyset.$$

Choose a rational homotopy fixed point

$$r \in X\,(hK)$$

such that

$$\mathrm{loc}\,(r) = h((x_\nu)).$$

Recall $\mathbf{N}_{\bar{Y}}$ of Definition 9.16 and the discussion prior to and including

Corollary 9.106 where we have described $\mathbf{N}_{\check{Y}}$ as the nerve of the groupoid $\mathcal{Y}$ of definition 9.104.

Consequently we have defined a map of simplicial Γ_K-sets

$$c_Y : \mathbf{N}_{\check{Y}} \longrightarrow \mathbf{B}G$$

inducing a map (denoted by the same name)

$$c_Y : \mathbf{N}_{\check{Y}}(hK) \longrightarrow H^1(K, G).$$

For each $\alpha \in H^1(K, G^\alpha)$, we will denote by

$$c_{Y^\alpha} : \mathbf{N}_{\check{Y}^\alpha}(hK) \longrightarrow H^1(K, G^\alpha)$$

the corresponding map for the twist Y^α. We will denote by

$$b : X(hK) \longrightarrow \mathbf{N}_{\check{Y}}(hK),$$

$$b_\alpha : X(hK) \longrightarrow \mathbf{N}_{\check{Y}^\alpha}(hK).$$

the natural projection map.

The strategy of the proof will be as following:

1. We will show that for each $r \in X(hK)$ there exists an α such

$$c_{Y^\alpha}(b_\alpha(r)) \in H^1(K, G^\alpha)$$

is the neutral element.

2. We will show that if $c_{Y^\alpha}(b_\alpha(r))$ is neutral and $\mathrm{loc}(r) = h((x_\nu))$ then

$$(f^\alpha)^{-1}((x_\nu)) \cap Y^\alpha(\mathbb{A})^h \neq \emptyset.$$

Proposition 9.142 *Let $f : Y \longrightarrow X$ be a torsor under a finite K-group G. Let $r \in X(hK)$ be a rational homotopy fixed point. Then there exists an $\alpha \in H^1(K, G)$ such that*

$$c_{Y^\alpha}(b_\alpha(r)) \in H^1(K, G^\alpha)$$

is the neutral element.

Proof For each $\alpha \in H^1(K, G)$ we will denote by

$$\tau_\alpha : H^1(K, G) \longrightarrow H^1(K, G^\alpha)$$

the inverse of twisting by α (hence sending $[\alpha] \in H^1(K, G)$ to the neutral element in $H^1(K, G^\alpha)$). We will prove the theorem by showing that for each $\alpha \in H^1(K, G)$ one has

$$c_{Y^\alpha} \circ b_\alpha = \tau_\alpha \circ c_Y \circ b. \qquad (\heartsuit)$$

Then if we simply choose $\alpha = c_Y(b(r))$ we will get that $c_{Y^\alpha}(b_\alpha(r))$ is the neutral element. We will now proceed to prove ($\heartsuit$).

Let $\alpha \in H^1(K, G^\alpha)$ be an element and let H be a big enough finite quotient of Γ_K such that the action of Γ_K on G and the element α factor through H. Then we can represent α by a 1-cocycle

$$u : H \longrightarrow G.$$

Let L/K be the finite Galois extension corresponding to H. Recall the étale hypercovering

$$\check{X}_L \longrightarrow X$$

from Definition 9.83. By choosing some identification of the geometric connected components of X_L with H we obtain an isomorphism of simplicial Γ_K-sets

$$d : \mathbf{N}_{\check{X}_L} \xrightarrow{\simeq} \mathbf{E}H.$$

Consider the following hypercoverings over X:

$$Y_L = Y \times_X X_L,$$

$$Y_L^\alpha = Y^\alpha \times_X X_L.$$

In Lemma 9.111 we have constructed an isomorphism of étale coverings over X

$$T_\alpha : Y_L \xrightarrow{\simeq} Y_L^\alpha$$

inducing an isomorphism of simplicial Γ_K-sets

$$(T_\alpha)_* : \mathbf{N}_{\check{Y}_L} \xrightarrow{\simeq} \mathbf{N}_{\check{Y}_L^\alpha}.$$

Note that over $\overline{K}$ the variety Y_L decomposes as the disjoint union of $|H|$ copies of Y. It is then not hard to see that the projections

$$Y_L \longrightarrow Y,$$

$$Y_L \longrightarrow X_L,$$

induce an isomorphism of simplicial Γ_K-sets

$$\mathbf{N}_{\check{Y}_L} \xrightarrow{\simeq} \mathbf{N}_{\check{Y}} \times \mathbf{N}_{\check{X}_L}.$$

Similarly, we have a natural isomorphism of simplicial Γ_K-sets

$$\mathbf{N}_{\check{Y}_L^\alpha} \xrightarrow{\simeq} \mathbf{N}_{\check{Y}^\alpha} \times \mathbf{N}_{\check{X}_L}.$$

Now the desired equality ($\heartsuit$) will follow from the following lemma:

Lemma 9.143 *There exists a Γ_K-equivariant isomorphism of groupoids*

$$\mathcal{T}_\alpha : \mathcal{E}H \times \mathcal{B}G \longrightarrow \mathcal{E}H \times \mathcal{B}G^\alpha$$

such that:

1. *The following diagram in the category simplicial Γ_K-set commutes:*

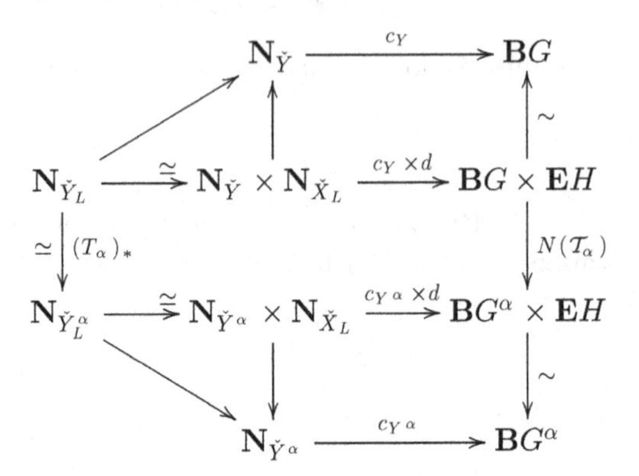

where all the unmarked maps are induced from the projections described above.

2. *All arrows in the right column of the diagram above are weak equivalences and the induced bijection*

$$\tau_\alpha : \mathbf{B}G(hK) = H^1(K, G) \longrightarrow H^1(K, G^\alpha) = \mathbf{B}G^\alpha(hK)$$

coincides with the inverse of twisting by α.

Proof We begin by defining

$$\mathcal{T}_\alpha : \mathcal{E}H \times \mathcal{B}G \longrightarrow \mathcal{E}H \times \mathcal{B}G^\alpha.$$

We define $\mathcal{T}_\alpha$ to be the identity on the objects. Now Let

$$(\tau, *), (\sigma, *) \in Ob(\mathcal{E}H \times \mathcal{B}G) = \mathcal{E}H \times \mathcal{B}G^\alpha.$$

One can naturally identify

$$\mathrm{Hom}_{\mathcal{E}H \times \mathcal{B}G}((\tau, *), (\sigma, *)) \cong G,$$

$$\mathrm{Hom}_{\mathcal{E}H \times \mathcal{B}G^\alpha}((\tau, *), (\sigma, *)) \cong G^\alpha.$$

Hence we need to define a map

$$\mathcal{T}_\alpha^{(\sigma,\tau)} : G \longrightarrow G^\alpha.$$

We shall take

$$\mathcal{T}_\alpha{}^{(\sigma,\tau)}(g) = u_\tau^{-1} g u_\sigma.$$

We leave it to the reader to verify that $\mathcal{T}_\alpha$ is indeed a Γ_K-equivariant isomorphism of groupoids.

Now one can easily verify the commutativity of the diagram in 1 above. For proving 2, note that since $\mathcal{T}_\alpha$ is an isomorphism so is $N(\mathcal{T}_\alpha)$. Thus since $\mathbf{E}H$ is contractible, all the maps in the right column are weak-equivalences. We leave to the reader to verify that the map

$$\tau_\alpha : \mathbf{B}G(hK) = H^1(K, G) \longrightarrow H^1(K, G^\alpha) = \mathbf{B}G^\alpha(hK)$$

is indeed the inverse of twisting by α □

This finishes the proof of Proposition 9.142. □

We now come to the second part of the proof of 9.141:

Proposition 9.144 *Let $\alpha \in H^1(K, G)$ be such that $c_{Y^\alpha}(b_\alpha(r))$ is the neutral element. Then*

$$(f^\alpha)^{-1}((x_\nu)) \cap Y^\alpha(\mathbb{A})^h \neq \emptyset.$$

Proof Note that without loss of generality we can replace Y by Y^α and α by the neutral element.

We want to show that there is a point in $f^{-1}((x_\nu))$ which is homotopy rational. For that it will be enough to show that every hypercovering $\mathcal{U}$ of Y and every $n \geq 1$ the subset of $f^{-1}((x_\nu))$ of points which are $(\mathcal{U}, n)$-homotopy rational is **non-empty** and **compact**. Then the standard compactness argument shows that

$$f^{-1}((x_\nu)) \cap Y(\mathbb{A})^h \neq \emptyset.$$

First, we will show that it is enough to prove this only for hypercoverings of a certain form.

Lemma 9.145 *Let $f : Y \longrightarrow X$ be a torsor under a K-group G. Let $\mathcal{U} \longrightarrow Y$ be a hypercovering. Then there exists a hypercovering $\mathcal{V} \longrightarrow X$ which admits a map*

$$\mathcal{V} \longrightarrow \check{Y}$$

of hypercoverings over X and a map

$$f^*(\mathcal{V}) \longrightarrow \mathcal{U}$$

of hypercoverings over Y.

Proof For a $g \in G$ we denote by $\mathcal{U}^g$ the hypercovering obtained from $\mathcal{U}$ by replacing each étale map

$$\mathcal{U}_n \longrightarrow Y$$

with the composition

$$\mathcal{U}_n \longrightarrow Y \xrightarrow{g} Y.$$

Define $\mathcal{U}^G$ to be the fiber product

$$\mathcal{U}^G = \prod_{g \in G} \mathcal{U}^g.$$

We have a natural action of G on $\mathcal{U}^G$ which is free and so we get a hypercovering

$$\mathcal{U}^G/G \longrightarrow X$$

of X. It is not hard to check that

$$\mathcal{U}^G = f^*\left(\mathcal{U}^G/G\right).$$

Now take the hypercovering

$$\mathcal{V} = \mathcal{U}^G/G \times_X \check{Y}$$

of X. Then one has a natural map over X

$$\mathcal{V} \longrightarrow \check{Y}$$

obtained by projection and since

$$f^*\mathcal{V} = \mathcal{U}^G \times_Y f^*\check{Y}$$

we have the composition of maps over Y

$$f^*\mathcal{V} \longrightarrow \mathcal{U}^G \longrightarrow \mathcal{U}$$

where the last one is obtained by projecting on the coordinate which corresponds to $1 \in G$. $\square$

In view of Lemma 9.145 it is enough to show that the set

$$f^{-1}((x_\nu)) \cap Y(\mathbb{A})^{f^*\mathcal{V},n}$$

is non-empty and compact for every $n \geq 1$ and every hypercovering $\mathcal{V} \longrightarrow X$ which admits a map

$$\phi : \mathcal{V} \longrightarrow \check{Y}$$

over X. Such hypercoverings have the following useful property

Lemma 9.146 *Let $f : Y \longrightarrow X$ be a torsor under a K-group G. Let $\mathcal{V} \longrightarrow X$ be a hypercovering and $\varphi : \mathcal{V} \longrightarrow \check{Y}$ a map of hypercoverings. Let*

$$p : \mathbf{N}_{f^*\mathcal{V}} \longrightarrow \mathbf{N}_{\mathcal{V}}$$

be the natural map. Then p is a principle G-covering. Furthermore, the composition

$$\mathbf{N}_{\mathcal{V}} \longrightarrow \mathbf{N}_{\check{Y}} \xrightarrow{c_Y} \mathbf{B}G$$

is the classifying map of p.

Proof We have a natural action of G on $\mathbf{N}_{f^*\mathcal{V}}$ such that $\mathbf{N}_{\mathcal{V}}$ is the quotient.

Now as in section §9.6.1 it will be enough to show that $\mathbf{N}_{f^*\mathcal{V}}$ fits in a commutative diagram

$$\begin{array}{ccc} \mathbf{N}_{f^*\mathcal{V}} & \xrightarrow{d} & \mathbf{E}G \\ \downarrow{\scriptstyle p} & & \downarrow \\ \mathbf{N}_{\mathcal{V}} & \longrightarrow & \mathbf{B}G \end{array}$$

such that d is G-equivariant. Since we have a map $\varphi : \mathcal{V} \longrightarrow \check{Y}$ it is enough to show this for $\mathcal{V} = \check{Y}$. This is done by explicit computation.

Consider the augmented functor $F \longrightarrow Id$ from groupoids to groupoids defined as follows. If C is a category then the objects of $F(C)$ are pairs (X, f) where $X \in Ob(C)$ and f is a morphism starting at X. The morphisms from (X, f) to (Y, g) are morphisms $h : X \longrightarrow Y$ in C such that $g \circ h = f$. The natural transformation from F to Id sends (X, f) to X.

Now recall the groupoid $\mathcal{Y}$ of Definition 9.104 such that $\mathbf{N}_{\check{Y}} \cong N(\mathcal{Y})$. It is then an easy verification to check that $\mathbf{N}_{f^*\check{Y}}$ is the nerve of the groupoid $F(\mathcal{Y})$ and that $\mathbf{E}G$ is the nerve of $F(\mathcal{B}G)$. Then the commutative diagram

$$\begin{array}{ccc} F(\mathcal{Y}) & \longrightarrow & F(\mathcal{B}G) \\ \downarrow & & \downarrow \\ \mathcal{Y} & \longrightarrow & \mathcal{B}G \end{array}$$

gives the desired commutative diagram

$$\begin{array}{ccc} \mathbf{N}_{f^*\check{Y}} & \longrightarrow & \mathbf{E}G \\ \downarrow & & \downarrow \\ \mathbf{N}_{\check{Y}} & \longrightarrow & \mathbf{B}G \end{array}$$

of simplicial sets. It is left to show that the map $\mathbf{N}_{f^*\check{Y}} \longrightarrow \mathbf{E}G$ is G equivariant. Note that in both $\mathcal{Y}$ and $\mathbf{B}G$ we have natural identification

$$\cup_Y \operatorname{Hom}(X, Y) = G$$

and the action of G on $F(\mathcal{Y})$ and $F(\mathcal{B}G)$ is given by multiplication

$$g(X, f) = (X, gf).$$

Hence it is clear that the map $F(\mathcal{Y}) \longrightarrow F(\mathcal{B}G)$ is G-equivariant and we are done. □

Now let $\mathcal{V} \longrightarrow X$ be a hypercovering which admits a map of hypercoverings

$$\phi : \mathcal{V} \longrightarrow \check{Y}$$

and let n be a natural number. We need to show that

$$f^{-1}((x_\nu)) \cap Y(\mathbb{A})^{f^*\mathcal{V},n}$$

is non-empty and compact. As above we will denote by

$$p : \mathbf{N}_{f^*\mathcal{V}} \longrightarrow \mathbf{N}_{\mathcal{V}}$$

the natural map which we have seen is a principal G-torsor. We will denote by

$$p_*^K : \mathbf{N}_{f^*\mathcal{V},n}(hK) \longrightarrow \mathbf{N}_{\mathcal{V},n}(hK)$$

$$p_*^{\mathbb{A}} : \mathbf{N}_{f^*\mathcal{V},n}(h\mathbb{A}) \longrightarrow \mathbf{N}_{\mathcal{V},n}(h\mathbb{A})$$

the induced maps.

We have a commutative diagram

$$\begin{array}{ccc} Y(\mathbb{A}) & \xrightarrow{h_{f^*\mathcal{V},n}} & \mathbf{N}_{f^*\mathcal{V},n}(h\mathbb{A}) \\ \downarrow f & & \downarrow p_*^{\mathbb{A}} \\ X(\mathbb{A}) & \xrightarrow{h_{\mathcal{V},n}} & \mathbf{N}_{\mathcal{V},n}(h\mathbb{A}) . \end{array}$$

By Lemma 9.146 the sequence

$$\mathbf{N}_{f^*\mathcal{V},n} \longrightarrow \mathbf{N}_{\mathcal{V},n} \longrightarrow P_n(\mathbf{B}G) = \mathbf{B}G$$

is a principle G-fibration sequence. In particular the first map is a normal covering with Galois group $G(\overline{K})$. Hence by Lemma 9.55 the sequence

$$\mathbf{N}_{f^*\mathcal{V},n}^{h\Gamma_K} \longrightarrow \mathbf{N}_{\mathcal{V},n}^{h\Gamma_K} \longrightarrow \mathbf{B}G^{h\Gamma_K} \qquad (\dagger)$$

is a fibration sequence.

From the spectral sequence that computes the homotopy groups of $\mathbf{B}G^{h\Gamma_K}$ we see that the connected component of the base point has vanishing homotopy groups in dimension greater than 1 and that its fundamental group is

$$H^0(\Gamma_K, G) = G^{\Gamma_K}$$

which we also denote by $G(K)$. Hence the trivial connected component of $\mathbf{B}G^{h\Gamma_K}$ is a model for $\mathbf{B}G(K)$.

Now from the fibration (†) we get the following sequence of sets/groups (obtained as the tail of the long exact sequence in homotopy groups):

$$G(K) \xrightarrow{\delta} \mathbf{N}_{f^*\mathcal{V},n}(hK) \xrightarrow{p_*^K} \mathbf{N}_{\mathcal{V},n}(hK) \xrightarrow{c_Y \circ \phi_*} H^1(K, G)$$

which is exact is the following sense:

1. A point $v \in \mathbf{N}_{\mathcal{V},n}(hK)$ can be lifted to $\mathbf{N}_{f^*\mathcal{V},n}(hK)$ if and only if $c_Y(\phi_*(v))$ is the neutral element.
2. If v does lift to $\mathbf{N}_{f^*\mathcal{V},n}(hK)$ then $G(K)$ acts transitively (although not necessarily faithfully) on $f_*^{-1}(v)$.

Note that we can do the same for $\mathbb{A}$ instead of K and get the following diagram:

$$\begin{array}{ccccccc}
G(K) & \xrightarrow{\delta^K} & \mathbf{N}_{f^*\mathcal{V},n}(hK) & \xrightarrow{p_*^K} & \mathbf{N}_{\mathcal{V},n}(hK) & \xrightarrow{c_Y \circ \phi_*} & H^1(K,G) \\
\downarrow{\scriptstyle \mathrm{loc}} & & \downarrow{\scriptstyle \mathrm{loc}} & & \downarrow{\scriptstyle \mathrm{loc}} & & \downarrow{\scriptstyle \mathrm{loc}} \\
G(\mathbb{A}) & \xrightarrow{\delta^{\mathbb{A}}} & \mathbf{N}_{f^*\mathcal{V},n}(h\mathbb{A}) & \xrightarrow{p_*^{\mathbb{A}}} & \mathbf{N}_{\mathcal{V},n}(h\mathbb{A}) & \xrightarrow{c_Y \circ \phi_*} & H^1(\mathbb{A},G) \\
\| & & \uparrow{\scriptstyle h_{f^*\mathcal{V},n}} & & \uparrow{\scriptstyle h_{\mathcal{V},n}} & & \| \\
G(\mathbb{A}) & \longrightarrow & Y(\mathbb{A}) & \xrightarrow{f} & X(\mathbb{A}) & \longrightarrow & H^1(\mathbb{A},G).
\end{array}$$

Note that Lemma 9.108 guarantees that the last row fits commutatively into the diagram. We now proceed to finish the proof of Proposition 9.144.

Let

$$A = \{v \in \mathbf{N}_{f^*\mathcal{V},n}(hK) \,|\mathrm{loc}\,(f_*^K(v)) = h_{\mathcal{V},n}(x_\nu)\}.$$

We wish to show that it is non-empty and finite. Let

$$r_{\mathcal{V},n} \in \mathbf{N}_{\mathcal{V},n}(hK)$$

be the image of r in $\mathbf{N}_{\mathcal{V},n}(hK)$. The map ϕ_* sends $r_{\mathcal{V},n}$ to the image of r in $\mathbf{N}_{\check{Y},n}(hK)$ and thus

$$c_Y(\phi_*(r_{\mathcal{V},n}))$$

is the neutral element. This means that there exists a

$$v \in \mathbf{N}_{f^*\mathcal{V},n}(hK)$$

such that

$$p_*^K(v) = r_{\mathcal{V},n}$$

and so we have that

$$\mathrm{loc}\,(p_*^K(v)) = \mathrm{loc}\,(r_{\mathcal{V},n}) = h_{\mathcal{V},n}(x_\nu)$$

implying that $A \neq \emptyset$.

On the other hand by Lemma 9.65 the set of all elements $s \in \mathbf{N}_{\mathcal{V},n}(hK)$ such that $\mathrm{loc}\,(s) = h_{\mathcal{V},n}(x_\nu)$ is finite and since $G(K)$ is finite so is A.

From the compactness of $G(\mathbb{A})$ and the commutativity of the diagram we see that the subspaces

$$B = \left(p_*^{\mathbb{A}}\right)^{-1}(h_{\mathcal{V},n}(x_\nu)) \subseteq \mathbf{N}_{f^*\mathcal{V},n}(h\mathbb{A})$$

and

$$f^{-1}(x_\nu) \subseteq Y(\mathbb{A})$$

are compact and that the continuous map (see Lemma 9.40)

$$h_{f^*\mathcal{V},n} : Y(\mathbb{A}) \longrightarrow \mathbf{N}_{f^*\mathcal{V},n}(h\mathbb{A})$$

restricts to a continuous map

$$h_B : f^{-1}(x_\nu) \longrightarrow B.$$

From the exactness of the rows we get that h_B is also **surjective**.

Now from commutativity we see that

$$\mathrm{loc}\,(A) \subseteq B$$

and furthermore if (y_ν) is in $f^{-1}(x_\nu)$ then $h_B((y_\nu)) \in B$ is rational if and only if $h_B((y_\nu)) \in \mathrm{loc}(A)$. Hence we get that

$$f^{-1}((x_\nu)) \cap Y(\mathbb{A})^{f^*\mathcal{V},n} = h_B^{-1}(\mathrm{loc}(A)).$$

Now since $\mathrm{loc}(A)$ is non-empty and finite and since h_B is continuous and surjective we have that $h_B^{-1}(\mathrm{loc}(A))$ is non-empty and closed in the compact space $f^{-1}((x_\nu))$. Hence

$$f^{-1}((x_\nu)) \cap Y(\mathbb{A})^{f^*\mathcal{V},n}$$

is non-empty and compact. □

This completes the proof of Proposition 9.141. □

9.12 Applications

In this section we shall present some applications of the theory developed in the paper.

Theorem 9.147 *Let K be number field and X, Y two smooth geometrically connected K-varieties. Then*

$$(X \times Y)(\mathbb{A})^{fin,\mathrm{Br}} = X(\mathbb{A})^{fin,\mathrm{Br}} \times Y(\mathbb{A})^{fin,\mathrm{Br}}.$$

Proof Let $\tau : K \hookrightarrow \mathbb{C}$ be an embedding. Then it is known that the pro-object $\acute{E}t(X)$ is isomorphic to the pro-finite completion of the topological space $(X \otimes_\tau \mathbb{C})(\mathbb{C})$. Since pro-finite completion commutes with products this means that the natural map

$$\acute{E}t(\overline{X} \times \overline{Y}) \longrightarrow \acute{E}t(\overline{X}) \times \acute{E}t(\overline{Y})$$

is an isomorphism. Recalling Proposition 9.19, the result then follows from Theorem 9.87. □

Theorem 9.148 *Let K be a number field and X a smooth geometrically connected variety over K. Assume further that*

$$\pi_2^{\acute{e}t}(\overline{X}) = 0,$$

then

$$X(\mathbb{A})^{fin} = X(\mathbb{A})^{fin,\mathrm{Br}}.$$

Proof In this case the map $\acute{E}t^2(\overline{X}) \longrightarrow \acute{E}t^1(\overline{X})$ is an isomorphism in $\mathrm{Pro\,Ho}\left(\mathrm{Set}^{\Delta^{op}}\right)$. By Proposition 9.19 and Theorem 9.87, this means that

$$X(\mathbb{A})^{h,1} = X(\mathbb{A})^{h,2}.$$

By Theorems 9.103 & 9.136, and Corollary 9.68 we then have

$$X(\mathbb{A})^{fin} = X(\mathbb{A})^{h,1} = X(\mathbb{A})^{h,2} = X(\mathbb{A})^{h} = X(\mathbb{A})^{fin,\mathrm{Br}}.$$

□

Corollary 9.149 *Let K be a number field and X a smooth geometrically connected proper variety over K such that $\pi_2^{\acute{e}t}(\overline{X}) = 0$. Then*

$$X(\mathbb{A})^{desc} = X(\mathbb{A})^{fin}.$$

Corollary 9.150 *Let C be a smooth curve over K, then*

$$C(\mathbb{A})^{fin} = C(\mathbb{A})^{fin,\mathrm{Br}}.$$

If C is also projective then

$$C(\mathbb{A})^{fin} = C(\mathbb{A})^{desc}.$$

Proof If $C \cong \mathbb{P}^1$ over $\overline{K}$ then

$$C(\mathbb{A})^{fin} = C(\mathbb{A}) = C(\mathbb{A})^{\mathrm{Br}} = C(\mathbb{A})^{fin,\mathrm{Br}}.$$

Otherwise, C is geometrically a $K(\pi, 1)$ and one can verify that $\pi_1(C(\mathbb{C}))$ is good in the sense of section 6 of [AMa69]. Hence

$$\pi_2^{\acute{e}t}(\overline{C}) = 0$$

and then the results follows from Theorem 9.148 and Corollary 9.149. □

Now let us recall the following definition (see [AMa69]):

Definition 9.151 Let $\overline{X}$ be a variety over an algebraically closed field $\overline{K}$. Let $\acute{E}t(\overline{X}) = \{X_\alpha\}_{\alpha \in I}$ be the étale homotopy type. Then we define the étale homology pro-groups of $\overline{X}$ to be

$$H_i^{\acute{e}t}(\overline{X}) = \{H_i(X_\alpha)\}_{\alpha \in I}.$$

Theorem 9.152 *Let X and K be as above. If $H_2^{\acute{e}t}(\overline{X}) = 0$ then*

$$X(\mathbb{A})^{\mathrm{Br}} = X(\mathbb{A})^{fin-ab}.$$

Proof In this case we get from 9.87 that

$$X(\mathbb{A})^{\mathbb{Z}h,2} = X(\mathbb{A})^{\mathbb{Z}h,1}$$

and so the result follows from Theorem 9.116, Corollary 9.68 and Theorem 9.103. □

Theorem 9.153 *Let X and K be as above and assume that $\pi_1^{\acute{e}t}(\overline{X})$ is abelian and that $\pi_2^{\acute{e}t}(\overline{X}) = 0$ (e.g. X is an abelian variety or an algebraic torus). Then*

$$\begin{array}{ccc} X(\mathbb{A})^{fin-ab} & = & X(\mathbb{A})^{fin} \\ \| & & \| \\ X(\mathbb{A})^{\mathrm{Br}} & = & X(\mathbb{A})^{fin,\mathrm{Br}}. \end{array}$$

Proof Applying Theorem 9.148 and using the fact that the fundamental group is abelian one gets

$$X(\mathbb{A})^{fin-ab} = X(\mathbb{A})^{fin} = X(\mathbb{A})^{fin,\mathrm{Br}}$$

and so in particular $X(\mathbb{A})^{fin-ab} = X(\mathbb{A})^{fin,\mathrm{Br}}$. Note that in general one has inclusions

$$X(\mathbb{A})^{fin,\mathrm{Br}} \subseteq X(\mathbb{A})^{\mathrm{Br}} \subseteq X(\mathbb{A})^{fin-ab}$$

and so in this case both inclusions are equalities. □

Theorem 9.154 *Let X and K be as above and assume that $\pi_1^{\acute{e}t}(\overline{X})$ is abelian and the Hurewicz map*

$$\pi^{\acute{e}t}(\overline{X})_2 \longrightarrow H_2^{\acute{e}t}(\overline{X})$$

is an isomorphism. Then

$$X(\mathbb{A})^{fin,Br} = X(\mathbb{A})^{\mathrm{Br}}.$$

Proof In this case we get from 9.87 that

$$X(\mathbb{A})^{h,2} = X(\mathbb{A})^{\mathbb{Z}h,2}$$

and so the result follows from Theorem 9.136, Corollary 9.68 and Theorem 9.116. □

References

[AMa69] Artin, M., Mazur, B., *Étale Homotopy*, Lecture Notes in Mathematics, 100, 1969.

[Bro73] Brown, K. S., Abstract homotopy theory and generalized sheaf cohomology, *Transactions of the American Mathematical Society*, 186, 1973, p. 419-458.

[BSc11] Barnea, I., Schlank, T. M., A projective model structure on pro simplicial sheaves, and the relative étale homotopy type, preprint `http://arxiv.org/pdf/1109.5477.pdf`.

[BSe64] Borel, A., Serre, J. -P., Théorèmes de finitude en cohomologie galoisienne, *Commentarii Mathematici Helvetici*, 39 (1), 1964, p. 111-164.

[CTS80] Colliot-Thélène, J. -L., Sansuc, J. -J., La descente sur les variétés rationnelles, *Journées de Géométrie Algébrique d'Angers*, Juillet 1979, p. 223-237.

[CTS87] Colliot-Thélène, J. -L., Sansuc, J. -J., La descente sur les variétés rationnelles II, *Duke Mathematical Journal*, 54 (2), 1987, p. 375-492.

[De09a] Demarche, C., Obstruction de descente et obstruction de Brauer-Manin étale, *Algebra and Number Theory*, 3 (2), 2009, p. 237-254.

[De09b] Demarche, C., Théorèmes de dualité pour les complexes de tores, preprint, 2009, `http://arxiv.org/abs/0906.3453v1`.

[Dol58] Dold, A., Homology of symmetric products and other functors of complexes, *Annals of Mathematics*, 68 (1), 1958.

[EHa76] Edwards, D. A., Hastings H. M. *Cech and Steenrod homotopy theories with applications to geometric topology*, Lecture Notes in Mathematics, 542, Springer Verlag, 1976.

[FIs07] Fausk, H., Isaksen, D. C., Model structures on pro-categories, *Homology, Homotopy and Applications*, 9 (1), 2007, p. 367-398.

[Fri82] Friedlander, E. M., *Étale homotopy of simplicial schemes*, Annals of Mathematics Studies, 104, Princeton University Press, 1982.

[GJa99] Goerss P. G, Jardine J. F. *Simplicial Homotopy Theory*, Progress in Mathematics, 174, Birkhäuser, 1999.

[Goe95] Goerss, P. G., Homotopy fixed points for Galois groups, *Contemporary Mathematics*, 181, p. 187-224.

[Har02] Harari, D., Groupes algébriques et points rationnels, *Mathematische Annalen*, 322 (4), 2002, p. 811-826.

[HSk02] Harari, D., Skorobogatov, A. N., Non-abelian cohomology and rational points, *Compositio Mathematica*, 130 (3), 2002, p. 241-273.

[HSz05] Harari, D., Szamuely, T., Arithmetic duality theorems for 1-motives, *Journal für die reine und angewandte Mathematik*, 578, 2005, p. 93-128.

[Isk71] Iskovskikh, V. A., A counterexample to the Hasse principle for systems of two quadratic forms in five variables, *Mathematical Notes*, 10 (3), 1971, p. 575-577.

[Jar87] Jardine, J. F., Simplicial Presheaves, *Journal of Pure and Applied Algebra*, 47, 1987, p. 35-87.

[Jos09] Jossen, P., *The Arithmetic of 1-motives*, Thesis, Central European University, 2009.

[Lie08] Lieblich, M., Twisted sheaves and the period-index problem, *Compositio Mathematica*, 144, 2008, p. 1-31.

[Lin40] Lind, C. -E., *Untersuchungen über die rationalen Punkte der ebenen kubischen Kurven vom Geschlecht Eins*, Thesis, University of Uppsala, 1940.

[Man70] Manin, Y. I., Le groupe de Brauer-Grothendieck en géométrie diophantienne, *Actes du Congrès International Des Mathématiciens*, Tome 1, 1970, p. 401-411.

[Mil06] Milne, J. -S., *Arithmetic Duality Theorems*, 2nd ed., BookSurge, LLC., 2006.

[Pal10] Pál, A., Homotopy sections and rational points on algebraic varieties, `http://arxiv.org/abs/1002.1731`, preprint, 2010.

[Poo10] Poonen, B., Insufficiency of the Brauer-Manin obstruction applied to étale covers, *Ann. of Math.* 171, 2010, p. 2157-2169.

[Qui09] Quick, G., Continuous group actions on profinite spaces, *Journal of Pure and Applied Algebra*, 215, 2011, p. 1024-1039.

[Rei42] Reichardt, H., Einige im Kleinen überall lösbare, im Grossen unlösbare diophantische Gleichungen, *Journal für die reine und angewandte Mathematik*, 184, 1942, p. 12-18.

[Sch09] Schlank, T. -M., On the Brauer-Manin obstruction applied to ramified covers, preprint, 2009, `http://arxiv.org/abs/0911.5728`.

[SGA4] Artin, M., Grothendieck, A., Verdier, J. L., *Théorie des Topos et Cohomologie Étale des Schémas*, Lecture Notes in Math. 270, Springer, 1972.

[Sko99] Skorobogatov, A. N., Beyond the Manin obstruction, *Inventiones Mathematicae*, 135 (2), 1999, p. 399-424.

[Sko01] Skorobogatov, A. N., *Torsors and Rational Points*, Cambridge University Press, 2001.

[Sko09] Skorobogatov, A. N., Descent obstruction is equivalent to étale Brauer-Manin obstruction, *Mathematische Annalen*, 344 (3), 2009, p. 501-510.

[Spa88] Spaltenstein, N., Resolutions of unbounded complexes, *Compositio Mathematica*, 65 (2), 1988, p. 121-154.

[Sto07] Stoll, M., Finite descent obstructions and rational points on curves, *Algebra and Number Theory*, 1 (4), 2007, p. 349-391.

10
Factorially graded rings of complexity one

Jürgen Hausen
Mathematisches Institut, Universität Tübingen

Elaine Herppich
Mathematisches Institut, Universität Tübingen

Abstract We consider finitely generated normal algebras over an algebraically closed field of characteristic zero that come with a complexity one grading by a finitely generated abelian group such that the conditions of a UFD are satisfied for homogeneous elements. Our main results describe these algebras in terms of generators and relations. We apply this to write down explicitly the possible Cox rings of normal complete rational varieties with a complexity one torus action.

10.1 Statement of the results

The subject of this note are finitely generated normal algebras $R = \oplus_K R_w$ over some algebraically closed field $\mathbb{K}$ of characteristic zero graded by a finitely generated abelian group K. We are interested in the following homogeneous version of a unique factorization domain: R is called *factorially (K-)graded* if every homogeneous nonzero nonunit is a product of K-primes, where a *K-prime* element is a homogeneous nonzero nonunit $f \in R$ with the property that whenever f divides a product of homogeneous elements, then it divides one of the factors. For free K, the properties factorial and factorially graded are equivalent [3], but for a K with torsion the latter is more general. Our motivation to study factorially graded algebras is that the Cox rings of algebraic varieties are of this type, see for example [2].

We focus on effective K-gradings of complexity one, i.e., the $w \in K$ with $R_w \neq 0$ generate K and K is of rank $\dim(R) - 1$. Moreover, we suppose that the grading is pointed in the sense that $R_0 = \mathbb{K}$ holds. The case of a free grading group K and hence factorial R was treated in [5, Section 1]. Here we settle the more general case of factorial gradings allowing torsion. Our results enable us to write down explicitly the possible Cox rings of normal complete rational varieties with a complexity one torus action. This complements [6], where the Cox ring of a given variety was computed in terms of the torus action.

In order to state our results, let us fix the notation. For $r \geq 1$, let $A = (a_0, \ldots, a_r)$ be a sequence of vectors $a_i = (b_i, c_i)$ in $\mathbb{K}^2$ such that any pair (a_i, a_k) with $k \neq i$ is linearly independent, $\mathfrak{n} = (n_0, \ldots, n_r)$ a sequence of positive integers and $L = (l_{ij})$ a family of positive integers, where $0 \leq i \leq r$ and $1 \leq j \leq n_i$. For every $0 \leq i \leq r$, define a monomial

$$T_i^{l_i} \;:=\; T_{i1}^{l_{i1}} \cdots T_{in_i}^{l_{in_i}} \;\in\; S \;:=\; \mathbb{K}[T_{ij};\, 0 \leq i \leq r,\, 1 \leq j \leq n_i].$$

Moreover, for any two indices $0 \leq i, j \leq r$, set $\alpha_{ij} := \det(a_i, a_j) = b_i c_j - b_j c_i$ and for any three indices $0 \leq i < j < k \leq r$ define a trinomial

$$g_{i,j,k} \;:=\; \alpha_{jk} T_i^{l_i} \;+\; \alpha_{ki} T_j^{l_j} \;+\; \alpha_{ij} T_k^{l_k} \;\in\; S.$$

We define a grading of S by an abelian group K such that all the $g_{i,j,k}$ become homogeneous of the same degree. For this, consider the free abelian groups

$$F \;:=\; \bigoplus_{i=0}^{r} \bigoplus_{j=1}^{n_i} \mathbb{Z} \cdot f_{ij} \;\cong\; \mathbb{Z}^n, \qquad\qquad N \;:=\; \mathbb{Z}^r,$$

where we set $n := n_0 + \ldots + n_r$. Set $l_i := (l_{i1}, \ldots, l_{in_i})$. Then we have a linear map $P\colon F \to N$ defined by the $r \times n$ matrix

$$P = \begin{pmatrix} -l_0 & l_1 & \ldots & 0 \\ \vdots & \vdots & \ddots & \vdots \\ -l_0 & 0 & \ldots & l_r \end{pmatrix}.$$

Let $P^*\colon M \to E$ be the dual map, set $K := E/P^*(M)$ and let $Q\colon E \to K$ be the projection. Let (e_{ij}) be the dual basis of (f_{ij}) and define a

K-grading on S by $\deg(T_{ij}) := Q(e_{ij})$. Then all $g_{i,j,k}$ are homogeneous of the same degree and we obtain a K-graded factor algebra

$$R(A, \mathfrak{n}, L) := \mathbb{K}[T_{ij};\ 0 \le i \le r,\ 1 \le j \le n_i] \ /\ \langle g_{i,i+1,i+2};\ 0 \le i \le r-2 \rangle.$$

We say that the triple $(A, \mathfrak{n}, L)$ is *sincere*, if $r \ge 2$ and $n_i l_{ij} > 1$ for all i, j hold; this ensures that there exist in fact relations $g_{i,j,k}$ and none of these relations contains a linear term. Note that for $r = 1$ we obtain the diagonal complexity one gradings of the polynomial ring S.

Theorem 10.1 *Let $(A, \mathfrak{n}, L)$ be any triple as above.*

(i) *The algebra $R(A, \mathfrak{n}, L)$ is factorially K-graded; the K-grading is effective, pointed and of complexity one.*
(ii) *Suppose that $(A, \mathfrak{n}, L)$ is sincere. Then $R(A, \mathfrak{n}, L)$ is factorial if and only if the group K is torsion free.*

The second part of this theorem provides examples of factorially graded algebras which are not factorial; note that K is torsion free if and only if the numbers $l_i := \gcd(l_{i1}, \ldots, l_{in_i})$ are pairwise coprime.

Example 10.2 Let A consist of the vectors $(1,0)$, $(1,1)$ and $(0,1)$, take $\mathfrak{n} = (1,1,1)$ and take the family L given by $l_{01} = l_{11} = l_{21} = 2$. Then the matrix

$$\begin{pmatrix} -2 & 2 & 0 \\ -2 & 0 & 2 \end{pmatrix}$$

describes the map $P\colon \mathbb{Z}^3 \to \mathbb{Z}^2$. Thus the grading group is $K = \mathbb{Z} \oplus \mathbb{Z}/2\mathbb{Z} \oplus \mathbb{Z}/2\mathbb{Z}$. Concretely this grading can be realized as

$$\deg(T_{01}) = (1, \overline{0}, \overline{0}), \qquad \deg(T_{11}) = (1, \overline{1}, \overline{0}), \qquad \deg(T_{21}) = (1, \overline{0}, \overline{1}).$$

The associated algebra $R(A, \mathfrak{n}, L)$ is factorially K-graded but not factorial. It is explicitly given by

$$R(A, \mathfrak{n}, L) = \mathbb{K}[T_{01}, T_{11}, T_{21}]/\langle T_{01}^2 - T_{11}^2 + T_{21}^2 \rangle.$$

Obvious further examples of algebras with an effective pointed factorial grading are $R(A, \mathfrak{n}, L)[S_1, \ldots, S_m]$, graded by $K \times \mathbb{Z}^m$ via $\deg(S_i) := e_i$, where $e_i \in \mathbb{Z}^m$ denotes the i-th canonical basis vector. The methods of [6, Section 3] apply directly to our situation and show that there are no other examples, i.e., we arrive at the following.

Theorem 10.3 *Every finitely generated normal $\mathbb{K}$-algebra with an effective, pointed, factorial grading of complexity one is isomorphic to some $R(A, \mathfrak{n}, L)[S_1, \ldots, S_m]$.*

We turn to Cox rings. Roughly speaking, the Cox ring of a complete normal variety X with finitely generated divisor class group $\mathrm{Cl}(X)$ is given as

$$\mathcal{R}(X) := \bigoplus_{\mathrm{Cl}(X)} \Gamma(X, \mathcal{O}_X(D)),$$

see [2] for the details of the precise definition. As mentioned, our aim is to write down all possible Cox rings of normal rational complete varieties with a complexity one torus action. They are obtained from $R(A, \mathfrak{n}, L)[S_1, \ldots, S_m]$ by coarsening the $(K \times \mathbb{Z}^m)$-grading as follows. Let $0 < s < n + m - r$, consider an integral $s \times n$ matrix d, an integral $s \times m$ matrix d' and the block matrix

$$\dot{P} = \begin{pmatrix} P & 0 \\ d & d' \end{pmatrix},$$

where d and d' are chosen in such a manner that the columns of the matrix $\dot{P}$ are pairwise different primitive vectors in $\mathbb{Z}^{r+s}$ which generate $\mathbb{Q}^{r+s}$ as a cone. Consider the linear map of lattices $\dot{P} \colon \ddot{F} \to \dot{N}$, where

$$\ddot{F} \;:=\; F \oplus \mathbb{Z}f_1 \oplus \ldots \oplus \mathbb{Z}f_m, \qquad \dot{N} \;:=\; \mathbb{Z}^{r+s}.$$

Let $\dot{P}^* \colon \dot{M} \to \ddot{E}$ be the dual map and $\dot{Q} \colon \ddot{E} \to \dot{K}$ the projection, where $\dot{K} := \ddot{E}/\dot{P}^*(\dot{M})$. Denoting by e_{ij}, e_k the dual basis to f_{ij}, f_k, we obtain a $\dot{K}$-grading of $R(A, \mathfrak{n}, L)[S_1, \ldots, S_m]$ by setting

$$\deg(T_{ij}) \;:=\; \dot{Q}(e_{ij}), \qquad \deg(S_k) \;:=\; \dot{Q}(e_k).$$

Theorem 10.4 *In the above notation, the following holds.*

(i) *The $\dot{K}$-grading of $R(A, \mathfrak{n}, L)[S_1, \ldots, S_m]$ is effective, pointed and factorial. Moreover, T_{ij}, S_k define pairwise nonassociated $\dot{K}$-prime generators.*

(ii) *The $\dot{K}$-graded algebra $R(A, \mathfrak{n}, L)[S_1, \ldots, S_m]$ is the Cox ring of a $\mathbb{Q}$-factorial rational projective variety with a complexity one torus action.*

Theorem 10.5 *Let X be a normal rational complete variety with a torus action of complexity one. Then the Cox ring of X is isomorphic as a graded ring to some $R(A, \mathfrak{n}, L)[S_1, \ldots, S_m]$ with a $\dot{K}$-grading as constructed above.*

10.2 Proof of the results

A very first observation lists basic properties of the K-grading of $R(A, \mathfrak{n}, L)$. We denote by $\mathbb{T}^n := (\mathbb{K}^*)^n$ the standard n-torus. Moreover, we work with the diagonal action of the quasitorus $H := \operatorname{Spec} \mathbb{K}[K]$ on $\mathbb{K}^n$ given by $t \cdot z = (\chi_{ij}(t) z_{ij})$, where $\chi_{ij} \in \mathbb{X}(H)$ is the character corresponding to $Q(e_{ij}) = \deg(T_{ij})$. By definition, this action stabilizes the zero set

$$\widetilde{X} \; := \; V(\mathbb{K}^n;\, g_{i,i+1,i+2},\, 0 \leq i \leq r-2) \; \subseteq \; \mathbb{K}^n.$$

Proposition 10.6 *The algebra $R(A, \mathfrak{n}, L)$ is normal, the K-grading is pointed, effective and of complexity one.*

Proof Effectivity of the K-grading is given by construction, because the degrees $\deg(T_{ij}) = Q(e_{ij})$ generate K. From [5, Prop. 1.2], we infer that $R(A, \mathfrak{n}, L)$ is a normal complete intersection. Thus, we have

$$\dim(R(A, \mathfrak{n}, L)) \; = \; n - r + 1 \; = \; \dim(\ker(P)) + 1$$

which means that the K-grading is of complexity one. Now consider the action of the quasitorus $H := \operatorname{Spec} \mathbb{K}[K]$ on $\mathbb{K}^n$ given by the K-grading. Note that $H \subseteq \mathbb{T}^n$ is the kernel of the homomorphism of tori

$$\mathbb{T}^n \; \to \; \mathbb{T}^r, \qquad (t_0, \ldots, t_r) \; \mapsto \; \left(\frac{t_1^{l_1}}{t_0^{l_0}}, \ldots, \frac{t_r^{l_r}}{t_0^{l_0}} \right).$$

The set $\widetilde{X} \subseteq \mathbb{K}^n$ of common zeroes of all the $g_{i,i+1,i+2}$ is H-invariant and thus it is invariant under the one-parameter subgroup of H given by

$$\mathbb{K}^* \; \to H, \qquad t \; \mapsto \; (t^{\zeta_{ij}}), \qquad\qquad \zeta_{ij} \; := \; n_i^{-1} l_{ij}^{-1} \prod_k n_k \prod_{k,m} l_{km}.$$

Since all ζ_{ij} are positive, any orbit of this one-parameter subgroup in $\mathbb{K}^n$ has the origin in its closure. Consequently, every H-invariant function on $\widetilde{X}$ is constant. This shows $R(A, \mathfrak{n}, L)_0 = \mathbb{K}$. □

We say that a Weil divisor on $\widetilde{X}$ is *H-prime* if it is non-zero, has only multiplicities zero or one and the group H permutes transitively the prime components with multiplicity one. Note that the divisor $\operatorname{div}(f)$ of a homogeneous function $f \in R(A, \mathfrak{n}, L)$ on $\widetilde{X}$ is H-prime if and only if f is K-prime [4, Prop. 3.2]. The following is an essential ingredient of the proof.

Proposition 10.7 *Regard the variables T_{ij} as regular functions on $\widetilde{X}$.*

(i) *The divisors of the T_{ij} on $\widetilde{X}$ are H-prime and pairwise different. In particular, the T_{ij} define pairwise nonassociated K-prime elements in $R(A, \mathfrak{n}, L)$.*

(ii) *If the ring $R(A, \mathfrak{n}, L)$ is factorial and $n_i l_{ij} > 1$ holds, then the divisor of T_{ij} on $\widetilde{X}$ is even prime.*

Proof For (i), we exemplarily show that the divisor of T_{01} is H-prime. First note that by [5, Lemma 1.3] the zero set $V(\widetilde{X}; T_{01})$ is described in $\mathbb{K}^n$ by the equations

$$T_{01} \;=\; 0, \qquad \alpha_{s+1\,0} T_s^{l_s} + \alpha_{0s} T_{s+1}^{l_{s+1}} \;=\; 0, \quad 1 \le s \le r-1. \qquad (10.1)$$

Let $h \in S$ denote the product of all T_{ij} with $(i, j) \neq (0, 1)$. Then, in $\mathbb{K}^n_h$, the above equations are equivalent to

$$T_{01} \;=\; 0, \qquad -\frac{\alpha_{s+1\,0} T_s^{l_s}}{\alpha_{0s} T_{s+1}^{l_{s+1}}} \;=\; 1, \quad 1 \le s \le r-1.$$

Now, choose a point $\widetilde{x} \in \mathbb{K}^n_h$ satisfying these equations. Then $\widetilde{x}_{01}$ is the only vanishing coordinate of $\widetilde{x}$. Any other such point is of the form

$$(0,\, t_{02}\widetilde{x}_{02},\, \ldots,\, t_{rn_r}\widetilde{x}_{rn_r}), \qquad t_{ij} \in \mathbb{K}^*,\; t_s^{l_s} = t_{s+1}^{l_{s+1}},\; 1 \le s \le r-1.$$

Setting $t_{01} := t_{02}^{-l_{02}} \cdots t_{0n_0}^{-l_{0n_0}} t_1^{l_1}$, we obtain an element $t = (t_{ij}) \in H$ such that the above point equals $t \cdot \widetilde{x}$. This consideration shows

$$V(\widetilde{X}_h; T_{01}) = H \cdot \widetilde{x}.$$

Using [5, Lemma 1.4], we see that $V(\widetilde{X}; T_{01}, T_{ij})$ is of codimension at least two in $\widetilde{X}$ whenever $(i, j) \neq (0, 1)$. This allows to conclude

$$V(\widetilde{X}; T_{01}) = \overline{H \cdot \widetilde{x}}.$$

Thus, to obtain that T_{01} defines an H-prime divisor on $\widetilde{X}$, we only need that the equations (10.1) define a radical ideal. This in turn follows from the fact that their Jacobian at the point $\widetilde{x} \in V(\widetilde{X}; T_{01})$ is of full rank.

To verify (ii), let $R(A, \mathfrak{n}, L)$ be factorial. Assume that the divisor of T_{ij} is not prime. Then we have $T_{ij} = h_1 \cdots h_s$ with prime elements $h_l \in R(A, \mathfrak{n}, L)$. Consider their decomposition into homogeneous parts

$$h_l = \sum_{w \in K} h_{l,w}.$$

Plugging this into the product $h_1 \cdots h_s$ shows that $\deg(T_{ij})$ is a positive combination of some $\deg(T_{kl})$ with $(k, l) \neq (i, j)$. Thus, there is a vector

$(c_{kl}) \in \ker(Q) \subseteq E$ with $c_{ij} = 1$ and $c_{kl} \leq 0$ whenever $(k,l) \neq (i,j)$. Since $\ker(Q)$ is spanned by the rows of P, we must have $n_i = 1$ and $l_{ij} = 1$, a contradiction to our assumptions. $\square$

We come to the main step of the proof, the construction of a T-variety having $R(A, \mathfrak{n}, L)[S_1, \ldots, S_m]$ as its Cox ring. We will obtain X as a subvariety of a toric variety Z and the construction of Z is performed in terms of fans. As before, consider the lattice

$$\ddot{F} := F \oplus \mathbb{Z}f_1 \oplus \ldots \oplus \mathbb{Z}f_m.$$

Let $\ddot{\Sigma}$ be the fan in $\ddot{F}$ having the rays $\ddot{\varrho}_{ij}$ and $\ddot{\varrho}_k$ through the basis vectors f_{ij} and f_k as its maximal cones. Let $0 < s < n + m - r$, choose an integral $s \times n$ matrix d and an integral $s \times m$ matrix d'. Consider the (block) matrices

$$\begin{pmatrix} P & 0 \\ d & d' \end{pmatrix}, \qquad \begin{pmatrix} P & 0 \end{pmatrix}$$

and suppose that the columns of the first one are primitive, pairwise different and generate $\dot{N}_{\mathbb{Q}}$ as a cone, where $\dot{N} := \mathbb{Z}^{r+s}$. With $N := \mathbb{Z}^r$, we have the projection $B \colon \dot{N} \to N$ onto the first r coordinates and the linear maps $\dot{P} \colon \ddot{F} \to \dot{N}$ and $\ddot{P} \colon \ddot{F} \to N$ respectively given by the above matrices.

Let $\dot{\Sigma}$ be the fan in $\dot{N}$ with the rays $\dot{\varrho}_{ij} := \dot{P}(\ddot{\varrho}_{ij})$ and $\dot{\varrho}_k := \dot{P}(\ddot{\varrho}_k)$ as its maximal cones. The ray ϱ_{ij} through the ij-th column of P is given in terms of the canonical basis vectors $v_1, \ldots, v_r$ in $N = \mathbb{Z}^r$ as

$$\varrho_{ij} = \mathbb{Q}_{\geq 0} v_i, \quad 1 \leq i \leq r, \qquad \varrho_{0j} = -\mathbb{Q}_{\geq 0}(v_1 + \ldots + v_r).$$

For fixed i all ϱ_{ij} are equal to each other; we list them nevertheless all separately in a system of fans Σ having the zero cone as the common gluing data; see [1] for the formal definition of this concept. Finally, we have the fan Δ in $\mathbb{Z}^r$ with the rays $\mathbb{Q}_{\geq 0} v_i$ and $-\mathbb{Q}_{\geq 0}(v_1 + \ldots + v_r)$ as its maximal cones.

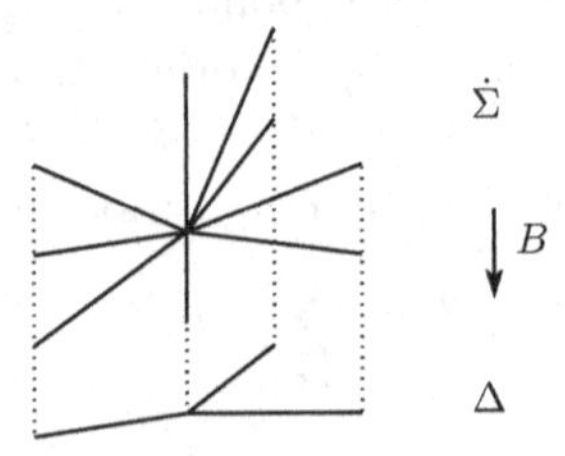

The toric variety $\ddot{Z}$ associated to $\ddot{\Sigma}$ has $\operatorname{Spec} \mathbb{K}[\ddot{E}] \cong \mathbb{T}^{n+m}$ as its acting torus, where $\ddot{E}$ is the dual lattice of $\ddot{F}$. The fan $\dot{\Sigma}$ in $\dot{N}$ defines a toric variety $\dot{Z}$ and the system of fans Σ defines a toric prevariety Z. The toric prime divisors corresponding to the rays $\ddot{\varrho}_{ij}, \ddot{\varrho}_k \in \ddot{\Sigma}$, $\dot{\varrho}_{ij}, \dot{\varrho}_k \in \dot{\Sigma}$ and $\varrho_{ij} \in \Sigma$, are denoted as

$$\ddot{D}_{ij}, \ddot{D}_k \subseteq \ddot{Z}, \qquad \dot{D}_{ij}, \dot{D}_k \subseteq \dot{Z}, \qquad D_{ij} \subseteq Z.$$

The toric variety associated to Δ is the open subset $\mathbb{P}_r^{(1)} \subseteq \mathbb{P}_r$ of the projective space obtained by removing all toric orbits of codimension at least two. The maps $\dot{P}$ and $\ddot{P}$ define toric morphisms $\dot{\pi}\colon \ddot{Z} \to \dot{Z}$ and $\ddot{\pi}\colon \ddot{Z} \to Z$. Moreover, $B\colon \dot{N} \to N$ defines a toric morphism $\beta\colon \dot{Z} \to Z$ and the identity $\mathbb{Z}^r \to \mathbb{Z}^r$ defines a toric morphism $\kappa\colon Z \to \mathbb{P}_r^{(1)}$. These morphisms fit into the commutative diagram

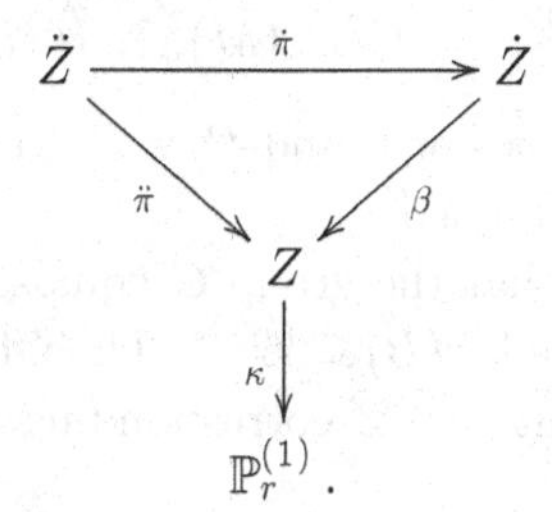

Note that $\kappa\colon Z \to \mathbb{P}_r^{(1)}$ is a local isomorphism which, for fixed i, identifies all the divisors D_{ij} with $1 \le j \le n_i$. Let $\dot{H} \subseteq \mathbb{T}^{n+m}$ and $\ddot{H} \subseteq \mathbb{T}^{n+m}$ be the kernels of the toric morphisms $\dot{\pi}\colon \ddot{Z} \to \dot{Z}$ and $\ddot{\pi}\colon \ddot{Z} \to Z$ respectively. Here are the basic features of our construction.

Proposition 10.8 *In the above notation, the following statements hold.*

(i) *With $\ddot{Z}_0 := \ddot{Z} \setminus (\ddot{D}_1 \cup \ldots \cup \ddot{D}_m)$, the restriction $\ddot{\pi}\colon \ddot{Z}_0 \to Z$ is a geometric quotient for the action of $\ddot{H}$ on $\ddot{Z}_0$.*
(ii) *The quasitorus $\dot{H}$ acts freely on $\ddot{Z}$ and $\dot{\pi}\colon \ddot{Z} \to \dot{Z}$ is the geometric quotient for this action.*
(iii) *The factor group $G := \ddot{H}/\dot{H}$ is isomorphic to $\mathbb{T}^s$ and it acts canonically on $\dot{Z}$.*
(iv) *The G-action on $\dot{Z}$ has infinite isotropy groups along $\dot{D}_1, \ldots, \dot{D}_m$ and isotropy groups of order l_{ij} along $\dot{D}_{ij}$.*
(v) *With $\dot{Z}_0 := \dot{Z} \setminus (\dot{D}_1 \cup \ldots \cup \dot{D}_m)$, the restriction $\beta\colon \dot{Z}_0 \to Z$ is a geometric quotient for the action of G on $\dot{Z}$.*

Proof The fact that $\ddot{\pi}\colon \ddot{Z}_0 \to Z$ and $\dot{\pi}\colon \ddot{Z} \to \dot{Z}$ are geometric quotients is due to known characterizations of these notions in terms of (systems of) fans, see e.g. [1]. As a consequence, $\beta\colon \dot{Z}_0 \to Z$ is a geometric quotient for the induced action of $G = \ddot{H}/\dot{H}$.

We verify the remaining part of (i). According to [2, Prop. II.1.4.2], the isotropy group of $\dot{H} = \ker(\dot{\pi})$ at a distinguished point $z_{\ddot{\varrho}} \in \ddot{Z}$ has character group isomorphic to

$$\ker(\dot{P}) \cap \mathrm{lin}_{\mathbb{Q}}(\ddot{\varrho}) \;\oplus\; (\dot{P}(\mathrm{lin}_{\mathbb{Q}}(\ddot{\varrho})) \cap \dot{N})/\dot{P}(\mathrm{lin}_{\mathbb{Q}}(\ddot{\varrho}) \cap \ddot{F}).$$

By the choice of d and d', the map $\dot{P}$ sends the primitive generators of the rays of $\ddot{\Sigma}$ to the primitive generators of the rays of $\dot{\Sigma}$. Thus we obtain that the isotropy of $z_{\ddot{\varrho}_{ij}}$ and $z_{\ddot{\varrho}_k}$ are all trivial.

We turn to (iii). With the dual lattices $\dot{M}$ of $\dot{N}$ and M of N, we obtain the character groups of $\dot{H}$ and $\ddot{H}$ and the factor group $\ddot{H}/\dot{H}$ as

$$\mathbb{X}(\dot{H}) \;=\; \ddot{E}/\dot{P}^*(\dot{M}),\; \mathbb{X}(\ddot{H}) \;=\; \ddot{E}/\ddot{P}^*(M),\; \mathbb{X}(\ddot{H}/\dot{H}) \;=\; \dot{P}^*(\dot{M})/\ddot{P}^*(M).$$

By definition of the matrices of $\dot{P}$ and $\ddot{P}$, we have $\ddot{P}^*(\dot{M})/\ddot{P}^*(M) \cong \mathbb{Z}^s$. This implies $G \cong \mathbb{T}^s$ as claimed.

To see (iv), note first that the group G equals $\ker(\beta)$ and hence corresponds to the sublattice $\ker(B) \subseteq \mathbb{Z}^{r+s}$. Thus, the isotropy group $G_{z_{\dot{\varrho}}}$ for the distinguished point $z_{\dot{\varrho}} \in \dot{Z}$ corresponding to $\dot{\varrho} \in \dot{\Sigma}$ has character group isomorphic to

$$\ker(B) \cap \mathrm{lin}_{\mathbb{Q}}(\dot{\varrho}) \;\oplus\; (B(\mathrm{lin}_{\mathbb{Q}}(\dot{\varrho})) \cap N)/B(\mathrm{lin}_{\mathbb{Q}}(\dot{\varrho}) \cap \dot{N}).$$

Consequently, for $\dot{\varrho} = \dot{\varrho}_k$ the isotropy group $G_{z_{\dot{\varrho}}}$ is infinite and for $\dot{\varrho} = \dot{\varrho}_{ij}$ it is of order l_{ij}. □

Now we come to the construction of the embedded variety. Let $\delta \subseteq \ddot{F}_{\mathbb{Q}}$ be the orthant generated by the basis vectors f_{ij} and f_k. The associated affine toric variety $\overline{Z} \cong \mathbb{K}^{n+m}$ is the spectrum of the polynomial ring

$$\mathbb{K}[\ddot{E} \cap \delta^\vee] = \mathbb{K}[T_{ij}, S_k;\; 0 \le i \le r,\; 1 \le j \le n_i,\; 1 \le k \le m] = S[S_1, \ldots, S_m].$$

Moreover, $\overline{Z}$ contains $\ddot{Z}$ as an open toric subvariety and the complement $\overline{Z} \setminus \ddot{Z}$ is the union of all toric orbits of codimension at least two. Regarding the trinomials $g_{i,j,k} \in S$ as elements of the larger polynomial ring $S[S_1, \ldots, S_m]$, we obtain an $\ddot{H}$-invariant subvariety

$$\overline{X} \;:=\; V(g_{i,i+1,i+2};\; 0 \le i \le r-2) \;\subseteq\; \overline{Z}.$$

Proposition 10.9 *Set $\ddot{X} := \overline{X} \cap \ddot{Z}$. Consider the images $\dot{X} := \dot{\pi}(\ddot{X}) \subseteq \dot{Z}$ and $Y := \beta(\dot{X}) \subseteq Z$.*

(i) *$\dot{X} \subseteq \dot{Z}$ is a normal closed G-invariant $s+1$ dimensional variety, $Y \subseteq Z$ is a closed non-separated curve and $\kappa(Y) \subseteq \mathbb{P}_r$ is a line.*

(ii) *The intersection* $\dot{C}_{ij} := \dot{X} \cap \dot{D}_{ij}$ *with the toric divisor* $\dot{D}_{ij} \subseteq \dot{Z}$ *is a single* G*-orbit with isotropy group of order* l_{ij}.

(iii) *The intersection* $\dot{C}_k := \dot{X} \cap \dot{D}_k$ *with the toric divisor* $\dot{D}_k \subseteq \dot{Z}$ *is a smooth rational prime divisor consisting of points with infinite* G*-isotropy.*

(iv) *For every point* $x \in \dot{X}$ *not belonging to some* $\dot{C}_{ij}$ *or to some* $\dot{C}_k$, *the isotropy group* G_x *is trivial.*

(v) *The variety* $\dot{X}$ *satisfies* $\Gamma(\dot{X}, \mathcal{O}) = \mathbb{K}$, *its divisor class group and Cox ring are given by*

$$\mathrm{Cl}(\dot{X}) \cong \dot{K}, \qquad \mathcal{R}(\dot{X}) \cong R(A, \mathfrak{n}, L)[S_1, \ldots, S_m].$$

The variables T_{ij} *and* S_k *define pairwise nonassociated* $\dot{K}$*-prime elements in* $R(A, \mathfrak{n}, L)[S_1, \ldots, S_m]$.

(vi) *There is a* G*-equivariant completion* $\dot{X} \subseteq X$ *with a* $\mathbb{Q}$*-factorial projective variety* X *such that* $\mathcal{R}(X) = \mathcal{R}(\dot{X})$ *holds.*

Proof By the definition of $\ddot{P}$ and $\ddot{H}$, the closed subvariety $\overline{X} \subseteq \overline{Z}$ is invariant under the action of $\ddot{H}$. In particular, $\ddot{X}$ is $\dot{H}$-invariant and thus the image $\dot{X} := \dot{\pi}(\ddot{X})$ under the quotient map is closed as well. Moreover, the dimension of $\dot{X}$ equals $\dim(\ddot{X}/\dot{H}) = s+1$. Analogously we obtain closedness of $Y = \ddot{\pi}(\ddot{X})$. The image $\kappa(Y) = \kappa(\ddot{\pi}(\ddot{X}))$ is given in $\mathbb{P}_r$ by the equations

$$\alpha_{jk}U_i + \alpha_{ki}U_j + \alpha_{ij}U_k = 0$$

with the variables $U_0, \ldots, U_r$ on $\mathbb{P}_r$ corresponding to the toric divisors given by the rays $\mathbb{Q}_{\geq 0}v_i$ and $-\mathbb{Q}_{\geq 0}(v_0 + \ldots + v_{r-1})$ of Δ; to see this, use that pulling back the above equations via $\kappa \circ \ddot{\pi}$ gives the defining equations for $\ddot{X}$. Consequently $\kappa(Y)$ is a projective line. This shows (i).

We turn to (ii). According to Proposition 10.7, the intersection $\ddot{X} \cap \ddot{D}_{ij}$ is a single $\ddot{H}$-orbit. Since $\dot{\pi}\colon \ddot{X} \to \dot{X}$ is a geometric quotient for the $\dot{H}$-action, we conclude that $\dot{C}_{ij} = \dot{\pi}(\ddot{D}_{ij})$ is a single G-orbit. Moreover, since $\dot{H}$ acts freely, the isotropy group of $G = \ddot{H}/\dot{H}$ along $\dot{C}_{ij}$ equals that of $\ddot{H}$ along $\ddot{D}_{ij}$ which, by Proposition 10.8 (iv), is of order l_{ij}.

For (iii) note first that the restrictions $\beta\colon \dot{D}_k \to Z$ and are isomorphisms onto the acting torus of Z. Moreover, the restricting κ gives an isomorphism of the acting tori of Z and $\mathbb{P}_r$. Consequently, β maps $\dot{C}_k$ isomorphically onto the intersection of the line Y with the acting torus of $\mathbb{P}_r$. Thus, $\dot{C}_k$ is a smooth rational curve. Proposition 10.8 (iv) ensures that $\dot{C}_k$ consists of fixed points. Assertion (iv) is clear.

We prove (v). From Proposition 10.6, we infer $\Gamma(\ddot{X}, \mathcal{O})^{\dot{H}} = \mathbb{K}$ which implies $\Gamma(\dot{X}, \mathcal{O}) = \mathbb{K}$. The next step is to establish a surjection $\dot{K} \to \mathrm{Cl}(\dot{X})$, where $\dot{K} := \ddot{E}/\dot{P}^*(\dot{M})$ is the character group of $\dot{H}$. Consider the push forward $\dot{\pi}_*$ from the $\dot{H}$-invariant Weil divisors on $\ddot{X}$ to the Weil divisors on $\dot{X}$ sending $\ddot{D}$ to $\dot{\pi}(\ddot{D})$. For every $\dot{w} \in \dot{K}$, we fix a $\dot{w}$-homogeneous rational function $f_{\dot{w}} \in \mathbb{K}(\dot{X})$ and define a map

$$\mu\colon \dot{K} \to \mathrm{Cl}(\dot{X}), \qquad \dot{w} \mapsto [\dot{\pi}_* \mathrm{div}(f_{\dot{w}})].$$

One directly checks that this does not depend on the choice of the $f_{\dot{w}}$ and thus is a well defined homomorphism. In order to see that it is surjective, note that due to Proposition 10.6, we obtain $\dot{C}_{ij}$ as $\dot{\pi}_*\mathrm{div}(T_{ij})$ and $\dot{C}_k$ as $\dot{\pi}_*\mathrm{div}(T_k)$. The claim then follows from the observation that removing all $\dot{C}_{ij}$ and $\dot{C}_k$ from $\dot{X}$ leaves the set $\dot{X} \cap \mathbb{T}^{r+s}$ which is isomorphic to $V \times \mathbb{T}^r$ with a proper open subset $V \subseteq \kappa(Y)$ and hence has trivial divisor class group.

Now [6, Theorem 1.3] shows that the Cox ring of $\dot{X}$ is $R(A, \mathfrak{n}, L)[S_1, \ldots, S_m]$ with the $\mathrm{Cl}(\dot{X})$-grading given by $\deg(T_{ij}) = [\dot{C}_{ij}]$ and $\deg(S_k) = [\dot{C}_k]$. Consequently, $R(A, \mathfrak{n}, L)[S_1, \ldots, S_m]$ is factorially $\mathrm{Cl}(\dot{X})$-graded and thus also the finer $\dot{K}$-grading is factorial. Since $\dot{H}$ acts freely on $\ddot{X}$, we can conclude $\mathrm{Cl}(\dot{X}) = \dot{K}$.

Finally, we construct a completion of $\dot{X} \subseteq X$ as wanted in (vi). Choose any simplicial projective fan $\dot{\Sigma}'$ in $\dot{N}$ having the same rays as Σ, see [7, Corollary 3.8]. The associated toric variety $\dot{Z}'$ is projective and it is the good quotient of an open toric subset $\ddot{Z}' \subseteq \overline{Z}$ by the action of $\dot{H}$. The closure X of $\dot{X}$ in $\dot{Z}'$ is projective and, as the good quotient of the normal variety $\overline{X} \cap \ddot{Z}'$, it is normal. By Proposition 10.7, the complement $X \setminus \dot{X}$ is of codimension at least two, which gives $\mathcal{R}(X) = \mathcal{R}(\dot{X})$. From [4, Cor. 4.13] we infer that X is $\mathbb{Q}$-factorial. □

Remark 10.10 We may realize any given $R(A, \mathfrak{n}, L)$ as a subring of the Cox ring of a surface: For every $l_i = (l_{i1}, \ldots, l_{in_i})$ choose a tuple $d_i = (d_{i1}, \ldots, d_{in_i})$ of positive integers with $\gcd(l_{ij}, d_{ij}) = 1$ and $d_{i1}/l_{i1} < \ldots < d_{in_i}/l_{in_i}$. Then take

$$\dot{P} := \begin{pmatrix} P & 0 & 0 \\ d & 1 & -1 \end{pmatrix}.$$

We are ready to verify the main results. As mentioned, Theorem 10.3 follows from [6, Sec. 3]. Moreover, the statements of Theorem 10.4 are contained in Proposition 10.9.

Proof of Theorem 10.1 According to Proposition 10.6, the algebra

$R(A, \mathfrak{n}, L)$ is normal and the K-grading is effective, pointed and of complexity one. By Proposition 10.9, the algebra $R(A, \mathfrak{n}, L)[S_1, \dots, S_m]$ is a Cox ring and hence it is factorially graded, see [4]. Clearly the graded subring $R(A, \mathfrak{n}, L)$ inherits the latter property.

Now suppose that $(A, \mathfrak{n}, L)$ is a sincere triple. If K is torsion free, then K-factoriality of $R(A, \mathfrak{n}, L)$ implies factoriality, see [3, Theorem 4.2]. Conversely, if $R(A, \mathfrak{n}, L)$ is factorial, then the generators T_{ij} are prime by Proposition 10.7. From [5, Lemma 1.5], we then infer that the numbers $\gcd(l_{i1}, \dots, l_{in_i})$ are pairwise coprime. This implies that $P\colon F \to N$ is surjective and thus K is torsion free. □

Proof of Theorem 10.5 According to [6, Theorem 1.3], the Cox ring $\mathcal{R}(X)$ is isomorphic to a ring $R(A, \mathfrak{n}, L)[S_1, \dots, S_m]$ with a grading by $\dot{K} := \mathrm{Cl}(X)$ such that the variables T_{ij} and S_k are homogeneous. In particular, X is the quotient by the action of $\dot{H} = \mathrm{Spec}\ \mathbb{K}[\dot{K}]$ on an open subset $\widehat{X}$ of

$$\overline{X} \;=\; V(g_{i,i+1,i+2};\ 0 \le i \le r-2) \;\subseteq\; \overline{Z}.$$

For $r < 2$, the variety X is toric, we may assume that T acts as a subtorus of the big torus and the assertion follows by standard toric geometry. So, let $r \ge 2$. By construction, the grading of $R(A, \mathfrak{n}, L)[S_1, \dots, S_m]$ by $\overline{K} := K \times \mathbb{Z}^m$ is the finest possible such that all variables T_{ij} and S_k are homogeneous. Consequently, we have exact sequences of abelian groups fitting into a commutative diagram

(10.2)

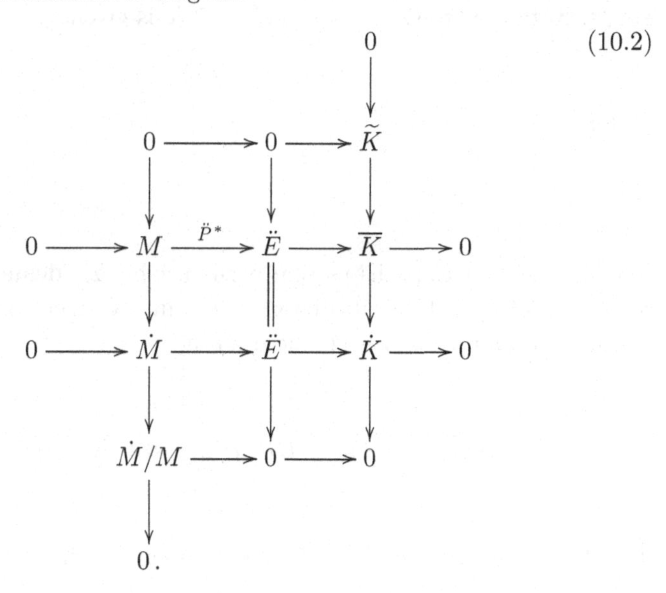

In particular we extract from this the following two commutative triangles, where the second one is obtained by dualizing the first one:

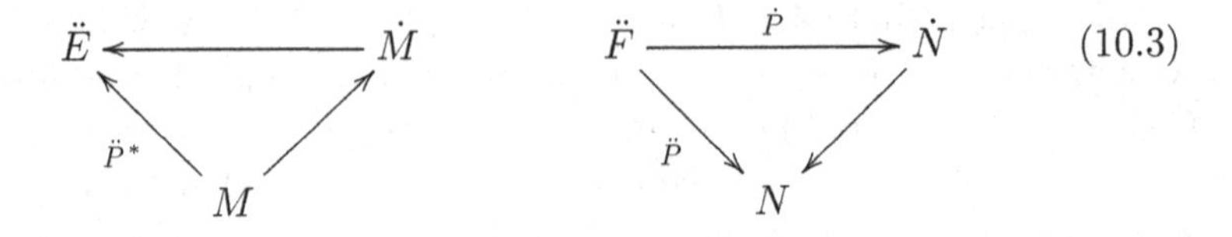

We claim that the kernel $\widetilde{K}$ is free. Consider $\overline{H} := \operatorname{Spec} \mathbb{K}[\overline{K}]$ and the isotropy group $\overline{H}_{ij} \subseteq \overline{H}$ of a general point $\widehat{x}(i,j) \in \widehat{X} \cap V(T_{ij})$. Then we have exact sequences

$$1 \longleftarrow \overline{H}/\overline{H}_{ij} \longleftarrow \overline{H} \longleftarrow \overline{H}_{ij} \longleftarrow 1$$

$$0 \longrightarrow \overline{K}(i,j) \longrightarrow \overline{K} \longrightarrow \overline{K}/\overline{K}(i,j) \longrightarrow 0$$

where the second one arises from the first one by passing to the character groups. Note that the subgroup $\overline{K}(i,j) \subseteq \overline{K}$ is given by

$$\overline{K}(i,j) = \operatorname{lin}_{\mathbb{Z}}(\deg T_{kl};\ (k,l) \neq (i,j)) + \operatorname{lin}_{\mathbb{Z}}(\deg T_p;\ 1 \leq p \leq m). \tag{10.4}$$

Now [6, Theorem 1.3] tells us that each variable T_{ij} defines a $\dot{K}$-prime element in $\mathcal{R}(X)$ and thus its divisor is $\dot{H}$-prime. Consequently, $\overline{H}/\dot{H}\overline{H}_{ij}$ is connected and has a free character group

$$\mathbb{X}(\overline{H}/\dot{H}\overline{H}_{ij}) \;=\; \widetilde{K}(i,j) \;:=\; \widetilde{K} \cap \overline{K}(i,j).$$

Mimicking equation (10.4), we define a subgroup $\dot{K}(i,j) \subseteq \dot{K}$ fitting into a commutative net of exact sequences

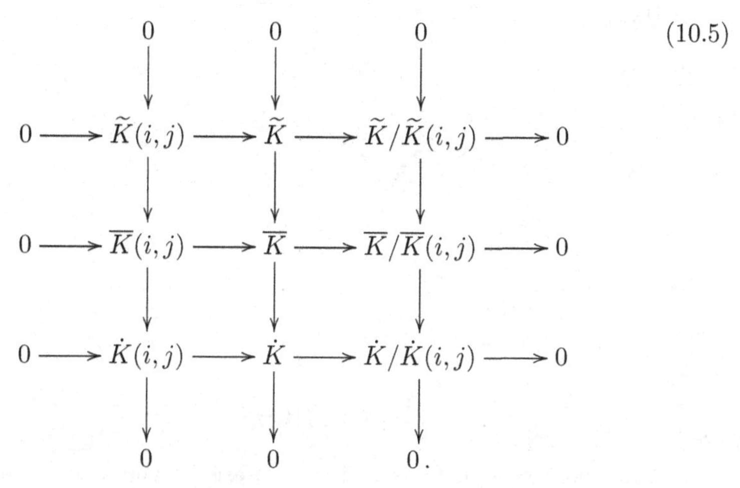

(10.5)

By general properties of Cox rings [4, Prop. 2.2], we must have $\dot{K}/\dot{K}(i,j) = 0$ and thus we can conclude

$$\widetilde{K}/\widetilde{K}(i,j) \;=\; \overline{K}/\overline{K}(i,j) \;\cong\; \mathbb{Z}/l_{ij}\mathbb{Z}. \tag{10.6}$$

Consider again $\widehat{x}(i,j) \in V(T_{ij}) \cap \widehat{X}$, set $x(i,j) = p_X(\widehat{x}(i,j))$ and let T denote the torus acting on X. Then [6, Prop. 2.6] and its proof provide us with a commutative diagram

$$\begin{array}{ccc} \widetilde{H} & \supseteq & \widetilde{H}_{ij} \\ \big\uparrow & & \big\uparrow{\scriptstyle\cong} \\ T & \supseteq & T_{x(i,j)} \end{array}$$

where $\widetilde{H} = \overline{H}/\dot{H}$ and $\widetilde{H}_{ij} = \overline{H}/\dot{H}\overline{H}_{ij}$. Using (10.6) and passing to the character groups we arrive at a commutative diagram

$$\begin{array}{ccccccccc} 0 & \longrightarrow & \widetilde{K}(i,j) & \longrightarrow & \widetilde{K} & \longrightarrow & \mathbb{Z}/l_{ij}\mathbb{Z} & \longrightarrow & 0 \\ & & & & \big\downarrow & & \big\uparrow{\scriptstyle\cong} & & \\ & & & & \mathbb{X}(T) & \longrightarrow & \mathbb{Z}/l_{ij}\mathbb{Z} & \longrightarrow & 0 \end{array}$$

with exact rows. As seen before, the group $\widetilde{K}(i,j)$ is free abelian. Consequently, also $\widetilde{K}$ must be free abelian.

Now the snake lemma tells us that $\dot{M}/M$ is free as well. In particular, the first vertical sequence of (10.2) splits. Thus, we obtain the desired matrix presentation of $\dot{P}$ from rewriting the second commutative triangle of (10.3) as

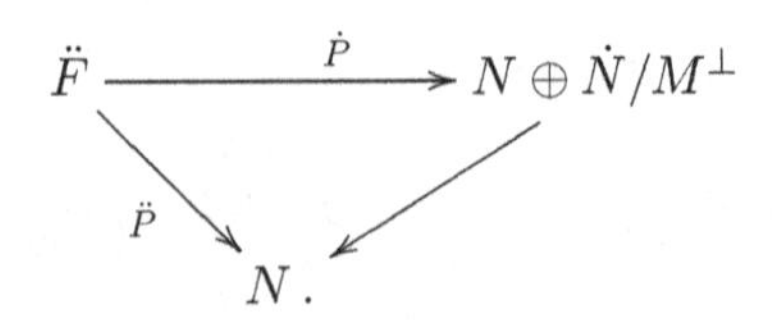

$$\ddot{F} \xrightarrow{\dot{P}} N \oplus \dot{N}/M^{\perp}, \quad \ddot{F} \xrightarrow{\ddot{P}} N, \quad N \oplus \dot{N}/M^{\perp} \to N.$$

□

References

[1] A. A'Campo-Neuen, J. Hausen: Toric prevarieties and subtorus actions. Geom. Dedicata 87 (2001), no. 1-3, 35–64.

[2] I. Arzhantsev, U. Derenthal, J. Hausen, A. Laface: Cox rings, arXiv:1003.4229.

[3] D.F. Anderson: Graded Krull domains. Comm. Algebra 7 (1979), no. 1, 79–106.

[4] J. Hausen: Cox rings and combinatorics II. Mosc. Math. J. 8 (2008), no. 4, 711–757.

[5] J. Hausen, E. Herppich, H. Süß: Multigraded factorial rings and Fano varieties with torus action. Documenta Math. 16 (2011), 71–109.

[6] J. Hausen, H. Süß: The Cox ring of an algebraic variety with torus action. Advances Math. 225 (2010), 977–1012.

[7] T. Oda, H.S. Park: Linear Gale transforms and Gelfand-Kapranov-Zelevinskij decompositions. Tohoku Math. J. (2) 43 (1991), no. 3, 375–399.

11
Nef and semiample divisors on rational surfaces

Antonio Laface
Departamento de Matemática, Universidad de Concepción

Damiano Testa
Mathematics Institute, University of Warwick

Abstract In this paper we study smooth projective rational surfaces, defined over an algebraically closed field of any characteristic, with pseudo-effective anticanonical divisor. We provide a necessary and sufficient condition in order for any nef divisor to be semiample. We adopt our criterion to investigate Mori dream surfaces in the complex case.

Introduction

Let X be a smooth projective rational surface defined over an algebraically closed field $\mathbb{K}$ of any characteristic. A problem that has recently attracted attention consists in finding equivalent characterizations of Mori dream surfaces, that is surfaces whose Cox ring is finitely generated (see [ADHL] for basic definitions), for instance, in terms of the Iitaka dimension of the anticanonical divisor. Indeed, if the Iitaka dimension of the anticanonical divisor is 2, then X is always a Mori dream surface, as shown in [TVAV]. If the Iitaka dimension of the anticanonical divisor is 1, then X admits an elliptic fibration $\pi\colon X \to \mathbb{P}^1$ and it is a

Mori dream surface if and only if the relatively minimal elliptic fibration of π has a Jacobian fibration with finite Mordell-Weil group, as shown in [AL11]. The authors are not aware of any previously known example of a Mori dream surface with Iitaka dimension of the anticanonical divisor equal to 0.

By [HK00, Definition 1.10], a smooth projective surface X is a Mori dream space if the irregularity of X vanishes and if the nef cone of X is the affine hull of finitely many semiample classes. In this paper we concentrate on giving necessary and sufficient conditions in order for any nef divisor to be semiample in the case in which X is a smooth projective rational surface with non-negative anticanonical Iitaka dimension. Our theorems hold in any characteristic; in the complex case we provide examples of Mori dream surfaces with vanishing anticanonical Iitaka dimension.

The paper is organized as follows. We introduce necessary the notation in Section 11.1. In Section 11.2 we prove some technical lemmas about the stable base locus of divisors on a smooth projective rational surface. We use these lemmas to prove Corollary 11.6 of Section 11.3. In Section 11.4 we prove that, if the anticanonical divisor is not effective, then every nef divisor is semiample. In Section 11.5 we consider the case where the anticanonical divisor is effective, providing a necessary and sufficient condition for every nef divisor to be semiample. At the end of this section we prove that the blow-up at 10 or more distinct points on a smooth cubic curve is never a Mori dream space. Finally, in Section 11.6 we give an example of a Mori dream surface with vanishing anticanonical Iitaka dimension.

Ackowledgments

The first author was supported by Proyecto FONDECYT Regular 2011, N. 1110096. The second author was supported by EPSRC grant number EP/F060661/1. It is a pleasure to thank Igor Dolgachev for the suggestion of looking at Coble surfaces with finite automorphism group to produce examples of Mori dream surfaces, and to thank Michela Artebani for helpful remarks and discussions. We would also like to thank Ana-Maria Castravet for pointing out a mistake in an older version of this paper. We have made large use of the computer algebra program Magma [BCP97] in Section 11.6 of this paper.

11.1 Notation

In this section we gather the basic definitions and the standard conventions that we use throughout the paper. Let X be a smooth projective surface with $\mathrm{H}^1(X, \mathcal{O}_X) = 0$. All the cones mentioned in this paper are contained in the rational vector space $\mathrm{Pic}_{\mathbb{Q}}(X) = \mathrm{Pic}(X) \otimes_{\mathbb{Z}} \mathbb{Q}$, see [ADHL]. The *effective cone* $\mathrm{Eff}(X)$ consists of the numerical equivalence classes of divisors admitting a positive multiple numerically equivalent to an effective divisor; the *pseudo-effective cone* $\overline{\mathrm{Eff}}(X)$ is the closure of the effective cone. The *nef cone* $\mathrm{Nef}(X)$ consists of the classes of divisors D such that $D \cdot C \geq 0$ for any curve C in X. The pseudo-effective cone and the nef cone are dual with respect to each other under the pairing induced by the intersection form of $\mathrm{Pic}(X)$. The *semiample cone* $\mathrm{SAmple}(X)$ consists of the classes of divisors admitting a positive multiple whose associated linear system is base point free. Clearly, every semiample divisor is also nef, and therefore there is an inclusion $\mathrm{SAmple}(X) \subset \mathrm{Nef}(X)$.

We specialize Definition 1.10 of [HK00] to our context.

Definition 11.1 Let X be a smooth projective surface. We say that X is a *Mori dream surface* if the following conditions hold:

(1) the group $\mathrm{H}^1(X, \mathcal{O}_X)$ vanishes;
(2) the nef cone of X is generated by a finite number of semiample classes.

Condition (2) means that the cones $\mathrm{Nef}(X)$ and $\mathrm{SAmple}(X)$ coincide and that both are polyhedral. The statement that the cone $\mathrm{Eff}(X)$ is polyhedral and that the cones $\mathrm{Nef}(X)$ and $\mathrm{SAmple}(X)$ coincide is also equivalent to Condition (2). If the cone $\mathrm{Eff}(X)$ is polyhedral and $\mathrm{rk}(\mathrm{Pic}(X)) \geq 3$, it is not difficult to prove that $\mathrm{Eff}(X)$ is spanned by classes of *negative curves*, that is integral curves C with $C^2 < 0$ (see [AL11, Proposition 1.1]). Examples of negative curves are $(-n)$-*curves* which are smooth rational curves of self-intersection $-n$. An *exceptional curve* is an alternative name for a (-1)-curve.

Recall that the *Iitaka dimension* $k(D)$ of a divisor D is one less than the Krull dimension of the ring

$$\bigoplus_{n \geq 0} \mathrm{H}^0(X, nD)$$

if a positive multiple of D is effective and $-\infty$ otherwise. Our initial approach is to study Mori dream surfaces according to their anticanonical Iitaka dimension $k(-K_X)$. Toric surfaces and *generalized del Pezzo surfaces*, that is rational surfaces with nef and big anticanonical divisor, have anticanonical Iitaka dimension 2. Rational elliptic surfaces have anticanonical Iitaka dimension 1.

11.2 Nef and semiample divisors

Let X be a smooth projective surface and let C be an effective divisor on X. Suppose that the matrix of pairwise intersections of the components of C is negative definite. It follows that there is an effective divisor B with support on C such that for every irreducible component D of C we have $B \cdot D < 0$.

Definition 11.2 We say that an effective divisor B with support in C is a block on C if for every irreducible component D of C the inequality $B \cdot D < \min\{0, -K_X \cdot D\}$ holds.

Often in our applications, the divisor C is the union of all the integral curves of X orthogonal to a big and nef divisor N on X; in this case the required negative definiteness follows by the Hodge Index Theorem. In such cases we say that a block on C is a block for N. Observe that from the definition of block it follows easily that if B is a block for C, then the supports of B and of C coincide.

Lemma 11.3 *Let X be a smooth projective surface. Let N be a big and nef divisor on X and let B be a block for N. Then the stable base locus of N is the stable base locus of the line bundle $\mathcal{O}_B(N)$. In particular, if R is a connected component of the block B such that the group $\mathrm{H}^1(R, \mathcal{O}_R)$ vanishes, then R is disjoint from the stable base locus.*

Proof To prove the lemma, we use some of the techniques of the proof of [Nak00, Theorem 1.1] Let E be an effective divisor on X; we establish below the following assertion:

($\star$) there is an effective divisor Z with the same support as $B \cup E$ such that, for sufficiently large n, the restriction map

$$\mathrm{H}^0(X, \mathcal{O}_X(nN)) \longrightarrow \mathrm{H}^0(X, \mathcal{O}_Z(nN))$$

is surjective.

Assuming ($\star$) we deduce that, if E is an integral curve with $N \cdot E > 0$, then the restriction of nN to Z admits global sections not vanishing identically on E, and hence that E is not contained in the stable base locus of N. Therefore the support of the stable base locus of N is contained in the support of B. Finally, from the proof of ($\star$) we obtain that when $E = 0$ we can choose $Z = B$, establishing the main statement of the lemma.

We now turn to the proof of ($\star$). Replacing B if necessary by a positive multiple, we obtain a block B' for N such that for every irreducible component D of B we have $B' \cdot D < -(E + K_X) \cdot D$; note that, in the case in which $E = 0$, we may choose $B' = B$. We show that the divisor $Z = B' + E$ satisfies ($\star$). Let A be an ample divisor on X. Since the divisor N is big, for large enough n the divisor $F = nN - Z - K_X - A$ is effective and hence the divisor $nN - Z - K_X = A + F$ is big. Moreover, if D is an integral curve on X such that $D \cdot (nN - Z - K_X) \leq 0$, then D is a component of F and it is not a component of B'. It follows that the inequality $N \cdot D > 0$ holds, and since F has only finitely many components, we obtain that for large enough n the divisor $nN - Z - K_X$ has positive intersection with every curve on X; since it is also big, we deduce from the Nakai-Moishezon criterion that the divisor $nN - Z - K_X$ is ample.

Thus, for every sufficiently large integer n the Kodaira Vanishing Theorem implies that the group $\mathrm{H}^1(X, \mathcal{O}_X(nN - Z))$ vanishes and ($\star$) follows from the long exact cohomology sequence associated to the sequence

$$0 \longrightarrow \mathcal{O}_X(nN - Z) \longrightarrow \mathcal{O}_X(nN) \longrightarrow \mathcal{O}_Z(nN) \longrightarrow 0,$$

establishing ($\star$).

For the last assertion, by the assumptions it follows that the restriction of N to R is trivial; in particular the line bundle $\mathcal{O}_R(N)$ is globally generated and the result follows from what we already proved. □

Lemma 11.4 *Let X be a smooth projective surface with $q(X) = 0$ and let B be an effective divisor on X. If the divisor $K_X + B$ is not effective, then the group $\mathrm{H}^1(B, \mathcal{O}_B)$ vanishes. In particular, if there is a nef divisor N such that the inequality $N \cdot (K_X + B) < 0$ holds, then the group $\mathrm{H}^1(B, \mathcal{O}_B)$ vanishes.*

Proof From the sequence

$$0 \longrightarrow \mathcal{O}_X(-B) \longrightarrow \mathcal{O}_X \longrightarrow \mathcal{O}_B \longrightarrow 0$$

we deduce that it suffices to show that the group $\mathrm{H}^2(X, \mathcal{O}_X(-B))$ vanishes. The first part follows by Serre duality. The second part is an immediate special case of the first and the lemma follows. □

11.3 The K_X-negative part of the nef cone

Lemma 11.5 *Let X be a smooth projective rational surface. If the divisor N on X is nef and it satisfies $-K_X \cdot N > 0$, then N is semiample. If N is not big, then it is base point free.*

Proof Suppose first that N is big and let B be a block for N. The result follows by Lemmas 11.3 and 11.4 since the inequality $N \cdot (K_X + B) < 0$ holds.

Suppose now that N is not big, and hence that $N^2 = 0$. The divisors $K_X - N$ and $K_X + N$ are not effective since $N \cdot (K_X - N) < 0$ and $N \cdot (K_X + N) < 0$. Therefore, by the Riemann-Roch formula, we deduce that the divisor N is linearly equivalent to an effective divisor. To avoid introducing more notation, we assume that N itself is effective. We deduce that the base locus of N is contained in N and that the group $\mathrm{H}^1(N, \mathcal{O}_N)$ vanishes by Lemma 11.4. In particular the line bundle $\mathcal{O}_N(N)$ is trivial and from the exact sequence

$$0 \longrightarrow \mathcal{O}_X \longrightarrow \mathcal{O}_X(N) \longrightarrow \mathcal{O}_N(N) \longrightarrow 0$$

and the vanishing of the group $\mathrm{H}^1(X, \mathcal{O}_X)$, we deduce that the base locus of $|N|$ is empty. □

As a consequence of what we proved so far, we deduce the following result (see also [AL11, Theorem 3.4] and [TVAV]).

Corollary 11.6 *Let X be a smooth projective rational surface and suppose that there are a positive integer k and $\mathbb{Q}$-divisors P and E such that P is non-zero and nef, E is effective, $-K_X = P + E$ and kP is integral. Every nef divisor on X not proportional to P is semiample; moreover P itself is semiample if P is big or kP is base point free.*

Proof Let N be a non-zero nef divisor on X. If $-K_X \cdot N > 0$, then we conclude that N is semiample by Lemma 11.5. Suppose that $-K_X \cdot N = 0$. In particular we deduce that $P \cdot N = 0$ and hence that P and N are not big and that N is proportional to P. Again Lemma 11.5 allows us to conclude. □

11.4 The anticanonical divisor is not effective

Let X be a smooth projective rational surface. In this section we assume that the anticanonical divisor $-K_X$ is not linearly equivalent to an effective divisor, but a positive multiple of $-K_X$ is. Let e be the least positive integer such that the linear system $|-eK_X|$ is not empty and let E be an element of $|-eK_X|$.

Theorem 11.7 *Every nef divisor on X is semiample.*

Proof If N is trivial, then the result is clear; we therefore suppose that N is non-trivial. Note that the divisor $K_X - N$ is not effective, as it is the negative of a non-zero pseudo-effective divisor. Thus by the Riemann-Roch formula and Serre duality we deduce that N is linearly equivalent to an effective divisor; to avoid introducing more notation, we assume that N itself is effective. If N is big, then let B denote a block for N; if N is not big, then let $B = N$. In both cases the base locus of N is contained in B.

Suppose first that $h^1(B, \mathcal{O}_B) = 0$. By Lemma 11.3 we deduce that if N is big, then it is semiample. Thus we reduce to the case in which N is not big; the line bundle $\mathcal{O}_N(N)$, having non-negative degree on each irreducible component of its support, is globally generated. From the sequence

$$0 \longrightarrow \mathcal{O}_X \longrightarrow \mathcal{O}_X(N) \longrightarrow \mathcal{O}_N(N) \longrightarrow 0$$

and the vanishing of the group $\mathrm{H}^1(X, \mathcal{O}_X)$ we deduce that also in this case the divisor N is semiample.

Suppose now that $h^1(B, \mathcal{O}_B) \geq 1$. By Lemma 11.4 we obtain that the divisor $K_X + B$ is linearly equivalent to the effective divisor B'. We deduce that the divisor eB is linearly equivalent to the divisor $eB' + E$. If the effective divisor eB were equal to the effective divisor $eB' + E$, it would follow that every prime divisor appears with multiplicity divisible by e in E; in particular, the divisor E would be a multiple of an effective divisor linearly equivalent to $-K_X$. Since we are assuming that $-K_X$ is not effective, it follows that eB and $eB' + E$ are not equal and we deduce that the dimension of $|eB|$ is at least one. In particular, N is not big, since otherwise the matrix of pairwise intersections of the components of B would be negative definite and hence no multiple of B would move. Therefore we have $B = N$; write $eN = M + F$ where $M \neq 0$ is the moving part of $|eN|$ and F is the fixed part. Since the divisor N is nef and not big it follows that $M \cdot F = F^2 = 0$ and by the Hodge Index

Theorem we obtain that F is proportional to M. Since M is semiample, the result follows. □

11.5 The anticanonical divisor is effective

Let X be a smooth projective rational surface. In this section we assume that X admits an effective anticanonical divisor. We let E be an element of the linear system $|-K_X|$ and we denote by ω_E the dualizing sheaf of E. Since E is an effective divisor on a smooth surface, its dualizing sheaf is invertible and by the adjunction formula we have $\omega_E \simeq \mathcal{O}_E$. Moreover from the sequence

$$0 \longrightarrow \mathcal{O}_X(-E) \longrightarrow \mathcal{O}_X \longrightarrow \mathcal{O}_E \longrightarrow 0$$

we deduce that $h^0(E, \mathcal{O}_E) = 1$; in particular the divisor E is connected and has arithmetic genus equal to one.

Let $\iota\colon E \to X$ denote the inclusion and let

$$\iota^*\colon \operatorname{Pic}(X) \longrightarrow \operatorname{Pic}(E)$$

be the pull-back map induced by ι. The map ι^* is a homomorphism of abelian groups whose image is a finitely generated subgroup of $\operatorname{Pic}(E)$, since $\operatorname{Pic}(X) \simeq \mathbb{Z}^{10-(K_X)^2}$. If N is a divisor on X, then we denote by $N_E \in \operatorname{Pic}(E)$ the class of the line bundle $\mathcal{O}_E(N)$.

We let $\Gamma \subset \operatorname{Pic}(E)$ be the image of the lattice $\operatorname{Nef}(X) \cap \langle K_X \rangle^\perp$ under the homomorphism ι^*; equivalently Γ is the abelian subgroup of $\operatorname{Pic}(E)$ generated by the classes N_E, where N ranges among the nef divisors such that $K_X \cdot N = 0$. In particular, Γ is a finitely generated abelian group. We shall prove in Theorem 11.11 that the equality of semiample and nef divisors is equivalent to the finiteness of the group Γ.

Lemma 11.8 *Let N be a nef divisor satisfying $N \cdot K_X = 0$. If N is semiample, then $N_E \in \operatorname{Pic}(E)$ is a torsion element.*

Proof Let r be a positive integer such that the linear system $|rN|$ is base point free. Since $N \cdot K_X = 0$ we deduce that the linear system $|rN|$ contains a divisor disjoint from E and hence $(rN)_E$ represents the trivial line bundle and $N_E \in \operatorname{Pic}(E)[r]$. □

Lemma 11.9 *Let N be a big and nef divisor satisfying $N \cdot K_X = 0$. The support of the stable base locus of N is contained in E. If moreover N_E is a torsion element of $\operatorname{Pic}(E)$, then N is semiample.*

Proof First, we reduce to the case in which all the exceptional curves F on X such that $N \cdot F = 0$ are contained in E.

Let $F \subset X$ be an exceptional curve such that $F \cdot N = 0$ and F is not contained in E, and let $b\colon X \to X'$ be the contraction of F. By construction, the divisor N is the pull-back of a big and nef divisor N' on X'; moreover since $E \cdot F = 1$ and F is not contained in E, it follows that F meets E transversely at a single smooth point. Therefore the restriction of b to E is an isomorphism to its image E', and the restriction of $\mathcal{O}_{X'}(N')$ to E' is isomorphic to the restriction of $\mathcal{O}_X(N)$ to E. Finally, if N' is semiample, then N is also semiample. Repeating if necessary the above construction starting from X', N' and E' we conclude the reduction step. Observe that the process described above terminates, since at each stage we contract an exceptional curve, thereby reducing the rank of the Picard group of X.

Thus, we assume from now on that the exceptional curves F on X such that $F \cdot N = 0$ are contained in E.

Let B be a block for N. Let F be a component of B not contained in E. Thus we have $F^2 < 0$ and $F \cdot E \geq 0$; it follows from the adjunction formula that F is a smooth rational curve and either $F \cdot K_X = -1$ and F is an exceptional curve, or $F \cdot K_X = 0$ and F is a (-2)-curve disjoint from E. By our reduction, we conclude that every irreducible component of B not contained in E is disjoint from E; we thus have $B = B' + E'$ where the support of E' is contained in the support of E and the support of B' is the union of the components of B disjoint from E.

We now show that no component of B' is contained in the stable base locus of N; for this it suffices to show that the divisor $K_X + B'$ is not effective by Lemmas 11.3 and 11.4. Since the matrix of pairwise intersections of the components of B' is negative definite, the divisor B' is the unique element of the linear system $|B'|$. If the divisor $B' - E$ were linearly equivalent to an effective divisor B'', then the linear system $|B'|$ would contain the divisor $B'' + E$ contradicting the fact that $|B'| = \{B'\}$. Thus the divisor $K_X + B'$ is not effective, as we wanted to show, and the stable base locus of N is contained in E'.

To prove the second part, we show that no component of E is contained in the stable base locus of N. Let r be a positive integer such that $(rN)_E$ is trivial. From the Kawamata-Viehweg Vanishing Theorem (see [Xie10, Corollary 1.4] for the positive characteristic case) we deduce that the group $\mathrm{H}^1(X, \mathcal{O}_X(rN - E))$ vanishes and therefore using the exact sequence

$$0 \longrightarrow \mathcal{O}_X(rN - E) \longrightarrow \mathcal{O}_X(rN) \longrightarrow \mathcal{O}_E(rN) \longrightarrow 0$$

we deduce that there are sections of $\mathcal{O}_X(rN)$ that are disjoint from E and the lemma is proved. □

Lemma 11.10 *Let N be a nef non big divisor satisfying $N \cdot K_X = 0$. If N_E is a torsion element of* Pic(E)*, then N is semiample.*

Proof If the divisor N is trivial, then the result is clear. Suppose that N is non-zero and let r be a positive integer such that $(rN)_E$ is trivial. Observe that $h^2(X, \mathcal{O}_X(rN - E)) = h^0(X, \mathcal{O}_X(-rN)) = 0$, since for every ample divisor A on X we have $A \cdot N > 0$ by Kleiman's criterion and therefore the divisor $-rN$ cannot be effective. Thus from the exact sequence

$$0 \longrightarrow \mathcal{O}_X(rN - E) \longrightarrow \mathcal{O}_X(rN) \longrightarrow \mathcal{O}_E(rN) \longrightarrow 0$$

and the fact that $h^1(E, \mathcal{O}_E) = 1$ we deduce that $h^1(X, \mathcal{O}_X(rN)) \geq 1$ and thus that the dimension of the linear system $|N|$ is at least one, by the Riemann-Roch formula. We may therefore write $N = M + F$, where $M \neq 0$ is the moving part and F is the divisorial base locus of $|N|$. Since N is nef and not big we deduce that $M \cdot F = F^2 = 0$ and thus by the Hodge Index Theorem we conclude that F is proportional to M and the lemma follows. □

Theorem 11.11 *Let X be a smooth projective rational surface and suppose that E is an effective divisor in the anticanonical linear system. The subgroup Γ of* Pic(E) *is finite if and only if every nef divisor on X is semiample.*

Proof Suppose that the group Γ is finite and let N be a nef divisor on X. If $-K_X \cdot N > 0$, then the result follows from Lemma 11.5. Suppose therefore that $-K_X \cdot N = 0$. It follows that N_E is contained in Γ and by assumption it has finite order. The conclusion follows by using Lemmas 11.9 and 11.10.

For the converse, suppose that every nef divisor on X is semiample. Let N be a nef divisor on X such that $K_X \cdot N = 0$. By Lemma 11.8 we deduce that N_E has finite order. Therefore Γ is finite, being a finitely generated abelian group all of whose elements have finite order; the result follows. □

Example 11.12 Let X be a smooth projective rational surface such that a positive multiple of the anticanonical divisor is effective.

Suppose that the anticanonical divisor of X is not linearly equivalent to an effective divisor, and let e be the least among the positive integers n such that the linear system $|-nK_X|$ is not empty. In view of Theorem 11.7, every nef divisor on X is semiample. We first show that if E is an element of the linear system $|-eK_X|$, then the cohomology group $\mathrm{H}^1(E, \mathcal{O}_E)$ vanishes. Indeed, from the long exact cohomology sequence associated to

$$0 \longrightarrow \mathcal{O}_X(eK_X) \longrightarrow \mathcal{O}_X \longrightarrow \mathcal{O}_E \longrightarrow 0$$

we deduce that the group $\mathrm{H}^1(E, \mathcal{O}_E)$ is a subgroup of

$$\mathrm{H}^2(X, \mathcal{O}_X(eK_X)) \simeq \mathrm{H}^0(X, \mathcal{O}_X(-(e-1)K_X)),$$

which vanishes by the definition of e. Next, if N is a nef divisor on X such that $N \cdot K_X = 0$, then it follows that the restriction of N to E is trivial, since the divisor N has degree zero on each irreducible component of E. We see that in this case "there is no space" for nef divisors that are not semiample: the subgroup of $\mathrm{Pic}(E)$ generated by the classes of the restrictions of the nef divisors to E is trivial.

Suppose that the anticanonical divisor on X is linearly equivalent to an effective divisor, let E be an element of the anticanonical linear system on X, and let $\langle E \rangle \subset \mathrm{Pic}(X)$ be the sublattice generated by the irreducible components of E. We analyze the possibilities for the group Γ in terms of the anticanonical Iitaka dimension of X and of the signature of the quadratic form on $\langle E \rangle$.

The anticanonical Iitaka dimension of X equals two if and only if the quadratic form $\langle E \rangle$ represents positive values. In this case the anticanonical divisor of X is big and the group Γ is trivial, since a nef divisor that is orthogonal to a big divisor vanishes. The class of rational surfaces with big anticanonical divisor is analyzed in [TVAV].

If the anticanonical Iitaka dimension of X equals 1, then the quadratic form $\langle E \rangle$ is negative semidefinite and it is not definite. Let M denote the moving part of a positive multiple of the anticanonical divisor on X with positive dimensional linear system. A nef divisor on X orthogonal to the anticanonical divisor must be orthogonal to M, and, by the Hodge Index Theorem, it is therefore proportional to M. We deduce that Γ is cyclic and torsion. The class of rational surface with anticanonical Iitaka dimension 1 is analyzed in [AL11].

If the anticanonical Iitaka dimension of X equals 0, then the quadratic form $\langle E \rangle$ is negative semidefinite. If the quadratic form on $\langle E \rangle$ is not definite, then there is an effective nef divisor F with support contained in the support of E, whose class generates the kernel of the intersection form on $\langle E \rangle$. In this case, the group Γ is again cyclic, contains the group generated by F with finite index, but Γ is not torsion, since otherwise a positive multiple of F would be base point free, contradicting the fact that the anticanonical Iitaka dimension vanishes. If the quadratic form on $\langle E \rangle$ is negative definite, then there are easy examples in which the group Γ is infinite and others in which it is finite. We analyze one such example after proving the following result.

Proposition 11.13 *Let X and $\langle E \rangle$ be as above. Suppose that the quadratic form on $\langle E \rangle$ is negative definite. Then every nef divisor of X is semiample if and only if, for every $[D] \in \langle E \rangle^\perp$, the line bundle $\mathcal{O}_E(D)$ represents a torsion element of* Pic(E).

Proof The irreducible components of E span an extremal face of the effective cone and therefore there are nef divisors N on X such that if F is an integral curve on X with $N \cdot F = 0$, then F is a component of E. More precisely, the set of nef divisors with the above property is a non-empty open subset of $\langle E \rangle^\perp$. In particular, there is a basis of the space $\langle E \rangle^\perp$ consisting of nef divisors, and the conclusion follows by Theorem 11.11. □

Example 11.14 Let $\overline{E} \subset \mathbb{P}^2$ be a smooth plane cubic curve and let $p_1, \ldots, p_r$ be distinct points lying on $\overline{E}$. Denote by X the blow-up of $\mathbb{P}^2$ at the points $p_1, \ldots, p_r$ by E the strict transform in X of $\overline{E}$. We denote by ℓ a line in $\mathbb{P}^2$ and by ℓ_E the effective divisor on E induced by ℓ; note that the degree of the divisor ℓ_E equals 3.

If $r \leq 8$, then the quadratic form on $\langle E \rangle$ is positive definite and every nef divisor is semiample (Lemma 11.5).

In case $r = 9$ the intersection form on $\langle E \rangle$ is negative semidefinite and is not definite since $E^2 = 0$. We observe that E, which is nef, is semiample if and only if the line bundle $\mathcal{O}_E(E) \simeq \mathcal{O}_E\big(3\ell_E - (p_1 + \cdots + p_9)\big)$ is torsion in Pic(E) (Corollary 11.6).

Consider now the case $r \geq 10$, that is when the quadratic form on $\langle E \rangle$ is negative definite. Let $E_1, \ldots, E_r$ be the exceptional divisors of the blow up morphism $X \to \mathbb{P}^2$. The classes of $E_1 - E_r$, $E_2 - E_r$, ..., $E_{r-1} - E_r$ and $\ell - 3E_r$ span $\langle E \rangle^\perp$. Moreover these classes restrict to the classes of $p_1 - p_r, p_2 - p_r, \ldots, p_{r-1} - p_r$ and $\ell_E - 3p_r$ on Pic(E). Thus we conclude by Proposition 11.13 that any nef divisor on X is semiample if and only if each of the above classes is torsion in Pic(E).

11.6 Examples of rational Mori dream surfaces with $k(-K_X) = 0$

We begin by recalling the following definition given by Dolgachev and Zhang in [DZ01].

Definition 11.15 A *Coble surface* X is a smooth projective rational surface with $-K_X$ not effective and $-2K_X$ effective.

In this section we provide an example of a Coble surface $\tilde{Y}$ with finitely generated Cox ring and $k(-K_{\tilde{Y}}) = 0$. Here we consider the following construction. Inside the moduli space of Enriques surfaces there are exactly two 1-dimensional families of Enriques surfaces whose general element has finite automorphism group. These have been classified in [Kon86]. To describe the general element Y_t of one such family, we recall a construction given in [Kon86, §3]. Let ϕ be the involution of $\mathbb{P}^1 \times \mathbb{P}^1$ defined by

$$\phi([x_0, x_1], [y_0, y_1]) = ([x_0, -x_1], [y_0, -y_1]).$$

Let $\{C_t\}_{t \in \mathbb{P}^1}$ be the pencil of curves of degree $(2, 2)$ in $\mathbb{P}^1 \times \mathbb{P}^1$ given by

$$C_t := V\big((2x_0^2 - x_1^2)(y_0^2 - y_1^2) + (2ty_0^2 + (1 - 2t)y_1^2)(x_1^2 - x_0^2)\big)$$

and define

$$L_1 := V(x_0 - x_1),\ L_2 := V(x_0 + x_1),\ L_3 := V(y_0 - y_1),\ L_4 := V(y_0 + y_1).$$

An elementary calculation shows that C_t is smooth irreducible for $t \neq 1, \frac{1}{2}, \frac{3}{2}, \infty$. Moreover, the curve C_t has an ordinary double point, for $t \in \{\frac{1}{2}, \frac{3}{2}\}$, it is the union of two irreducible curves of degree $(1, 1)$, for $t = 1$, and is the union of $L_1, \ldots, L_4$ for $t = \infty$. Note that the base locus of the pencil C_t consists of the four points

$$([1, 1], [1, 1]), \quad ([1, 1], [1, -1]), \quad ([1, -1], [1, 1]), \quad ([1, -1], [1, -1]),$$

and also note that each of these points is contained in two of the curves $L_1, \ldots, L_4$. For each $s \in \mathbb{P}^1$ the birational map $\pi_s \colon S_s \to \mathbb{P}^1 \times \mathbb{P}^1$ is obtained by first blowing up the four base points of the pencil C_t and then blowing up the 12 points of intersections of any exceptional divisor with the strict transform of C_s and of $L_1, \ldots, L_4$. Thus the surface S_s is rational with Picard group of rank 18.

For each $t \in \mathbb{P}^1$, the reducible curve $B_t \subset \mathbb{P}^1 \times \mathbb{P}^1$ defined by

$$B_t := C_t + L_1 + L_2 + L_3 + L_4$$

is a divisor of degree $(4,4)$, invariant with respect to the involution ϕ. Dropping the subscript t from the morphisms to simplify the notation, this leads us to the diagram

$$\begin{array}{ccccc} X_t & \xrightarrow[2:1]{\varphi} & S_t & \xrightarrow{\pi} & \mathbb{P}^1 \times \mathbb{P}^1 \\ {\scriptstyle \psi}\big\downarrow{\scriptstyle 2:1} & & & & \\ Y_t & & & & \end{array}$$

where the morphism π is birational, the surface S_t is rational with Picard group of rank 18, the morphism φ is a double cover, the branch locus of φ consists of the strict transform B_t' of B_t together with a union Γ of disjoint curves in the exceptional locus of π, and finally ψ is the double cover $X_t \to Y_t = X_t/\langle\sigma\rangle$, where $\sigma \in \operatorname{Aut}(X_t)$ is the involution induced by ϕ.

Kondo proves that Y_t is an Enriques surface for all t different from $1, \frac{1}{2}, \frac{3}{2}, \infty$. Moreover he proves that $\operatorname{Aut}(Y_t)$ is a finite group for any value of t (see [Kon86, §3 and Corollary 5.7] for precise statements). Let $t_0 \in \mathbb{P}^1$ be such that C_{t_0}' is irreducible and has an ordinary double point and let $X_0 := X_{t_0}$ be the corresponding surface. Then X_0 has a singularity of type $\mathbf{A}_1$ at a point p. Since p is the only singular point of X_0, it must be stable with respect to the involution σ, and thus Y_0 is singular at one point q as well. This leads to the commutative diagram

$$\begin{array}{ccc} \tilde{X} & \xrightarrow{\pi_p} & X_0 \\ {\scriptstyle \tilde{\psi}}\big\downarrow & & \big\downarrow{\scriptstyle \psi} \\ \tilde{Y} & \xrightarrow[\pi_q]{} & Y_0 \end{array}$$

where π_p and π_q are minimal resolution of singularities, $\tilde{X}$ is a K3-surface and $\tilde{\psi}$ is a double cover branched along the exceptional divisor E of π_q.

Proposition 11.16 *The surface $\tilde{Y}$ is a Coble surface.*

Proof Let $\sigma \in \operatorname{Aut}(X_0)$ be the involution induced by ϕ. As we noted before, $\sigma(p) = p$ so that σ lifts to an involution $\tilde{\sigma}$ of $\tilde{X}$. The involution $\tilde{\sigma}$ is non-symplectic since σ is non-symplectic and that the exceptional divisor $R := \pi_p^{-1}(p)$ is fixed by this involution. Thus the quotient surface $\tilde{Y} = \tilde{X}/\langle\tilde{\sigma}\rangle$ is a smooth projective rational surface. Observe that the branch divisor of $\tilde{\psi}$ is contained in $|-2K_{\tilde{Y}}|$, and in particular $-2K_{\tilde{Y}}$ is effective. Moreover, since ψ does not ramify along a divisor, the branch divisor of $\tilde{\psi}$ is exactly E. Since X_0 has an $\mathbf{A}_1$-singularity at p, it follows that $F = \pi_p^{-1}(p)$ is a (-2)-curve on $\tilde{X}$. In particular, the curve $E = \tilde{\psi}(F)$ is irreducible and reduced with $E^2 < 0$. Thus $-K_{\tilde{Y}}$ cannot be effective, since otherwise $E \in |-2K_{\tilde{Y}}|$ would be non-reduced. □

We are now ready to prove the following.

Theorem 11.17 *The Cox ring of $\tilde{Y}$ is finitely generated.*

Proof We want to use [Kon86, Corollary 5.7] to prove that the group $\mathrm{Aut}(Y_0)$ is finite, even though Y_0 is a rational surface and not an Enriques surface. The reason we can still apply Kondo's result is that the Néron-Severi lattice of Y_0 is isomorphic to the Néron-Severi lattice of an Enriques surface, and moreover the minimal resolution of the limit X_0 of the K3-surfaces X_t is still a K3-surface and therefore it still corresponds to a point in the period space. Moreover Y_0 is Kawamata log terminal, klt for short, since

$$K_{\tilde{Y}} = \pi_q^* K_{Y_0} - \frac{1}{2}E$$

and π_q is a resolution of singularities of Y_0. Observe that indeed Y_0 is a klt Calabi-Yau surface since its canonical divisor is numerically trivial. Let $\Delta := \frac{1}{2}E$. Since the pair $(\tilde{Y}, \Delta)$ is the terminal model of Y_0, it follows that $\mathrm{Aut}(Y_0) = \mathrm{Aut}(\tilde{Y}, \Delta)$. In particular the last group is finite so that the image of the map $\mathrm{Aut}(X, \Delta) \to \mathrm{GL}\big(\mathrm{Pic}_{\mathbb{Q}}(X)\big)$ is finite as well. Hence the Cox ring of $\tilde{Y}$ is finitely generated by [Tot10, Corollary 5.1]. □

Remark 11.18 In general a Coble surface is not a Mori dream space. To see this let $p_1, \ldots, p_9$ denote the nine intersection points of two general plane cubic curves C_1 and C_2. Let $q \in C_1$ be such that the divisor class of $p_9 - q$ is a 2-torsion point of $\mathrm{Pic}(C_1)$. Then the blow up Z of the plane at $p_1, \ldots, p_8, q$ admits an elliptic fibration defined by $|-2K_Z|$. Due to the generality assumption on C_1 and C_2, it is easy to see that the linear system $|-2K_Z|$ does not contain reducible elements. Equivalently Z does not contain (-2)-curves, so that it is not a Mori dream space by [AL11]. By an Euler characteristic calculation the fibration induced by $|-2K_Z|$ contains 12 nodal curves. The blow up Y of Z at one of these nodes p is a Coble surface since $|-K_Y|$ is empty, but $|-2K_Y|$ contains the strict transform of the fiber through p. Moreover Y is not a Mori dream surface since Z is not a Mori dream surface.

In the following remark, we obtain a description of the Coble surface $\tilde{Y}$ as an iterated blow up of $\mathbb{P}^2$.

Remark 11.19 The Coble surface $\tilde{Y}$ constructed above as the minimal resolution of a limit of a family of Mori dream Enriques surfaces contains (at least) the configuration of (-2)-curves of the general element of such family. The intersection graph of this surface is given in Figure 11.1 (see [Kon86, Example 1], where double edges mean that the corresponding curves have intersection equal to 2.

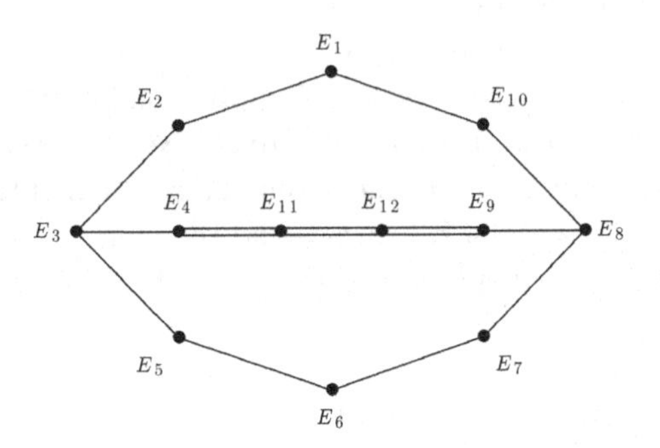

Figure 11.1 Configuration of (-2)-curves of $\tilde{Y}$.

It is not difficult to see that $F := E_1+E_2+E_3+E_5+E_6+E_7+E_8+E_{10}$ has $F^2 = 0$ and $|2F|$ is an elliptic pencil on $\tilde{Y}$ with two reducible fibers one of type I_8 and one of type I_2. Thus $\tilde{Y}$ is the blow-up of a Mori dream surface Z with Picard group of rank 10, admitting an elliptic fibration whose dual graph of singular fibers contains I_8 and I_2. The surface Z' defined by the jacobian fibration of that induced by $|-2K_Z|$ has finite Mordell-Weil group by [AL11]. Moreover Z' is unique up to isomorphism and its unique elliptic fibration, given by $|-K_{Z'}|$, admits exactly four singular fibers of type I_8, I_2, I_1, I_1, by [Dui10, Proposition 9.2.19]. This is the same configuration of singular fibers of the elliptic pencil defined by the linear system $|-2K_Z|$. Explicitly, let x, y, z be homogeneous coordinates on $\mathbb{P}^2$ and let

$$c_0 := 4x^2(x^2 - yz)^2$$

and

$$c_\infty := (2x^2y - x^2z + y^3 - 2y^2z + yz^2)(2x^2y + x^2z - y^3 - 2y^2z - yz^2).$$

The surface Z is the rational elliptic surface associated to the pencil generated by the two plane sextics $C_0 = V(c_0)$ and $C_\infty = V(c_\infty)$. The fiber corresponding to the curve C_0 has multiplicity two and is of type I_8.

The fiber corresponding to the curve C_∞ is reduced and is of type I_2. The only two remaining singular fibers are both of type I_1 and correspond to the plane sextics $V(c_0 + c_\infty)$ and $V(c_0 - c_\infty)$. To construct the jacobian surface Z' associated to Z, let

$$d_0 := 4x(yz - x^2) \qquad \text{and} \qquad d_\infty := z(y+z)^2 - x^2(y+2z);$$

the surface Z' is the rational elliptic surface determined by the pencil generated by the plane cubics $D_0 = V(d_0)$ and $D_\infty = V(d_\infty)$. As before, the fiber corresponding to D_0 is of type I_8; the fiber corresponding to D_∞ is of type I_2; the only remaining singular fibers are of type I_1 and correspond to the cubics $V(d_0 + d_\infty)$ and $V(d_0 - d_\infty)$. Observe that the singularity of the pencil defining Z' at the point $[0, 1, -1]$ is resolved only after two successive blow ups. To obtain Z from Z' it suffices to contract the exceptional curve lying above the point $[0, 1, -1]$ (the one introduced by the second blow up) and to blow up the point $[0, 1, 1]$. Finally, the Coble surface $\tilde{Y}$ is obtained from Z by blowing up the singular point of one of the two I_1 fibers. The two surfaces obtained by the choice of the last blown up point are isomorphic: the substitution $([x, y, z], t) \mapsto ([ix, y, -z], -t)$ in the pencil $c_0 + tc_\infty$ defining Z determines an automorphism of order 4 on Z exchanging the two I_1 fibers.

References

[AL11] M. Artebani and A. Laface, Cox rings of surfaces and the anticanonical Iitaka dimension, *Advances in Mathematics* **226** (2011), 5252–5267.

[ADHL] I. Arzhantsev, U. Derenthal, J. Hausen and A. Laface, *Cox Rings*, in preparation.

[BCP97] W. Bosma, J. Cannon and C. Playoust, The Magma algebra system. I. The user language, *J. Symbolic Comput.* **24**(2011), 235–265.

[DZ01] I. Dolgachev and D.-Q. Zhang, Coble rational surfaces *Amer. J. Math.* **123** (2011), 79–114.

[Dui10] J. Duistermaat, *Discrete Integrable Systems*, Springer Monographs in Mathematics, Springer, New York, 2010.

[HK00] Y. Hu and S. Keel, Mori dream spaces and GIT, *Michigan Math. J.* **48** (2000), 331–348.

[Kon86] S. Kondō, Enriques surfaces with finite automorphism groups, *Japan. J. Math. (N.S.)* **12** (1986), 191–282.

[Nak00] M. Nakamaye, Stable base loci of linear series, *Math. Ann.*, **318** (2000), 837–847.

[TVAV] D. Testa, A. Várilly-Alvarado and M. Velasco, Big rational surfaces, *Math. Ann.* **351** (2011), 95–107.

[Tot10] B. Totaro, The cone conjecture for Calabi-Yau pairs in dimension 2, *Duke Math. J.* **154** (2010), 241–263.

[Xie10] Q. Xie, Strongly liftable schemes and the Kawamata-Viehweg vanishing in positive characteristic, *Math. Res. Lett.* **17** (2010), 563–572.

12

Example of a transcendental 3-torsion Brauer-Manin obstruction on a diagonal quartic surface

Thomas Preu

Institut für Mathematik, Universität Zürich

Abstract We give an explicit example of a diagonal quartic surface defined over the rationals for which a necessarily transcendental 3-torsion element of its Brauer group gives rise to an obstruction to weak approximation. We also show that no non-constant 2-primary element exists in its Brauer group, demonstrating the need to consider all of the Brauer group for obstructions on diagonal quartics over the rationals and not just the 2-torsion part. We give a concrete number theoretic application of our construction.

12.1 Motivation and introduction

By variety we mean a geometrically reduced irreducible scheme separated and of finite type over a field. Let X be a smooth projective variety defined over $\mathbb{Q}$. We associate to it a well-known sequence of inclusions

$$X(\mathbb{Q}) \subset X(\mathbb{A}_{\mathbb{Q}})^{\mathrm{Br}} \subset X(\mathbb{A}_{\mathbb{Q}}).$$

All terminology and the underlying theory is standard and can be found in text books, e.g., in [14, Ch. 5]. If $X(\mathbb{A}_{\mathbb{Q}})^{\mathrm{Br}} \neq X(\mathbb{A}_{\mathbb{Q}})$ we say we have a Brauer-Manin obstruction to weak approximation.

Diagonal quartic surfaces are a case of special interest: they have relatively rich arithmetic behavior and still have a simple enough representation to be tackled by explicit computations. They are smooth projective varieties defined by a single equation of the form

$$a_0 x_0^4 + a_1 x_1^4 + a_2 x_2^4 + a_3 x_3^4 = 0, \quad a_i \in \mathbb{Q}^{\times}.$$

For background on the geometry and arithmetic of such a surface we refer to [1]. Currently the only known Brauer-Manin obstructions for such surfaces are mediated by 2-torsion elements of the Brauer group of X, cf. [1] for an algebraic one or [6] for a transcendental one. In [1] and [7] one finds bounds on the algebraic respectively transcendental Brauer group:

$$\#(\mathrm{Br}_1(X)/\mathrm{Br}_0(X)) \mid 2^5, \quad \#(\mathrm{Br}(X)/\mathrm{Br}_1(X)) \mid 2^{20}3^2 5^2.$$

We restate a sufficient criterion for triviality of transcendental prime torsion from [6, Theorem 5.2.] and [7, Theorem 3.2.]:

Proposition 12.1 *Let X be a diagonal quartic over $\mathbb{Q}$ as above. Denote by M the subgroup in $\mathbb{Q}^\times/\mathbb{Q}^{\times 4}$ generated by a_i/a_j for $i, j \in \{0, 1, 2, 3\}$ and -4. Let M' be the subgroup generated by M and -1. Then*

1. $(\mathrm{Br}(X)/\mathrm{Br}_1(X))[2] \cong 0$ *if* $2 \notin M$,
2. $(\mathrm{Br}(X)/\mathrm{Br}_1(X))[3] \cong 0$ *if* $3 \notin M'$,
3. $(\mathrm{Br}(X)/\mathrm{Br}_1(X))[5] \cong 0$ *if* $5 \notin M'$.

Given the above results a natural question is whether (necessarily transcendental) 3-torsion elements are of any interest for Brauer-Manin obstruction for the class of diagonal quartic surfaces. Our main result Theorem 12.8 answers this affirmatively by an explicit example. For other types of surfaces obstructions necessarily mediated by algebraic 3-torsion elements are known, e.g., cubic surfaces as shown in [2].

12.2 The candidate surface

We look at the diagonal quartic surface X over the rationals in $\mathbb{P}^3_{\mathbb{Q}}$ defined by the equation

$$x_0^4 + 3x_1^4 - 4x_2^4 - 9x_3^4 = 0. \tag{12.1}$$

Searching for rational points with a computer yields the following list of small primitive solutions:

```
[ 1, 1, 1, 0 ],          [ 7, 3, 5, 2 ],            [ 95, 63, 73, 36 ],
[ 43, 167, 155, 42 ], [ 101, 201, 157, 130 ], [ 331, 393, 103, 310 ],
[ 715, 39, 307, 398 ], [ 299, 581, 65, 444 ], [ 767, 273, 85, 448 ].
```

By inspection we see that the relation $3 \mid x_1x_3$ holds in all the found points, while there are 3-adic solutions which do not satisfy this relation, e.g., $(\sqrt[4]{10}, 1, 1, 1)$. This hints at a failure of weak approximation possibly explained by a Brauer-Manin obstruction.

Using Bright's `MAGMA`-script, which is available at `www.boojum.org.uk/maths/quartic-surfaces/index.html`, we see that the algebraic part is trivial. Alternatively we can read this off from the appendix in [1, p.143] as X falls into case `A282`. Combining this with the criterion of Proposition 12.1 we get that $\mathrm{Br}(X)/\mathrm{Br}_0(X) \leq (\mathbb{Z}/3\mathbb{Z})^2$ and any Brauer-Manin obstruction must come from a transcendental 3-torsion element.

12.3 Constructing a 3-torsion element

In this section we construct a 3-torsion element in the Brauer group of a surface related to X which is used in the next section to get a Brauer element on X.

The inspiration for the construction as follows is from [9], where the approach to compute transcendental Brauer-Manin obstruction effectively with the help of fibrations is systematically studied. Without further mention, we use a result of Grothendieck [5, V.2.2] that the Brauer group and the cohomological Brauer group of a quasi-projective smooth surface are canonically isomorphic.

The fibration f that we use is a rational map with locus of indeterminancy given by $x_0 = x_1 = 0$.

$$X \dashrightarrow \mathbb{P}^1_{\mathbb{Q}}, [x_0 : x_1 : x_2 : x_3] \mapsto [x_0 : x_1].$$

We will relate X to a product of elliptic curves and find a connection between some of their Brauer group elements. All maps involved are generically finite of degree a power of 2. For a clearer exposition we give the diagram first and elaborate on it afterward.

$$\begin{array}{ccccccccccccc}
X & \xleftarrow{bl} & X' & \xleftarrow{h} & X'_k & \xleftarrow{h'} & X'' & \xleftarrow{bl'} & X''' & \xrightarrow{h'''} & E \times F & \longrightarrow & F. \\
 & \searrow^{f} & \downarrow & & \downarrow & & \downarrow & & \downarrow & \swarrow & & & \\
 & & \mathbb{P}^1_{\mathbb{Q}} & \longleftarrow & \mathbb{P}^1_k & \xleftarrow{g} & E & = & E & & & &
\end{array}$$

(In the diagram, f is a dashed arrow from X to $\mathbb{P}^1_{\mathbb{Q}}$, and the unlabelled diagonal arrow goes from $E \times F$ to E.)

We do not carry out the computations for the blowups, desingularizations, loci of indeterminacy and generic degrees. The reader should easily reconstruct those standard computations. We will however give some illustrative explicit equations.

First by blowing up the locus of indeterminacy of the fibration f via bl, we get a smooth projective variety X' and a morphism from X' to $\mathbb{P}^1_{\mathbb{Q}}$, which is a fibration.

Then we perform two base extensions. The first is given by a Galois field extension $k/\mathbb{Q}$ of degree 16. We give a definition of k below where it will become clear why we perform it.

Then we extend $\mathbb{P}^1_k$ to an elliptic curve E by a degree 2 cover g ramified in 4 points. E is defined as a variety in weighted projective space, and X'' is the fiber product $X'_k \times_{\mathbb{P}^1_k} E$.

$$\mathbb{P}_k(1,1,2) \supset E := \{s^4 + 3t^4 - u^2 = 0\} \to \mathbb{P}^1_k, [s:t:u] \mapsto [s:t]. \quad (12.2)$$

Explicitly X'' can be given as a subvariety of $\mathbb{P}^3_k \times \mathbb{P}_k(1,1,2)$ by the equations in (12.1), (12.2) and $tx_0 = sx_1$. It has, geometrically, 4 singular points determined by $x_2 = x_3 = u = 0$ which define a subscheme $W'' \subset X''$. Its image $W'_k := h''(W'') \subset X'_k$ also consists of 4 geometric points and we have by restriction of h' a flat morphism $\tilde{h}' : \tilde{X}'' := X'' \setminus W'' \to X'_k \setminus W'_k =: \tilde{X}'_k$ which is finite of degree 2 and ramified at 16 geometric lines. We also have a rational map from X'' to another elliptic curve F sending $((x_1, x_2, x_3), (t, u))$ to $[x_2 : x_3 : u]$ on the affine chart defined by $x_0 = 1 = s$:

$$F := \{4y^4 + 9z^4 - v^2 = 0\} \subset \mathbb{P}_k(1,1,2).$$

Its locus of indeterminacy coincides exactly with the singularities of X''. By blowing up via bl' we can remove the singular locus and get a smooth projective variety X''' with a morphism to F.

This induces a morphism $h''' : X''' \to E \times F$, which turns out to be generically finite of degree 2. The locus where the morphism is not finite is the exceptional locus of the blowup bl'.

We have automorphisms on E, F and X''' respectively each denoted by ϕ which should not cause any confusion, since each is given by either $u \mapsto -u$ or $v \mapsto -v$ which behave compatibly. These morphisms induce automorphisms on $k(E), k(F), k(X''')$ and $\mathrm{Br}(X''')$, each denoted ϕ^*.

Lemma 12.2 *Let $q_1, q_2 \in k(X''')^\times$ be as defined below. Then $(q_1, q_2)_3$ represents a class in $\mathrm{Br}(k(X'''))[3]$ which is the restriction of a unique element $\mathcal{A} \in \mathrm{Br}(X''')[3]$.*

Proof First we construct an explicit 2-cocycle $(\hat{q}_1, \hat{q}_2)_3$ representing a nontrivial element in $\mathrm{Br}(E \times F)[3]$. Essentially we use the cup product pairing (cf. [10, VI Corollary 8.13.])

$$H^1_{ét}(E, \mathbb{Z}/3\mathbb{Z}) \times H^1_{ét}(F, \mathbb{Z}/3\mathbb{Z}) \to H^2_{ét}(E \times F, \mathbb{Z}/3\mathbb{Z}).$$

For a smooth projective variety Y over a field of characteristic prime to 3 like E, F or $E \times F$ the Kummer sequence yields $\alpha_Y : H^1_{ét}(Y, \mu_3) \twoheadrightarrow \mathrm{Pic}(Y)[3]$ and $\beta_Y : H^2_{ét}(Y, \mu_3) \twoheadrightarrow \mathrm{Br}(Y)[3]$. We illustrate α_Y (cf. [10, p. 125]) by computing preimages: for $D \in \mathrm{Div}(Y)$ with $[D] \in \mathrm{Pic}(Y)[3]$ we choose a trivialization $\mathcal{O}_Y(3D) \xrightarrow{\sim} \mathcal{O}_Y$ inducing the structure of an algebra on $\mathcal{M} := \mathcal{O}_Y \oplus \mathcal{O}_Y(D) \oplus \mathcal{O}_Y(2D)$. We get a μ_3-Galois cover $Y' := \mathrm{Spec}(\mathcal{M}) \to Y$ representing an element of $H^1_{ét}(Y, \mu_3)$. If $q \in k(Y)^\times$ satisfies $\mathrm{div}(q) = 3D$, then $k(Y') \cong k(Y)[T]/(T^3 - q)$, and the element in $H^1_{ét}(\mathrm{Spec}(k(Y)), \mu_3)$ represented by q is the restriction of an element in $H^1_{ét}(Y, \mu_3)$ which maps to $[D]$ via α_Y.

Now k is chosen as the smallest number field over which the full 3-torsion in the Picard group of E and F is defined. Using division polynomials we get $k := \mathbb{Q}(\zeta_{12}, \rho)$, where $\rho := \sqrt[4]{3 + 2\zeta_{12} - 3\zeta_{12}^2 - 4\zeta_{12}^3}$. Since k contains a 3rd root of unity we can fix an isomorphism $\mathbb{Z}/3\mathbb{Z} \cong \mu_3$ over k. We have $[k : \mathbb{Q}] = 16$ and it is a splitting field of $x^{16} - 6x^{12} + 39x^8 + 18x^4 + 9$, of which we denote a chosen root in k by η.

By consequences of the Kummer sequence stated above a pair $([D_E], [D_F]) \in \mathrm{Pic}(E)[3] \times \mathrm{Pic}(F)[3]$ of divisor classes given by divisors D_E, D_F give rise to an element $\widehat{\mathcal{A}}$ of $\mathrm{Br}(E \times F)[3]$. First, let $\hat{q}_1 \in k(E)^\times, \hat{q}_2 \in k(F)^\times$ represent preimages of $[D_E], [D_F]$ by α_E, α_F. By the cup product the pair $(\hat{q}_1, \hat{q}_2)$ represents an element in $H^2_{ét}(E \times F, \mu_3)$. Via $\beta_{E\times F}$ the class of the cyclic algebra $(\hat{q}_1, \hat{q}_2)_3$ represents $\widehat{\mathcal{A}} \in \mathrm{Br}(E \times F)$. Here we used the inclusion of Brauer groups of smooth varieties into Brauer groups of their function fields implicitly (cf. [10, IV Corollary 2.6.]) and continue to do so.

By construction of k we have that $E(k)[3] \cong (\mathbb{Z}/3\mathbb{Z})^2$. To compute $\hat{q}_1$ we fix a base point O_1 on E. This allows us to write $D_E = P_1 - Q_1$ for 3-torsion points P_1, Q_1. We additionally require $P_1 = -Q_1$ in our construction and normalize $\hat{q}_1(O_1) = 1$. This ensures $\hat{q}_1$ to be anti-invariant under the map given by $u \mapsto -u$, i.e., $\phi^*(\hat{q}_1) = \frac{1}{\hat{q}_1}$ as required in the proof of Lemma 12.3. We compute $\hat{q}_1$ by effective Riemann-Roch as implemented in MAGMA. We proceed analogously for $\hat{q}_2$ and F. We write q_1, q_2 for the pullbacks to X''' giving rise to the desired class $\mathcal{A} = h'''^*\widehat{\mathcal{A}}$.

Let V be the affine chart of X'' defined by $s = 1 = x_0$. On $U := V \backslash W''$ which is at the same time an open subset of X''' we get for some choice of the P_i, Q_i:

$$
\begin{aligned}
q_1 :=\\
&((21\eta^{14}+33\eta^{12}-169\eta^{10}-195\eta^{8}+1131\eta^{6}+1287\eta^{4}-1572\eta^{2}+828)t^3+\\
&(8\eta^{14}-24\eta^{12}-39\eta^{10}+156\eta^{8}+273\eta^{6}-936\eta^{4}+261\eta^{2}-198)t^2+\\
&(64\eta^{15}+28\eta^{13}-429\eta^{11}-143\eta^{9}+2769\eta^{7}+897\eta^{5}-603\eta^{3}+1635\eta)ut+\\
&(-95\eta^{14}+43\eta^{12}+585\eta^{10}-273\eta^{8}-3783\eta^{6}+1833\eta^{4}-1242\eta^{2}-396)t+\\
&3(3\eta^{15}-26\eta^{13}+156\eta^{9}-1014\eta^{5}+821\eta^{3}-468\eta)u+\\
&3(30\eta^{14}+2\eta^{12}-195\eta^{10}+1261\eta^{6}-45\eta^{2}+452))/\\
&((36\eta^{14}+93\eta^{12}-299\eta^{10}-546\eta^{8}+1911\eta^{6}+3510\eta^{4}-2667\eta^{2}+2259)t^3+\\
&6(16\eta^{14}-21\eta^{12}-91\eta^{10}+130\eta^{8}+585\eta^{6}-858\eta^{4}+561\eta^{2}-183)t^2+\\
&(-70\eta^{14}+21\eta^{12}+429\eta^{10}-156\eta^{8}-2769\eta^{6}+936\eta^{4}-\\
&909\eta^{2}-207)t+156),
\end{aligned}
$$

$$
\begin{aligned}
q_2 :=\\
&(2(4\eta^{14}+25\eta^{12}-39\eta^{10}-130\eta^{8}+195\eta^{6}+1014\eta^{4}-279\eta^{2}-213)x_2^3+\\
&2(-72\eta^{14}+39\eta^{12}+468\eta^{10}-234\eta^{8}-3042\eta^{6}+1638\eta^{4}+108\eta^{2}+351)x_2^2x_3+\\
&(47\eta^{15}+15\eta^{13}-286\eta^{11}-117\eta^{9}+1950\eta^{7}+585\eta^{5}+417\eta^{3}+270\eta)ux_2+\\
&3(-80\eta^{14}-15\eta^{12}+507\eta^{10}+78\eta^{8}-3315\eta^{6}-234\eta^{4}-1089\eta^{2}-153)x_2x_3^2+\\
&(-34\eta^{15}+18\eta^{13}+195\eta^{11}-117\eta^{9}-1131\eta^{7}+351\eta^{5}-729\eta^{3}+675\eta)ux_3+\\
&9(-6\eta^{12}+13\eta^{10}-39\eta^{6}-39\eta^{2}-264)x_3^3)/\\
&(4(-\eta^{14}-5\eta^{12}+65\eta^{8}-78\eta^{6}-273\eta^{4}+99\eta^{2}+66)x_2^3+\\
&12(12\eta^{14}-11\eta^{12}-78\eta^{10}+78\eta^{8}+468\eta^{6}-390\eta^{4}-18\eta^{2}-81)x_2^2x_3+\\
&18(15\eta^{14}+2\eta^{12}-91\eta^{10}-13\eta^{8}+585\eta^{6}+117\eta^{4}+\\
&192\eta^{2}+75)x_2x_3^2+1404x_3^3).
\end{aligned}
$$

□

12.4 Analysis of the geometry

Lemma 12.3 *There is a $\mathcal{B} \in \mathrm{Br}(X'_k)[3]$ such that $(h' \circ bl')^*\mathcal{B} = \mathcal{A} \in \mathrm{Br}(X''')[3]$.*

Unfortunately h' and X'' are not nice enough to conclude this from standard theorems. We have to consider open subvarieties missing the singular locus to transfer $\mathcal{A}$ to a Brauer class on X'_k.

Proof By [10, IV Corollary 2.6.] the inclusion of the generic point of a smooth variety induces an inclusion between the Brauer groups; by functoriality and properties of injective maps, an open inclusion between smooth varieties also induces an inclusion between Brauer groups. We get

$$\mathrm{Br}(X''')[3] \hookrightarrow \mathrm{Br}(\tilde{X}'')[3] \hookrightarrow \mathrm{Br}(k(X'''))[3] \qquad (12.3)$$

and we may identify the corresponding function fields and their Brauer groups.

The finite flat morphism $\tilde{h}' : \tilde{X}'' \to \tilde{X}'_k$ induces a Galois extension on function fields with Galois group $H \cong \mathbb{Z}/2\mathbb{Z}$. Explicitly the nontrivial Galois automorphism ϕ^* on $k(X''')/k(X'_k)$ is induced by $u \mapsto -u$. Also $\tilde{h}'$ fits into the following commutative diagram:

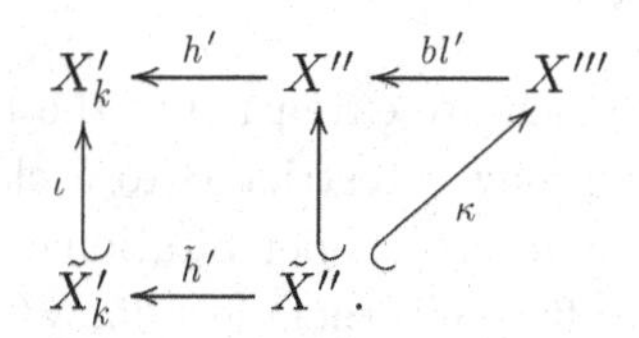

Recall that $\mathcal{A}$ is given by $(q_1, q_2)_3$. By construction we had $\phi^*(q_i) = \frac{1}{q_i}$ for $i \in \{1, 2\}$ which yields $\phi^*[(q_1, q_2)_3] = [(q_1, q_2)_3]$, i.e., Galois invariance. This implies $\mathcal{A}'' := \kappa^*\mathcal{A} \in \mathrm{Br}(\tilde{X}'')[3]^H$. Restriction gives an isomorphism $\mathrm{Br}(\tilde{X}'_k)[3] \xrightarrow{\sim} \mathrm{Br}(\tilde{X}''))[3]^H$ (cf. [7, Theorem 1.4]) and we have $\mathcal{A}'' = \tilde{h}'^*\mathcal{A}'$ for a unique $\mathcal{A}' \in \mathrm{Br}(\tilde{X}'_k)[3]$.

Let $\mathcal{B} \in \mathrm{Br}(X'_k)[3]$ be the class associated to $\mathcal{A}'$ via the restriction $\mathrm{Br}(X'_k)[3] \xrightarrow{\sim} \mathrm{Br}(\tilde{X}'_k)[3]$ which is an isomorphism by the purity theorem [5, VI.6.1 and VI.6.2] since W'_k has codimension 2 in X'_k.

We get

$$(h' \circ bl' \circ \kappa)^*\mathcal{B} = (\iota \circ \tilde{h}')^*\mathcal{B} = \mathcal{A}''.$$

Since $\mathcal{A}''$ and $\mathcal{A}$ are equal in $\mathrm{Br}(k(X'''))[3]$ by (12.3) we get our claim. □

Next we apply corestriction to Brauer groups and explain some of its properties first. By the first half of the proof of [7, Theorem 1.4] we get for a finite flat morphism $\Gamma : Z \to Y$ of noetherian schemes whose Brauer group and cohomological Brauer group are canonically isomorphic a corestriction map $\mathrm{cores}_{Z/Y} : \mathrm{Br}(Z) \to \mathrm{Br}(Y)$ (this argument – valid in this more general context compared to that in loc. cit. – uses the existence of a norm map of structure sheaves, which is standard). If we fit Γ into a fiber diagram of noetherian schemes as on the left in (12.4) we get a commutative square as on the right. This is by compatibility of the norm map with base change, for such Γ, of the Leray spectral sequence with pull backs (which uses [10, II. Corollary 3.6.], i.e., $R^i\Gamma_* = 0$ for Γ finite and $i > 0$) and of the Leray spectral sequence with change of the coefficient sheaf which all carries over to cohomology. These facts are also used in the following section.

$$\begin{array}{ccc} Z & \xleftarrow{G'} & Z' \\ \downarrow{\scriptstyle\Gamma} & & \downarrow{\scriptstyle\Gamma'} \\ Y & \xleftarrow[G]{} & Y' \end{array} \qquad \begin{array}{ccc} \mathrm{Br}(Z) & \xrightarrow{G'^*} & \mathrm{Br}(Z') \\ {\scriptstyle \mathrm{cores}_{Z/Y}}\downarrow & & \downarrow{\scriptstyle \mathrm{cores}_{Z'/Y'}} \\ \mathrm{Br}(Y) & \xrightarrow[G^*]{} & \mathrm{Br}(Y')\,. \end{array} \tag{12.4}$$

Lemma 12.4 *The class $\mathcal{B}$ corestricts to a class in $\mathrm{Br}(X')$ which is the pull back of a class $\mathcal{C} \in \mathrm{Br}(X)$.*

Proof Since h is induced by a finite extension of the base field, it is a finite and flat morphism. We may corestrict $\mathcal{B}$ to a class $\mathcal{B}' \in \mathrm{Br}(X')$ (see above). Point blowups as bl are proper birational morphisms, and we can therefore apply a result of Grothendieck [5, VI.7.3.] for smooth projective varieties over fields of characteristic 0 to get an isomorphism $bl^* : \mathrm{Br}(X) \xrightarrow{\sim} \mathrm{Br}(X')$, which yields $\mathcal{C} \in \mathrm{Br}(X)$. □

12.5 Analysis of the arithmetic

In this section we analyze local invariants for $\mathcal{C} \in \mathrm{Br}(X)$.

We denote by $\Omega_{\mathbb{Q}}$ the set of places of $\mathbb{Q}$, for $p \in \Omega_{\mathbb{Q}}$ and $c \in X(\mathbb{Q}_p)$ let $\mathrm{inv}_p(\mathcal{C}(c))$ be the invariant of the $\mathbb{Q}_p$-Brauer class obtained by restricting $\mathcal{C}$ to c and let inv_K denote the invariant map for a local field K.

We start with a preparatory lemma, in which $p \in \{2, 3\}$ and k_p denotes the completion of k at the unique prime over the non-split prime p. For a k-scheme such as X''' we let $X'''(k_p)$ denote the set of k_p-points of the k-scheme, and call a point $a \in X'''(k_p)$ a lift of some $c \in X(\mathbb{Q}_p)$ if the following diagram commutes:

$$\begin{array}{ccc} X''' & \xleftarrow{a} & \mathrm{Spec}(k_p) \\ \downarrow & & \downarrow \\ X & \xleftarrow{c} & \mathrm{Spec}(\mathbb{Q}_p)\,. \end{array} \tag{12.5}$$

Lemma 12.5 *For $p \in \{2,3\}$, every $c \in X(\mathbb{Q}_p)$ can be written as $c = [1 : y_1 : y_2 : y_3]$ with $y_1, y_2, y_3 \in \mathbb{Z}_p$ and there are exactly two lifts $a_\pm \in X'''(k_p)$ with $\mathrm{inv}_p(\mathcal{C}(c)) = \mathrm{inv}_{k_p}(\mathcal{A}(a_+)) = \mathrm{inv}_{k_p}(\mathcal{A}(a_-))$.*

Proof For both primes divisibility arguments from elementary number theory yield that any $c \in X(\mathbb{Q}_p)$ can be represented in coordinates by $[1 : y_1 : y_2 : y_3]$ with $y_1, y_2, y_3 \in \mathbb{Z}_p$, and thus there is a unique $b' \in X'(\mathbb{Q}_p)$ with $c = bl(b')$. If not stated otherwise we assume $y_0 = 1$ for any point given by coordinates. Because p is non-split there is only a single lift of b' to a point $b \in X'_k(k_p)$. Furthermore there are exactly two points (see below) of the form

$$a_\pm = ([1 : y_1 : y_2 : y_3], [1 : t : u]) = ([1 : y_1 : y_2 : y_3], [1 : y_1 : \pm\sqrt{1 + 3y_1^4}]) \\ \in U(k_p) \subset X'''(k_p)$$

mapping to b under $h' \circ bl'$. All lifts of $\mathbb{Q}_p$-points of X avoid the exceptional locus of bl' and can indeed be considered in $U(k_p)$ which allows us to use simpler coordinates.

For $p = 2$ explicit calculations show that any solution of $1 + 3y_1^4 = 4y_2^4 + 9y_3^4$ in $\mathbb{Z}/2^5\mathbb{Z}$ has the property that $1 + 3y_1^4$ is a square in $(\mathbb{Z}/2^5\mathbb{Z})[\sqrt{-3}]$. Let $\mathrm{val}_{\mathbb{Q}_2(\sqrt{-3})}$ be the normalized valuation. We apply a version of Hensel's lemma for $\mathrm{val}_{\mathbb{Q}_2(\sqrt{-3})}(\frac{\partial r}{\partial y}) > 0$ to the polynomial $r := y^2 - (1 + 3y_1^4)$ over $\mathbb{Q}_2(\sqrt{-3}) \subset k_2$. Thus above every $\mathbb{Q}_2$-point of X are two points in $U(k_2)$. For $p = 3$ we have $1^2 = 1 + 3y_1^4 \in \mathbb{Z}/3\mathbb{Z}$ is a square and applying standard Hensel's lemma to r shows that any $\mathbb{Q}_3$-point of X lifts to two points of $U(k_3)$.

For a finite separable extension of local fields L/K we have by [12, (7.1.4) Corollary] the commutative diagram:

$$\begin{array}{ccc} \mathrm{Br}(L) & \xrightarrow{\mathrm{inv}_L} & \mathbb{Q}/\mathbb{Z} \\ {\scriptstyle \mathrm{cores}_{L/K}} \downarrow & & \| \\ \mathrm{Br}(K) & \xrightarrow[\mathrm{inv}_K]{} & \mathbb{Q}/\mathbb{Z}. \end{array} \tag{12.6}$$

Functoriality of the Brauer group, i.e., compatibility with restriction shows $\mathrm{inv}_p(\mathcal{C}(c)) = \mathrm{inv}_p(\mathcal{B}'(b'))$ and $\mathrm{inv}_{k_p}(\mathcal{B}(b)) = \mathrm{inv}_{k_p}(\mathcal{A}(a_+)) = \mathrm{inv}_{k_p}(\mathcal{A}(a_-))$. We apply (12.4) to a diagram like (12.5) with b, b' replacing a, c and (12.6) with $L = k_p, K = \mathbb{Q}_p$ to conclude $\mathrm{inv}_p(\mathcal{B}'(b')) = \mathrm{inv}_{k_p}(\mathcal{B}(b))$, and thus get:

$$\mathrm{inv}_p(\mathcal{C}(c)) = \mathrm{inv}_{k_p}(\mathcal{A}(a_+)).$$

□

Proposition 12.6 *For $p \in \Omega_{\mathbb{Q}} \setminus \{3\}$ the constant* $\mathrm{inv}_p(\mathcal{C}(c'))$ *is independent of $c' \in X(\mathbb{Q}_p)$, every $c \in X(\mathbb{Q}_3)$ can be written as $c = [1 : y_1 : y_2 : y_3]$ with $y_1, y_2, y_3 \in \mathbb{Z}_3$ and* $\mathrm{inv}_3(\mathcal{C}(c)) = 0$ *if and only if* $3 \mid y_1 y_3$.

Proof Applying a result of Colliot-Thélène and Skorobogatov [3, Proposition 2.4. and Corollary 3.3.] shows that at all places except possibly 2 and 3 the invariant is constant. So we are reduced to considering $p \in \{2, 3\}$. We apply Lemma 12.5 and are left with computing invariants of $\mathcal{A}$ for one of the two lifts a of c.

Let us consider 2-adic points first. We show that q_1 evaluated at a is a cube in k_2 and so $\mathrm{inv}_{k_p}(\mathcal{A}(a)) = 0$ (cf. [13, XIV§2. Proposition 4.]): we apply Hensel's lemma to $y^3 - q_1(a)$. MAGMA calculates all solutions modulo 2, evaluates q_1 at these points and checks, if it is a 3rd root in k_2. This turns out to be the case for any modulo 2 solution. This concludes the proof of the first part of the proposition.

For $p = 3$ we use the explicit formulas for the local Hilbert symbol (sometimes called norm residue symbol) to compute local invariants as given in [11, V. (3.7) Theorem] (see [4, VII.] for more general formulas). The formulas involve computing Laurent series associated to $q_1(a), q_2(a)$ and taking Laurent series residues of that series, followed by a mod 3-residue. The associated Laurent series have to be computed only up to the precision where the mod 3-residue of the -1-power coefficient of the resulting series stabilizes. In [16] one can find bounds which apply in our case.

Since by Hensel's lemma any solution over $\mathbb{Z}/3^n\mathbb{Z}$ for $n \geq 1$ of (12.1) induces a solution c over $\mathbb{Z}_3$, we may work mod 3^n where n is big enough such that the Hilbert symbols of associated pairs $(q_1(a), q_2(a))$ are constant on the mod 3^n-classes of $\mathbb{Z}_3$-solutions of (12.1). It suffices to take n big enough such that $q_1(a)k_3^{\times 3} = q_1(a')k_3^{\times 3}$ and $q_2(a)k_3^{\times 3} = q_2(a')k_3^{\times 3}$ at suitable lifts a, a' of any two different integral 3-adic points $c, c' \in X(\mathbb{Q}_3)$ with $c \equiv c' \bmod 3^n$. If we take $n = 2$, then the coordinates $y_0 = s = 1, u = \pm\sqrt{1 + 3y_1^4}, t = y_1, y_2, y_3$ associated to a, a' coincide mod 3^2 and moreover y_2 always has valuation 0. This together with valuation analysis of the coefficients of q_1, q_2 yields that the quotients $q_1(a)/q_1(a')$ and $q_2(a)/q_2(a')$ are of the form $d := 1 + 3^2 \cdot d'$ for some $d' \in \mathfrak{o}_{k_3}$. As $\mathrm{val}_{k_3}(3) = 8$ for the normalized valuation the proof of [4, I. Proposition (5.9)] tells us that $d \in k_3^{\times 3}$.

We compute all solutions mod 3^2 by iterating over y_1, y_3 mod 3^2 and setting $y_2 = \pm\sqrt[4]{(1 + 3y_1^4 - 9y_3^4)/4} \in \mathbb{Z}_3$. With MAGMA we can go through all the 162 cases and evaluate the invariant of $(q_1, q_2)_3$ at the lifts using the local Hilbert symbol. The correlation between invariants and residue classes $r \equiv y_1 y_3 \bmod 3$ is:

	$\mathrm{inv}_3 = 0$	$\mathrm{inv}_3 = 1/3$	$\mathrm{inv}_3 = 2/3$
$r = 0$	90	0	0
$r = 1$	0	18	18
$r = 2$	0	18	18

□

Since there are some $\mathbb{Q}$-rational points in the class with $r = 0$ and since the invariant is constant at all other places besides 3 we conclude that any $c \in X(\mathbb{Q})$ for the surface X satisfies $\sum_{p \in \Omega_{\mathbb{Q}} \setminus \{(3)\}} \mathrm{inv}_p(\mathcal{C}(\iota_p(c))) = 0$ where $\iota_p : X(\mathbb{Q}) \to X(\mathbb{Q}_p)$ is the canonical inclusion. Thus the local invariant at 3 must be 0 for any $\mathbb{Q}$-point c. We deduce by the above table that $3 \mid y_1 y_3$, and this divisibility relation is completely explained by an obstruction to weak approximation by a transcendental 3-torsion element in $\mathrm{Br}(X)$.

Remark 12.7 1. The invariants at each prime for the Brauer element $\mathcal{C}$ agree for all rational points. This is wrong in general as [15, proof of Proposition 7.1.] shows.

2. That a 3-torsion element obstructs at the place 3 is only a coincidence (see, e.g., [8, Example 8.] for a 2-torsion element obstructing at 17).
3. Whether all points with $r = \mathrm{inv}_3 = 0$ can be approximated is unclear to us. Using a list of all 57 positive primitive solutions with coordinates less than 10^7 we can however, varying the signs of the coordinates, cover all 90 non-obstructed classes mod 3^2.

12.6 Conclusion

We summarize the above calculations and arguments.

Theorem 12.8 1. *For the diagonal quartic surface* $X = \{x_0^4 + 3x_1^4 - 4x_2^4 - 9x_3^4 = 0\} \subset \mathbb{P}^3_{\mathbb{Q}}$ *there is a transcendental* $\mathcal{C} \in \mathrm{Br}(X)[3]$ *giving rise to a Brauer-Manin obstruction to weak approximation.*

2. *We have* $\mathbb{Z}/3\mathbb{Z} \leq \mathrm{Br}(X)/\mathrm{Br}_0(X) \cong \mathrm{Br}(X)/\mathrm{Br}_1(X) \leq (\mathbb{Z}/3\mathbb{Z})^2$, *so there are neither algebraic nor 2-primary Brauer elements that could account for the obstruction.*

3. *Every integral solution* $c = [y_0, y_1, y_2, y_3]$ *to the defining equation, e.g.,* $[1,1,1,0]$ *obeys the congruence* $3 \mid y_1 y_3$*, while this is not apparent from 3-adic analysis as* $[\sqrt[4]{10}, 1, 1, 1]$ *illustrates. We have* $X(\mathbb{Q}) \subset X(\mathbb{Q}_3)$ *is not dense and* $c' \in X(\mathbb{A}_\mathbb{Q})$ *satisfies* $c' \in X(\mathbb{A}_\mathbb{Q})^{\mathcal{C}}$ *if and only if* $3 \mid z_1 z_3$ *holds for the* $\mathbb{Z}_3$*-coordinate representation* $[1, z_1, z_2, z_3]$ *of the* $\mathbb{Q}_3$*-component of* c'*.*

Remark 12.9 As the first referee indicated the argument gets simpler by discarding most of Proposition 12.6 and only computing $\mathrm{inv}_3(\mathcal{C}(\iota_3(c'))) \neq \mathrm{inv}_3(\mathcal{C}(c''))$ explicitly for, e.g., $c' = [1:1:1:0] \in X(\mathbb{Q})$ and $c'' = [\sqrt[4]{10} : 1 : 1 : 1] \in X(\mathbb{Q}_3) \setminus X(\mathbb{Q})$ if one is only interested in the first two parts of Theorem 12.8.

The `MAGMA`-scripts which were used for the computations are part of the authors PhD thesis are available from the author on request.

An open problem for future research is to determine if 5-torsion in Brauer groups of diagonal quartic surfaces plays any role for arithmetic obstructions.

12.7 Acknowledgment

I thank Andrew Kresch for his guidance through my thesis and the many colleges from Zurich and various workshops for moral support and helpful discussions, especially the organizers and participants of the workshop on torsors in Edinburgh, 2011. I also thank Andreas-Stephan Elsenhans for providing a list of positive integral solutions. The author was supported in part by a grant from the SNF.

References

[1] Bright M.J., *Computations on diagonal quartic surfaces*, PhD thesis, University of Cambridge, 2002, `www.boojum.org.uk/maths/quartic-surfaces/thesis.pdf`

[2] Colliot-Thélène J.-L., D. Kanevsky, Sansuc J.-J., *Arithmètique des surfaces cubiques diagonales*, Ed. Wüstholz G., *Diophantine Approximation and Transcendence Theory*, Springer-Verlag, LNM 1290, Berlin, Heidelberg, New York, 1987, pp. 1 - 108

[3] Colliot-Thélène J.-L., Skorobogatov A.N., *Good reduction of the Brauer-Manin obstruction*, to appear in Trans. Amer. Math. Soc., 365, 2013, pp. 579 - 590

[4] Fesenko I.B., Vostokov S.V., *Local Fields and Their Extensions*, Amer. Math. Soc., Providence, 2002

[5] Giraud J., Grothendieck A., Kleiman S.L., Raynaud M., Tate J., *Dix exposés sur la cohomologie des schémas*, North-Holland Publishing Company, Amsterdam, 1968

[6] Ieronymou E., *Diagonal quartic surfaces and transcendental elements of the Brauer group*, J. Inst. Math. Jussieu 9, 2010, pp. 769 - 798

[7] Ieronymou E., Skorobogatov A.N., Zarhin Yu.G., *On the Brauer group of diagonal quartic surfaces*, J. London Math. Soc. 83, 2011, pp. 659 - 672

[8] Kresch A., Tschinkel Yu., *On the arithmetic of Del Pezzo surfaces of degree* 2, Proc. London Math. Soc. (3), 89, 2004, pp. 545 - 569

[9] Kresch A., Tschinkel Yu., *Effectivity of Brauer-Manin obstructions on surfaces*, Adv. in Math. 226, 2011, pp. 4131 - 4144

[10] Milne J.S., *Étale Cohomology*, Princeton Univ. Press, Princeton, NJ, 1980

[11] Neukirch J., *Algebraische Zahlentheorie*, Springer, Berlin, Heidelberg, 2007

[12] Neukirch J., Schmidt A., Wingberg K., *Cohomology of Number Fields*, Springer, GmW 323, Berlin, Heidelberg, New York, 1999

[13] Serre J.P., *Corps locaux*, Hermann, Paris, 1962

[14] Skorobogatov A., *Torsors and Rational Points*, Cambrige Univ. Press, CTM 144, Cambridge, 2001

[15] Várilly-Alvarado A., *Weak approximation on Del Pezzo surfaces of degree 1*, Adv. in Math. 219, 2008, pp. 2123 - 2145

[16] Yamamoto K., *An explicit formula of the norm residue symbol in a local number field* Science reports of Tokyo Woman's Christian College, 24-28, 1972, pp. 302 - 334

Printed in the United States
by Baker & Taylor Publisher Services

Printed in the United States
by Baker & Taylor Publisher Services